Biomaterials Engineering

Biomaterials Engineering

Edited by
Shay Fisher

Willford Press
www.willfordpress.com

Published by Willford Press,
118-35 Queens Blvd., Suite 400,
Forest Hills, NY 11375, USA

ISBN: 978-1-68285-330-6

Cataloging-in-publication Data

Biomaterials engineering / edited by Shay Fisher.
p. cm.
Includes bibliographical references and index.
ISBN 978-1-68285-330-6
1. Biomedical engineering. 2. Bioengineering. 3. Tissue engineering. 4. Biomedical materials. I. Fisher, Shay.
R856 .B56 2017
610--dc23

For information on all Willford Press publications
visit our website at www.willfordpress.com

WILLFORD PRESS

Printed in China.

Contents

Preface

Biomaterial is any material that interacts with a biological system primarily for a medical purpose. This field of study is interdisciplinary in nature because it combines other fields of study like medicine, biology, chemistry, tissue engineering and material sciences. This book strives to provide a fair idea about this discipline and to help develop a better understanding of the latest advances within this field. The text is an important source of knowledge as it provides a comprehensive study of the various applications of biomaterial engineering. From theories to research to practical applications, case studies related to all contemporary topics of relevance to this field have been included in this book. Some of the diverse topics covered herein address the varied branches that fall under this category. Researchers and students in this field will be assisted by this book.

The main aim of this book is to educate learners and enhance their research focus by presenting diverse topics covering this vast field. This is an advanced book which compiles significant studies by distinguished experts. This book addresses successive solutions to the challenges arising in the area of application, along with it; the book provides scope for future developments.

It was a great honour to edit this book, though there were challenges, as it involved a lot of communication and networking between me and the editorial team. However, the end result was this all-inclusive book covering diverse themes in the field.

Finally, it is important to acknowledge the efforts of the contributors for their excellent chapters, through which a wide variety of issues have been addressed. I would also like to thank my colleagues for their valuable feedback during the making of this book.

Editor

1

The Braid Index of Complicated DNA Polyhedral Links

Xiao-Sheng Cheng[1,2], Xian'an Jin[2]*

1 Department of Mathematics, Huizhou University, Huizhou, Guangdong, P. R. China, 2 School of Mathematical Sciences, Xiamen University, Xiamen, Fujian, P. R. China

Abstract

The goal of this paper is to determine the braid index of two types of complicated DNA polyhedral links introduced by chemists and biologists in recent years. We shall study it in a more broad context and actually consider so-called Jaeger's links (more general Traldi's links) which contain, as special cases, both four types of simple polyhedral links whose braid indexes have been determined and the above two types of complicated DNA polyhedral links. Denote by $b(L)$ and $c(L)$ the braid index and crossing number of an oriented link L, respectively. Roughly speaking, in this paper, we prove that $b(L)=\frac{c(L)}{2}+1$ for any link L in a family including Jaeger's links and contained in Traldi's links, which is obtained by combining the MFW inequality and an Ohyama's result on upper bound of the braid index. Our result may be used to to characterize and analyze the structure and complexity of DNA polyhedra and entanglement in biopolymers.

Editor: Laurent Kreplak, Dalhousie University, Canada

Funding: XSC was supported by grants from the National Natural Science Foundation of China (number 11101174), Natural Science Foundation of Guangdong Province of China (number S2011040003984) and Foundation for Distinguished Young Talents in Higher Education of Guangdong, China (number LYM11120). XJ was supported by grants from the National Natural Science Foundation of China (numbers 10831001 and 11271307), Natural Science Foundation of Fujian Province of China (number 2012J01019) and the Fundamental Research Funds for the Central Universities (number 2010121007). The funders had no role in study design, data collection and analysis, decision to publish, or preparation of the manuscript.

Competing Interests: The authors have declared that no competing interests exist.

* E-mail: xajin@xmu.edu.cn

Introduction

The braid index of links has some applications in chemistry and molecular biology. For example, representing knotted hydrocarbon complexes as closed braids [1] can facilitate the study of their properties. In recent 20 years, many DNA polyhedral links such as DNA cube [2], DNA tetrahedron [3], DNA octahedron [4], DNA truncated octahedron [5], DNA bipyramid [6], DNA dodecahedron [7], DNA icosahedron and buckyballs [8] etc, have been synthesized in laboratories or schemes for synthesizing them have been designed by chemists and biologists. In addition, several novel types of polyhedral links have also been posed by mathematicians as a potential object to be synthesized. For four types of simple polyhedral links, their braid indexes have been determined in [9]. The purpose of this paper is to determine braid index of another two types of more complicated polyhedral links appeared in [8,10,11,12,13] and [14], respectively. We shall study it in a more broad context and actually consider so-called Jaeger's links and more general Traldi's links.

In [15], Jaeger associated to a plane graph an oriented link diagram by replacing each edge of the graph by an oriented clasp as illustrated in (*a*) of Figure 1. Then he established a relation between the Tutte polynomial [16] of the graph and the HOMFLY polynomial [17,18] of the associated oriented link. Moreover, about one year later, both the construction and the relation were extended by Traldi in [19], Traldi constructed his oriented link diagram based on a plane graph via replacing each edge by one of four types of oriented clasps (*a*),(*b*),(*c*) and (*d*), as illustrated in Figure 1. By assigning four different weights to edges of the plane graph according to the four types of oriented clasps he built a relation between the weighted dichromatic polynomial [19] of the graph and the HOMFLY polynomial of the associated oriented link.

Formally, let L be an oriented link, we call L to be a Jaeger's link if L, or its reverse, or its mirror image, or its inverse (i.e. the composition of the reverse and the mirror image), has an oriented diagram which can be obtained from a connected plane graph by replacing each edge of the graph by an oriented clasp (*a*) in Figure 1. According to the above definition, Jaeger's links will be all alternating. Analogously, we call L to be a Traldi's link if L, or its reverse, has an oriented diagram which can be obtained from a connected plane graph by replacing each edge of the graph by one of four types of oriented clasps in Figure 1. Note that different edges may be replaced by different types of oriented clasps. Traldi's links may not be alternating, say, the link $L8n7$ in the Thistlethwaite Link Table [20]. Clearly, Traldi's links contain Jaeger's links as special cases.

The braid index of an oriented link is the minimum number of strings among all closed braid representatives for the given oriented link. We point out that the braid index depends on orientations of links. See Page 215 of [21]. Let L be an oriented link. We denote by $b(L)$ the braid index of L. It is, in general, very hard to determine this geometric and numerical invariant $b(L)$ [21]. Let $P_L(v,z)$ be the HOMFLY polynomial of the oriented link L. In [22] and [23], Franks, Williams and Morton gave independently a lower bound for the braid index $b(L)$ of an oriented link L in terms of $span_v P_L(v,z)$ as follows:

$$\frac{1}{2}span_v P_L(v,z)+1\leq b(L), \qquad (1)$$

where $span_v P_L(v,z) = \max \deg_v P_L(v,z) - \min \deg_v P_L(v,z)$, and $\max \deg_v P_L(v,z)$ and $\min \deg_v P_L(v,z)$ denote, respectively, the maximal degree and minimal degree of v in the polynomial $P_L(v,z)$. This inequality (1) is usually called MFW inequality. The MFW inequality was the first known result relating the braid index to the HOMFLY polynomial. To date, we have known that the MFW inequality for many link families is sharp. For example, the inequality is sharp for all torus links, closed positive n-braids with a full twist [22], 2-bridge links [24], alternating fibred links [24] and a certain family of closed positive braids [25]. In this paper we shall prove the sharpness of the MFW inequality for a family of links including Jaeger's links and contained in Traldi's links. We shall make use of the following result obtained by Ohyama in 1993 [26] which states that for a non-splittable oriented link L,

$$b(L) \leq 1 + \frac{c(L)}{2}, \tag{2}$$

where $c(L)$ is the crossing number of L. In fact, you will see this upper bound and the MFW lower bound coincide for a family of links including Jaeger's links, but do not coincide for general Traldi's links.

Note that many types of polyhedral links are, in fact, Jaeger's links defined above. Four types of simple polyhedral links considered in [9] are actually all Jaeger's links constructed from subdivisions of (convex) polyhedral graphs or their duals. Note that the dual of a polyhedral graph is also a polyhedral graph and Steinitz's theorem states that a polyhedral graph is a 3-connected simple (i.e. no loops and no multiple edges) planar graph. Thus the result in [9] can not be used to deal with general Jaeger's links.

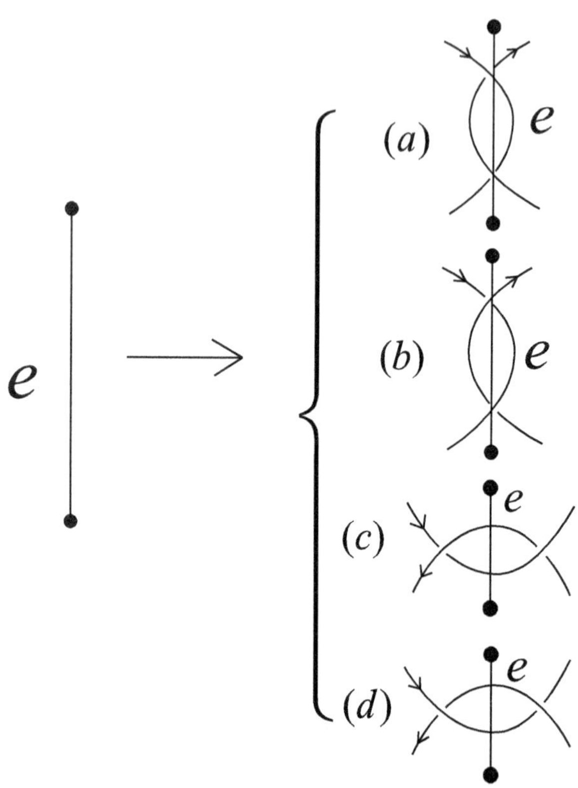

Figure 1. Replace an edge by four types of oriented clasp.

Roughly speaking, in this paper we prove that $b(L) = \frac{c(L)}{2} + 1$ for any link L in a family including Jaeger's links and contained in Traldi's links and thus extend the result in [9]. This extension enables us to determine the braid index of another two types of more complicated polyhedral links in [8,10] and [14] as well as four types of polyhedral links in [9]. It is probable that our result be used to determine braid indexes of new types of polyhedral links to be synthesized in the future. Our research demonstrates that using the braid index or crossing index to describe the complexity of some polyhedral links are equivalent and it may open a door to characterize, analyze the structure and complexity of DNA polyhedra and entanglement in biopolymers.

Analysis

Before analyzing the braid index of polyhedral links, we give some preliminary knowledge on links, the HOMFLY polynomial, graphs and the dichromatic polynomial.

1. Links

A **knot** is a simple closed curve in $R^3 \subset S^3$, i.e. the image of the embedding of S^1 into R^3. A **link** L of n components is a disjoint union of n knots: $L = K_1 \cup \cdots \cup K_n$. A knot is a link with one component. An oriented link is a link with each of its component assigned an orientation. A **link diagram** is a (regular) projection of a link onto a plane with the under-strand specified by using gaps at each crossing. When the link is oriented, the link diagram will inherit the orientation of the link in a natural way and called oriented link diagram. The **crossing number** or **crossing index** of a link L, denoted by $c(L)$, is the least number of crossings that occur in any diagram of the link. A diagram with fewest number of crossings for a given link is called a **minimal** diagram for the link.

An **n-string braid** b is a set of n strings in a 3-dimensional cube $I \times I \times I$, where $I = [0,1]$, all of which are attached to a horizontal bar at the top $\{\frac{1}{2}\} \times I \times \{1\}$ and at the bottom $\{\frac{1}{2}\} \times I \times \{0\}$ such that each $I \times I \times x$, $x \in S^1$, meets the n strings in exactly n points. A **closed n-string braid** $\hat{b}$ is a set of n strings embedded in $I \times I \times S^1$ such that each $I \times I \times x$, $x \in S^1$, meets the n strings in exactly n points. An example is illustrated in Figure 2.

It is clear that a closed n-string braid $\hat{b}$ is an oriented link with the convention that we always orient b from the top to the bottom. Conversely, Alexander [27], in 1923, showed that every oriented link in S^3 can be represented as a closed n-string braid. The **braid**

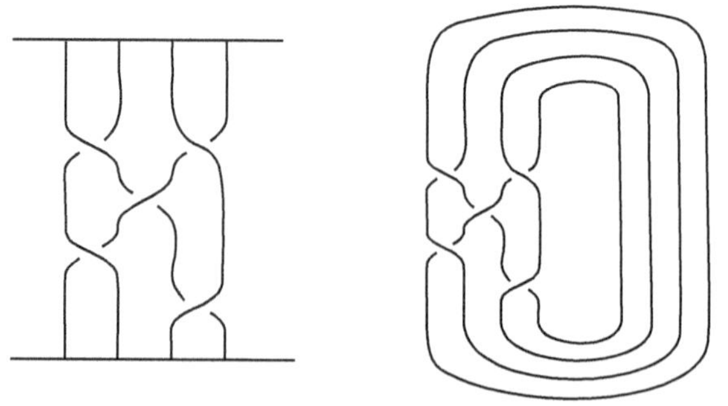

Figure 2. A braid b and its corresponding closed braid $\hat{b}$.

index $b(L)$ for an oriented link L is the smallest positive integer n such that L can be represented as a closed n-string braid.

Let L be an oriented link. We denote by L^r and L^m the reverse (reversing the orientation of each component of L) and the mirror image of L, respectively. Then it is obvious that $b(L)=b(L^r)=b(L^m)$.

2. The HOMFLY polynomial

The HOMFLY polynomial is an invariant of oriented links, introduced in [17] and [18] independently. Now we recall the definition of the HOMFLY polynomial. The HOMFLY polynomial of an oriented link L, denoted by $P_L(v,z)$, can be defined by the three following axioms [28].

(1) $P_L(v,z)$ is invariant under ambient isotopy of L.
(2) If L is the trivial knot then $P_L(v,z)=1$.
(3) It satisfies the skein relation: $v^{-1}P_{L_+}(v,z)-vP_{L_-}(v,z)=zP_{L_0}(v,z)$, where L_+, L_- and L_0 are link diagrams which are identical except near one crossing where they are as in Figure 3 and are called a skein triple.

It possesses the following basic properties [28], which imply that $span_v P_L(v,z)=span_v P_{L^m}(v,z)=span_v P_{L^r}(v,z)$.

(a) If L^r is the reverse of L, then $P_{L^r}(v,z)=P_L(v,z)$.
(b) If L^m is the mirror image of L, then $P_{L^m}(v,z)=P_L(-v^{-1},z)$.

3. Graphs

Let $G=\langle V(G),E(G)\rangle$ be a graph. We denote by $|V(G)|$ and $|E(G)|$ the numbers of vertices and edges, respectively, of the graph G. A graph G is said to be **connected** if it has a path with two distinct vertices u and v whenever $u,v\in V(G)$ and G is said to be disconnected otherwise. A **component** of a graph G is a subgraph that is connected and is not properly contained in any other connected subgraph of G. We denote by $k(G)$ the number of components of the graph G. An edge e of G is said to be a **bridge** if $k(G-e)>k(G)$. If G has no bridges, G is said to be bridgeless. An edge e of G is said to be a **loop** if its two end-vertices are the same. Iff G has no loops, G is said to be loopless. The **rank** $r(G)$ and **nullity** $n(G)$ of the graph G is defined to be $|V(G)|-k(G)$ and $|E(G)|-|V(G)|+k(G)$, respectively. It is well known that $r(G)$ and $n(G)$ are exactly the dimensions of cycle space and cut space, respectively, of the graph G and hence, are both non-negative integers. See, for example, [29].

Given a graph G and an edge e of G, we write $G-e$ for the graph obtained from G by deleting the edge e, and G/e for the graph obtained from G by contracting e, i.e. deleting e firstly and then identifying its two endvertices. A **planar** graph is a graph which can be embedded in the plane or equivalently, the sphere S^2. A specific embedding of a planar graph is called a **plane** **graph**. For undefined notations and terminologies on graph theory, we refer the readers to [29].

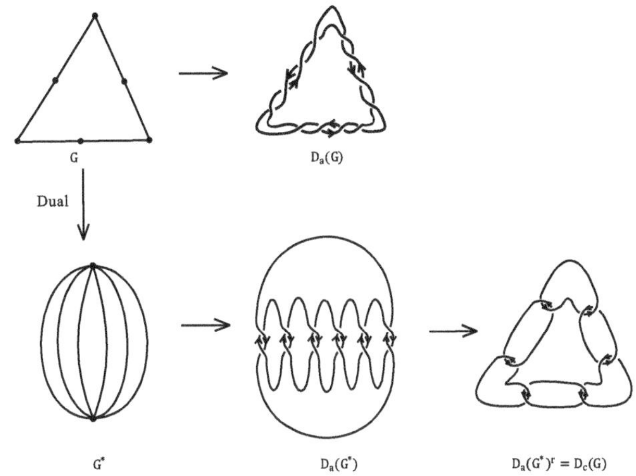

Figure 4. An example.

4. The dichromatic polynomial

A weighted graph is a graph G together with a function w mapping $E(G)$ into some commutative ring R with unity 1. If e is an edge of the weighted graph G, then $w(e)$ is the weight of the edge e. The dichromatic polynomial for weighted graphs was introduced by Traldi in [19], which is a generalization of the Tutte polynomial for signed graphs introduced by Kauffman in [30]. We point out there are actually several weighted versions of the Tutte polynomial.

The dichromatic polynomial $Q(t,z)$ of a weighted graph G can be defined as

$$Q(t,z)=\sum_{F\subseteq E(G)}\left(\prod_{f\in F} w(f)\right)t^{k\langle F\rangle}z^{n\langle F\rangle}, \qquad (3)$$

where the summation is over all edge subsets, F, of G, $k\langle F\rangle$ and $n\langle F\rangle$ are the number of components and the nullity, respectively, of the spanning subgraph $\langle F\rangle$, induced by F, of G. It can also be defined by using the following recursive relations:

(1) If G is an edgeless graph with $n\geq 1$ vertices $Q(G)=t^n$.
(2) If e is a loop of G, then $Q(G)=(1+w(e)z)Q(G-e)$.
(3) If e is not a loop of G, then $Q(G)=Q(G-e)+w(e)Q(G/e)$.

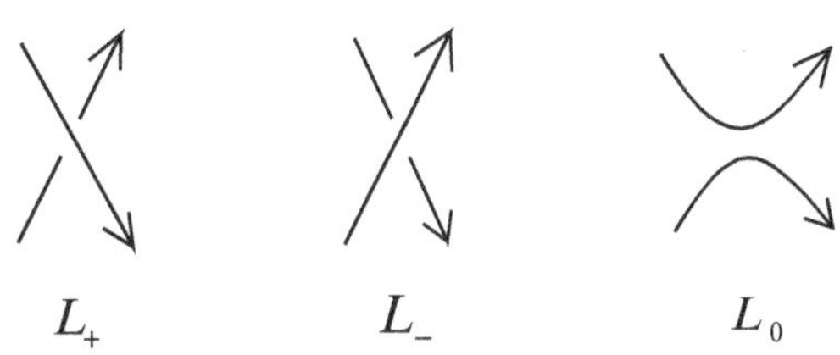

Figure 3. L_+, L_- **and** L_0**.**

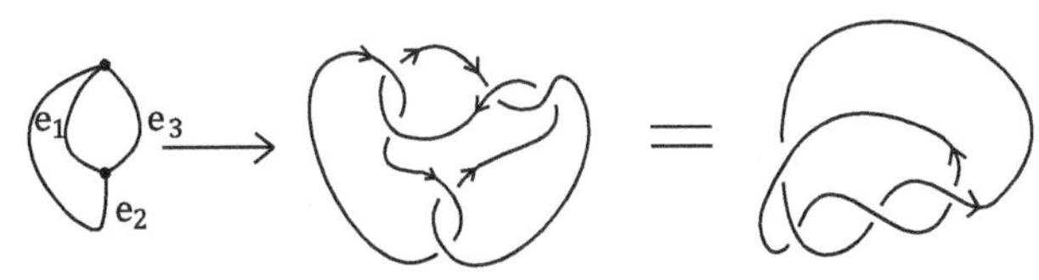

Figure 5. An alternating Traldi's link upper and lower bounds of whose braid index are not equal.

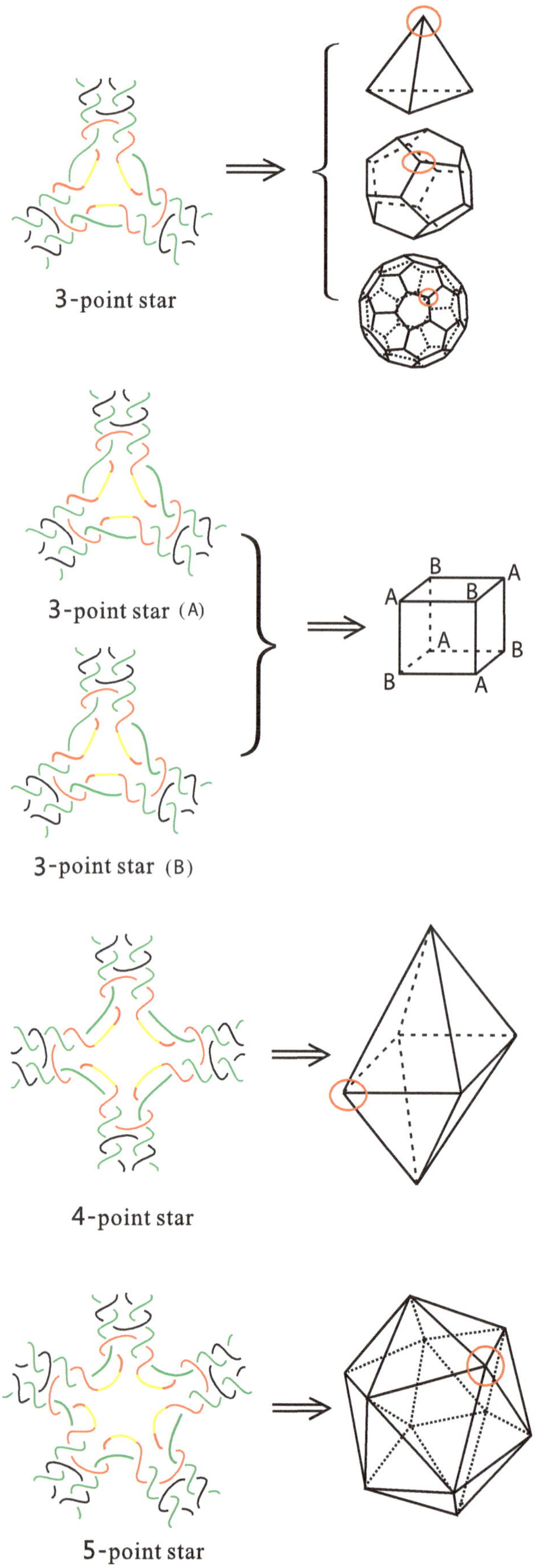

Figure 6. 3-point star motifs: DNA tetrahedron, cube, dodecahedron, buckyball; 4-point star motifs: DNA octahedron; 5-point star motifs: DNA icosahedron.

Just as the Tutte polynomial, it also has a spanning tree expansion and we do not explain it here.

Now we start to analyze the braid index of polyhedral links. We first consider a family of Jaeger's links.

Let G be a connected plane graph. Let $D_a(G)$ be the oriented link diagram constructed based on G by covering each edge of G with oriented clasp (a) in Figure 1. Oriented link diagrams $D_b(G)$,$D_c(G)$ and $D_d(G)$ can be defined similarly. Let $D_{a,d}(G)$ be the oriented link diagram constructed based on G by covering some edges of G with oriented clasp (a) and other edges with oriented clasp (d) in Figure 1. Oriented link diagrams $D_{b,c}(G)$ can be defined similarly. Note that $D_{a,d}(G)$ (resp. $D_{b,c}(G)$) contains $D_a(G)$,$D_d(G)$ (resp. $D_b(G)$,$D_c(G)$) as special cases. More generally, $D_{a,b,c,d}(G)$ is defined to be the oriented link diagram obtained from G by covering each edge with one of four types of clasps (a),(b),(c) and (d). In [19], Traldi obtained the following general formula.

Theorem 1. *Let G be a plane graph. Let $D(G)=D_{a,b,c,d}(G)$ be the Traldi's oriented link diagram. Let the weight of the edge replaced by oriented clasp (a),(b),(c) and (d) be $\frac{v}{z}$, $-\frac{1}{vz}$, $\frac{z}{v}$ and $-vz$, respectively. Then*

$$P_{D(G)}(v,z)=\frac{z}{v^{-1}-v}(vz)^a\left(-\frac{z}{v}\right)^b v^{2c-2d}Q_G\left(\frac{v^{-1}-v}{z},\frac{v^{-1}-v}{z}\right), \quad (4)$$

where a,b,c and d are numbers of edges of G which are replaced by oriented clasp (a),(b),(c) and (d), respectively.

The following theorem is a main theoretical tool for determining the braid index of polyhedral links.

Theorem 2. *Let G be a connected bridgeless and loopless plane graph. Let $L_{a,d}(G)$ (resp. $L_{b,c}(G)$) be the oriented link that $D_{a,d}(G)$ (resp. $D_{b,c}(G)$) represents. Then*

$$b(L_{a,d}(G))=|E(G)|+1=\frac{c(L_{a,d}(G))}{2}+1, \quad (5)$$

$$b(L_{b,c}(G))=|E(G)|+1=\frac{c(L_{b,c}(G))}{2}+1. \quad (6)$$

Proof. Note that $L_{b,c}(G)$ is the mirror image of some $L_{a,d}(G)$ and taking the mirror image does not change the braid index and the crossing index. It suffices for us to prove Eq. (5). When G is bridgeless and loopless, it is not hard to observe that $D_{a,d}(G)$ is a connected alternating oriented link diagram without nugatory crossings. According to the famous result [31,32,33] due to Kauffman, Murasugi and Thistlethwaite: A connected alternating link diagram without nugatory crossings is minimal, we obtain that

$$c(L_{a,d}(G))=2|E(G)|, \quad (7)$$

where $L_{a,d}(G)$ is the link that the link diagram $D_{a,d}(G)$ represents. In addition, by a result [34] due to Menasco that states that a link with an alternating diagram will be non-splittable if and only if the diagram is connected, we know that $L_{a,d}(G)$ is non-splittable. Thus, by Eq. (2), we have

$$b(L_{a,d}(G)) \leq |E(G)|+1. \tag{8}$$

Now we consider the lower bound of $b(L_{a,d}(G))$. For the oriented link $L_{a,d}(G)$, as a special case of Theorem 1, we obtain

$$\begin{aligned}
P_{L_{a,d}(G)}(v,z) &= \frac{z}{v^{-1}-v}(vz)^a v^{-2d} Q_G\left(\frac{v^{-1}-v}{z},\frac{v^{-1}-v}{z}\right)\Big|_{w_a=vz,w_d=-vz}\\
&=\left(\frac{z}{v^{-1}-v}\right)v^{a-2d}z^a \sum_{F\subseteq E(G)}\left(\prod_{f\in F} w(f)\right)\left(\frac{v^{-1}-v}{z}\right)^{k\langle F\rangle}\left(\frac{v^{-1}-v}{z}\right)^{n\langle F\rangle}\\
&=\left(\frac{z}{v^{-1}-v}\right)v^{a-2d}z^a \sum_{F\subseteq E(G)}\left(\prod_{f\in F} w(f)\right)\left(\frac{v^{-1}-v}{z}\right)^{|F|-|V(G)|+2k\langle F\rangle}\\
&=\left(\frac{z}{v^{-1}-v}\right)v^{a-2d}z^a \sum_{F\subseteq E(G)}\left(\left(\frac{v}{z}\right)^{a(F)}(-vz)^{d(F)}\right)\left(\frac{v^{-1}-v}{z}\right)^{|F|-|V(G)|+2k\langle F\rangle} \qquad (9)\\
&= v^{a-2d}z^a\sum_{F\subseteq E(G)}(-1)^{d(F)}v^{|F|}z^{d(F)-a(F)}\left(\frac{v^{-1}-v}{z}\right)^{|F|-|V(G)|+2k\langle F\rangle-1}\\
&= \sum_{F\subseteq E(G)}(-1)^{d(F)}v^{a-2d+|V(G)|-2k\langle F\rangle+1}(1-v^2)^{|F|-|V(G)|+2k\langle F\rangle-1}\\
&\qquad z^{a+d(F)-a(F)-|F|+|V(G)|-2k\langle F\rangle+1},
\end{aligned}$$

where $a(F)$ and $d(F)$ are numbers of edges in F covered by clasps of types (a) and (d), respectively. It is clear that $a(F)+d(F)=|F|$.

Now we compute the highest and lowest degrees in v of Eq. (9). Note that the power of $1-v^2$ is $|F|-|V(G)|+2k\langle F\rangle-1 = n\langle F\rangle+k\langle F\rangle-1\geq 0$. Thus we have

$$\begin{aligned}
\max \deg_v P_{L_{a,d}(G)}(v,z) &= \max_F\{(a-2d+|V(G)|-2k\langle F\rangle+1)\\
&\quad +2(|F|-|V(G)|+2k\langle F\rangle-1)\}\\
&= \max_F\{a-2d+|V(G)|-1+2n\langle F\rangle\}\\
&= a-2d+|V(G)|-1+2n\langle E(G)\rangle\\
&= a-2d+|V(G)|-1+2|E(G)|-2|V(G)|+2\\
&= a-2d+2|E(G)|-|V(G)|+1.
\end{aligned}$$

The third equation holds since $n\langle F\rangle$ attains maximality if and only if $F=E(G)$ because of G is bridgeless. Similarly, we have

$$\begin{aligned}
\min \deg_v P_{L_{a,d}(G)}(v,z) &= \min_F\{a-2d+|V(G)|-2k\langle F\rangle+1\}\\
&= a-2d+|V(G)|-2k\langle\varnothing\rangle+1\\
&= a-2d-|V(G)|+1.
\end{aligned}$$

The second equation holds since $k\langle F\rangle$ attains maximality if and only if $F=\varnothing$ since G is loopless. Thus,

$$\begin{aligned}
\mathrm{span}_v P_{L_{a,d}(G)}(v,z) &= \max \deg_v P_{L_a(G)}(v,z)-\min \deg_v P_{L_a(G)}(v,z)\\
&= (a-2d+2|E(G)|-|V(G)|+1)-(a-2d-|V(G)|+1)\\
&= 2|E(G)|.
\end{aligned}$$

Hence, by Eq. (1), we obtain

$$b(L_{a,d}(G)) \geq |E(G)|+1. \tag{10}$$

Combining Eq. (8) with Eq. (10), we show that Eq. (5) holds. □

Corollary. *(1). Let G be a connected bridgeless plane graph. Then*

$$b(L_a(G)) = |E(G)|+1 = \frac{c(L_a(G))}{2}+1, \tag{11}$$

$$b(L_b(G)) = |E(G)|+1 = \frac{c(L_b(G))}{2}+1. \tag{12}$$

(2). Let G be a connected loopless plane graph. Then

$$b(L_c(G)) = |E(G)|+1 = \frac{c(L_c(G))}{2}+1, \tag{13}$$

$$b(L_d(G)) = |E(G)|+1 = \frac{c(L_d(G))}{2}+1. \tag{14}$$

As a result,

$$b(L) = \frac{c(L)}{2}+1 \tag{15}$$

for any Jaeger's link L.

Proof. By checking the proof of Theorem 2, we find that the condition "loopless" is actually not necessary for the special case $L_a(G)$. Hence, Eq. (11) holds. Eq. (12) follows from Eq. (11) and the fact $L_b(G)=L_a^m(G)$. Let G be a connected loopless plane graph and G^* be the planar dual of G. Then G^* is connected and

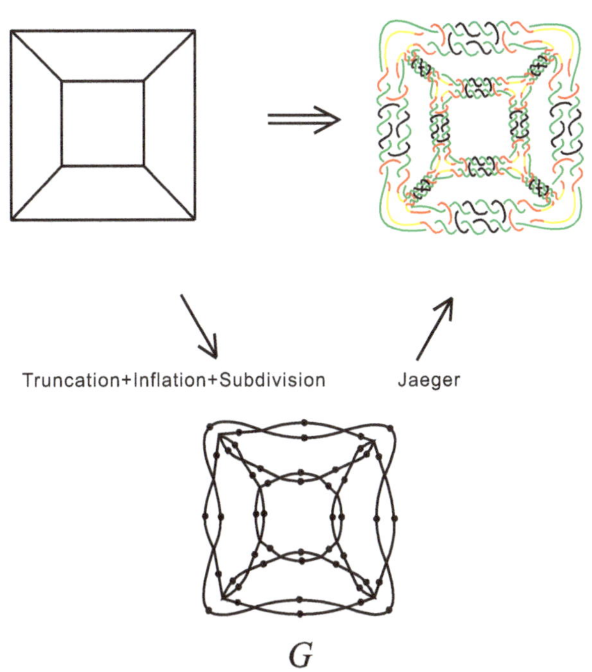

Figure 7. Planar form of double crossover DNA cube and its associated plane graph *G*.

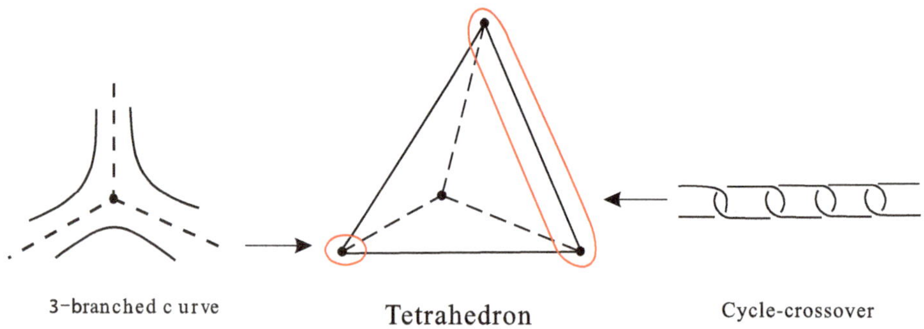

Figure 8. 3-**branched curve, cycle-crossover and the construction of cycle-crossover tetrahedral link.**

bridgeless. Note that $L_c(G)=L_a^r(G^*)$ (see Figure 4 for an example) and $L_d(G)=L_a^{rm}(G^*)$ and reversing orientations do not change braid index and crossing index. Hence, Eqs. (13) and (14) hold. By the definition of a Jaeger's link L, we know that there is a connected plane graph G such that L is equivalent to one of the four links: $L_a(G)$, $L_a^m(G)$, $L_a^r(G)$ and $L_a^{mr}(G)=L_a^{rm}(G)$. It will suffice to prove that $b(L_a(G))=\frac{c(L_d(G))}{2}+1$ for any connected plane graph G. Firstly, $b(L_a(G))=|E(G)|+1=\frac{c(L_d(G))}{2}+1$ for any connected bridgeless plane graph G. Secondly, if G has bridges, let G' be the connected plane graph obtained from G by contracting all bridges of G. Note that $L_a(G)$ is equivalent to $L_a(G')$ via untwisting oriented clasps covering bridges. Hence, we have $b(L_a(G))=b(L_a(G'))=\frac{c(L_a(G'))}{2}+1=\frac{c(L_d(G))}{2}+1$.

□

Remark. *Eq. (15) generalizes Theorem 4.3 in [9] from subdivisions of polyhedral graphs to any connected plane graphs.*

To conclude the section, we point out that the braid index of general Traldi's links (even for alternating Traldi's links) can not be determined similarly since the upper and lower bounds are not always equal in general. An example of alternating Traldi's links, denoted by L, is shown in Figure 5 which is constructed from a theta graph by replacing edges e_1,e_2 by oriented clasp (a) and e_3 by clasp (c). On the one hand, by using the software KnotGTK, we obtain

$$P_L(v,z)=z^{-1}(-v^5+v^3)+z(-v^5+3v^3)+z^3v^3 \quad (16)$$

and hence, $\text{span}_v P_L(v,z)=2$. On the other hand, we have $c(L)=4$. Therefore,

$$2=\frac{\text{span}_v P_L(v,z)}{2}+1\le b(L)\le\frac{c(L)}{2}+1=3. \quad (17)$$

Results

It is obvious that by the Corollary the braid index of four types of polyhedral links (see Fig. 2 of [9]) can be determined. In this section, we determine the braid index of two types of complicated DNA polyhedral links. However, braid indexes of these two new types of polyhedral links can not be obtained by the previous result in [9]. You will see these new types of polyhedral links are all alternating and hence their crossing numbers are easily obtained. By the Corollary, in order to obtain the braid index, we only need to show that these links are Jaeger's links, and equivalently to find their associated plane graphs. Note that two strands of DNA have antiparallel orientations which are consistent with Jaeger's orientations. For simplicity, we do not draw orientations in the following any more.

1. Double crossover DNA polyhedral links

Recent years, in [8,10,11,12,13], the authors, in laboratory, designed and synthesized some fancy double crossover DNA polyhedra, such as DNA tetrahedron, cube, octahedron, dodecahedron, icosahedron and buckyball, by covering each vertex of degree n of the polyhedron by "n-point star motif (tiles)" and through sticky-end association between the tiles. The "n-point star motif" has an n-fold rotational symmetry and consists of $2n+1$ single strands: a long repetitive central DNA strand (colored red and yellow), n identical medium DNA strands (colored green) and n identical short DNA strands (colored black). The colored yellow part at the center of the motif represents n unpaired DNA single-strands whose lengths can be adjusted to change bending degree of the whole structure. See Figure 6.

We point out in Figure 6 the tiles are a little different in the ends of tiles from the actual assemblies of such DNA polyhedra. For

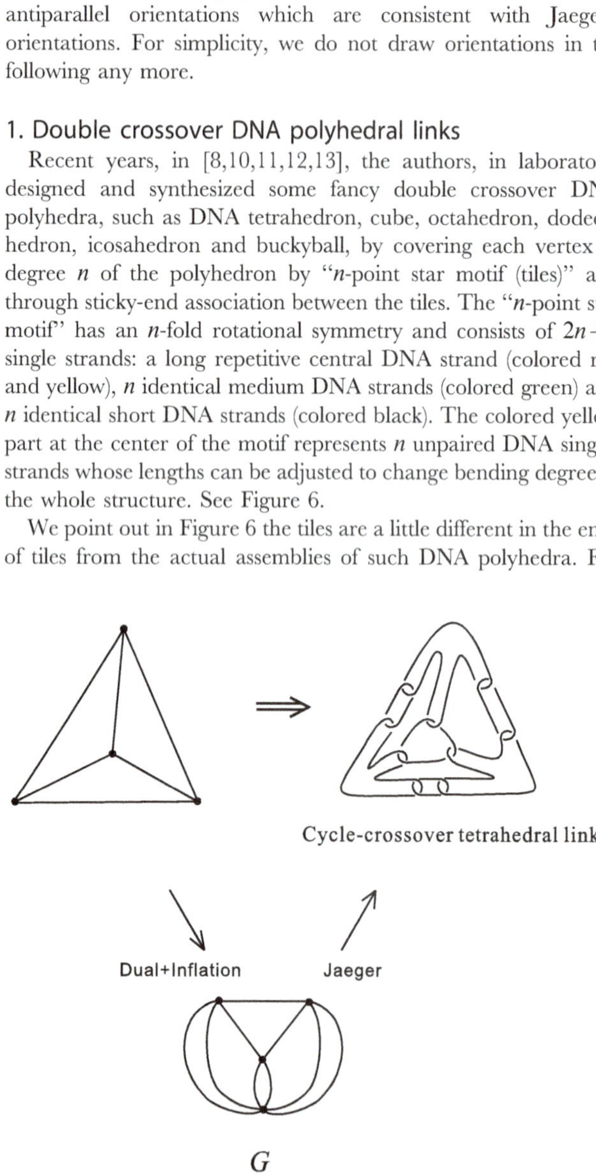

Figure 9. Planar form of cycle-crossover DNA tetrahedron and its associated plane graph G.

actual assemblies, we refer the reader to Fig. 1 of [8] for DNA tetrahedron, dodecahedron and buckyball, Fig. 4 of [11] for DNA cube, Fig. 1 of [13] for DNA octahedron and Fig. 4(f) of [10] and Fig. 1 of [12] for DNA icosahedron. However, one will obtain the same polyhedron links based on our drawings as actual assemblies, and hence do not affect the analysis of their braid indexes.

Now we take double crossover DNA cube as an example to show double crossover DNA polyhedra are all Jaeger's links. The planar form of double crossover DNA cube link and its associated plane graph are shown in Figure 7 (upper and right) and (lower), respectively. In general, the associated plane graph G can be obtained from the polyhedron P by the following steps:

Step 1: Truncating the polyhedron P, i.e. cutting each corner of the polynhedron P, we obtain a polyhedral graph Q.

Step 2: Inflating edges of Q corresponding to edges of P, i.e. converting a single edge into two parallel edges, we obtain a plane graph H.

Step 3: Subdividing each edge of H by inserting a vertex, we obtain the associated graph G.

Let P be a polyhedron with m edges. Then, by simple calculation, the associated plane graph G has $8m$ edges and thus the braid index of the corresponding double crossover DNA polyhedral link is $8m+1$.

2. Cycle-crossover polyhedral links

This type of polyhedral links was introduced in [14], designed as a potential object for synthesising, which can be constructed from a polyhedron by replacing each edge of the polyhedron by a "cycle-crossover" and replacing each vertex of degree n by an n-branched curve. See Figure 8. The planar form of the cycle-crossover cube link is shown in Figure 9 (upper and right) and its associated plane graph is shown in Figure 9 (lower).

In general, cycle-crossover polyhedral links are all Jaeger's links. The associated plane graph G can be obtained from the polyhedron P by the following steps:

Step 1: Computing the planar dual of P, we obtain a plane graph H

Step 2: Inflating edges of H, i.e. converting a single edge into several parallel edges, we obtain the associated graph G.

In this case suppose that the total lengths of cycle-crossovers used to construct cycle-crossover polyhedral links are m. Then the associated plane graph G has m edges and thus the braid index of the corresponding cyle-crossover DNA polyhedral link is $m+1$.

Discussion

Firstly, note that the two associated plane graphs in Figures 7 and 9 are both not subdivisions of polyhedral graphs and for both types of polyhedral links. This means that the method used in [9] can not be used here. Secondly, it is easy to find that the braid index of two types of complicated polyhedral links does not depend on the structure of the polyhedron. Thirdly, as the referee points out that there are two types of DNA cube: 4 turns and 4.5 turns. See Figure 4 and Figure 1, respectively, of [11], we only consider the 4-turn DNA cube in this paper. It deserves to study the braid index of 5-turn DNA cube and more general 5-turn DNA polyhedra. Finally, in fact Theorem 2 is more powerful than it has been used, and hence may be used to determine the braid index of more complicated polyhedral links in the future.

Author Contributions

Conceived and designed the experiments: XSC XJ. Performed the experiments: XSC XJ. Analyzed the data: XSC XJ. Contributed reagents/materials/analysis tools: XSC XJ. Wrote the paper: XSC XJ.

References

1. Cox MA, Hughes TS, Ellis-Monaghan JA, Mondanaro KR (2008) Hydrocarbon links in an octet truss. J Math Chem 43: 874–891.
2. Chen J, Seeman NC (1991) Synthesis from DNA of a molecule with the connectivity of a cube. Nature 350: 631–633.
3. Goodman RP, Berry RM, Turberfield AJ (2004) The single-step synthesis of a DNA tetrahedron. Chem Commun 12: 1372–1373.
4. Shih WM, Quispe JD, Joyce GF (2004) A 1.7-kilobase single-stranded DNA that folds into a nanoscale octahedron. Nature 427: 618–621.
5. Zhang Y, Seeman NC (1994) The construction of a DNA truncated octahedron. J Am Chem Soc 116: 1661–1669.
6. Erben CM, Goodman RP, Turberfield AJ (2007) A Self-Assembled DNA Bipyramid. J Am Chem Soc 129: 6992–6993.
7. Zimmermann J, Cebulla MPJ, Mönninghoff S, Kiedrowski GV (2008) Self-Assembly of a DNA Dodecahedron from 20 Trisoligonucleotides with C3h Linkers. Angew Chem Int Ed 47: 3626–3630.
8. He Y, Ye T, Su M, Zhang C, Ribbe AE, et al. (2008) Hierarchical self-assembly of DNA into symmetric supramolecular polyhedra. Nature 452: 198–202.
9. Cheng XS, Jiang X, Dai H (2012) The braid index of polyhedral links. J Math Chem 50: 1386–1397.
10. Lin C, Liu Y, Yan H (2009) Designer DNA Nanoarchitectures. Biochemistry 48(8): 1663–1674.
11. Zhang C, Ko SH, Su M, Leng Y, Ribbe AE, et al. (2009) Symmetry Controls the Face Geometry of DNA Polyhedra. J Am Chem Soc 131:1413–1415.
12. Zhang C, Su M, He Y, Zhao X, Fang P, et al. (2008) Conformational flexibility facilitates selfassembly of complex DNA nanostructures. Proc Natl Acad Sci USA 105: 10665–10669.
13. He Y, Su M, Fang P, Zhang C, Ribbe AE, et al. (2010) On the chirality of self-assembled DNA octahedra. Angew Chem Int Ed 49: 748–751.
14. Cheng XS, Zhang HP, Hu G, Qiu WY (2010) The architecture and Jones polynomials of cycle-crossover polyhedral links. MATCH Commun Math Comput Chem 63:637–635.
15. Jaeger F (1988) On Tutte polynomials and link polynomials. Proc Amer Math Soc 103: 647–654.
16. Tutte WT (1954) A contribution to the theory of chromatic polynomials. Canad J Math 6: 80–91.
17. Freyd P, Yetter D, Hoste J, Lickorish WBR, Millett K, et al. (1985) A new polynomial invariant of knots and links. Bull Amer Math Soc (NS) 12: 239–246.
18. Przytycki JH, Traczyk P (1987) Invariants of links of Conway type. Kobe J Math 4: 115–139.
19. Traldi L (1989) A dichromatic polynomial for weighted graphs and link polynomials. Proc Amer Math Soc 106: 279–286.
20. Math.toronto.edu. Available: http://katlas.math.toronto.edu/wiki/The Thistle-thwaite Link Table. Accessed 2012 Oct 11.
21. Murasugi K (1996) Knot theory and its applications. Boston: Birkhauser. 341 p.
22. Franks J, Williams RF (1987) Braids and the Jones polynomial. Trans Amer Math Soc 303: 97–108.
23. Morton HR (1986) Seifert circles and knot polynomials. Math. Proc. Cambridge Philos Soc 99: 107–109.
24. Murasugi K (1991) On the braid index of alternating links. Trans Amer Math Soc 326: 237–260.
25. Nakamura T (2004) Notes on the braid index of closed positive braids. Topology Appl 135: 13–31.
26. Ohyama Y (1993) On the minimal crossing number and the braid index of links. Canad J Math 45: 117–131.
27. Alexander JW (1923) A lemma on systems of knotted curves. Proc Natl Acad Sci USA 9: 93–95.
28. Cromwell PG (2004) Knots and Links. Cambridge University Press. 328 p.
29. Bondy JA, Murty USR (1976) Graph theory with applications. London and Basingstoke: The Macmillan Press Ltd. 264 p.
30. Kauffman LH (1989) A Tutte polynomial for signed graphs. Discrete Appl Math 25: 105–127.
31. Kauffman LH (1987) State models and the Jones polynomial. Topology 26: 395–407.
32. Thistlethwaite MB (1987) A spanning tree expansion of the Jones polynomial. Topology 26: 297–309.
33. Murasugi K (1987) Jones polynomial and classical conjectures in knot theory. Topology 26: 187–194.
34. Menasco W (1984) Closed incompressible surfaces in alternating knot and link complements. Topology 23(1): 37–44.

2

Rapidly-Deposited Polydopamine Coating via High Temperature and Vigorous Stirring: Formation, Characterization and Biofunctional Evaluation

Ping Zhou[1,2], Yi Deng[1,2], Beier Lyu[3], Ranran Zhang[4], Hai Zhang[4,5], Hongwei Ma[3], Yalin Lyu[4]*, Shicheng Wei[1,2]*

1 Department of Oral and Maxillofacial Surgery, School and Hospital of Stomatology, Peking University, Beijing, China, **2** Center for Biomedical Materials and Tissue Engineering, Academy for Advanced Interdisciplinary Studies, Peking University, Beijing, China, **3** Suzhou Institute of Nano-Tech and Nano-Bionics, Chinese Academy of Sciences, Suzhou, China, **4** Department of Stomatology, Beijing Anzhen Hospital, Capital Medical University, Beijing, China, **5** Department of Restorative Dentistry, School of Dentistry, University of Washington, Washington, United States of America

Abstract

Polydopamine (PDA) coating provides a promising approach for immobilization of biomolecules onto almost all kinds of solid substrates. However, the deposition kinetics of PDA coating as a function of temperature and reaction method is not well elucidated. Since dopamine self-polymerization usually takes a long time, therefore, rapid-formation of PDA film becomes imperative for surface modification of biomaterials and medical devices. In the present study, a practical method for preparation of rapidly-deposited PDA coating was developed using a uniquely designed device, and the kinetics of dopamine self-polymerization was investigated by QCM sensor system. It was found that high temperature and vigorous stirring could dramatically speed up the formation of PDA film on QCM chip surface. Surface characterization, BSA binding study, cell viability assay and antibacterial test demonstrates that the polydopamine coating after polymerization for 30 min by our approach exhibits similar properties to those of 24 h counterpart. The method has a great potential for rapid-deposition of polydopamine films to modify biomaterial surfaces.

Editor: Jie Zheng, University of Akron, United States of America

Funding: This work was supported by the National Natural Science Foundation (81371697), www.nsfc.gov.cn, to SCW; Beijing Natural Science Foundation (7132124), www.bjnsf.org, to SCW; Peking University's 985 Grant, www.pku.edu.cn, to SCW. The funders had no role in study design, data collection and analysis, decision to publish, or preparation of the manuscript.

Competing Interests: The authors have declared that no competing interests exist.

* Email: sc-wei@pku.edu.cn (SCW); lvyalin@vip.sina.com (YPL)

Introduction

Inspired by the composition of adhesive proteins in mussels, Messersmith et al [1] (2007) employed the innate self-polymerization attribute of dopamine to form thin and surface-adherent polydopamine (PDA) film, which later on has been widely applied as a surface modification agent. Polydopamine possesses covalent and non-covalent bonding capabilities for a broad range of organic, inorganic and metallic substrates [2], harboring potential applications in the challenging field of antibacterial [3], antifungal [4], antifouling [5,6], drug delivery vehicle [7], biosensor [8], cell culture [9], tissue engineering [10–12] and so on. However, to author's knowledge, dopamine solution is usually polymerized for at least several hours prior to surface modification/deposition. It is too long from practical point of view and it is crucial to develop a rapid PDA film formation method.

Although PDA coating has been used for numerous years and tremendous effort has been invested in understanding its structure, it has yet to be unambiguously determined. It is known that the process of dopamine-polymerization first involves oxidation of a catechol to a benzoquinone [13] and PDA film is a complex network with free catechol groups available for further chemical surface modification [14]. However, recent research suggests that PDA is a supra-molecular aggregate of monomers rather than a covalent polymer [15].

Despite the wide application and numerous structure analyses for polydopamine coating, much less effort has been focused on factors that would affect the deposition kinetics of polydopamine film, which has become an obstacle for further optimization. Dopamine self-polymerizes to polydopamine usually under slightly alkaline condition, a pH typical of marine environments (2 $g{\cdot}L^{-1}$ dopamine in 10 mM Tris-HCl, pH = 8.5). The thickness of PDA film is growing as a function of time in 24 h detected by spectroscopic ellipsometry [16]. Zhao, C et al [17] found that addition of oxidizing agents increased the rate of PDA formation under basic conditions. Moreover, the effect of pH and concentration on the deposition kinetics of dopamine solution has also been studied by spectroscopic ellipsometry, AFM [18] and surface plasmon resonance (SPR) [19]. A constant increase in the maximal film mass, thickness and roughness was observed as an augment in the dopamine concentration from 0.1 $g{\cdot}L^{-1}$ to 10 $g{\cdot}L^{-1}$, and a pH value of 8.5 was demonstrated to be the best pH for Tris-HCl buffer solution [18,19]. Furthermore, Xu,Y et al [20] observed that the self-polymerization speed of dopamine grew as temperature increased from 20°C to 60°C. Besides these influence factors, the reaction method could also play a key role in the deposition kinetics of PDA. There is, however, barely any effort devoted in the subject area, which results in inconsistent reaction temperature and time reported in a variety of research articles [21–26]. Currently, there are three major dopamine polymerization processes: static, shake, and stir. With regard to the static method, substrates are submerged in dopamine solution for 16 h [10], 18 h [3], overnight [21,22] or 24 h [12] at room temperature. In shake method, dopamine solutions are shaken to achieve uniform PDA film on the substrate at quite different reaction temperature and time (such as 1 h at ambient temperature [23], 24 h at ambient temperature [24] and 8 h at 30°C [25]). Stir method is popularly used recently, which protects substrates from the particle deposition in the solution [16]. Unfortunately, like static and shake method, the stirring time for dopamine-polymerization is undefined and varies from several hours [26] to 24 h [18]. Undefined conditions in all three methods generate PDA films of various thickness, mass and quality, and make it impossible to compare studies between different labs.

The main aim of this study is to investigate the deposition kinetics of PDA film as a function of temperature and stirring, and to develop a method for rapidly-deposited PDA coating under standard environment (2 $g{\cdot}L^{-1}$ dopamine in 10 mM Tris-HCl, pH = 8.5).In this investigation, the impact of temperature and reaction method on the deposition kinetics of PDA coating was evaluated by quartz crystal microbalance (QCM) sensor system, which is commonly used to monitor changes in mass at the quartz crystal sensing surface and capable of real-time *in situ* detection in solution [27]. It could be a powerful analytical tool to study the deposition kinetics and properties of PDA coating. In addition, a unique device was designed for preparation of rapidly-deposited PDA film, in which QCM chip substrates were vertically placed in dopamine solution under vigorous stirring (300 $r{\cdot}min^{-1}$) at 60°C and QCM sensor system was used to monitor the mass change of PDA coating on chip surface. PDA films were characterized using a variety of analytical tools and bio-performance of these films was also evaluated. Results show that using high temperature and vigorous stirring could dramatically speed up dopamine-polymerization and thirty minutes reaction time was found appropriate for surface modification. Our method has demonstrated great potential of being a rapid and practical approach for surface modification of biomaterials and medical devices using self-polymerized PDA coating.

Materials and Methods

Materials

Tris (hydroxymethyl) aminomethane (Tris) was purchased from Beijing Chemical Reagents Company (Beijing, China). Dopamine hydrochloride was obtained from Sigma (Missouri, USA). Bovine serum albumin (BSA) fraction V was purchased from Aladdin (Shanghai, China). Commercial pure titanium (Ti), grade 2, was provided from Northwest Institute for Non-ferrous Metal Research (Xi'an, China). Pure polyetheretherketone (PEEK) was purchased from Hua-jun Special Engineering Plastic Products co., LTD (Changzhou, China). Cefotaxime sodium (CS, $C_{16}H_{16}N_5O_7S_2Na$) was provided by Amresco Inc. (Ohio, USA).All reagents were of analytical grade. All aqueous solutions were prepared with distilled water.

Quartz crystal microbalance and chips

The QCM and chips were purchased from Dongwei Biological Technology Co., LTD (Hangzhou, China).The chips were AT-cut planar silicon oxide-coated QCM crystals (14 mm in diameter) with a 5 MHz (4.95 MHz±50 KHz) nominal resonance frequency. Prior to use, all QCM chips were cleaned in an UV/ozone Tip-Cleaner (BioForce Nanosciences, Ames, IA,USA) for 30 min and thoroughly rinsed with distilled water and ethanol for three times, then dried in a stream of nitrogen. Reaction solution was pumped through sensor with QCM chip by a peristaltic pump (BT100K, Baoding, China). The system and reaction solution were conditioned to a required temperature prior to experiment.

The QCM is based on the piezoelectric influence, where a deposited mass is registered as changes in frequency of an oscillating QCM crystal. The adsorbed mass can then be calculated by the Sauerbrey equation [28]:

$$\Delta m = -C\Delta f/n$$

where, Δm equals change in mass per unit surface area, C is the instrument sensitivity constant (17.7 $ng{\cdot}cm^{-2}$ Hz^{-1}), Δf is the frequency change of the specific harmonic and n denotes the number of overtone. The fundamental frequency of our study was simultaneously acquired at of 15 MHz (n = 3).

The influence of temperature and reaction method on the deposition kinetics and characteristics of polydopamine films

The effect of temperature on the polymerization kinetics of PDA coating was detected *in-situ* by QCM. The clean QCM chip was placed onto the QCM sensor system. QCM sensor system was conditioned at 25°C, 37°C and 60°C respectively prior to experiment. After the baseline became stable, 500 μL 2 $g{\cdot}L^{-1}$ dopamine solution (10 mM Tris-HCl buffer, pH = 8.5) was placed on the sensor surface to study the influence of temperature on dopamine-polymerization *in situ* and in real-time. Throughout this investigation, all solutions were preheated in a Thermostatic Water Bath (HWSX-650/T, Jinhua, China) at the same temperature to the QCM sensor system.

To investigate the influence of reaction method on dopamine-polymerization, the mass change of chips decorated by PDA films formed by conditions of static, shaking and stirring were measured. For the static method, 3 ml 2 $g{\cdot}L^{-1}$ dopamine solution (10 mM Tris-HCl buffer, pH = 8.5) was added into each well of 12-well plate (Corning, New York, USA) with cleaned chips and reacted at 60°C for 1 h. In the shaking method, cleaned chips were placed in a 12-well plate. 3 ml standard dopamine solution was injected into each well for 1 h at 60°C with shaking (300 $r{\cdot}min^{-1}$). In the stirring method, on the other hand, cleaned chips were vertically placed into 200 ml standard dopamine solution at 60°C with stirring (300 $r{\cdot}min^{-1}$) for 1 h. 300 $r{\cdot}min^{-1}$ was chosen as the maximal speed that would not affect the vertically placed QCM chips in dopamine solution to investigate the effect of stirring to dopamine-polymerization. The plate and buffer were preheated prior to experiment. After rinsed with distilled water and ethanol, each chip was dried in a stream of nitrogen and the mass change was investigated by QCM sensor system. Each group had three chips and their average values were calculated.

Preparation of rapidly deposited polydopamine coating

Figure 1 illustrates the formation of polydopamine coating on QCM chips surface by normal method and rapidly-deposited method. In the rapidly-deposited method, 200 ml 10 mM Tris-

HCl solution (pH = 8.5) was preheated in a constant temperature magnetic stirrer (SHJ-11, Jinhua, China) at 60°C, and the clean QCM chips were vertically placed into the solution in a uniquely designed device, which was made up of an retort stand, a wire loop and ten small clips. The chips substrates were fixed by the clips with the same interval and their surfaces were in a circle. For PDA coating, 0.4 g dopamine was dissolved and polymerized with vigorous stirring (300 r·min^{-1}) for 5 min, 10 min, 20 min,30 min, 1 h, 2 h, 4 h and 8 h respectively. PDA formed by our method named stir-PDA (sPDA). The mass changes of QCM chips decorated by PDA formed at 37°C with gently shaking at the speed of 70 r·min^{-1} for 24 h was studied as the control, named normal method (nPDA). Briefly, the cleaned chips were placed in 12-well plates (Corning, New York, USA). 3 ml 2 g·L^{-1} dopamine solution (10 mM Tris-HCl buffer, pH = 8.5) was injected into each well for 24 h at 37°C with shaking (70 r·min^{-1}). The plates and buffer were preheated prior to experiment. After rinsed with distilled water and ethanol, each chip was dried in a stream of nitrogen and the mass change was monitored by QCM sensor. To totally clear away the physically adsorbed PDA particles that not possess covalent bonding capability on substrate surface, PDA coated chips were cleaned in the distilled water by ultrasonic (KQ-5200DE, Kunshan, China) for different periods (1 min, 2 min, 5 min and 10 min) and the mass changes were investigated by QCM.

Moreover, the film thickness of sPDA coated chips with various polymerization times and 24 h-nPDA coated chips that had ultrasonically cleaned for 10 min was further investigated by spectroscopic ellipsometry (MD2000D, J.A.Woollam, Nebraska, USA).

Characterization

Particles of sPDA and nPDA that acquired from previously reacted dopamine solution were used to evaluate the chemical properties. Briefly, dopamine solutions polymerized by our newly developed method and normal method were filtered by 0.22 um filter respectively, and the obtained PDA particles were dried in a vacuum oven (SDZF-6050, Shanghai, China) at 37°C for 1 h. Fourier-transform infrared spectrometry (FTIR) (Magna-IR 750, Nicolet, USA) was used to identify the functional groups of sPDA and nPDA in the form of pellets (KBr pellet).The spectra were recorded from 4000 cm^{-1} to 400 cm^{-1}. The chemical constituents of sPDA and nPDA particles were analyzed by X-ray photoelectronic spectroscopy (XPS) (Kratos, Manchester, UK).The surface hydrophilicity and surface energy (SE) of QCM chips modified by sPDA or nPDA that hat had ultrasonically cleaned for 10 min were characterized by water contact angle measurements (Dataphysics Instrument, Stuttgart, Germany). Three measurements were performed for each chip at ambient temperature based on the sessile drop method, and the mean value was taken as the reported result. The SE value was determined according to OWRK method using distilled water. The surface topographies of sPDA and nPDA films on the surface of QCM chips that had ultrasonically cleaned for 10 min were analyzed by an atomic force microscope (AFM) (Buker, Massachusetts, USA). AFM images were obtained in contact mode. The scan range was 1.0 um × 1.0 um, and scan rate was 1 Hz. Ra was used to evaluate the surface roughness on the basis of scan area.Before AFM measurement, the chips were rinsed with distilled water and ethanol, and allowed to air dry.

BSA immobilization study

BSA was selected as the characteristic protein to investigate the adsorbing properties of sPDA film with various polymerization times (5 min, 10 min, 20 min, 30 min, 1 h, 2 h, 4 h and 8 h), and to compare with that of 24 h nPDA film. Briefly, sPDA and nPDA film decorated QCM chips that had ultrasonically cleaned for 10 min were, respectively, put onto the flow chamber, which was then continuously flushed with 0.01 M PBS buffer (pH = 7.4). After the baseline became stable, the solution was replaced by fresh 5 mg/ml BSA solution in PBS buffer. The reaction was conducted at 25°C by temperature monitor and all used solutions were preheated.

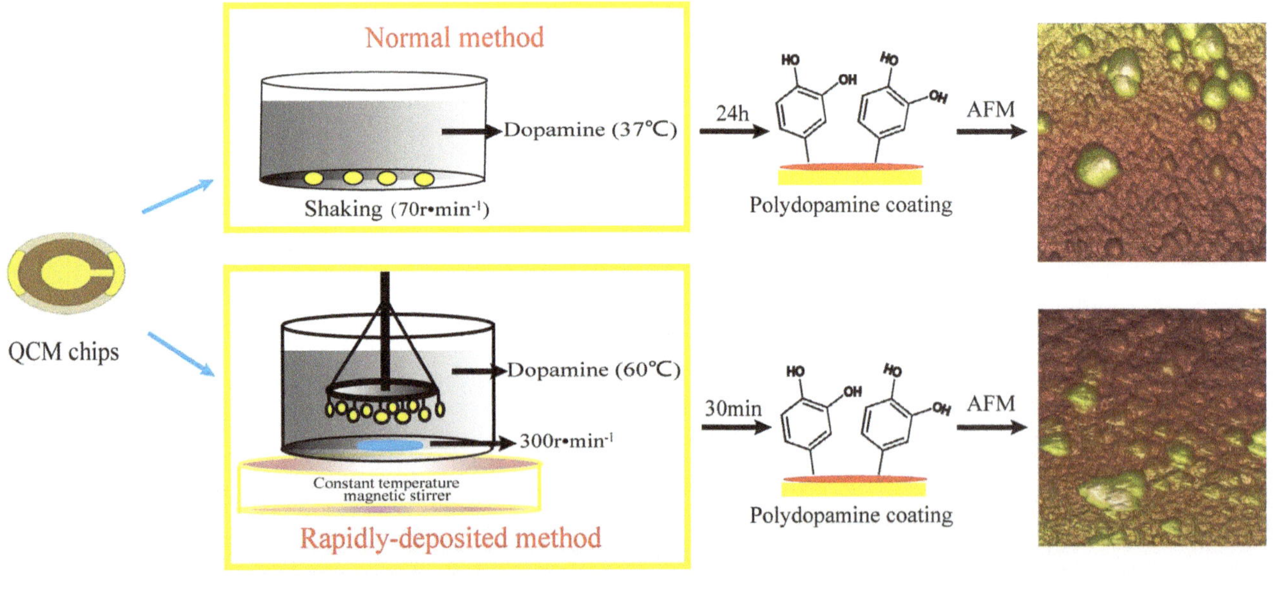

Figure 1. Scheme of polydopamine coating process on QCM chips surfaces by normal method and rapidly-deposited method. Substrates are vertically placed in standard dopamine solution at the temperature of 60°C with stirring (300 r·min^{-1}) using our uniquely designed device, which can significantly improve the deposition of polydopamine on QCM chips.

Cytotoxicity assay

The pristine QCM chips, 30 min-sPDA film coated chips and 24 h-nPDA film coated chips were sterilized under UV irradiation for 30 min, placed in a 24-well plate. The influence of 30 min-sPDA coating and 24 h-nPDA coating on cell viability was evaluated by CCK-8 assay (CCK-8, Dojindo, Kumamoto, Japan), using MG-63 osteoblast cells obtained from the American Type Culture Collection. Cells were grown in Dulbecco's modified Eagle medium (DMEM) (Hyclone, UT,USA), supplemented with 10% fetal bovine serum (Hyclone, UT,USA), 100 $U \cdot mL^{-1}$ penicillin (Amresco, Cleveland, USA), and 0.1 $mg \cdot mL^{-1}$ streptomycin (Amresco, Cleveland, USA) under standard cell culture conditions (37°C, 100% humidity, 95% air and 5% CO_2).

After cell counting, MG-63 osteoblast cell was seeded into 24-well culture plates with sterilized PDA film modified QCM chips at a density of 2×10^4 cells/well. The medium was replaced with fresh one every 24 h.

The control experiments were carried out using the complete growth culture medium without substrates (non-toxic control). After incubating for 24 h and 72 h, respectively, 500 µL medium containing 50 µL of CCK-8 reagent was added into each well and incubation went on for two more hours. Then, 200 µL of supernatant from each well was transferred to a new 96-well cell culture plate. The absorbance value (OD value) was measured at 450 nm with a microplate reader (Model 680, Bio-Rad, Hercules, CA). Three specimens were tested for each incubation period, and each test was performed in triplicate.

Evaluation of antibiotics-binding ability

To study the ability of sPDA film for the immobilization of biomolecules, the inhibition of bacterial adhesion on CS decorated Ti and PEEK surfaces by 30 min-sPDA and 24 h-nPDA was compared. The method for antibacterial activity evaluation is the same as previous reported [29]. Briefly, prior to surface modification, Ti and PEEK were cut into discs of 15 mm in diameter and 2 mm in thickness. These disc samples were polished with a series of SiC abrasive papers (400, 1000, 1500 and 2000 grit). They were then ultrasonically cleaned in acetone, anhydrous ethanol and distilled water, respectively, for 10 min and dried in a stream of nitrogen. 30 min-sPDA coated Ti and PEEK, 24 h-nPDA coated Ti and PEEK were prepared as described previously. 3 ml fresh 50 $mg \cdot mL^{-1}$ CS in distilled water was injected into each well of 12-well plate with modified Ti and PEEK at 37°C with shaking (70 $r \cdot min^{-1}$) for 24 h. They were then rinsed with distilled water and dried in a stream of nitrogen. 30 min-sPDA-CS decorated Ti, 24 h-nPDA-CS decorated Ti, 30 min-sPDA-CS decorated PEEK and 24 h-nPDA-CS decorated PEEK were sterilized under UV irradiation for 30 min on each side. *Escherichia coli* (*E. Coli* 1.1369, obtained from China General Microbiological Culture Collection Center, China) and *Streptococcus mutans* (*S. Mutans* UA159, obtained from American type Culture Collection, USA) were selected as the experimental strains. *E.Coli* and *S.Mutans* were cultured in Luria-Bertani (LB) medium and Brian Heart Infusion (BHI) medium respectively. Bacterial adhesion on the surfaces of pristine Ti, pristine PEEK, 30 min-sPDA-CS modified Ti, 30 min-sPDA-CS modified PEEK, 24 h-nPDA-CS modified Ti and 24 h-nPDA-CS modified PEEK for various period (4 h, 24 h and 72 h) was assessed *via* Microbial Viability Assay Kit-WST (Dojindo, Kumamoto, Japan) and a microplate reader (Elx808, Bio-tek, Vermont, USA).Moreover, the activity of bacterial in the medium was also measured to evaluate the release action of CS on those samples. Minimum bactericidal concentration of CS was added into the mixture of bacterial solution and culture medium as the control.

Statistical analysis

The results were expressed as mean ± standard deviations (SD) derived from experiment and assessed statistically using Student's t-test. Statistical analysis was carried out with software SPSS13.0 for Windows. Statistical significance was accepted at $p<0.05$.

Results and Discussion

Dopamine-polymerization as a function of temperature and reaction method

The influence of temperature on the deposition kinetics of PDA coating was *in-situ* and real-time monitored for the first time by QCM (Figure 2A). The Δm signals of PDA films on the sensor surface markedly augment with an increase in the polymerization temperature between 25°C and 60°C, indicating temperature has significant impact on dopamine-polymerization and subsequent deposition. According to the thermodynamic principle, higher temperature would promote the oxidation of dopamine as well as the deposition process of PDA onto substrate surface. This result is consistent with investigation of dopamine polymerization under various temperatures measured by spectroscopic ellipsometry [20].

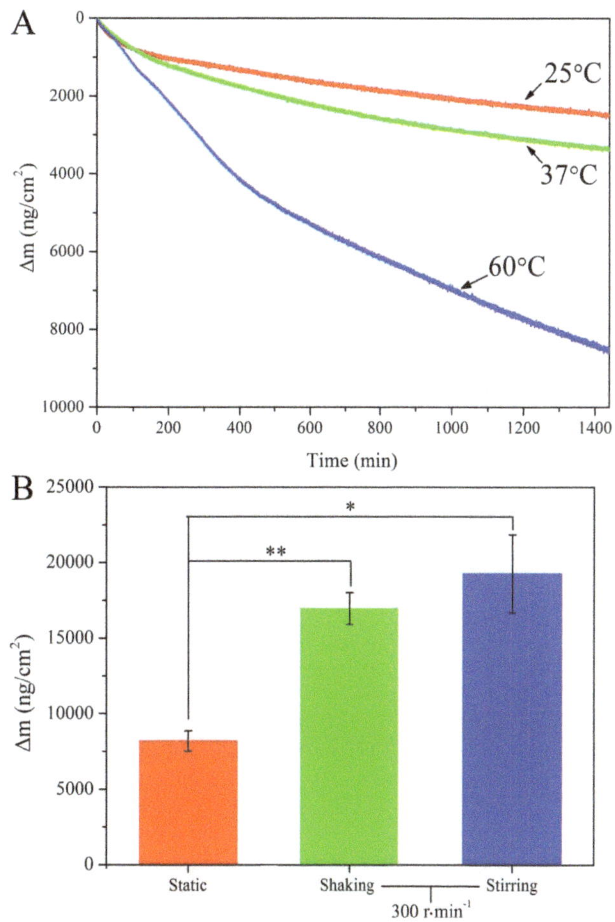

Figure 2. The influence of temperature and reaction method on the deposition kinetics of polydopamine films. (A): Representative mass change vs. time (min) curves as a function of temperature for polymerization of dopamine solution. (B): The mass change of QCM chips coated by PDA formed by methods of static, shaking (300 $r \cdot min^{-1}$) and stirring (300 $r \cdot min^{-1}$) for 1 h respectively. *Represents $p<0.05$ and ** Represents $p<0.01$ compared with chips coated by PDA formed at static group, n = 3.

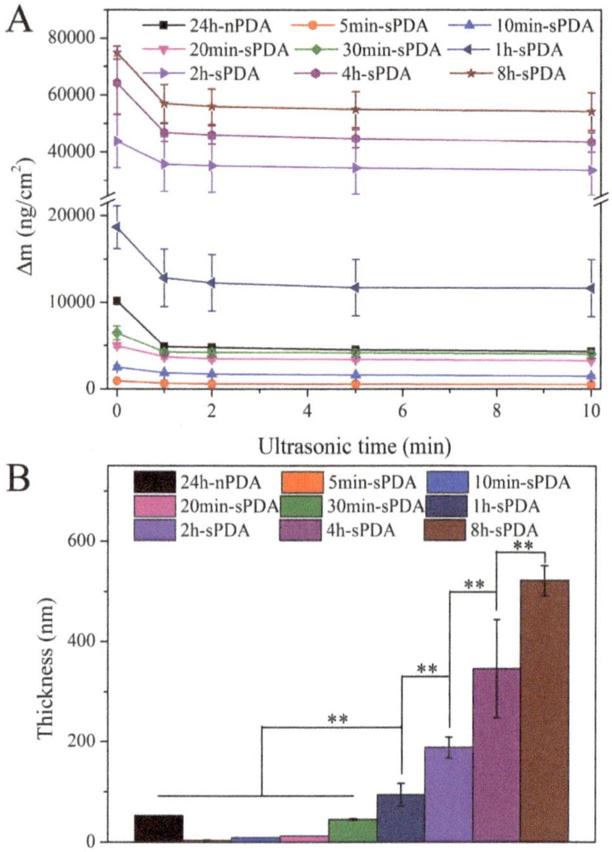

Figure 3. The mass and thickness measurement of sPDA films and nPDA films. (A): The mass change of QCM chips decorated by sPDA films of varying polymerization times (5 min, 10 min, 20 min, 30 min, 1 h, 2 h, 4 h and 8 h) as a function of ultrasonic cleaning time, and compared with that of 24 h-nPDA film coated chips. (B): The film thickness of QCM chips coated by sPDA of different polymerization times (5 min, 10 min, 20 min, 30 min, 1 h, 2 h, 4 h and 8 h) and 24 h-nPDA. Those chips had ultrasonically cleaned in distilled water for 10 min before measurement. ** Represents $p<0.01$, $n=3$.

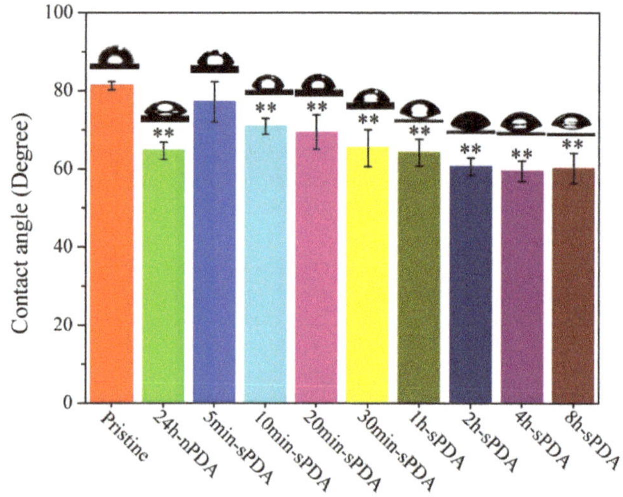

Figure 4. Water contact angles of QCM chips modified by sPDA films and nPDA films. The Contact angles of pristine QCM chips and QCM chips modified by coating of 24 h-nPDA, 5 min-sPDA, 10 min-sPDA, 20 min-sPDA 30 min-sPDA, 1 h-sPDA, 2 h-sPDA, 4 h-sPDA and 8 h-sPDA. All samples were ultrasonically cleaned for 10 min and dried. ** Represents $p<0.01$ compared with the pristine chips group, $n=3$.

The mass change of QCM chips decorated by PDA films formed from methods of static, shaking and stirring was measured to evaluate the effect of reaction method on the polymerization behavior of PDA coating (Figure 2B). The mass change of chips decorated by PDA film polymerized for 1 h at 60°C with stirring method (19275 ± 2575 ng·cm^{-2}) is a little higher than that of PDA film from shaking method (16981 ± 1037 ng·cm^{-2}) ($p>0.05$), and both are significantly higher than that of PDA film formed at static condition (8202 ± 665 ng·cm^{-2}) ($p<0.05$), indicating vigorous shaking and stirring could dramatically increase the formation speed of PDA film on the sensor surface. That is presumably due to more fresh oxygen in air dissolved in solution with shaking or stirring, which contributes to the self-oxidation of dopamine. However, some disadvantages, such as solution will be splashed out and the chips may turned over to the back side, were observed for dopamine-polymerization with vigorous shaking. This may be the reason why most reported papers we know had used shaking method at low speed for the polymerization of dopamine [23,25,29]. Therefore, vigorous stirring possess good potential to be the reaction method to facilitate the polymerization of dopamine solution.

Mass and thickness of polydopamine film

For surface modification, dopamine solution is generally polymerized for several hours in the presence of substrates, which is too long for any practical applications. Knowing that both high temperature and vigorous stirring could promote the polymerization of dopamine, a unique device was designed, which uses a constant temperature magnetic stirrer for the formation of rapidly-deposited PDA coating. The mass change of chips decorated by sPDA films with varying polymerization times (5 min, 10 min, 20 min,30 min,1 h, 2 h, 4 h and 8 h) was measured by QCM. To better evaluate PDA coating formed by our method, dopamine solution polymerization under normal condition was conducted as control. As mentioned in the introduction, the dopamine-polymerization process is quite different in a variety of research articles [21–26]. In our study, condition of 37°C with gently shaking at the speed of 70 r·min^{-1} for 24 h was used as the normal methodfor the reasons as follows: 1). 37°C is the most used reaction temperature; 2). Gentle shaking is popularly applied for the polymerization of dopamine solution [23,25,29]; 3). 24 h is the longest reported reaction time reported for PDA coating at which the mass and film thickness will get equilibrium [1].

It has been recognized that excess physically adsorbed PDA particles on the surface of substrate do not have the covalent bonding capability. It is crucial to eliminate the potentially detrimental effect from physically adsorbed PDA particles on the cell functions. PDA films on the substrates are usually washed with distilled water for three times [30,31]. Compared with conventional water rinsing procedure, ultrasonic cleaning may be more efficient and stable. Unfortunately, various ultrasonic times, such as 5 min [32],10 min thrice [12,33] and 15 min [34], were reported to clean out physically adsorbed PDA particles on substrate surface, and the cleaning efficiency had not been clearly demonstrated. In the present study, we assessed the mass changes of PDA film decorated chips undergone ultrasonic cleaning for 1 min, 2 min, 5 min and 10 min, respectively. A great amount of physically adsorbed PDA was obliterated after ultrasonic cleaning for 5 min, and mass change of PDA-coated chips reached

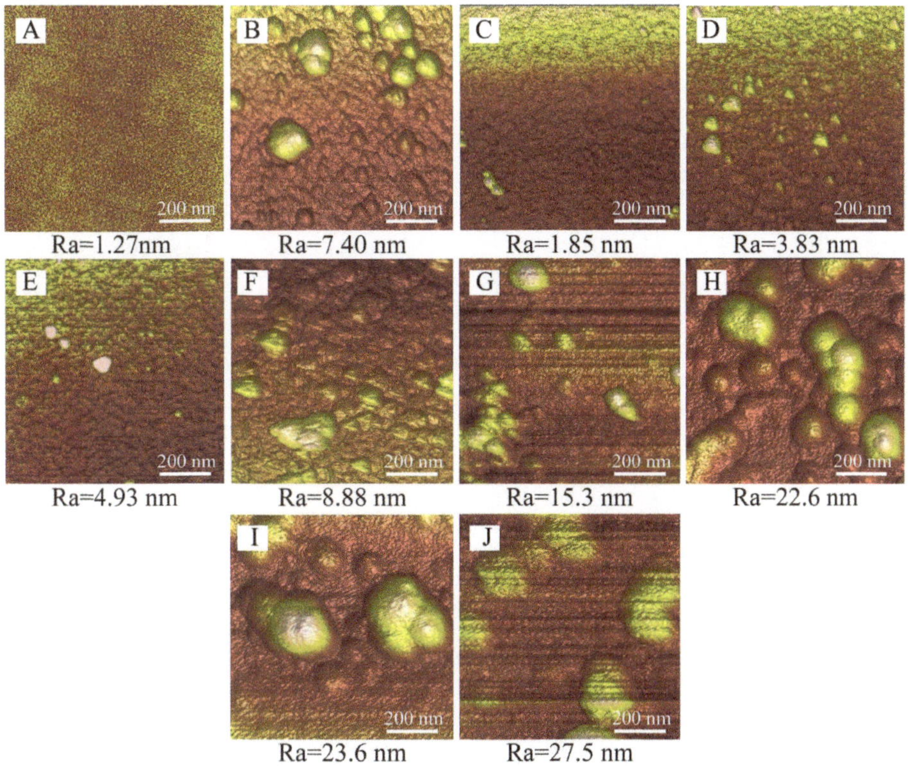

Figure 5. The AFM surface topographies of QCM chips modified by sPDA films and nPDA films. (A): Pristine chip. (B) 24 h-nPDA decorated chip. (C): 5 min-sPDA decorated chip. (D): 10 min-sPDA decorated chip. (E): 20 min-sPDA decorated chip. (F): 30 min-sPDA decorated chip. (G): 1 h-sPDA decorated chip. (H): 2 h-sPDA decorated chip. (I): 4 h-sPDA decorated chip. (J): 8 h-sPDA decorated chip. All those samples had ultrasonically cleaned in distilled water for 10 min. Scale bar: 200 nm.

equilibrium with the extension of time (Figure 3A), suggesting that 5 min treatment is enough to get rid of excessive PDA particles on the sensor surface. This is in consistent with the ultrasonic cleaning time on the surface of silicon wafers reported by Junfei Ou et al [32].

After ultrasonically cleaned, it was found that the mass change of chips modified by PDA coating attained by our method was dramatically increased with the augment of the polymerization time, and reached a plateau after reacted for 8 h (Figure S1 in File S1). After polymerization for 1 h, the mass change of chips decorated by sPDA films is much bigger than that of 24 h-nPDA ($p<0.01$). Interestingly, the mass change of 30 min-sPDA coated chips (6448±788 ng·cm^{-2}) was much less than that of 24 h-nPDA (10186±453 ng·cm^{-2}) prior to cleaning (Figure 3A). However, after ultrasonically cleaned, the mass change of chips modified by 30 min-sPDA films (3992±454 ng·cm^{-2}) is in close vicinity to that of 24 h-nPDA (4324±178 ng·cm^{-2}) (Figure 3A). These results indicate fewer physically absorbed sPDA particles existed on the chip surface than that of nPDA, since vigorous stirring could protect substrates from the particle aggregation in dopamine solution. Besides, there is no PDA on the surface of substrate at the first 30 min by common method [35], But evident mass change was observed when dopamine-polymerization went on for 5 min by our method.

The film thickness of chips coated by sPDA films with different polymerization times (5 min, 10 min, 20 min,30 min,1 h, 2 h, 4 h and 8 h) and 24 h-nPDA that had been ultrasonically cleaned for 10 min was further investigated by spectroscopic ellipsometry. In accordance with the mass change, the film thickness of sPDA on the sensor surfaces increases with the increase in polymerization time and get equilibrium after reacted for 8 h (Figure S1 in File S1). As shown in Figure 3, part B, the thickness of 30 min-sPDA film (44.95±1.15 nm) on the surface of chips is slightly under that of 24 h-nPDA (53.55±0.25 nm) ($p>0.05$), which is close to 50 nm that reported by Messersmith et al [1]. However, the thickness of sPDA films after polymerized for 1 h are much higher than that of 24 h-nPDA ($p<0.01$) (Figure 3B). As shown in Figure S3 in File S1, a good correlation was observed for the curve of sPDA mass versus time and that thickness of versus time, indicating the average molecular weight and density of PDA exhibit no significant difference with the augment of polymerization time, since PDA particles is presumed to be hierarchical aggregation comprised with plate-like aggregates at a size of 1–2 nm through π–π stacking [36].

Those mass and film thickness measurements of PDA coating prove that our method using thermal-dynamic treatment of high temperature and vigorous stirring could not only dramatically promote the polymerization speed of dopamine solution, but also deposit much more PDA onto the substrate surface. Moreover, the mass and film thickness of PDA coating after polymerization for 30 min by our approach was similar to those of 24 h counterpart, possess good prospects to be the condition for rapid-deposition of PDA films to modify biomaterial surfaces.

Characterization of PDA films

The chemical attributes of sPDA and nPDA were probed using Fourier-transform infrared spectrometry (FTIR), X-ray photoelectronic spectroscopy (XPS) and water contact angle measurement. FTIR spectra (KBr) of sPDA and nPDA powders are similar

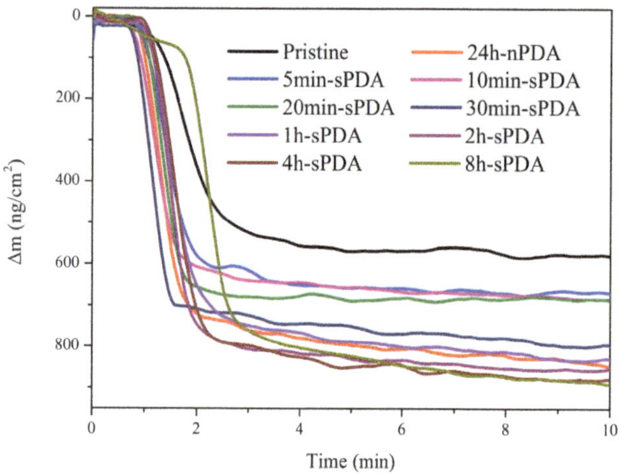

Figure 6. BSA binding evaluation of sPDA films and nPDA films. Representative mass change vs. time (min) curves of pristine chip, 24 h-nPDA coated chip, 5 min-sPDA coated chip, 10 min-sPDA coated chip, 20 min-sPDA coated chip, 30 min-sPDA coated chip, 1 h-sPDA coated chip, 2 h-sPDA coated chip, 4 h-sPDA coated chip and 8 h-sPDA coated chip reaction with 5 mg·ml^{-1} BSA solution.

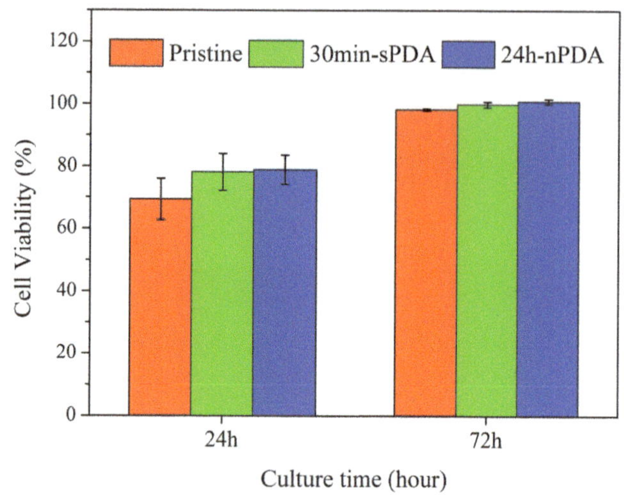

Figure 7. Cytotoxicity assay of QCM chips modified by 30 min-sPDA films and 24 h-nPDA films. Viability of MG63 osteoblasts incubated with pristine QCM chips, 30 min-sPDA decorated chips and 24 h-nPDA decorated chips for 24 h and 72 h. Viability is expressed as a percentage relative to the result obtained with the non-toxic control (MG63 osteoblasts incubated without substrates). n = 3.

(Figure S2 in File S1). According to Daniel R. Dreyer et al's previous study [15], the peaks at 1510 cm^{-1} and 1600 cm^{-1} are consistent with the indole or indoline structures. The large peak at 3376 cm^{-1} is in accordance with the presence of hydroxyl structures as well as water. No evident difference in chemical element content of C, N and O was observed between sPDA powder and nPDA powder in X-ray photo-electronic spectroscopy (XPS) (Figure S3A in File S1).Moreover, the high-resolution C 1 s spectra of sPDA powder and nPDA powder were deconvoluted into five different curves respectively and they are similar(Figure S3B in File S1). The binding energies centered at 284.60 eV, 285.59 eV, 286.01 eV,287.71 eV and 291.43 eV were assigned to the carbon skeleton (-C-C-/-C-H-), amino group (-C-N-), hydroxyl group (-C-OH), carbonyl group (-C = O and -C(O)O-) respectively [29,37].These results indicate that rapidly-deposited PDA film formed by the treatment of high temperature and vigorous stirring, compared with nPDA, exhibits no chemical alteration. Figure 4 shows the water contact angles of the pristine QCM chip, sPDA film modified chips with different polymerization times and 24 h-nPDA film modified chips that hat had ultrasonically cleaned for 10 min. The pristine chip is hydrophobic, with a contact angle of 81.30±1.02°. After coated by 5 min-sPDA film, the contact angle of chips presents no evident change. However, significant decrease was observed for the contact angle of chips (70.95±2.03°) when the dopamine-polymerization time increased to 10 min. It remained in the close vicinity to those of 30 min-sPDA (65.34±4.96°), 1 h-sPDA film (64.18±3.40°)and 24 h-nPDA film coated chips (64.67±2.19°). The contact angles reduced to 60.66±2.18° when polymerization time increased to 2 h and reached an equilibrium plateau with the extension of time. These results indicate sPDA coating possesses good hydrophilicity.

The surface topographies of sPDA films of varying polymerization times (5 min, 10 min, 20 min,30 min,1 h, 2 h, 4 h and 8 h) and 24 h-nPDA films on chips that had ultrasonically cleaned for 10 min were analyzed by an atomic force microscope (AFM) (Figure 5). The Ra value of chip decorated by sPDA coating augments with the increase of polymerization time. For sPDA coating at 5 min of dopamine polymerization time by our method, partial aggregates were formed on the surface of QCM chips (Figure 5C). When the polymerization time increased to 10 min and 20 min, more PDA aggregates appeared on chip surface, but there was no significant change in particle size and Ra value, which is consistent with the thickness measurement (Figure 5D–E). The surface topography of 30 min-sPDA film modified chip (Ra = 8.88 nm) (Figure 5F) is similar to 24 h-nPDA film coated chip (Ra = 7.40 nm) (Figure 5B), with a significant higher roughness than that of 20 min-sPDA film (Ra = 4.93 nm). Much higher surface roughness is observed for chip coated by sPDA after polymerized for 1 h and the Ra value reaches up to 27.5 nm for 8 h-sPDA (Figure 5G–J). These results are in accordance with the mass change and film thickness measurements, and further confirms that high temperature and vigorous stirring could dramatically improve the polymerization speed of dopamine solution.

BSA binding ability evaluation

BSA was selected to be the characteristic protein in real-time evaluation of the protein adhesion properties on the sPDA film coated chips surface by QCM, and in comparison with that of 24 h-nPDA film coated chips (Figure 6). In accordance with the mass measurement, the Δm signals of sPDA films reaction with 5 mg·ml^{-1} BSA solution augment with an increase in polymerization time between 5 min and 8 h, and present higher mass change than that of the pristine chips. The curves of mass change versus time for chips coated by 5 min-sPDA, 10 min-sPDA, 20 min-sPDA reacted with BSA are similar. When the dopamine polymerization time of our method increases to 30 min, much more BSA is bonded onto the chip surface, and the Δm reaches up to 800 ng·cm^{-2}, which is slightly less than that of 24 h-nPDA decorated chip (840 ng·cm^{-2}).The curve of mass change versus time for chip coated by 1 h-sPDA reacted with BSA is almost overlap with that of 24 h-nPDA. The mass changes of chips coated by 2 h-sPDA (859 ng·cm^{-2}), 4 h-sPDA (876 ng·cm^{-2}) and 8 h-sPDA (891 ng·cm^{-2}) reacted with BSA are similar. The small difference among sPDA coating at a time scale of twenty minutes in film thickness and surface topography contributes to the slow growth of Δm signals reacted with BSA solution. In the same way,

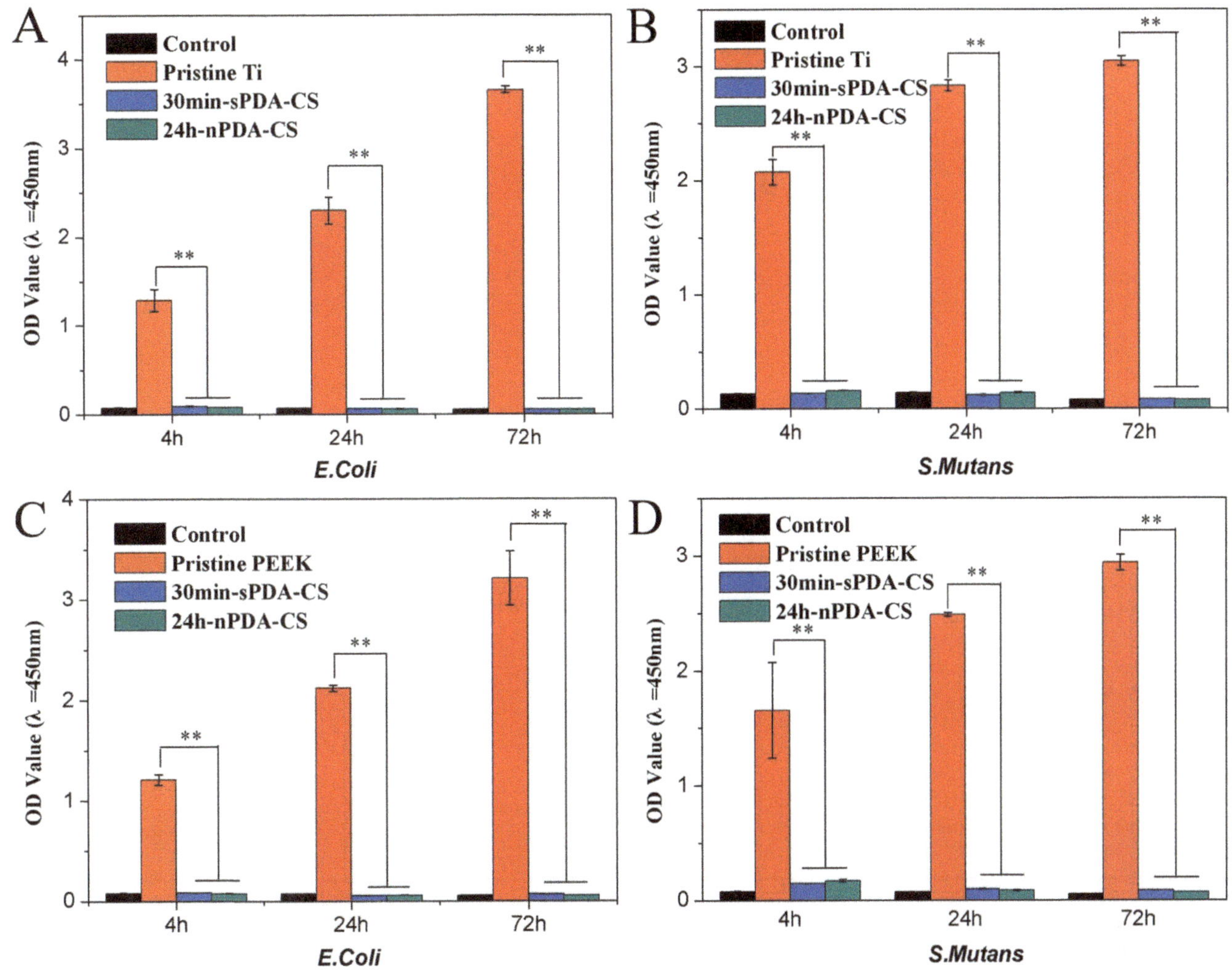

Figure 8. Bacterial adhesion on the surface of 30 min-sPDA-CS and 24 h-nPDA-CS decorated Ti and PEEK. Number of living *E.Coli* and *S.Mutans* adhered on decorated Ti and PEEK surfaces after exposed to bacterial suspension for 4 h, 24 h and 72 h. ** represents $p<0.01$ compared with the pristine group, $n=3$.

the film thickness and surface topography is significantly changed after polymerized for 30 min and a much higher Δm signal was detected. Even though evident augment was observed for the mass changes and film thickness of chips coated by sPDA with the polymerization time increased from 1 h to 8 h, their BSA absorbing ability possess no big difference. The reason is that the surface of sPDA decorated chips has been covered with PDA film after polymerized for 30 min, and there is no significant change in surface coverage with the extension of reaction time.

Cytotoxicity assay

Previous results indicate coating of sPDA and nPDA exhibit same chemical property. The mass change, film thickness and BSA binding ability of 30 min-sPDA coating is in close vicinity to that of 24 h-nPDA. Moreover, 30 min-sPDA coating and 24 h-nPDA possess similar surface morphology and hydrophilicity. Therefore, 30 min was chosen as the standard polymerization time for our rapidly-deposited PDA coating in follow-up cytotoxicity study and surface functional modification.

The effect of 30 min-sPDA film and 24 h-nPDA film on cell viability was evaluated by CCK-8 assay with MG-63 osteoblasts (Figure 7). After incubated for 24 h, the cell viability of 30 min-sPDA coated chips is slightly higher than that of pristine chips, and similar to that of 24 h-nPDA coated chips. The cell viability of all samples is significantly lower than that of the culture plate ($p<0.01$). However, MG63 osteoblasts on all samples exhibit almost the same cell viability when incubated for 72 h. Our study suggest that both sPDA film and nPDA film possess good cytocompatibility, which is consistent with previous report [25].

Antibiotics evaluation

Antibiotic property is one of the most important properties for biomaterials, especially for implant applications. In order to explore the practical use of our newly-developed method, inhibition of bacterial adhesion on the surface of 30 min-sPDA-CS and 24 h-nPDA-CS decorated Ti and PEEK for up to 72 h was investigated and the results are shown in Figure 8. For the amount of adhered viable *E.Coli* and *S.Mutans*, no statistically significant difference was found between 30 min-sPDA-CS modified Ti and 24 h-nPDA-CS modified Ti for 4 h, 24 h and 72 h ($p>0.05$). Moreover, their bacteria adhesion amounts are dramatically lower than that of the pristine Ti ($p<0.01$) (Figure 8A–B). Inhibition of bacterial adhesion on the surface of 30 min-sPDA-CS decorated PEEK and 24 h-nPDA-CS decorated

PEEK, on the other hand, is the same to that of Ti (Figure 8C–D). Furthermore, both for Ti and PEEK, the number of *E.Coli* and *S.Mutans* in the medium with 30 min-sPDA-CS modified substrates is similar to that of 24 h-nPDA-CS modified substrates for up to 72 h ($p>0.05$), which is significantly lower than that of the pristine ($p<0.01$) (Figure S4 in File S1). These results demonstrate superior bioactivity of PDA films formed by our method, which has great potential to be universal approach for surface modification.

Conclusions

In this study, a new method for the formation of rapidly-deposited PDA film was developed, in which substrates are vertically placed in standard dopamine solution at the temperature of 60°C with vigorous stirring (300 $r \cdot min^{-1}$) using an uniquely designed device. The influence of temperature and stirring on the deposition kinetics of PDA film was investigated by the quartz crystal microbalance (QCM). Our results indicate that high temperature and vigorous stirring both can dramatically speed up the rate of dopamine-polymerization. Physical treatment of high temperature and vigorous stirring would result in no chemical changes for polydopamine. The mass, thickness and BSA binding ability of PDA film formed by our approach for 30 min are comparable to those of shaking counterpart for 24 h. Moreover, cell viability assay and antibacterial test demonstrate that our rapidly-deposited PDA coating possesses equally excellent surface modification properties compared with PDA formed by normal method. Furthermore, substrates placed vertically in our approach could not only reduce the deposited PDA aggregates from solution, but also be decorated with PDA film on all surfaces. Our method for rapid deposition of polydopamine film may be used as a potential approach for the surface modification of biomaterials and medical devices.

Supporting Information

File S1 Contains the following files: Figure S1, Mass change versus time and film thickness versus time for sPDA decorated chips. The black curve shows the mass changes versus time of QCM chips coated by sPDA with various polymerization time (5 min, 10 min, 20 min, 30 min, 1 h, 2 h, 4 h and 8 h) that had ultrasonically cleaned for 10 min. The red curve shows the film thickness versus time of the same samples. n = 3. Figure S2, FTIR spectra analysis. The FTIR spectra (KBr) of nPDA powder and sPDA powder. Figure S3, XPS survey scan spectra of nPDA powder and sPDA powder. (A): XPS wide spectra of nPDA powder and sPDA powder, insert table shows the contents of C, N and O. (B): High-resolution spectrum of carbon peaks (C 1 s) for nPDA powder and sPDA powder respectively. Figure S4, Bacterial adhesion in the medium with 30 min-sPDA-CS and 24 h-nPDA-CS decorated Ti and PEEK. Number of living *E.Coli* and *S.Mutans* in the medium cultured with modified Ti and PEEK after exposed to bacterial suspension for 4 h, 24 h and 72 h. ** represents $p<0.01$ compared with the pristine group, n = 3.

Acknowledgments

We thank Mrs. Chao Xu from State Key Laboratory of Advanced Optical Communication Systems & Networks for the help of AFM measurement in this manuscript.

Author Contributions

Conceived and designed the experiments: PZ HZ HWM YPL SCW. Performed the experiments: PZ YD BRL RRZ. Analyzed the data: PZ YD BRL. Contributed reagents/materials/analysis tools: PZ BRL HWM YPL SCW. Wrote the paper: PZ YPL SCW.

References

1. Lee H, Dellatore SM, Miller WM, Messersmith PB (2007) Mussel-Inspired surface chemistry for multifunctional coatings. Science 318: 426–430.
2. Brubaker CE, Messersmith PB (2012) The present and future of biologically inspired adhesive interfaces and materials. Langmuir 28: 2200–2205.
3. Sileika TS, Kim HD, Maniak P, Messersmith PB (2011) Antibacterial performance of polydopamine-modified polymer surfaces containing passive and active components. ACS Appl Mater Interfaces 3: 4602–4610.
4. Paulo CSO, Vidal M, Ferreira LS (2010) Antifungal nanoparticles and surfaces. Biomacromolecules 11: 2810–2817.
5. Ding X, Yang C, Lim TP, Hsu LY, Engler AC, et al. (2012) Antibacterial and antifouling catheter coatings using surface grafted PEG-b-cationic polycarbonate diblock copolymers. Biomaterials 33: 6593–6603.
6. Cho JH, Shanmuganathan K, Ellison CJ (2013) Bioinspired catecholic copolymers for antifouling surface coatings. ACS Appl Mater Interfaces 5: 3794–3802.
7. Cui J, Yan Y, Such GK, Liang K, Ochs CJ, et al. (2012) Immobilization and intracellular delivery of an anticancer drug using mussel-inspired polydopamine capsules. Biomacromolecules 13: 2225–2228.
8. Peng HP, Liang RP, Zhang L, Qiu JD (2013) Facile preparation of novel core-shell enzyme-Au-polydopamine-Fe_3O_4 magnetic bionanoparticles for glucose sensor. Biosens Bioelectron 42: 293–299.
9. Rim NG, Kim SJ, Shin YM, Jun I, Lim DW, et al. (2012) Mussel-inspired surface modification of poly (L-lactide) electrospun fibers for modulation of osteogenic differentiation of human mesenchymal stem cells. Colloids Surf B - Biointerfaces 91: 189–197.
10. Ku SH, Park CB (2010) Human endothelial cell growth on mussel-inspired nanofiber scaffold for vascular tissue engineering. Biomaterials 31: 9431–9437.
11. Tsai WB, Chen WT, Chien HW, Kuo WH, Wang MJ (2011) Poly(dopamine) coating of scaffolds for articular cartilage tissue engineering. Acta Biomater 7: 4187–4194.
12. Zhou YZ, Cao Y, Liu W, Chu CH, Li QL (2012) Polydopamine-induced tooth remineralization. ACS Appl Mater Interfaces 4: 6901–6910.
13. Lynge ME, van der Westen R, Postma A, Stadler B (2011) Polydopamine–a nature-inspired polymer coating for biomedical science. Nanoscale 3: 4916–4928.
14. Herlinger E, Jameson RF, Linert W (1995) Spontaneous autoxidation of dopamine. Journal of the Chemical Society, Perkin Transactions 2: 259–263.
15. Dreyer DR, Miller DJ, Freeman BD, Paul DR, Bielawski CW(2012) Elucidating the structure of poly(dopamine). Langmuir 28: 6428–6435.
16. Kang K, Choi IS, Nam Y (2011) A biofunctionalization scheme for neural interfaces using polydopamine polymer. Biomaterials 32: 6374–6380.
17. Wei Q, Zhang F, Li J, Li B, Zhao C (2010) Oxidant-induced dopamine polymerization for multifunctional coatings. Polym Chem 1: 1430.
18. Vincent B, Doriane DF, Valérie T, David R (2012) Kinetics of polydopamine film deposition as a function of pH and dopamine concentration: Insights in the polydopamine deposition mechanism. Colloid Interface Sci 386: 366–372.
19. Li H, Cui D, Cai H, Zhang L, Chen X, et al. (2013) Use of surface plasmon resonance to investigate lateral wall deposition kinetics and properties of polydopamine films. Biosens Bioelectron 41: 809–814.
20. Jiang J, Zhu L, Zhu L, Zhu B, Xu Y (2011) Surface characteristics of a self-polymerized dopamine coating deposited on hydrophobic polymer films. Langmuir 27: 14180–14187.
21. Bhang SH, Kwon SH, Lee S, Kim GC, Han AM, et al. (2013) Enhanced neuronal differentiation of pheochromocytoma 12 cells on polydopamine-modified surface. Biochem Biophys Res Commun 430: 294–1300.
22. Wang LQ, Jeong KJ, Chiang HH, Zurakowski D, Behlau I, et al. (2011) Hydroxyapatite for keratoprosthesis biointegration. INVEST OPHTH VIS SCI 52: 7392–7399.
23. Lee YB, Shin YM, Lee JH, Jun I, Kang JK, et al. (2012) Polydopamine-mediated immobilization of multiple bioactive molecules. Biomaterials 33: 8343–8352.
24. Wei Q, Li B, Yi N, Su B, Yin Z, et al. (2011) Improving the blood compatibility of material surfaces via biomolecule-immobilized mussel-inspired coatings. J Biomed Mater Res A 96: 38–45.
25. Zhu LP, Jiang JH, Zhu BK, Xu YY (2011) Immobilization of bovine serum albumin onto porous polyethylene membranes using strongly attached polydopamine as a spacer. Colloids Surf B Biointerfaces 86: 111–118.
26. Zheng L, Liu Q, Xiong L, Li Y, Han K, et al. (2012) Controlled preparation of titania nanofilm by a template of polydopamine film and its reversible wettability. Thin Solid Films 520: 2776–2780.

27. Fatisson J, Azari F, Tufenkji N (2011) Real-time QCM-D monitoring of cellular responses to different cytomorphic agents. Biosens Bioelectron 26: 3207–3212.
28. G S (1959) Verwendung von schwingquarzen zur wagung dunner schichten und zur mikrowagung. Z Angew Phys 155: 206–222.
29. He S, Zhou P, Wang L, Xiong X, Zhang Y, et al. (2014) Antibiotic-decorated titanium with enhanced antibacterial activity through adhesive polydopamine for dental/bone implant. J R Soc Interface 11: 20140169.
30. Lynge ME, Ogaki R, Laursen AO, Lovmand J, Sutherland DS, et al. (2011) Polydopamine/liposome coatings and their interaction with myoblast cells. ACS Appl Mater Interfaces 3: 2142–2147.
31. Shin YM, Lee YB, Shin H (2011) Time-dependent mussel-inspired functionalization of poly (L-lactide-co-varepsilon-caprolactone) substrates for tunable cell behaviors. Colloids Surf B Biointerfaces 87: 79–87.
32. Junfei Ou, Jinqing Wang, Sheng Liu, Zhou J, Yang S (2009) Self-assembly and tribological property of a novel 3-layer organic film on silicon wafer with polydopamine coating as the Interlayer. J Phys Chem C 113: 20429–20434.
33. Luo R, Tang L, Zhong S, Yang Z, Wang J, et al. (2013) In vitro investigation of enhanced hemocompatibility and endothelial cell proliferation associated with quinone-rich polydopamine coating. ACS Appl Mater Interfaces 5: 1704–1714.
34. Pop-Georgievski O, Verreault D, Diesner MO, Proks V, Heissler S, et al. (2012) Nonfouling poly (ethylene oxide) layers end-tethered to polydopamine. Langmuir 28: 14273–14283.
35. Bernsmann F, Ponche A, Ringwald C, Hemmerle J, Raya J, et al. (2009) Characterization of dopamine-melanin growth on silicon oxide. J Phys Chem C 113: 8234–8242.
36. Lynge ME, van der Westen R, Postma A, Städler B (2011) Polydopamine—a nature-inspired polymer coating for biomedical science. Nanoscale 3: 4916–4928.
37. Delpeux S, Beguin F, Benoit R, Erre R, Manolova N, et al. (1998) Fullerene core star-like polymers-1. Preparation from fullerenes and monoazidopolyehers. Eur Polym J 34: 905–915.

3

High Calcium Bioglass Enhances Differentiation and Survival of Endothelial Progenitor Cells, Inducing Early Vascularization in Critical Size Bone Defects

Karam Eldesoqi[¶], Caroline Seebach*[¶], Christina Nguyen Ngoc, Simon Meier, Christoph Nau, Alexander Schaible, Ingo Marzi, Dirk Henrich

Department of Trauma-, Hand- and Reconstructive Surgery, Hospital of the Goethe- University, Frankfurt/Main, Germany

Abstract

Early vascularization is a prerequisite for successful bone healing and endothelial progenitor cells (EPC), seeded on appropriate biomaterials, can improve vascularization. The type of biomaterial influences EPC function with bioglass evoking a vascularizing response. In this study the influence of a composite biomaterial based on polylactic acid (PLA) and either 20 or 40% bioglass, BG20 and BG40, respectively, on the differentiation and survival of EPCs *in vitro* was investigated. Subsequently, the effect of the composite material on early vascularization in a rat calvarial critical size defect model with or without EPCs was evaluated. Human EPCs were cultured with β-TCP, PLA, BG20 or BG40, and seeding efficacy, cell viability, cell morphology and apoptosis were analysed *in vitro*. BG40 released the most calcium, and improved endothelial differentiation and vitality best. This effect was mimicked by adding an equivalent amount of calcium to the medium and was diminished in the presence of the calcium chelator, EGTA. To analyze the effect of BG40 and EPCs *in vivo*, a 6-mm diameter critical size calvarial defect was created in rats ($n = 12$). Controls ($n = 6$) received BG40 and the treatment group ($n = 6$) received BG40 seeded with 5×10^5 rat EPCs. Vascularization after 1 week was significantly improved when EPCs were seeded onto BG40, compared to implanting BG40 alone. This indicates that Ca^{2+} release improves EPC differentiation and is useful for enhanced early vascularization in critical size bone defects.

Editor: Sanjoy Bhattacharya, Bascom Palmer Eye Institute, University of Miami School of Medicine;, United States of America

Funding: This work was supported by DAAD (grant # A109234), https://www.daad.de/de/index.html. The funders had no role in study design, data collection and analysis, decision to publish, or preparation of the manuscript.

Competing Interests: The authors have declared that no competing interests exist.

* E-mail: caroline.seebach@kgu.de

¶ These authors contributed equally to this work.

Introduction

Biomaterial-based strategies in bone tissue engineering combine principles of biology and engineering to develop functional substitutes for damaged bone and improve bone regeneration. Designing and fabricating composite biomaterials for bone regeneration from different synthetic biodegradable polymers and bioactive materials is an essential step in engineering bone tissue [1–5].

Biomaterials serve as matrices for tissue formation and surface properties promoting cell adhesion, proliferation and differentiation, desirable mechanical properties, non-cytotoxicity and osteoconductivity are all essential [6,7]. Bioglass (BG), a calcium silicate (CaO-SiO_2), is similar to the natural inorganic bone component and has been shown to stimulate the formation, precipitation and deposition of calcium phosphates from physiological solution and can result in enhanced bone-matrix interface strength [8].

Recently, composite biomaterials comprising a bioactive bioglass and a biodegradable polymeric component have been developed for bone tissue engineering scaffolds [9,10]. The composite materials were designed to mimic bone-forming components, to elicit specific cellular responses due to the release of highly reactive components and provide an ideal environment for bone formation [11,12].

Reparative cells are also an important aspect of bone tissue engineering and they require a biomaterial scaffold, which positively influences cell adhesion, morphology, proliferation and differentiation of neighbouring cells [13,14]. Relatively little is known about the effect of composite materials containing a bioglass fraction on cells with regenerative capabilities. Previous studies have predominantly focussed on marrow stromal cells (MSC). It has been demonstrated that bioglass materials are well tolerated by MSC, improving their function and differentiation, probably due to calcium ion release [15,16].

However, depending on the size of the bone defect, the survival and growth of MSC seeded onto a biomaterial influence the in-growth of endogenous bone-forming cells and may be limited due to lacking vascularization and insufficient nutritional bone graft support. Thus, early vascularization of the composite material in the bone defect is a crucial step for in-growth of osteogenic reparative cells in regenerating bone *in vivo* [17]. Improvement of early vascularization has been achieved through the use of endothelial progenitor cells (EPC) [18,19], isolated from diverse starting populations [20].

In the current investigation "endothelial-like cells" or "early EPCs" were used. These cells are presumably derived from monocytic/dendritic cells coexpressing some endothelial markers together with leukocyte markers and demonstrating a high VEGF synthesis [21–24]. In the following 'early EPCs' will be referred to as EPCs.

Early EPCs are presumably derived from monocytic/dendritic precursors, express the common leukocyte marker CD45 and some investigators therefore designate them as endothelial like differentiated PBMCs. These cells can be generated in a sufficient amount within 3 to 5 days from a reasonable volume of blood. Early EPCs are potent producers of vascular endothelial growth factor (VEGF) [21–24].

Outgrowth EPCs or late EPCs are characterized by a broad spectrum of endothelial markers including VEGF-R2 and UEA-I-Lectin. They express CD34, lack myeloid markers (CD45) and can be expanded *in vitro*. It is likely that these cells are generated from bone-marrow derived CD133+ cells [25,26]. The culture period of late EPCs is much longer, compared to that of early EPCs. Single colonies of late EPCs appear after 3–4 weeks [27], whereas early EPCs require only 3–5 days [28,29].

There is wide spread evidence that EPCs can be cultured to be functionally active on ceramic and natural biomaterials, alone and in combination with marrow stromal cells (MSC) [19,30,31]. It has also been demonstrated that EPCs seeded on ceramic biomaterials are capable of increasing the vascularization in bone defects within one week after implantation [19,32,33].

Recently, the combination of "bone marrow-derived EPCs" which require a culture period of at least 12 days, with polylactic acid (PLA) and a bioglass was analyzed *in vitro* [34]. This composite material promoted EPC mobilisation, differentiation and angiogenesis, likewise probably due to ionic calcium and mechanical cues [34]. No data is available about the effect of PLA-bioglass biomaterial combined with "early EPCs" on early vascularization in a critical size bone defect *in vivo*.

Therefore, the current investigation was designed to elucidate the effect of a composite material consisting of polylactic acid and increasing fractions of bioglass on early EPCs *in vitro*. The amount of calcium released from the composite material was measured and morphological changes and apoptosis of EPCs *in vitro* were recorded. Subsequently, EPCs were incubated with equivalent calcium concentrations and morphological changes, vitality and apoptosis of the EPCs were recorded. The effect of the composite material combined with EPCs on early vascularization (after 1 week) *in vivo* in a critical size calvarian defect model of the rat was also investigated.

Materials and Methods

Ethics

Human early EPCs were isolated from buffy coat (Red Cross Blood Donor Service, Frankfurt, Germany). The use of anonymous buffy coat for research purposes was approved by the local ethics committee (Ethik-Kommission des Fachbereichs Medizin der Johann Wolfgang Goethe-Universität, Project No. 329/10) and the donors signed informed consent.

All animal experiments were approved and performed in accordance with regulations set forth by our institution's animal care and oversight committee located at the Regierungspräsidium Darmstadt (Regierungspräsidium Darmstadt – Veterinärdezernat - Tierschutzkommission, Darmstadt, Germany, Project No. F3/22), in accordance with German law. All surgery was performed under general anesthesia, which was administered intraperitoneally as a mixture of Ketavet and Rompun. All efforts were made to minimize suffering. The animals were sacrificed with an overdose of pentobarbital (150 mg/kg i.p.).

Biomaterial Characterisation: PLA, BG20, BG40 and β-TCP

The composite biomaterials consisted of a PLA-component supplemented with either 20 or 40% bioglass (BG). Tetraethyl orthosilicate (TEOS, ≥99%) and nitric acid 65% were supplied by Merck Chemicals KgaA (Darmstadt, Germany). Calcium nitrate [Ca $(NO_3)_2$ $4H_2O$, ≥99%], poly (L-lactide) and chloroform ($CHCl_3$, ≥99.4%) were purchased from Sigma-Aldrich (Steinheim, Germany). All chemicals were reagent grade and used without further purification. For bioglass synthesis (CaO-SiO_2 - SiO_2 80 mol-%, CaO 20 mol-%), a low viscosity gel was obtained by mixing 31 mL of tetraethyl orthosilicate (TEOS) and 8.6 g of $Ca(NO_3)_2.4H_2O$ in a solution of 5.5 mL of HNO_3 2 M, used as catalyst, in 31.5 mL H_2O. The initial pH was 0.5. The bioglass was cast at room temperature in a Teflon container (Thermo Scientific Nalgene, Germany) until the gel had hardened. Aging was performed at 60°C for 3 days. Drying was carried out at 120°C. The glass was collected in a porcelain crucible (Haldenwanger GmbH, Waldkraiburg, Germany) and heated in a muffle furnace (Nabertherm GmbH, Lilienthal, Germany) at a rate of 3°C/1 minute to 700 C° and maintained at 700°C for 3 hours. The glass particles were ground to powder in a small porcelain mortar (Haldenwanger GmbH, Waldkraiburg, Germany). The particles were then sieved to achieve a size ranging from 106 μm to 125 μm by sieves with a mesh size of 106 μm and 125 μm (Retsch GmbH, Haan, Germany). Composite biomaterials were prepared by mixing polymer [poly(L-lactide)(PLA)] and bioglass with 10 ml chloroform as follows: PLA, PLA/BG 20% and PLA/BG 40% biomaterial. The bioglass content was 0%, 20% and 40% by weight. These biomaterials will be referred to as PLA, BG20 and BG40, respectively. Disc shaped specimens with a diameter of 5 mm and a thickness of 1 mm were cut and stored at room temperature under sterile conditions until use (Fig. 1A). The synthetic β-tricalcium phosphate (β-TCP) Chronos® (Synthes, Dübendorf, Switzerland) was employed as a reference material [19,31,32,33]. The β-TCP particles have a size of 0.7–1.4 mm, porosity 60%, pore size 100–500 mm, low mechanical stability and moderate biodegradability.

Scanning Electron Microscopy

Surface topography, roughness and morphology of the biomaterial, as well as adherent EPCs, were assessed by scanning electron microscopy (SEM). Untreated biomaterials and biomaterials seeded with EPCs were fixed 5 days after EPC seeding with glutaraldehyde for 30 min and subsequently dehydrated by 15 min incubation in a 5-step ethanol gradient (25%, 50%, 75%, 96%, 100% ethanol). Biomaterials were then incubated overnight in 1,1,1,3,3,3-hexamethyldisilazane (Merck-Schuchardt, Hohenbrunn, Germany) and drained. Afterwards the samples were sputtered with gold (3×60 s, Agar Sputter Coater, Agar Scientific Ltd., UK) using a Hitachi FE-SEM S4500 (Hitachi, Dusseldorf, Germany) with a voltage of 5 kV. The images were digitally recorded using the Digital Image Processing System 2.6 (Point Electronic, Halle, Germany). The public domain software *ImageJ* (http://rsb.info.nih.gov/ij/) was used to evaluate pore size, pore density and fiber diameter.

Calcium Release

30 mg of cell free BG20, BG40, control PLA or β-TCP were placed in individual wells of 24 well plates and incubated over a period of 2, 6 or 10 days with 1 mL endothelial basal medium (EBM, Cambrex, Verviers, Belgium) supplemented with endothe-

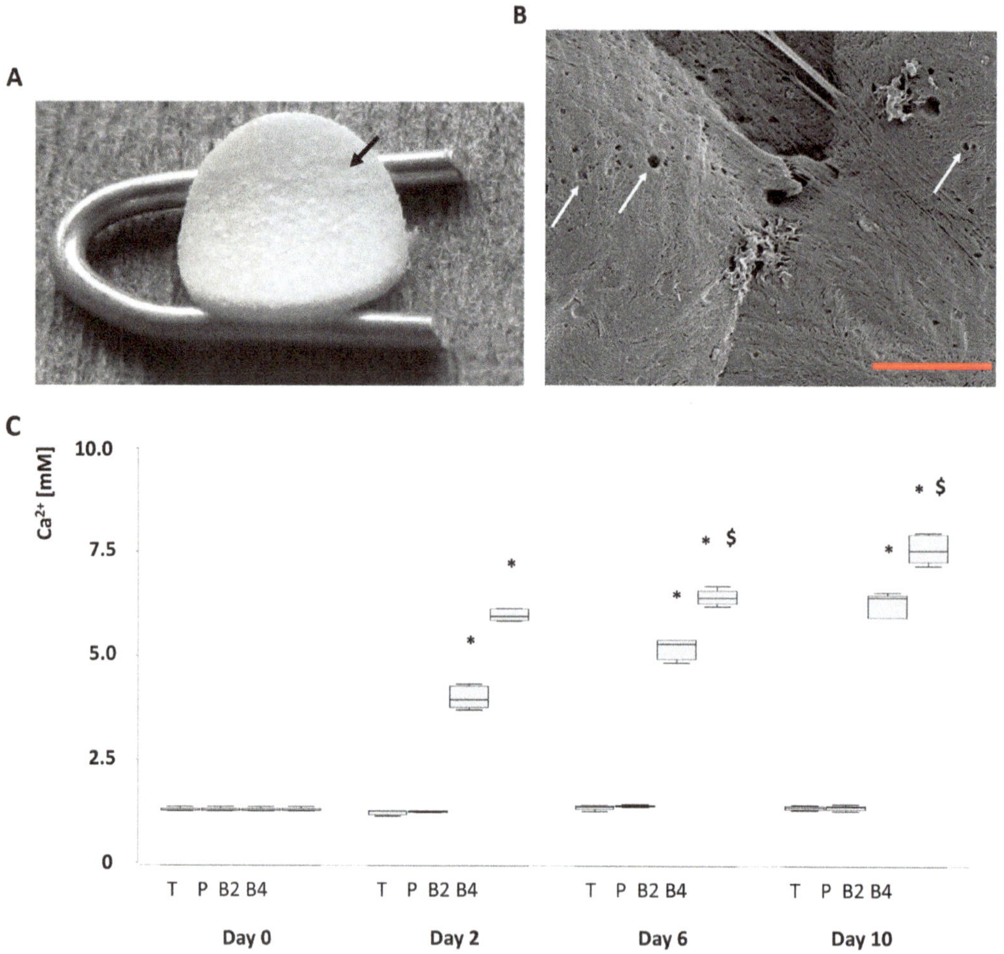

Figure 1. Surface structure and calcium release of the composite scaffold BG40. Disc shaped polymer/bioglass scaffold BG40, with a diameter of 5 mm and a thickness of 1 mm (A). A representative SEM image of BG40 is shown in (B). The composite demonstrates a relatively smooth surface with a thin fibrous texture and small pores (arrows). Scale bar indicates 6 µm. The release of ionic calcium to medium by BG40 (B4), BG20 (B2), PLA (P) and β-TCP (T) is shown in C. The most ionic calcium was released within the first two days. * = $p<0.05$ BG20, BG40 vs b-TCP, PLA; $ = $p<0.05$ BG40 vs BG20, n = 5.

lial growth medium-2MV SingleQuot (Lonza, Basel, Switzerland). At the indicated time points Ca^{2+} in the medium was measured (ABL 800 Flex; Radiometer GmbH, Willich, Germany). The calcium content of untreated EBM-medium served as control.

Cell Preparation and Culture

Isolation of early EPCs from buffy coat and rat spleen. Due to accessibility, early human EPCs were employed for all *in vitro* experiments. To avoid host versus graft reactions EPCs isolated from rat spleen were used in the animal study.

Buffy coat was subjected to density gradient centrifugation (30 min, 900 g) with Ficoll (1.077 g/ml, Biochrom, Berlin, Germany). Recovered mononucleated cells were washed twice with cold $PBS^{w/o}$ (10 min, 900 g) and 2×10^6 cells/cm^2 were cultivated on fibronectin coated (10 µg/ml, Sigma, Deisenhofen, Germany) culture dishes with endothelial basal medium (EBM, Cambrex, Verviers, Belgium) supplemented with EGM SingleQuot at 37°C, 5% CO_2. After 48 hours, non- and weakly-adherent cells were removed, the medium was changed and the cells were cultivated for an additional 72 hrs. A parallel preparation was performed to evaluate the percentage of endothelial differentiated cells. EPCs were identified by staining with 1,1′-dioctadecyl-3,3,3′,3′-tetramethylindo-carbocyanine-labeled acetylated low density lipoprotein (DiLDL, Cell-Systems, St. Katharinen, Germany) in EBM supplemented with 20% FCS. Cells were fixed with 2% paraformaldehyde for 10 min and after washing with $PBS^{+/+}$ FITC-labeled Ulex europaeus agglutinin-1 [10 µg/ml] (lectin, Sigma, Deisenhofen, Germany) were incubated for 1 h. Cells with double-positive fluorescence were considered to be EPCs [28,29]. Only preparations with a percentage of endothelial like differentiated cells greater than 80% were used. For experiments the cells were detached by accutase treatment (10 min) (PAA-laboratories, Linz, Austria), washed once with EBM+supplements (Cell-Systems, St. Katharinen, Germany) and subsequently adjusted to a density of 5×10^5 cells in 50 µl. Rat EPCs were isolated from the spleen of homozygous male SD-rats. The spleen was cut in small pieces (approximately 3 mm) and gently mashed using syringe plungers. The cell suspension was filtered through a 100 µm mesh, washed once with PBS and layered onto a ficoll density gradient. The subsequent following isolation, culture and identification were identical to the procedures described for human EPCs.

Biomaterial Biocompatibility Assessment

Seeding efficacy of early EPCs on composite biomaterial. Composite biomaterial chips (BG20, BG40) and PLA chips (Fig. 1a) were individually placed in single wells of a 96-well plate (Nunc, Wiesbaden, Germany) using sterile forceps. Granules of β-TCP were placed in a dense monolayer in single wells of a 96-well plate as additional controls.

5×10^4 EPCs in a volume of 50 µl were dripped onto the biomaterials and incubated for 10 min at 37°C. After incubation the medium containing non-adherent cells was removed and dripped once again over the biomaterials, followed by incubation as indicated above. This procedure was repeated three times. The bioglass chips or β-TCP granules were then gently transferred to another well containing 100 µl *EBM2+EGM2 Singlequot.* The remaining cells in the supernatant and at the bottom of the initial seeding well were isolated, counted and the percentage of adherent cells was calculated ((initial cell number - remaining cell number)/ initial cell number)*100. The experiment was performed in duplicate and was repeated 5 times.

Effect of Composite Biomaterials on EPC Vitality, Morphology, Apoptosis and Gene Expression

Twenty-four well plates were coated with fibronectin as described above. 1×10^5 EPCs in 1 mL EBM2+EGM2SQ were placed in each well. Transwell inserts with a pore size of 8 µm, containing equal amounts of biomaterials [30 mg of BG20, BG40, control PLA or β-TCP], were placed in the wells. The pores allowed diffusion of soluble biomaterial components to the cell compartment. After 2, 5 and 10 days the morphology of the cells was assessed by phase contrast microscopy at 100 fold magnification. Cells were photographed and the length of the EPCs was evaluated using the software Axiovison 4.7 (Zeiss). An increase in cell length is a cellular characteristic of early EPCs [21,35].

The funcionality of the cells was assessed by DiL-ac-LDL staining on day 5. The amount of DiL-ac-LDL uptake reflects EPC metabolic status and can be measured by image analysis as mean pixel brightness/cell [19,30]. DiL-ac-LDL-stained cells were photographed at 100 fold magnification with equal exposure (200 mS]. In each group (control, calcium [10 mM], BG40, BG40+ EGTA [3 mM]) the mean pixel brightness of 20 cells was analyzed (ImageJ (http://rsbweb.nih.gov/ij/).

Detection of necrotic EPCs on the biomaterials. Necrotic EPCs were detected by staining nuclei with DAPI (2-(4-Amidinophenyl)-6-indolecarbamidine, Sigma-Aldrich, Deisenhofen, Germany). DAPI passes less efficiently through the membrane of living cells and thus the staining of living cells is much lower [36].

Each 5×10^4 DiL-ac-LDL prestained EPCs were then seeded onto chips of PLA, BG20 and BG40 that were placed in individual wells of 24-well plates. The seeding procedure was performed as described above. The medium was removed on day 3 and day 5 and a DAPI-solution [final concentration 1 µg/ml in PBS] was added to each well. The DAPI solution was removed after 3 min followed by three washings with 1 mL PBS for each wash. The samples were then subjected to fluorescence microscopy. Five random high power fields were recorded per scaffold at 100 fold magnification. Cells with red fluorescence (DiLac-LDL) were judged as vital cells, cells with red and blue (DAPI) fluorescence were deemed necrotic. Results are presented as number of adhering cells per microscopic field of view (FOV) at 100 fold magnification as well as percentage of necrotic cells per FOV.

To evaluate the effect of Ca^{2+}-ions on EPC function, $CaCl_2$ was added to EPC cultured in fibronectin coated 24-well plates in the a concentration measured in the aforementioned experiments [10 mM].

Morphological changes and viability of the EPCs were assessed on day 5 after adding $CaCl_2$, as previously described. Additionally, EPC apoptosis was assessed on day 3 after $CaCl_2$ addition. The cells were detached by *Accutase* treatment and apoptosis was assessed by means of Annexin-V-staining (Annexin V apoptosis detection Kit, BD-Biosciences, Heidelberg, Germany) and flow cytometry (FACScalibur, BD-Biosciences), according to the manufacturer's instructions. All experiments described in this paragraph were performed with 5 different EPC-preparations.

Expression of endothelial marker genes by Real time RT-PCR. 5×10^4 EPCs were seeded onto PLA, BG20 or BG40 chips and incubated for 3 or 5 subsequent days. EPCs subjected to RNA-isolation on the day of seeding served as day zero.

Total RNA was isolated using the *RNeasy*-system (Qiagen, Hilden, Germany) following the manufacturer's instructions, with the following exception. Each chip of biomaterial (PLA, BG20, BG40) was individually incubated with approximately 100 µl *RLT* buffer for 3 min, the mixture was gently vortexed and the supernatant was subjected to the RNA isolation procedure. The quality and quantity of RNA was determined using a *NanoDrop* ND-1000 device (Nanodrop technologies, Wilmington, Delaware, USA). Contaminating genomic DNA was removed by digestion with the *RNase free DNase-Kit* following the manufacturer's protocol (Qiagen).

75 ng of RNA was reverse transcribed using an *Affinity script QPCR-cDNA synthesis kit* (Stratagene, La Jolla, CA, USA), following the manufacturer's instructions.

Real time RT-PCR was performed on a *Stratagene MX3005p qPCR system* (Stratagene, La Jolla, CA, USA). PCR was performed using the primer assays for human vascular endothelial growth factor (VEGF, NM_003376.4, catalog number PPH00251B) and von Willebrandt factor (vWF, NM_000552.3, catalog number PPH02567E), all purchased from Biomol (SuperArray, Frederick, MD, USA). The expression of glyceraldehyde-3-phosphate dehydrogenase (GAPDH, NM_002046.3, catalogue number PPH00150E) was measured as reference gene.

A melting curve analysis was applied to ensure the specificity of the PCR reaction. Relative quantification of the mRNA levels of the target genes was determined using the comparative CT (threshold cycle values) method ($2^{-\Delta CT}$ method). The results are presented as fold change to GAPDH gene expression. This experiment was performed with 4 independent EPC preparations.

Animals and Cell Transplantation

Eight-week old male Sprague-Dawley rats ($n = 12$, Charles River, Germany), weighing approximately 350–400 g were housed, four animals per cage under standardized conditions: 15–21°C, air flow and light (12 h day/12 h night), rat food and water ad libitum.

The rats were randomly allocated to the control ($n = 6$) or experimental group ($n = 6$). Control animals were implanted with BG40 biomaterial (disc shaped, 5 mm diameter, 1 mm thickness), experimental animals received BG40 seeded with 5×10^5 rat EPCs 1 h before implantation. General anesthesia was administered intraperitoneally as a mixture of Ketavet and Rompun. All efforts were made to minimize suffering. To create a critical size defect (CSD) the head was shaved and cleaned with antiseptic fluid. A lateral longitudinal incision over the head was made under aseptic conditions. The skull cortex was drilled using a 6 mm bit so that a circular critical calvarial bone defect of 6 mm was created. The biomaterials were implanted into the defect zone and their position controlled. The wound was then closed with continuous subcutaneous stiches using a 4/0-monofilament nylon suture. Animals had free access to food and water and were monitored daily in the postoperative period for any complications or abnormal behaviour.

The animals were sacrificed with an overdose of pentobarbital (150 mg/kg i.p.) and weighed after 1 week. The skull bone was dissected free and removed. Bones were wrapped in gauze moistened with physiologic NaCl-solution and stored at −80°C until preparation for immuno-histological examination.

von Willebrand Factor (vWF) as Marker of Early Vascularization

Skull bones were fixed in 4% *Zinc-Formal-Fixx, 4%* (Thermo Electron, Pittsburgh, USA) for 20 hrs, decalcified over 7 days in a 10% Tris buffered EDTA-solution under continuous stirring and embedded in paraffin. Sections (5 μm) of the decalcified specimens parallel to the long axis of the head were stained with hematoxylin and eosin or incubated with a mouse anti vWF-antibody, which crossreacts with both human and rat vWF (8 μg/mL, Biomol, Hamburg; Germany). An isotype identical (IgG_1) non specific mouse antibody served as negative control (Dako, Hamburg, Germany). As secondary antibody, a polymer HRP-conjugated goat anti mouse antibody (Histofine simple stain rat Max Po (M), Nichirei Biosciences, Tokyo, Japan) was applied and the sections were incubated with 3-amino-9-ethylcarbazole (AEC, Lab Vision, Dreieich, Germany). Finally, a counterstain with hematoxylin was performed. One slide per animal was analyzed using light microscopy (Axioobserver Z1, Zeiss, Gottingen, Germany) at a magnification of 100x in combination with a computer-supported imaging picture analysis system (Axiovision 4.7; Zeiss). Von Willebrandt factor positive blood vessels with a lumen were counted in 6 non overlapping images/slide/animal sourrounding the defect area and the mean number of vessels was calculated. These means were subsequently used for statistical anlysis. Cells positive for von Willebrandt factor were not considered. Imaging and blood vessel counting were performed with blinded specimens examined in random order by an independent observer, blinded to the group setup.

Statistics

Results are presented as box-plots of the median in figures, 25% and 75% quartiles ((M (25%q/75%q)) in the text and tables. Nonparametric Kruskal–Wallis test and multiple Conover-Iman test were consequently used, and a Bonferroni-Holm corrected $p<0.05$ was used to indicate statistical significance. Statistics were calculated using the software *Bias* 10.03 (Epsilon-Verlag, Darmstadt, Germany).

Results

Biomaterial Characterization and Calcium Release

Scanning electron microscopy revealed a relatively smooth surface of the composite biomaterial BG40. Higher magnification revealed fibrous structure with a fiber diameter of approximately 0.1 μm. The fibres were partly arranged in parallel structures but also significant areas of irregularly arranged fibres were found. Putatively not interconnecting micropores with a diameter ranging from 0.2 μm up to 1.2 μm were evenly distributed on the surface at a median density of 58 micropores/1000 μm^2 (Figure 1B).

The release of ionic calcium by BG20 as well as BG40 was significantly elevated in comparison to β-TCP during the whole 10 day observation period. The amount of released calcium ions through BG40 was significantly higher, compared to BG20 (d6, d10). The majority of calcium was released between days 1 and 2 (BG20 57% of total d10 release); BG40 74% of total d10 release, Figure 1C).

Seeding Efficacy of Early EPCs on the Composite Biomaterial

Seeding efficacies of EPCs were comparable between β-TCP and BG40. The percentage of initially adhering EPCs was significantly decreased on PLA, compared to β-TCP (Figure 2A).

Vitality and Function of EPCs Cultured on the Composite Biomaterial

Scanning electron microscopy performed on the first day after EPC seeding revealed evenly distributed EPCs on PLA and BG40. EPCs demonstrated a rounded phenotype on BG40 and a more flattened appearance when seeded on PLA (Fig. 2B). EPCs cultured 5 days on PLA demonstrated a necrotic phenotype, whereas EPCs cultured for the same time period on BG40 had a vital appearance (Fig. 2B).

Additional fluorescence microscopy analysis was performed for confirmation. Dil-ac-LDL prestained EPC were seeded onto the biomaterials. The number of adhering EPC was significantly decreased on PLA in comparison to BG20 and BG40 on day 3 after cell seeding but those differences were levelled on day 5, though (Fig. 2C). Necrotic EPC were identified through an additional DAPI-staining as indicated in the Materials and Methods-section. The percentage of necrotic EPC was significantly increased on PLA compared to EPCs that have been cultured on BG20 and BG40 (Fig. 2C, 2D).

The gene expression of VEGF and vWF was evaluated to determine the endothelial function of EPCs cultured on the composite materials over a period of 5 days. Gene expression is presented as fold change of the housekeeping gene expression (GAPDH). VEGF gene expression remained stable on BG40 on day 3 and day 5 in comparison to day 0, whereas VEGF gene expression decreased singificantly from day 0 to day 5 on BG20 and PLA. No significant differences were found between the PLA, BG20 and BG40 on day 3 and day 5 however a trend towards higher values were seen for BG40 vs PLA on day 5 ($p=0.1$) (Table 1). The gene expression of vWF increased significantly from day 0 to day 5 for all biomaterials. No significant differences between PLA, BG20 and BG40 were found albeit a trend for increased vWF-gene expression on BG40 in comparison to PLA on day 5 ($p=0.1$, Table 1).

Effect of Composite Scaffolds on EPC Morphology, Function and Survival

The co-incubation of EPCs with BG40, and to a lesser extent with BG20, resulted in significant EPC elongation compared to control (medium), PLA and β-TCP. These changes became prominent on day 6 and day 10 after incubation started (Figure 3A, B).

To test the hypothesis that calcium released by the composite biomaterials is critical, calcium was added to the medium in a final concentration of 10 mM. This concentration was comparable to the calcium released by BG40 in the experimental setting (Figure 1C). The elongation of the cells was assessed five days later. In line with the aforementioned results a statistically significant increase in cell length was observed (Figure 4A, B). As final proof that calcium derived from bioglass induces EPC elongation, EPCs were incubated over a period of five days with medium conditioned with BG40 (over 2 days) in the presence or absence of the specific calcium chelator EGTA (ethylene glycol tetraacetic acid). The addition of EGTA lowered the Ca^{2+}

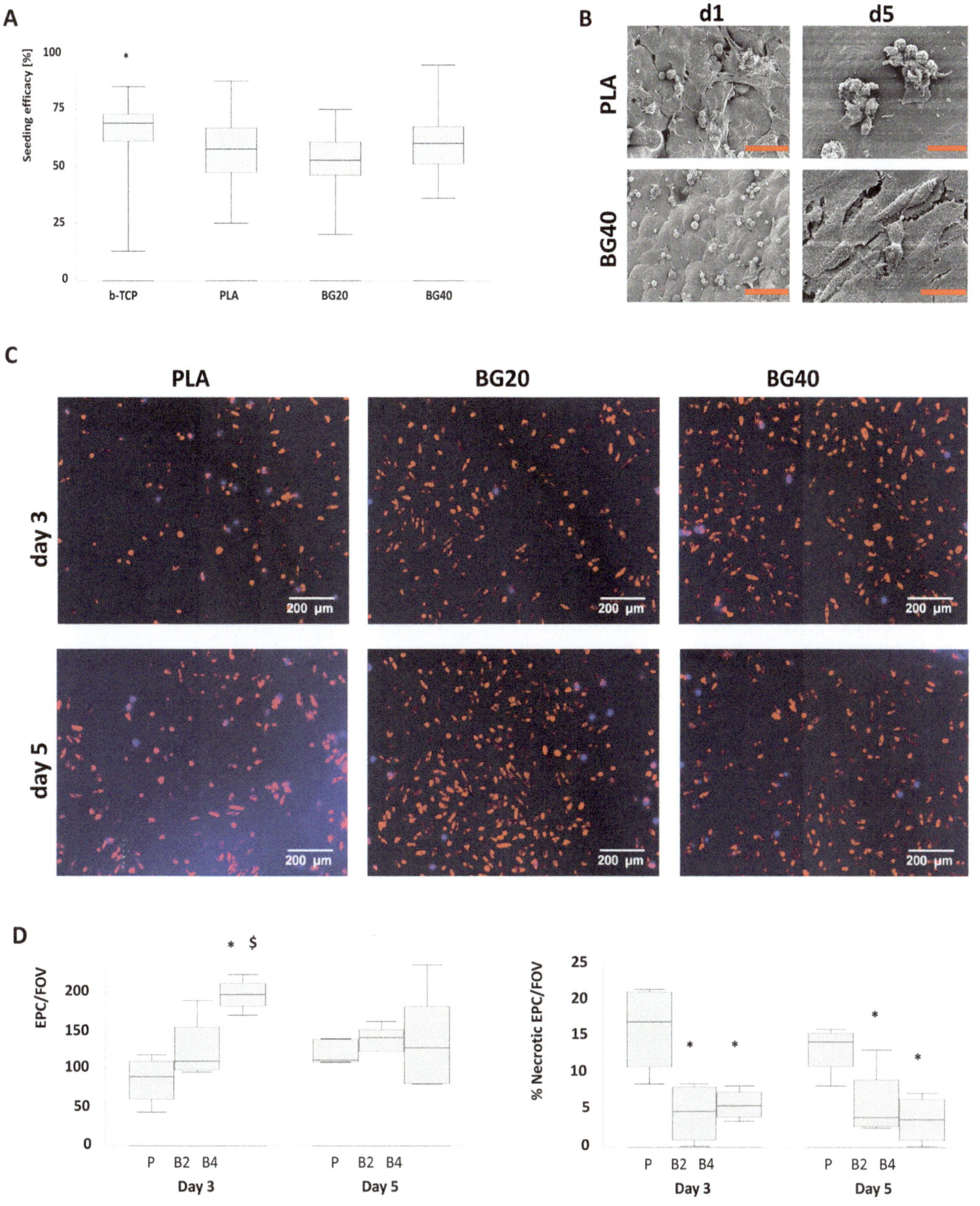

Number of adhering DiL-ac-LDL+ EPC/FOV

Percentage of necrotic EPC (DiL-ac-LDL+ DAPI+ / FOV)

Figure 2. Seeding efficacy and EPC phenotype on the composite scaffold. Seeding efficacy of EPCs on BG40, BG20, PLA and β-TCP (A). The percentage of initially adhering EPCs on β-TCP was significantly increased compared to PLA. Appearance of EPCs seeded on BG40 on day 1 and day 5 (B). Representative SEM images are shown. EPCs demonstrated a spherical structure on the first day after seeding on BG40 whereas a flattened appearance was seen when EPCs were seeded on PLA. On day 5 EPCs presented a necrotic phenotype when seeded on PLA, whereas they appear vital on BG40. Scale bars indicate 60 μm (day 1: PLA, BG40) respectively 20 μm (day 5: PLA, BG40). *: p = 0.045 β-TCP vs. PLA, n = 5. Confirmation of increased necrosis of EPC on PLA compared to BG20 and BG40, representative fluorescence microscopy images were shown in (C) at an original magnification of 100 fold. DiL-ac-LDL prestained EPCs were seeded onto the scaffolds. DAPI was added over a period of 3 min to discriminate between live (red cytoplasm, no blue nucleus) and necrotic EPCs (red cytoplasm, blue nucleus) on day 3 or 5 after seeding. The quantitative evaluation of this experiment is depicted in (D). The number of EPC was significantly lowered on PLA on day 3 (p = 0.02 vs BG20, p = 0.01 vs BG40) and the percentage of necrotic EPCs differed significantly between PLA and BG40 on day 3 (p = 0.02) and day 5 (p = 0.02). *: p<0.05 vs. PLA, $: p = 0.045 BG40 vs. BG20; P: PLA; B2: BG20; B4: BG40, n = 5. Scale bar indicates 200 μm.

concentration from 10 to approximately 2 mM. A significant decline of EPC length in the presence of EGTA was apparent (Figure 4A). A Dil-ac-LDL-staining performed in parallel proved the vitality and endothelial function of the EPCs (Figure 4A, B, C). The mean DiL-ac-LDL uptake per cell was comparable in all groups (Control: 31 (25/44); 10 mM $CaCl_2$:27 (22/43); BG40:36 (26/46); BG40+EGTA: 32 (27/44), not significant).

Next, the influence of Ca^{2+} ions added to the medium on EPC survival was investigated. In the control experiment the number of EPC/field of view (FOV) declined significantly between day 1 and day 5 (99 (78/130) vs 51 (46/57), p<0.05). In contrast, EPCs/FOV remained stable between days 1 and 5 in the presence of BG40 (Day 1:106 (93/120) vs day 5:93 (86/99), not significant) or $CaCl_2$ [10 mM, added to medium] (Day 1:104 (91/111) vs day 5:85 (75/105), not significant). Additionally, cell numbers in the presence of BG40 or $CaCl_2$ in the medium were significantly elevated, compared to the control at day 5. To investigate whether EPC maintenance was due to $CaCl_2$ mediated apoptosis inhibition, EPC apoptosis over a period of three days with and without 10 mM $CaCl_2$ was followed by apoptosis measurement. A significantly decreased rate of apoptotic EPCs under the influence of 10 mM $CaCl_2$, compared to controls was observed (Control: 30.4% (28.1/33.2) vs $CaCl_2$:8.6% (8.5/10.6); p<0.05). Representative dotblots of the measurement are depicted in figure 5.

Table 1. Gene expression of VEGF and vWF in EPC cultured on PLA, BG20 and BG40 over a period of 5 days.

Gene	Biomaterial	Day 0	Day 3	Day 5
VEGF	PLA	0.0079	0.0026	0.0014
		(0.0045/0.0107)	(0.0016/0.0066)	(0.0002/0.0027)
				↓ p = 0.03 vs. day 0
	BG20	0.0079	0.0008	0.0013
		(0.0045/0.0107)	(0.0005/0.002)	(0.0007/0.0256)
			↓ p = 0.05 vs. day 0	
	BG40	0.0079	0.0425	0.0075
		(0.0045/0.0107)	(0.0004/0.35)	(0.0025/0.014)
				↑ p = 0.1 vs PLA day 5
vWF	PLA	0.0056	0.0136	0.0153
		(0.0017/0.0072)	(0.0102/0.0184)	(0.0105/0.0211)
				↑ p = 0.04 vs day 0
	BG20	0.0056	0.018	0.0141
		(0.0017/0.0072)	(0.0073/0.013)	(0.0111/0.1902)
				↑ p = 0.05 vs day 0
	BG40	0.0056	0.266	0.0728
		(0.0017/0.0072)	(0.0233/1.017)	(0.0276/0.1224)
			↑ p = 0.04 vs day 0	↑ p = 0.04 vs day 0
				↑ p = 0.1 vs PLA day 5

The values are given as fold change to the expression of the GAPDH gene which served as housekeeping gene (median value (25% quartile/75% quartile). Day 0 demonstrates the gene expression of the EPC before seeding on the biomaterials. Generally, gene epression of VEGF declined over the time whereas gene expression of vWF increased during the observation period. The experiment was performed with 4 independent EPC preparations.
↑ = increased gene expression; ↓ = decreased gene expression.

Effect of BG40 and BG40+EPC on Early Vascularization in vivo

Areas of neovascularization in the critical calvarial bone defect of rats treated with BG40 or BG40+ EPC were detected after 1 week by vWF staining (Fig. 6A, 6B).

The histomorphometric analysis demonstrated significantly higher blood vessel density surrounding the defect area in samples obtained from animals receiving BG40 loaded with EPCs, compared to samples from animals treated with BG 40 alone (Fig. 6E, p<0.015). Besides the increased amount of vWF positive structures, more large vessels in the newly formed tissue were detectable (Fig. 6A, 6B) under the influence of BG40+ EPCs, compared to BG40 alone.

Discussion

This study demonstrates that EPCs can successfully be cultured on a PLA/bioglass composite biomaterial and that calcium released from the bioglass fraction leads to significant elongation and significantly improved survival of EPCs *in vitro*. *In vivo*, EPCs seeded onto BG40 induce significantly improved early vascularization to a critical size skull defect in the rat one week after implantation.

Physical Properties

Ideal biomaterials for bone reconstruction should fulfill requirements including mechanical stability, osteoinductivity, osteoconductivity and support of revascularization. Currently available single component materials do not meet all these requirements. Hence, more sophisticated biomaterials are needed and combining different biodegradable biomaterials with complementary properties may circumvent individual shortcomings.

In the present investigation a novel composite biomaterial based on a PLA carrier and a bioglass fraction, consisting of calcium oxide and silicium dioxide, was investigated. PLA is highly biocompatible with better thermal processability than other biopolymers. The main limitations of PLA are poor toughness, slow degradation and hydrophobic properties resulting in low cell affinity [37]. Pure bioglass is hard and brittle but offers a surface

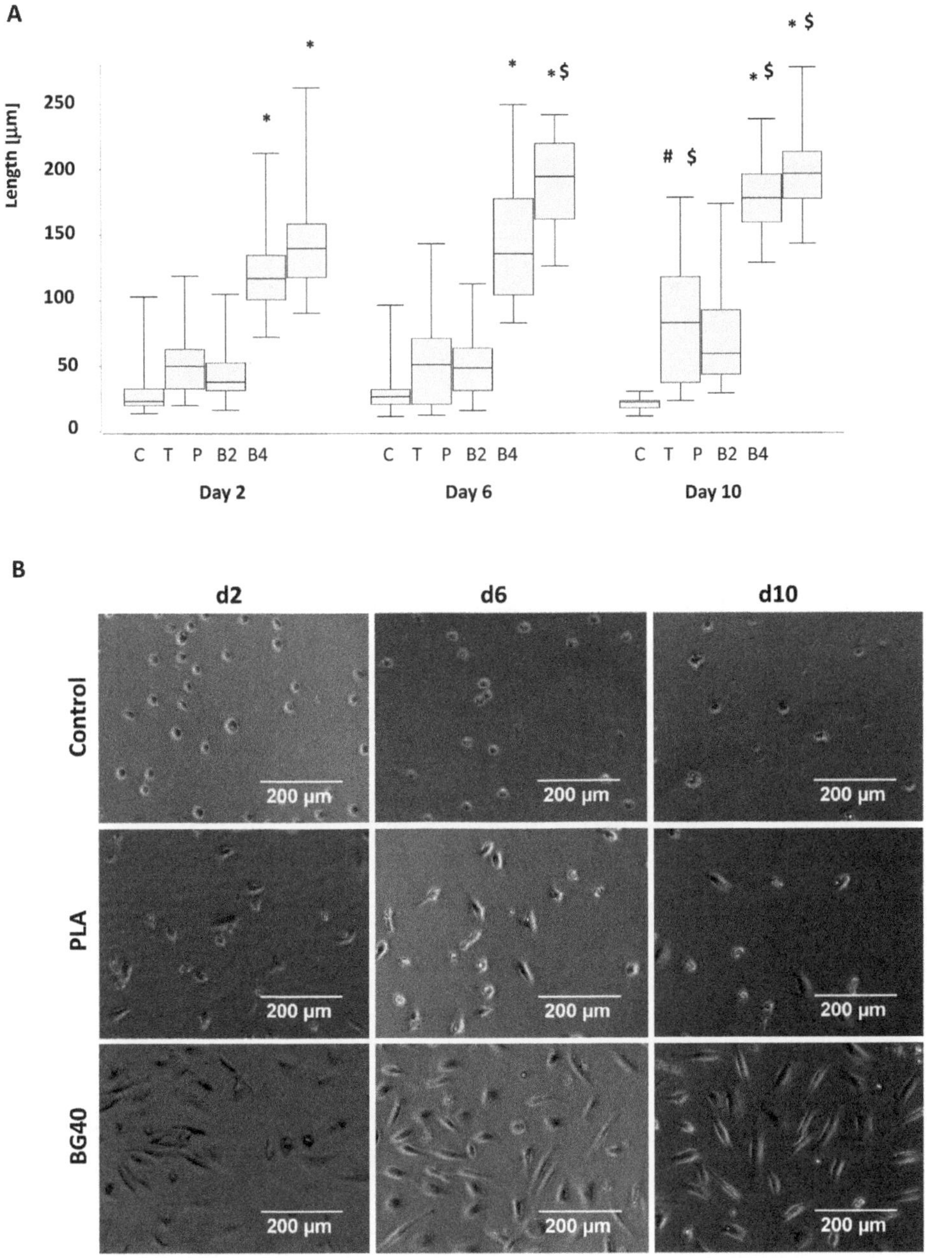

Figure 3. Coincubation of EPCs with BG40 and BG20 leads to EPC elongation. Incubation with BG20 (B2) and BG40 (B4) lead to a significantly sustained increase in EPC length compared to EPCs incubated with β-TCP (T), PLA (P) or the medium control (C) during the whole observation period (A). The length of EPCs incubated with BG20 and BG40 increased significantly from day 2 to day 10 (A). Representative micrographs of EPCs incubated with either medium, PLA or BG40 demonstrated EPC elongation (B). * = $p<0.05$ BG20, BG40 vs control, β-TCP, PLA; $ = $p<0.05$ d6, d10 vs d2; # = β-TCP vs control, n = 5.

suitable for cell attachment. It is highly biodegradable and influences the local environment by releasing bioactive ions such as ionic calcium [38], which may lead to improved cellular responses at the implantation site [39].

The composite BG40 demonstrated a relatively smooth surface with fibrous structures in the submicrometer scale, and small pores were present in high density. The occurrence, size and form of distinct micro- and macroporous structures on PLA−/bioglass based biomaterials depend on the fabrication process [37]. Apart from structural aspects due to different fabrication processes, composites containing a bioglass component share the common feature of releasing ions such as calcium.

In line with previous reports a significant Ca^{2+} release was observed with the bioglass composites employed in this investigation. The vast majority of ionic calcium was set free within the first two days, but further calcium release to the medium occurred over the following eight days. Similar release dynamics have been obtained using PLA with a bioglass fraction [34] and PGLA composites supplemented with bioglass [40]. However, the amount of calcium released has differed among studies, possibly due to the different fractions of bioglass within the biomaterials

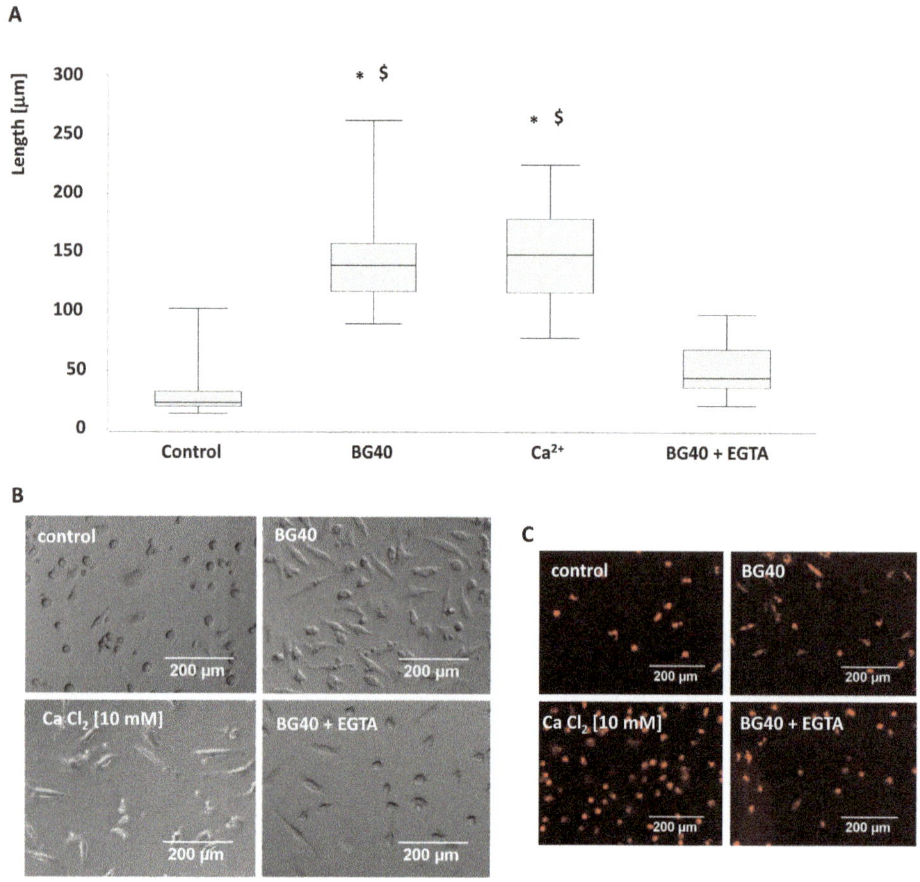

Figure 4. Calcium released by BG40 induces EPC elongation. EPCs were incubated in the presence of either 30 mg BG40/mL, 10 mM Ca^{2+} or BG40 conditioned medium+EGTA [3 mM]. Cell length was determined on day five (A) by means of phase contrast microscopy and subsequent histomorphometric evaluation. In (B) corresponding micrographs of EPCs are shown. The elongation of the EPCs was significantly lower in the presence of the calcium chelator EGTA [3 mM]. The uptake of DiL-ac-LDL demonstrated the endothelial phenotype and EPC vitality. The mean DiL-ac-LDL amount per cells was not significantly altered among the groups (control, 10 mM calcium, BG40, BG40+EGTA) (C). * = $p<0.05$ vs control; $ = $p<0.05$ vs BG40+EGTA, n = 5. Scale bars indicate 200 µm.

and/or different amounts of biomaterials used in the calcium release experiments.

It should be noted that the form of the biomaterials influences the properties of the biomaterial [41]. In the current investigation film like discs of the composite material were employed. Granules of BG20 and BG40 might have accelerated calcium release due to the higher surface area compared to disc shaped biomaterials. A greater surface area might also allow higher loading with regenerative cells.

Bioglass Effect on Cells and Vascularization in vitro and in vivo

The initial adhesion of EPCs on the bioglass composites BG20/BG40 and on β-TCP was comparable but significantly lower, compared to former studies [30–32]. These deviating results might be explained by alterations in the seeding procedure, which was performed in 96 well plates with comparatively low cell numbers in the present investigation. In former studies an EPC seeding efficacy of over 90% on β-TCP has been demonstrated [19]. Seeding efficacy on BG20/BG40 may be equally good if the same seeding conditions are applied.

The *in vitro* investigation demonstrated that BG40 induced the best EPC differentiation and survival, which hypothetically could result in better vascularization of a bone critical size defect *in vivo*.

To test this, the composite biomaterial BG40 with or without EPCs, was implanted into a critical size calvarial defect and early vascularization was analyzed. Much better vascularization was observed in animals receiving BG40 and EPCs. The experimental setup did not provide an evaluation of the effect of BG40 alone on early vascularization since BG40 without EPCs was not tested.

Two factors possibly play a dominant role in the increased *in vivo* vascularization. First, it has been demonstrated in different animal models that early EPCs increase vascularization [32,33,42]. Second, the beneficial effects of calcium ions released by the bioglass component might also contribute to the increased vascularization in *vivo* by supporting EPC function and survival. Separating the effects of the bioglass from the effects of EPCs on vascularization in this *in vivo* model will require future investigation.

The contribution of early EPCs in forming blood vessels is a matter of debate. Crosby and colleagues have reported that 8.3% –11.2% of endothelial cells that developed in sponge induced granulation tissue over 1 month were derived from circulating hematopoietic progenitor cells [43]. Hence, it is has been proposed that early EPCs more likely act in a paracrine manner, secreting proangiogenic factors such as VEGF [22]. Previous findings, whereby human early EPCs were implanted in a rat critical size defect of the femur, support this hypothesis. When the animals

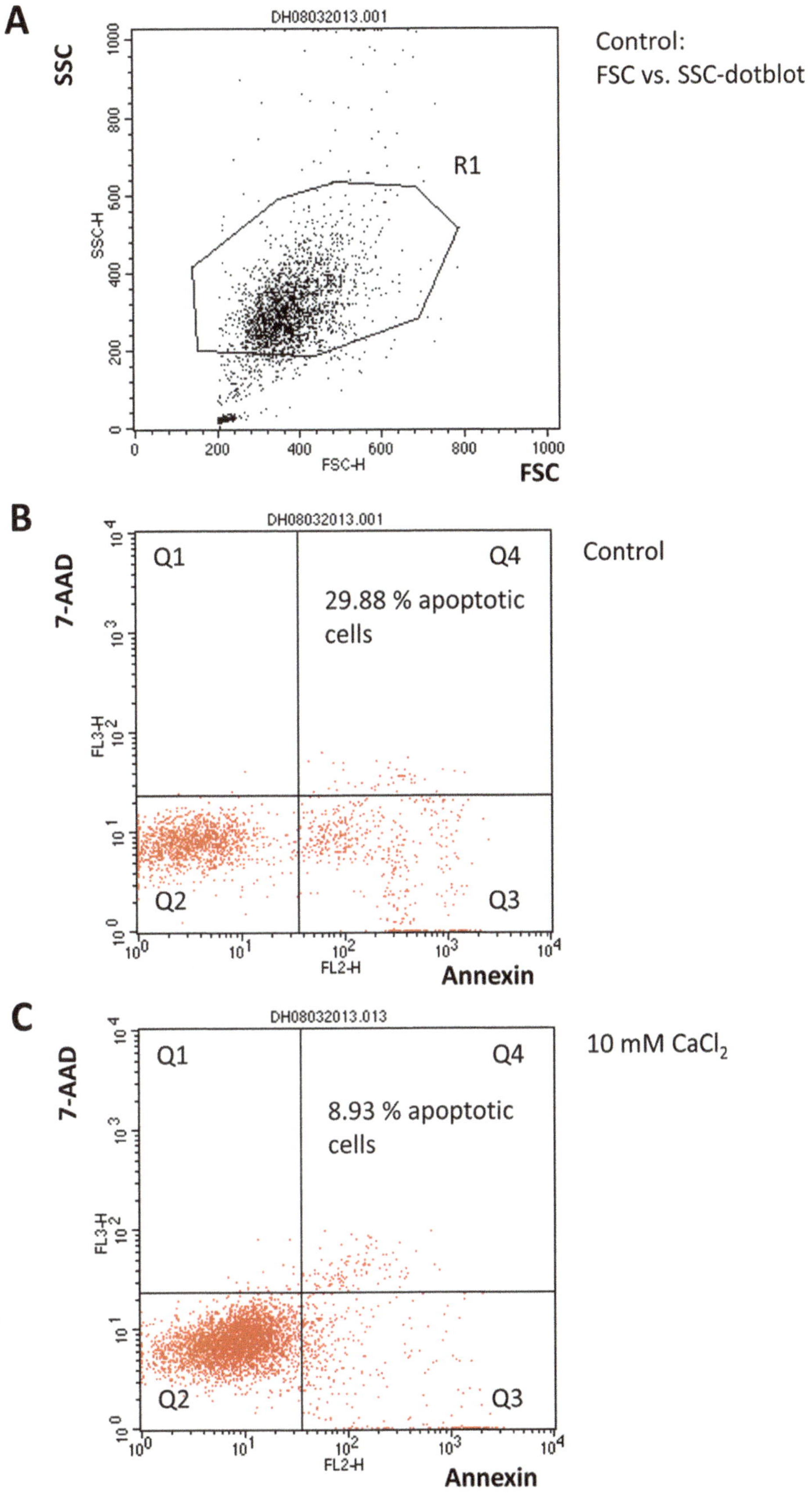

Figure 5. Decline of EPC apoptosis after 3 day incubation with Ca^{2+} [10 mM]. Apoptosis was measured by Annexin-staining, necrotic cells were stained with 7-AAD. Representative dot blots of the FACS-analysis are presented. Cells with appropriate forward/sideward scatter characteristics

were gated ("R1", A). The percentage of apoptotic cells was subsequently determined through quadrant analysis. Early and late apoptotic EPCs were located in quadrants Q3 and Q4, respectively. The percentage of apoptotic cells consists of the sum of early and late apoptotic cells. Apotosis of early EPCs in the control experiment is shown in (B) and in the presence of Ca^{2+} in (C).

were sacrificed one week after implantation and the VEGF expression and distribution of the human cells were analyzed a significantly elevated expression of VEGF in the defect area, compared to control animals without EPC, was observed. In contrast, the incorporation of the human cells in vWF-positive vessel structures was a rare event [33].

A stimulating effect of ionic products released by bioglass on osteogenic cell types has been reported in several investigations. Kaufmann et al. have reported an up-regulation of osteogenic

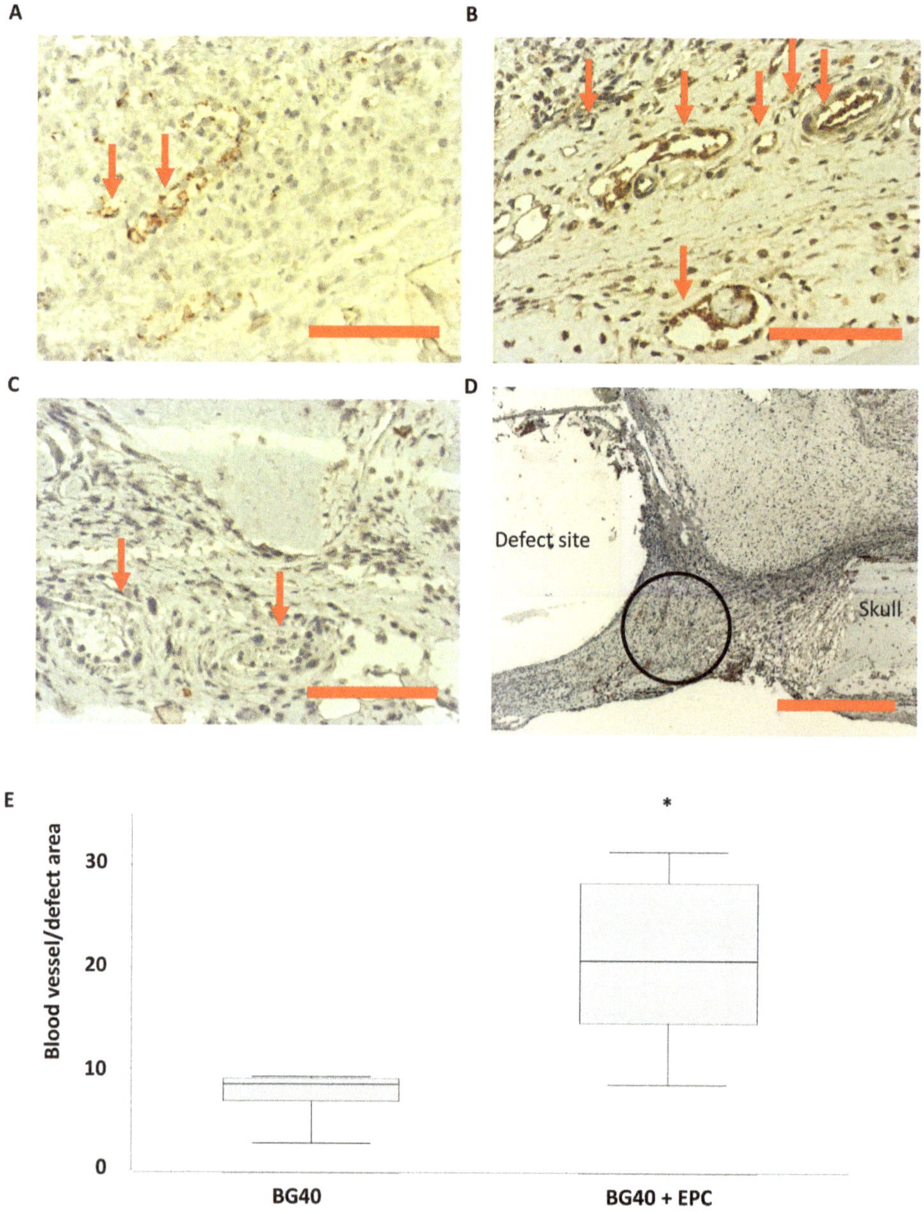

Figure 6. BG40+EPC increase vascularization one week after implantation. Representative vWF-stained histological sections of the defect site of male rats treated with BG40 (A, n=6) and BG40+EPC (B, n=6). The blood vessel density increased significantly in animals treated with BG40+EPC, compared to BG40 alone (B). vWF-positive structures appear brown (arrows). The isotype control is presented in figure 6C. The absence of staining in blood vessel structure (arrows) indicates the specificity of the vWF-staining (C). An overview of the defect area is provided in d (original magnification 50x). The BG40 implant could not be preserved during sample sectioning and left a gap corresponding to the BG area (defect site). The skull bone is located in the right corner (skull). The encircled area marks the region shown at higher magnification (original magnification 200x) in (A) and (B). The quantitative evaluation of vWF-positive blood vessel structures is depicted in (e). All vWF-positive, blood vessel like structures in the neighborhood of the whole defect site were counted. Red scale bars: 100 μm (A, B, C), 500 μm (D), *: p=0.015.

marker genes such as osteocalcin, osteonectin and osteopontin in osteoblasts cultured with bioglass [44]. Jell and coworkers have reported a significant up-regulation of osteogenic markers such as alkaline phosphatase and bone sialoprotein in osteoblasts incubated with bioglass conditioned medium [45].

More recently, different aspects of the proangiogenic potential of biomaterials containing bioglass have also been described. Day and colleagues have demonstrated enhanced synthesis of VEGF and basic fibroblast growth factor after incubating fibroblasts with bioglass [46]. Leu and Leach have shown that low concentrations of bioglass possess proangiogenic potential when they incubated human microvascular endothelial cells with bioglass conditioned medium. A dose dependent significant increase in cell proliferation, tube formation and VEGF-synthesis *in vitro* was observed [47]. The same group analyzed the angiogenic response to bioactive glass in an irradiated calvarial defect in rats. A bilateral calvarian defect was created and filled with a bioglass supplemented collagen sponge. An empty collagen sponge served as control. Significantly greater neovascularization within the defect in the presence of bioglass indicates a proangiogenic effect of this material [48].

The effect of a composite biomaterial consisting of a PLA-component and bioglass on EPC has recently been investigated by Aguirre et al. [34]. They report a significant increase of VEGF production and tube density if EPCs were seeded on a PLA-bioglass composite. However, the type of EPCs that were used in that study is not clearly defined. Based on the protocol for isolation and differentiation of the EPC it may be assumed that "late" EPCs (CD133 positive cells) were employed.

In the present study early EPCs were used [19,30–33]. In accordance to Aguirre et al. [34] an elongation of EPCs, which indicates sprouting activity [21,28,35], was observed after incubation with bioglass, correlating to the calcium concentration available in the medium.

A trend toward increased gene expression of VEGF and vWF was seen after incubation of EPC on BG40 in comparison to EPC cultured on PLA. This result is not as distinct as described by Aguirre et al who observed a significant up regulation of VEGF in EPC after incubation on a PLA-bioglass composite material [34]. This inconsistency might be due to differences in the mixture of the composite biomaterials, the sort of EPC and different measurement points (day 1 [34] vs. day 3, day 5 in this work). Calcium ions released by the bioglass might enhance the expression of both genes as it was discussed for VEGF by Aguirre et al [34]. The general increase of vWF gene expression of early EPC cultured on PLA and PLA-bioglass composite materials has not been described yet and might indicate a generally advanced endothelial differentiation of the EPC over the time [21].

The phenotypic changes of the early EPC were associated with a decline in EPC apoptosis in our experiments. Calcium mediates diverse cell processes and improved EPC survival may be due to a calcium dependent activation of PI3K, which activates the protein kinase Akt, a multifunctional regulator of EPC survival [28]. Cell shape alteration might occur over calcium mediated activation of rho-kinase (ROCK), a key regulator of cytoskeleton and cell polarity [49].

Conclusion

The present study demonstrates that the composite materials BG20 and BG40 support the function and survival of EPCs *in vitro* and that those effects are mediated by calcium released from the biomaterials. These observations suggest that improved EPC sprouting and delayed apoptosis should not only be seen as a pure *in vitro* phenomenon but may also be relevant to the improved vascularization in the skull defect model *in vivo*. It has become evident that understanding the effects of calcium ions on EPCs *in vitro* and *in vivo* requires further investigation. Elucidating underlying molecular mechanisms and subsequent optimization of the ionic components of bioglass may help to develop more effective biomaterials.

Author Contributions

Conceived and designed the experiments: KE CS DH. Performed the experiments: KE CS DH CNN AS. Analyzed the data: DH IM SM CN. Contributed reagents/materials/analysis tools: IM. Wrote the paper: CS KE DH.

References

1. Hong Z, Reis RL, Mano JF (2008) Preparation and in vitro characterization of scaffolds of poly(L-lactic acid) containing bioactive glass ceramic nanoparticles. Acta Biomater 4: 1297–1306. S1742-7061(08)00084-6 [pii];10.1016/j.actbio.2008.03.007 [doi].
2. Kim HW, Lee HH, Chun GS (2008) Bioactivity and osteoblast responses of novel biomedical nanocomposites of bioactive glass nanofiber filled poly(lactic acid). J Biomed Mater Res A 85: 651–663. 10.1002/jbm.a.31339 [doi].
3. Liu A, Hong Z, Zhuang X, Chen X, Cui Y, et al. (2008) Surface modification of bioactive glass nanoparticles and the mechanical and biological properties of poly(L-lactide) composites. Acta Biomater 4: 1005–1015. S1742-7061(08)00051-2 [pii];10.1016/j.actbio.2008.02.013 [doi].
4. Niemela T, Niiranen H, Kellomaki M (2008) Self-reinforced composites of bioabsorbable polymer and bioactive glass with different bioactive glass contents. Part II: In vitro degradation. Acta Biomater 4: 156–164. S1742-7061(07)00091-8 [pii];10.1016/j.actbio.2007.06.007 [doi].
5. Li X, Shi J, Dong X, Zhang L, Zeng H (2008) A mesoporous bioactive glass/polycaprolactone composite scaffold and its bioactivity behavior. J Biomed Mater Res A 84: 84–91. 10.1002/jbm.a.31371 [doi].
6. Porter JR, Ruckh TT, Popat KC (2009) Bone tissue engineering: a review in bone biomimetics and drug delivery strategies. Biotechnol Prog 25: 1539–1560. 10.1002/btpr.246 [doi].
7. Salgado AJ, Coutinho OP, Reis RL (2004) Bone tissue engineering: state of the art and future trends. Macromol Biosci 4: 743–765. 10.1002/mabi.200400026 [doi].
8. Jell G, Stevens MM (2006) Gene activation by bioactive glasses. J Mater Sci Mater Med 17: 997–1002. 10.1007/s10856-006-0435-9 [doi].
9. Brodie JC, Goldie E, Connel G, Merry J, Grant MH (2005) Osteoblast interactions with calcium phosphate ceramics modified by coating with type I collagen. J Biomed Mater Res A 73: 409–421. 10.1002/jbm.a.30279 [doi].
10. Oki A, Qiu XD, Alawode O, Foley B (2006) Synthesis of organic-inorganic hybrid composite and its thermal conversion to porous bioactive glass monolith. Mater Lett 60: 1–5.
11. Liao SS, Cui FZ, Zhang W, Feng QL (2004) Hierarchically biomimetic bone scaffold materials: nano-HA/collagen/PLA composite. J Biomed Mater Res B Appl Biomater 69: 158–165. 10.1002/jbm.b.20035 [doi].
12. Rezwan K, Chen QZ, Blaker JJ, Boccaccini AR (2006) Biodegradable and bioactive porous polymer/inorganic composite scaffolds for bone tissue engineering. Biomaterials 27: 3413–3431. S0142-9612(06)00123-2 [pii];10.1016/j.biomaterials.2006.01.039 [doi].
13. Fan X, Chen J, Ruan JM, Zhou ZC, Zou JP (2010) Preparation and Properties of Poly(lactide-co-glycolide)/Bone Meals Composite Materials. Polym Plast Technol 49: 309–315.
14. He Z, Xiong L (2009) Assessment of cytocompatibility of bulk modified ploy-L-lactic acid biomaterials. Polym Plast Technol 48: 1020–1024.
15. Vila OF, Bago JR, Navarro M, Alieva M, Aguilar E, et al. (2013) Calcium phosphate glass improves angiogenesis capacity of poly(lactic acid) scaffolds and stimulates differentiation of adipose tissue-derived mesenchymal stromal cells to the endothelial lineage. J Biomed Mater Res A 101: 932–941. 10.1002/jbm.a.34391 [doi].
16. Xu C, Su P, Chen X, Meng Y, Yu W, et al. (2011) Biocompatibility and osteogenesis of biomimetic Bioglass-Collagen-Phosphatidylserine composite scaffolds for bone tissue engineering. Biomaterials 32: 1051–1058. S0142-9612(10)01281-0 [pii];10.1016/j.biomaterials.2010.09.068 [doi].
17. Arkudas A, Beier JP, Heidner K, Tjiawi J, Polykandriotis E, et al. (2007) Axial prevascularization of porous matrices using an arteriovenous loop promotes survival and differentiation of transplanted autologous osteoblasts. Tissue Eng 13: 1549–1560. 10.1089/ten.2006.0387 [doi].

18. Asahara T, Murohara T, Sullivan A, Silver M, van der Zee R, et al. (1997) Isolation of putative progenitor endothelial cells for angiogenesis. Science 275: 964–967.
19. Henrich D, Seebach C, Kaehling C, Scherzed A, Wilhelm K, et al. (2009) Simultaneous cultivation of human endothelial-like differentiated precursor cells and human marrow stromal cells on beta-tricalcium phosphate. Tissue Eng Part C Methods 15: 551–560. 10.1089/ten.TEC.2008.0385 [doi];10.1089/ten.TEC.2008.0385 [pii].
20. Doyle B, Metharom P, Caplice NM (2006) Endothelial progenitor cells. Endothelium 13: 403–410. VUX6138862211824 [pii];10.1080/10623320601061656 [doi].
21. Fernandez PB, Lucibello FC, Zuzarte M, Lutjens P, Muller R, et al. (2001) Dendritic cells derived from peripheral monocytes express endothelial markers and in the presence of angiogenic growth factors differentiate into endothelial-like cells. Eur J Cell Biol 80: 99–110.
22. Rehman J, Li J, Orschell CM, March KL (2003) Peripheral blood "endothelial progenitor cells" are derived from monocyte/macrophages and secrete angiogenic growth factors. Circulation 107: 1164–1169.
23. Zhang SJ, Zhang H, Wei YJ, Su WJ, Liao ZK, et al. (2006) Adult endothelial progenitor cells from human peripheral blood maintain monocyte/macrophage function throughout in vitro culture. Cell Res 16: 577–584. 7310075 [pii];10.1038/sj.cr.7310075 [doi].
24. Zhang SJ, Zhang H, Hou M, Zheng Z, Zhou J, et al. (2007) Is it possible to obtain "true endothelial progenitor cells" by in vitro culture of bone marrow mononuclear cells? Stem Cells Dev 16: 683–690. 10.1089/scd.2006.0062 [doi].
25. Gehling UM, Ergun S, Schumacher U, Wagener C, Pantel K, et al. (2000) In vitro differentiation of endothelial cells from AC133-positive progenitor cells. Blood 95: 3106–3112.
26. Peichev M, Naiyer AJ, Pereira D, Zhu Z, Lane WJ, et al. (2000) Expression of VEGFR-2 and AC133 by circulating human CD34(+) cells identifies a population of functional endothelial precursors. Blood 95: 952–958.
27. Fuchs S, Hofmann A, Kirkpatrick C (2007) Microvessel-like structures from outgrowth endothelial cells from human peripheral blood in 2-dimensional and 3-dimensional co-cultures with osteoblastic lineage cells. Tissue Eng 13: 2577–2588. 10.1089/ten.2007.0022 [doi].
28. Dimmeler S, Assmus B, Hermann C, Haendeler J, Zeiher AM (1998) Fluid shear stress stimulates phosphorylation of Akt in human endothelial cells: involvement in suppression of apoptosis. Circ Res 83: 334–341.
29. Vasa M, Fichtlscherer S, Adler K, Aicher A, Martin H, et al. (2001) Increase in circulating endothelial progenitor cells by statin therapy in patients with stable coronary artery disease. Circulation 103: 2885–2890.
30. Henrich D, Wilhelm K, Warzecha J, Frank J, Barker J, et al. (2013) Human endothelial-like differentiated precursor cells maintain their endothelial characteristics when cocultured with mesenchymal stem cell and seeded onto human cancellous bone. Mediators Inflamm 2013: 364591. 10.1155/2013/364591 [doi].
31. Schultheiss J, Seebach C, Henrich D, Wilhelm K, Barker JH, et al. (2011) Mesenchymal stem cell (MSC) and endothelial progenitor cell (EPC) growth and adhesion in six different bone graft substitutes. Eur J Trauma Emerg Surg 37: 635–644.
32. Seebach C, Henrich D, Kahling C, Wilhelm K, Tami AE, et al. (2010) Endothelial progenitor cells and mesenchymal stem cells seeded onto beta-TCP granules enhance early vascularization and bone healing in a critical-sized bone defect in rats. Tissue Eng Part A 16: 1961–1970. 10.1089/ten.TEA.2009.0715 [doi].
33. Seebach C, Henrich D, Wilhelm K, Barker JH, Marzi I (2012) Endothelial progenitor cells improve directly and indirectly early vascularization of mesenchymal stem cell-driven bone regeneration in a critical bone defect in rats. Cell Transplant 21: 1667–1677. ct2189seebach [pii];10.3727/096368912X638937 [doi].
34. Aguirre A, Gonzalez A, Navarro M, Castano O, Planell JA, et al. (2012) Control of microenvironmental cues with a smart biomaterial composite promotes endothelial progenitor cell angiogenesis. Eur Cell Mater 24: 90–106. vol024a07 [pii].
35. Henrich D, Hahn P, Wahl M, Wilhelm K, Dernbach E, et al. (2004) Serum derived from multiple trauma patients promotes the differentiation of endothelial progenitor cells in vitro: possible role of transforming growth factor-beta1 and vascular endothelial growth factor165. Shock 21: 13–16. 10.1097/01.shk.0000101669.49265.50 [doi].
36. Zink D, Sadoni N, Stelzer E (2003) Visualizing chromatin and chromosomes in living cells. Methods 29: 42–50. S104620230200289X [pii].
37. Rasal RM, Janokar AV, Hirt DE (2010) Poly(lactic acide) modifications. Prog Polymer Sci 33: 338–356.
38. Jones JR, Lin S, Yue S, Lee PD, Hanna JV, et al. (2010) Bioactive glass scaffolds for bone regeneration and their hierarchical characterisation. Proc Inst Mech Eng H 224: 1373–1387.
39. Gorustovich AA, Roether JA, Boccaccini AR (2010) Effect of bioactive glasses on angiogenesis: a review of in vitro and in vivo evidences. Tissue Eng Part B Rev 16: 199–207. 10.1089/ten.TEB.2009.0416 [doi].
40. Lu HH, Tang A, Oh SC, Spalazzi JP, Dionisio K (2005) Compositional effects on the formation of a calcium phosphate layer and the response of osteoblast-like cells on polymer-bioactive glass composites. Biomaterials 26: 6323–6334. S0142-9612(05)00291-7 [pii];10.1016/j.biomaterials.2005.04.005 [doi].
41. Dorozhkin SV (2012) Calcium orthophosphate coatings, films and layers. Prog Biomater 1.
42. Sander AL, Jakob H, Henrich D, Powerski M, Witt H, et al. (2011) Systemic transplantation of progenitor cells accelerates wound epithelialization and neovascularization in the hairless mouse ear wound model. J Surg Res 165: 165–170. S0022-4804(09)00369-2 [pii];10.1016/j.jss.2009.07.003 [doi].
43. Crosby JR, Kaminski WE, Schatteman G, Martin PJ, Raines EW, et al. (2000) Endothelial cells of hematopoietic origin make a significant contribution to adult blood vessel formation. Circ Res 87: 728–730.
44. Kaufmann EA, Ducheyne P, Shapiro IM (2000) Effect of varying physical properties of porous, surface modified bioactive glass 45S5 on osteoblast proliferation and maturation. J Biomed Mater Res 52: 783–796. 10.1002/1097-4636(20001215)52:4<783::AID-JBM24>3.0.CO;2-J [pii].
45. Jell G, Notingher I, Tsigkou O, Notingher P, Polak JM, et al. (2008) Bioactive glass-induced osteoblast differentiation: a noninvasive spectroscopic study. J Biomed Mater Res A 86: 31–40. 10.1002/jbm.a.31542 [doi].
46. Day RM (2005) Bioactive glass stimulates the secretion of angiogenic growth factors and angiogenesis in vitro. Tissue Eng 11: 768–777. 10.1089/ten.2005.11.768 [doi].
47. Leu A, Leach JK (2008) Proangiogenic potential of a collagen/bioactive glass substrate. Pharm Res 25: 1222–1229. 10.1007/s11095-007-9508-9 [doi].
48. Leu A, Stieger SM, Dayton P, Ferrara KW, Leach JK (2009) Angiogenic response to bioactive glass promotes bone healing in an irradiated calvarial defect. Tissue Eng Part A 15: 877–885. 10.1089/ten.tea.2008.0018 [doi].
49. Amano M, Nakayama M, Kaibuchi K (2010) Rho-kinase/ROCK: A key regulator of the cytoskeleton and cell polarity. Cytoskeleton (Hoboken) 67: 545–554. 10.1002/cm.20472 [doi].

4

Collagen Self-Assembly on Orthopedic Magnesium Biomaterials Surface and Subsequent Bone Cell Attachment

Nan Zhao[1,2], Donghui Zhu[1,2]*

1 Department of Chemical, Biological and Bio-Engineering, North Carolina Agricultural and Technical State University, Greensboro, North Carolina, United States of America, 2 NSF Engineering Research Center-Revolutionizing Metallic Biomaterials, North Carolina Agricultural and Technical State University, Greensboro, North Carolina, United States of America

Abstract

Magnesium (Mg) biomaterials are a new generation of biodegradable materials and have promising potential for orthopedic applications. After implantation in bone tissues, these materials will directly interact with extracellular matrix (ECM) biomolecules and bone cells. Type I collagen, the major component of bone ECM, forms the architecture scaffold that provides physical support for bone cell attachment. However, it is still unknown how Mg substrate affects collagen assembly on top of it as well as subsequent cell attachment and growth. Here, we studied the effects of collagen monomer concentration, pH, assembly time, and surface roughness of two Mg materials (pure Mg and AZ31) on collagen fibril formation. Results showed that formation of fibrils would not initiate until the monomer concentration reached a certain level depending on the type of Mg material. The thickness of collagen fibril increased with the increase of assembly time. The structures of collagen fibrils formed on semi-rough surfaces of Mg materials have a high similarity to that of native bone collagen. Next, cell attachment and growth after collagen assembly were examined. Materials with rough surface showed higher collagen adsorption but compromised bone cell attachment. Interestingly, surface roughness and collagen structure did not affect cell growth on AZ31 for up to a week. Findings from this work provide some insightful information on Mg-tissue interaction at the interface and guidance for future surface modifications of Mg biomaterials.

Editor: Jie Zheng, University of Akron, United States of America

Funding: The authors acknowledge funding from the National Institute of Health (SC2NS082475) and support from NSF Engineering Research Center-Revolutionizing Metallic Biomaterials (ERC-RMB) (NSF-0812348) at North Carolina A&T State University. The funders had no role in study design, data collection and analysis, decision to publish, or preparation of the manuscript.

Competing Interests: The authors have declared that no competing interests exist.

* Email: dzhu@ncat.edu

Introduction

There is an increasing interest in magnesium (Mg)-based alloys as implantable orthopedic medical devices because of their biodegradability and good biocompatibility [1–11]. Compared with other metal biomaterials, e.g., stainless steel, titanium alloys, and cobalt-chromium alloys, Mg alloys have several advantages for orthopedic application. First, their physical and mechanical properties including density (1.74–2.0 g/cm^3), elastic modulus (41–45 GPa), and compressive yield strength (65–100 MPa), are much closer to that of natural bone, and therefore can avoid the stress shielding effect [12–14]. Second, Mg is an essential element for many biological activities including enzymatic reaction, formation of apatite, and bone cells adsorption [15]. Third, Mg alloys can eliminate the necessity of a second surgery to remove the permanent bone implants.

The success of an medical implant is largely dependent on the interaction between the surface of the implant and the surrounding tissues [16]. Both surface chemistry and topography of implants can affect biological activities such as osteoblasts metabolism, collagen synthesis, and alkaline phosphatase activity [17–20]. Cells often display distinctive morphological and metabolic properties when they are in contact with materials with different surface roughness [19]. It is a general consensus that cells cannot directly recognize bare metal materials in vitro or in vivo. It is the biomacromolecules absorbed on metal materials serve as a bridge connecting cells to the solid surface [20]. Therefore, the adsorption of ECM proteins and subsequent structure changes may lead to different cell fate. Collagen as the most abundant ECM protein is the major component of natural bone. It plays an important role in cell attachment, mechanical support, and apatite nucleation [21]. The mean weight percent of collagen in modern mammal bone is around 20.8%, and 90% of the organic matrix in bone is comprised of collagen [22,23]. Studies have been carried out in the past with respect to the self-assembly characteristic of collagen [24–26] and application of type I collagen as coating materials [27–29]. Fang et al. showed that different mica surfaces affect D-period during collagen self-assembly [30]. Nassif et al. reported that collagen-apatite matrix is necessary for organization of collagen fibrils into 3-D scaffolds and nucleation of hydroxyapatite [31]. However, the information on collagen and Mg biomaterial interaction is still missing in the literature.

Previous studies showed that biodegradable Mg alloys enhanced bone-implant strength and osseointegration compared to titanium alloys [2,32]. With the increasing orthopedic applications of Mg alloys, there is an urgent need to fill such a gap to understand how collagen molecules interact with the solid metal phase at the interface as well as the subsequent cell attachment. Evaluating the interaction between collagen and Mg implant in vivo could be very challenging currently owing to the complexity of biological system. Hence, an in vitro model was developed here to study type I collagen adsorption, assembly and osteoblasts adhesion on different Mg materials.

Materials and Methods

Material Preparation

High Purity Mg extrusion rod and MgAlZn (AZ31) extrusion rod with diameter of 10 mm were obtained from Goodfellow, USA. Materials were cut (Techcut 5, Allied High Tech Products, USA) into D10×1 mm disc and polished with SiC paper up to 1200 grit. All materials were supersonically cleaned (M2510 Ultrasonic cleaner, Branson, USA) in acetone (Fisher Scientific, USA) and washed 3 times with ethanol (Fisher Scientific, USA) followed by 30 minutes UV sterilization (1380 Biological Safety Cabinet, Thermo, USA). For each experiment, at least 3 replicates were used.

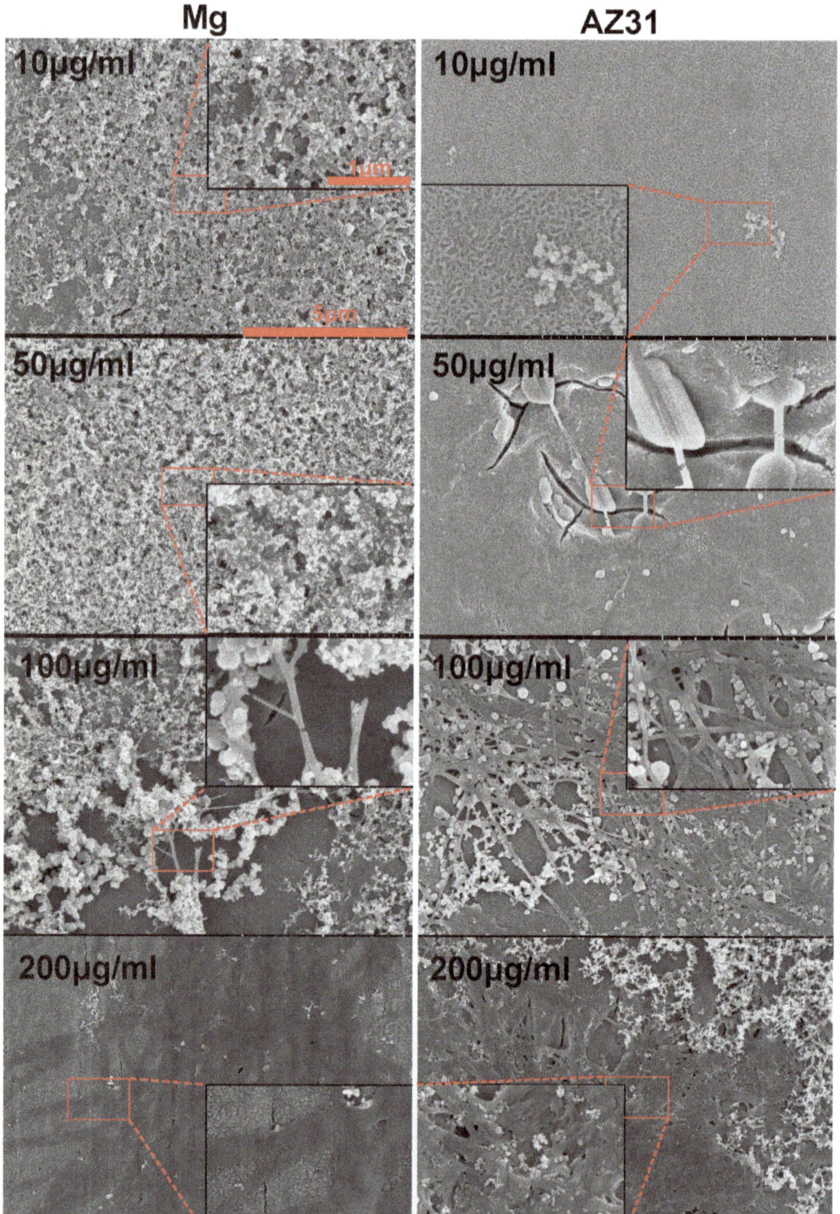

Figure 1. Representative SEM images of collagen self-assembly on Mg (left) and AZ31 (right). The concentration of collagen in each group is shown on the top left corner of each image.

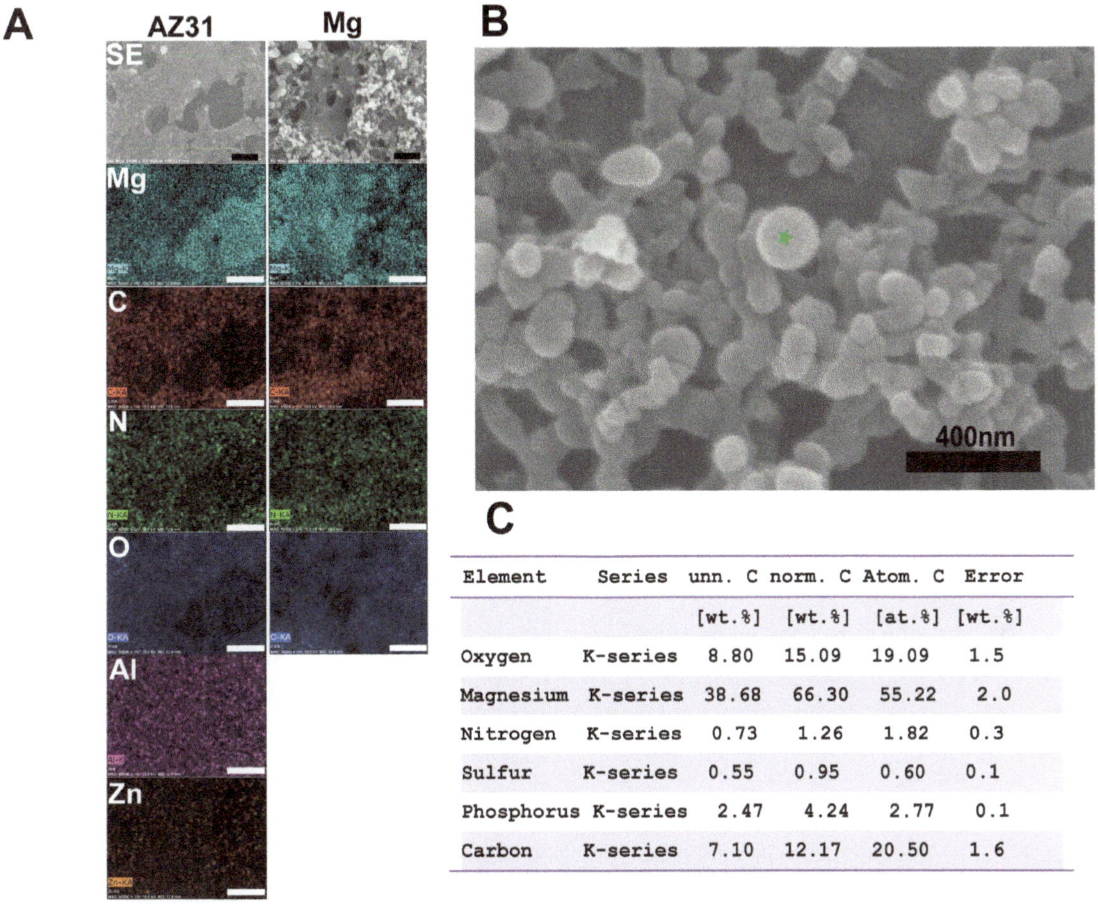

Element	Series	unn. C	norm. C	Atom. C	Error
		[wt.%]	[wt.%]	[at.%]	[wt.%]
Oxygen	K-series	8.80	15.09	19.09	1.5
Magnesium	K-series	38.68	66.30	55.22	2.0
Nitrogen	K-series	0.73	1.26	1.82	0.3
Sulfur	K-series	0.55	0.95	0.60	0.1
Phosphorus	K-series	2.47	4.24	2.77	0.1
Carbon	K-series	7.10	12.17	20.50	1.6

Figure 2. EDS analysis of collagen fibril and spherical particles on AZ31 and Mg. The corresponding element is shown on the top left corner. Scale bar is 400 nm. (A) EDS element mapping, (B) High magnification SEM image of nanoparticles attached on collagen fibril, (C) Elemental composition of a nanoparticle (green star in B) by EDS point analysis.

Collagen assembly on Mg surface

The effect of collagen concentration. Rat tail type I collagen solution (Life Technologies, USA) of 3 mg/ml was diluted by D-phosphate-buffered solution (DPBS, Invitrogen, USA) to 10, 50, 100, and 200 μg/ml, respectively. 50 μl final collagen solutions (pH of 7) were spread on testing materials surface followed by incubating under 37°C for 2 hours and then dehydrate with ethanol gradient. The morphology of final collagen structure was characterized by Scanning Electron Microscope (SEM, SU8000, Hitachi, USA). Mg materials treated with 200 μg/ml collagen solution were used for elemental analysis. Energy dispersive X-ray spectroscopy (EDS, Qantax200, Bruker, USA) element mapping was performed with high voltage of 15 kV and probe current of 15 mA. Quantitative spectrum analysis was carried out after the spectra acquisition.

Effect of pH

The pH of DPBS (7.49) solutions were adjusted to 7, 9, and 11 by 1 M NaOH (Sigma-Aldrich, USA) and 1 M HCl (Sigma-Aldrich, USA). Stock collagen solution was diluted by DPBS solutions with different pH to final concentration of 200 μg/ml. 50 μl final collagen solutions were spread on Mg and AZ31 surface (polished up to 1,200 grit SiC paper) for 2 hours of assembly, then followed by dehydration using ethanol gradient. The morphology of collagen was imaged by SEM.

Effect of assembly time

50 μl of 200 μg/ml DPBS diluted collagen solution (pH of 7) was spread on Mg and AZ31 surface and allowed to assemble for 4 h and 8 h. Then samples were dehydrated with ethanol gradient. The morphology of collagen was imaged by SEM.

Effect of surface roughness

Mg and AZ31 were divided into 3 groups and polished up to 180, 800 and 1,200 grit SiC paper, respectively. Materials polished up to 180 grit SiC paper were denoted as rough surface (RS); materials polished up to 800 grit SiC paper were denoted as semi-rough surface (SR); and materials polished up to 1,200 grit SiC paper were denoted as smooth surface (SS). Surface roughness was characterized by WYKO Optical Profiler (Veeco, USA). 50 μl 200 μg/ml DPBS diluted collagen solution (pH of 7) was spread on Mg and AZ31 surface and allowed to assemble for 2 h. The morphology of collagen fibril was imaged by SEM.

Collagen dynamic adsorption

RS, SR and SS materials were soaked into 1 ml 60 μg/ml diluted collagen solution (pH of 7) with only one side exposed to the solution. The amount of absorbed collagen was quantified by Sirius Red Assay (Abacam, USA) according the method described previously with minor modification [33,34]. In brief, materials were soaked into collagen solution for 0.5 h, 1 h, 2 h, 4 h, and

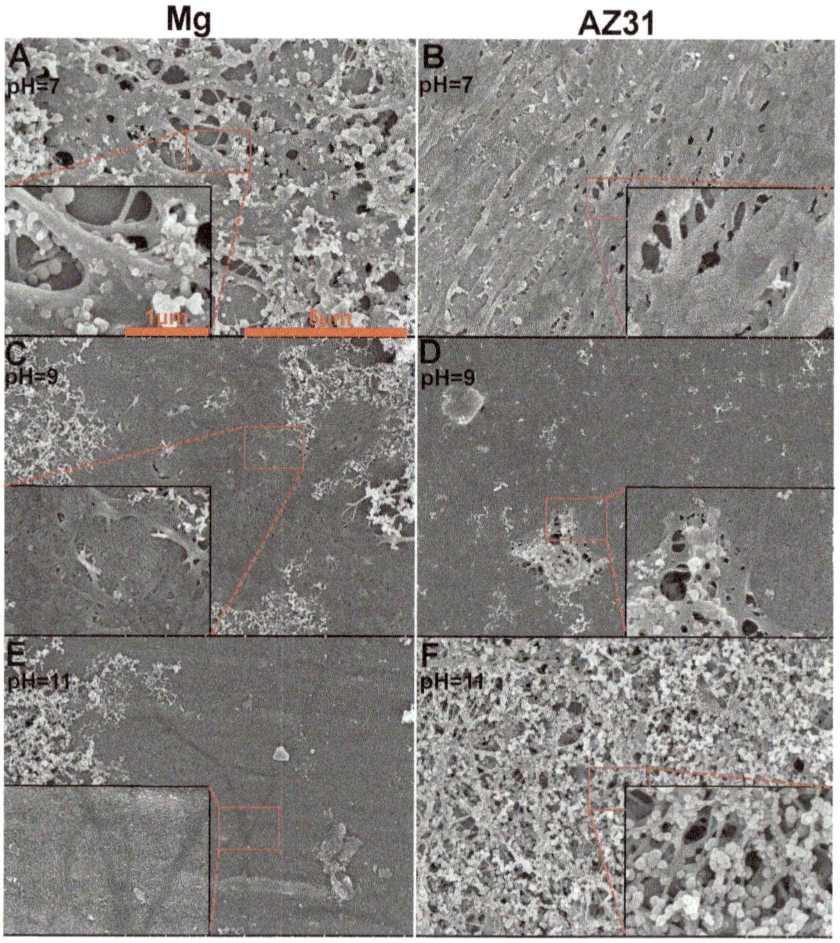

Figure 3. Representative SEM images of 200 μg/ml collagen self-assembly in DPBS with different pH values (A–B: 7, C–D: 9, E–F: 11) on Mg surface (A, C, E) and AZ31 (B, D, F).

8 h, respectively. At each time point, samples were removed and washed with DPBS for 3 times. Then the unattached collagen in solution was transferred to a new tube followed by incubation with Sirius Red for 1 hour. The solutions were centrifuged at 8,000 g for 15 min and the dye was eluted by 0.1 M NaOH. The absorbance was measured at 540 nm by 10S UV-Vis Spectrometer (Thermo, USA). Standard curve of a series of collagen solution (7.5, 15, 22.5, 30, 37.5, 45, 52.5, and 60 μg/ml) was obtained as the same procedures. Linear regression was performed in Prism 5 (GraphPad, USA). The attached collagen was calculated by subtraction of initial collagen by the collagen remained in the solution.

Cell attachment

Mouse osteoblasts (MC 3T3, ATCC, USA) was expanded in Minimum Essential Medium α (MEM, Life Technologies, USA) supplemented with 10% fetal bovine serum (Sciencell, USA), 100 U/ml penicillin (Sciencell, USA) and 100 μg/ml streptomycin (Sciencell, USA) in humidified incubator (Heracell 150 I, Thermo, USA) with 5% CO_2 as previously described [35]. 50 μl 200 μg/ml collagen solution (pH of 7) was allowed to self-assemble on Mg and AZ31 with RS, SR, and SS for 2 h in a 24-well culture plate (BD Bioscience, USA). Then these materials were gently rinsed by DPBS for 3 times. 50 μl cell solution with density of 10,000 cell/ml was dipped onto the surface of collagen treated materials. Cells were allowed to attach for 30 min and then samples were washed gently with DPBS for 3 times. After 4 hours, cells were fixed with 4% paraformaldehyde (Boston Bioproducts, USA) followed by ethanol gradient dehydration for 10 minutes. Samples were coated (E5400, Polaron Instruments, USA) with gold nanoparticles for 2 min, and imaged by SEM.

Cell proliferation

Mg and AZ31 with different surface roughness treated by collagen as described above were used to test cell proliferation. MC 3T3 Cells were seeded onto the collagen coated material surface with density of 10,000 cell/ml in a 24-well culture plate. At 1st, 4th, and 7th day, cell culture media were changed and cells were stained by Live/Dead kit (Invitrogen, USA). Culture media were centrifuged (Biofuge Stratos, Thermo, USA) at 8,000 g for 10 min and pH was measured by a pH meter (Eutech, USA). The fluorescent images were taken by a digital inverted light microscope (EVOS, Advanced Microscopy, USA).

Mg^{2+} concentration measurement

Mg^{2+} concentration was measured by xylidyl blue magnesium kit (Pointe Scientific, USA) as previously described [36]. In brief, 10 μl aliquot of test solution was added to 1.5 ml final xylidyl solution (0.1 mM xylidyl blue, 0.13 mM EGTA, 1.4 M DMSO, 0.02% potassium cyanide) and incubated for 10 min at room

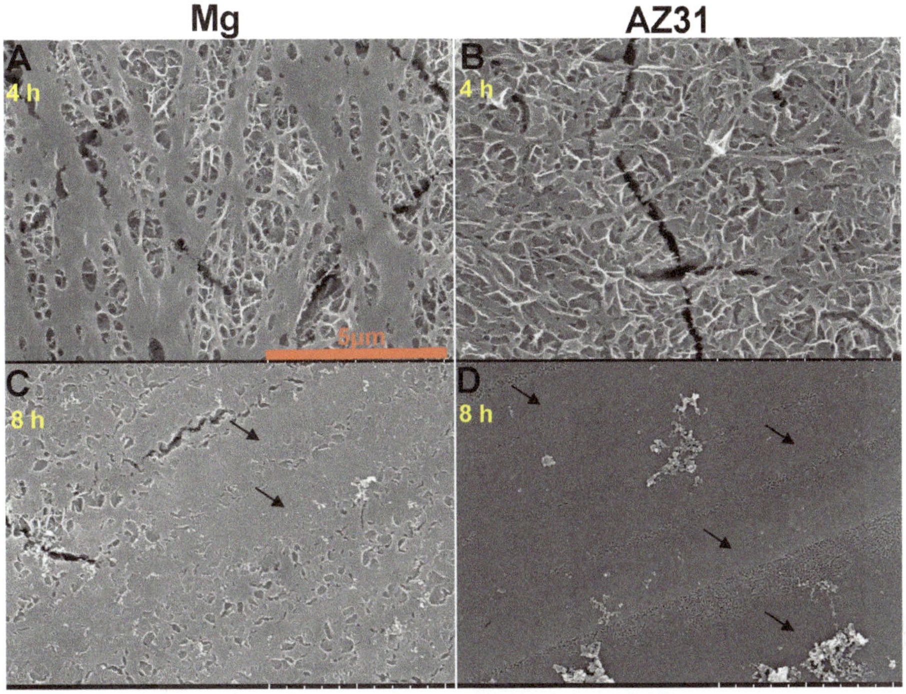

Figure 4. Representative SEM images of 200 μg/ml collagen self-assembly on Mg (A, C) and AZ31 (B, D) for 4 h (A, B) and 8 h (C, D). Arrows indicate the large collagen band.

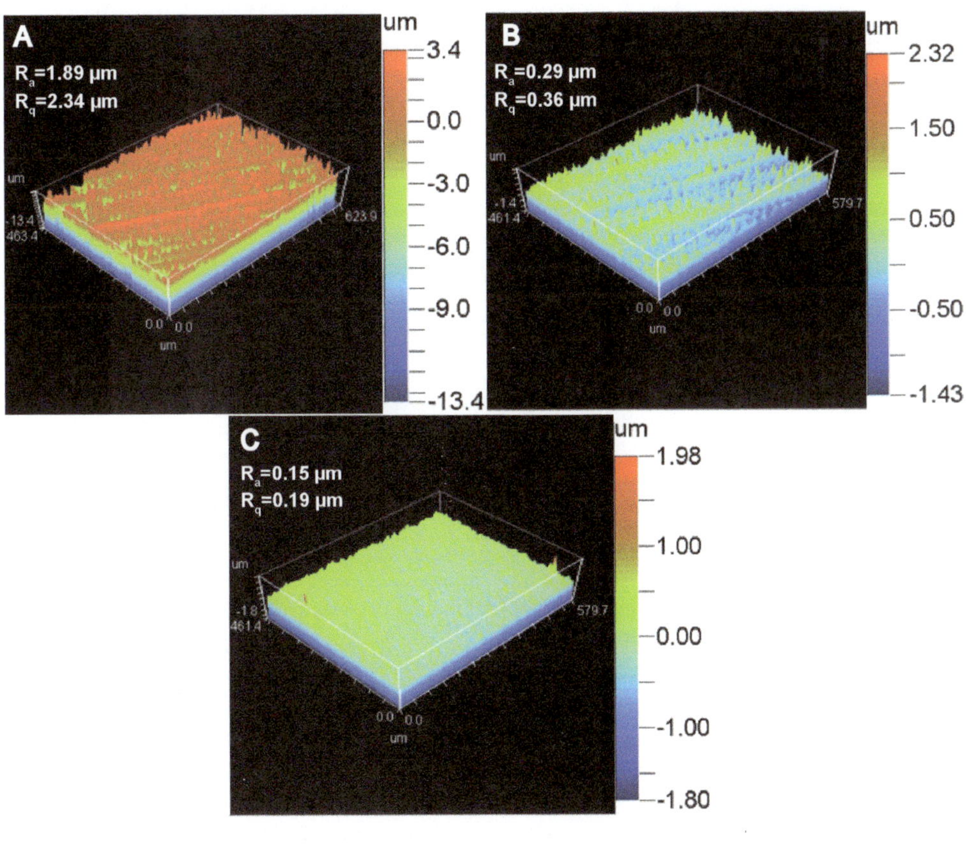

Figure 5. 3-D surface topography of Mg sample polished by different SiC paper (A: 180 grit; B: 800 grit; C: 1200 grit). The arithmetic average of surface roughness (R_a) and root mean square (R_q) calculated from 4 different samples are shown on the top left corner.

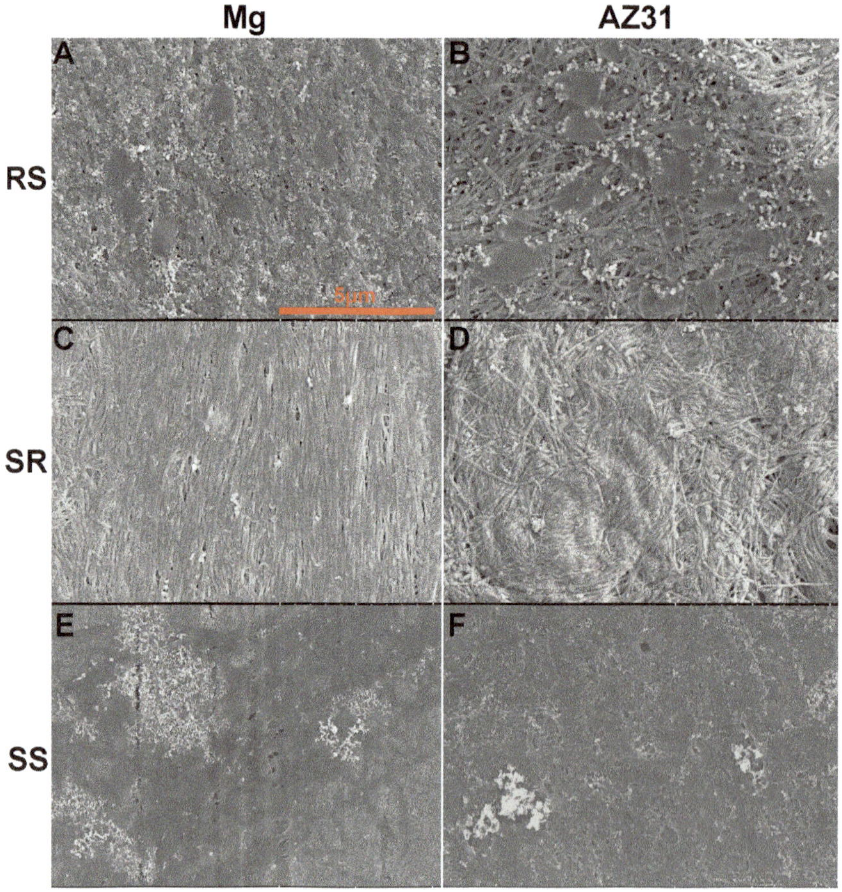

Figure 6. Representative SEM images of 200 μg/ml collagen self-assembly on Mg (A, C, E) and AZ31 (B, D, F) with different surface roughness (A–B: RS; C–D: SR; E–F: SS). RS: rough surface; SR: semi-rough surface; SS: smooth surface.

temperature. The absorbance of the mixtures was measured at 520 nm by a UV-Vis Spectrometer (Thermo, USA). Standard curve was obtained by using gradient $MgCl_2$ solution. Linear regression was performed in Prism 5 (GraphPad, USA). Mg^{2+} concentration was determined by the standard curve.

Statistics Analysis

Data was expressed as Mean ± SD in all figures unless otherwise specified. Statistical analysis was performed in Prism 5 software (GraphPad, USA). Unpaired two-tailed *t* test was used to compare difference between two groups. Multiple comparisons were performed by using one-way ANOVA followed by post hoc analysis. Two-way ANOVA was used to evaluate the effect of two

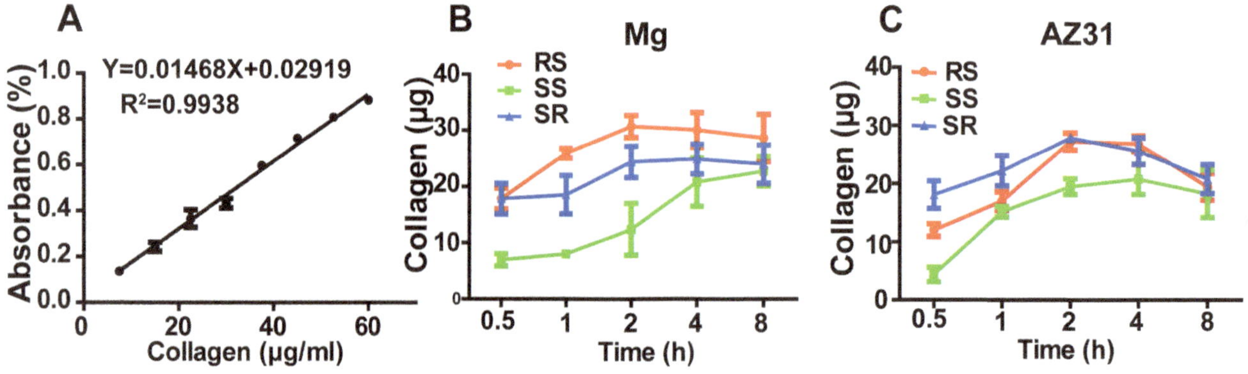

Figure 7. Collagen adsorption quantification. Standard curve for quantification of collagen by Picro-sirius red stain kit (A). The amount of collagen absorbed on Mg with different surface roughness at 0.5, 1, 2, 4, and 8 h (B). The amount of collagen absorbed on AZ31 with different surface roughness at 0.5, 1, 2, 4, and 8 h (C).

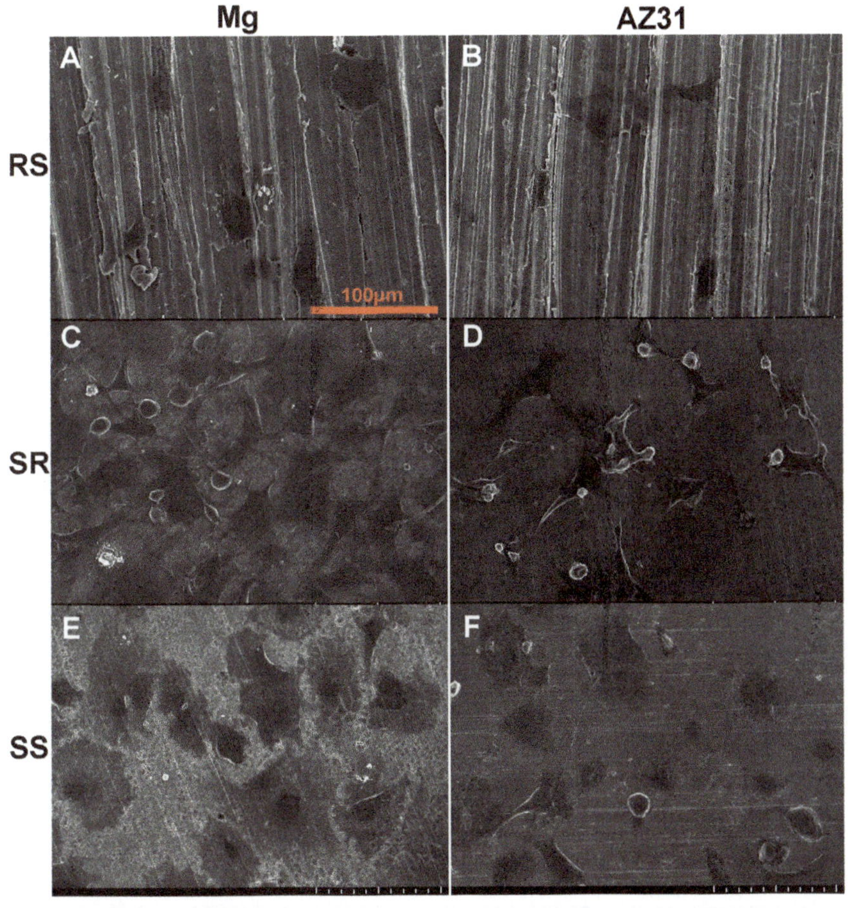

Figure 8. Representative SEM images of MC 3T3 cell attachment on collagen self-assembled at Mg (A, C, E) and AZ31 (B, D, F) with different surface roughness (A–B: RS; C–D: SR; E–F: SS).

independent factors. It is considered significantly different statistically if the $P<0.05$.

Results

Effect of collagen concentration

We first explored collagen self-assembly with different concentrations of collagen monomers on SS of Mg and AZ31 at neutral pH (Fig. 1). On pure Mg, collagen monomers agglomerated into a non-uniform structure and no long fibril was observed when the concentration was lower than 50 μg/ml. Some spherical particles were present on the surface of pure Mg in both 10 μg/ml and 50 μg/ml groups. When the collagen concentration reached to 100 μg/ml, a few fibrils started to appear. The structure changed from thin fibrils to wide bands as the initial concentration increased to 200 μg/ml. On AZ31 surface, long fibril started to appear as the collagen concentration reached to 50 μg/ml. In addition, sparsely dispersed woven structure composed of collagen fibers (fibril bundles) were observed in the 100 μg/ml group. Multiple-layer network structure with collagen ribbon of 100 nm width was the predominant structure in the 200 μg/ml group. Spherical particles in different sizes were also present in all the groups.

Next, we examined the chemical composition distribution of the fibril structure and the spherical particles by EDS (Fig. 2). Carbon (C), nitrogen (N) and oxygen (O) were mainly from the collagen fibrils structures on pure Mg and AZ31, which further demonstrated that they are not degradation products but collagen fibrils. In pure Mg group, less Mg was detected in some spherical particle dense zones. Al and Zn were evenly distributed on the collagen layer. A higher magnification image of the spherical particles (Fig. 2B) shows that the size of those particles is about one hundred nanometer. Their chemical composition is summarized in Fig. 2C.

Effect of pH

Effect of pH on collagen assembly was also investigated. SEM images of collagen self-assembly under different pH on SS materials for 2 h are shown in Fig. 3. At pH 7, collagen ribbons with width ~100 nm conjugated with other fibrils, forming a multiple layer network structure on pure Mg. A few nanofibril side chains connecting adjacent long collagen ribbon were also seen. On AZ31, some parallel collagen ribbons were connected with adjacent collagen ribbons by smaller branches and others merged with their proximal collagen, forming a uniform sheet. At pH 9, more collagen ribbons wove together spreading on Mg surface. In addition, some bare areas and crevices were observable. The whole surface of AZ31 was almost covered by a collage layer with some small holes. Few long collagen fibrils were present on pure Mg when pH rose to 11. On AZ31, thin fibrils randomly crossed with others, resulting in a network structure. A lot of nano-spherical particles attached to the collagen fibrils.

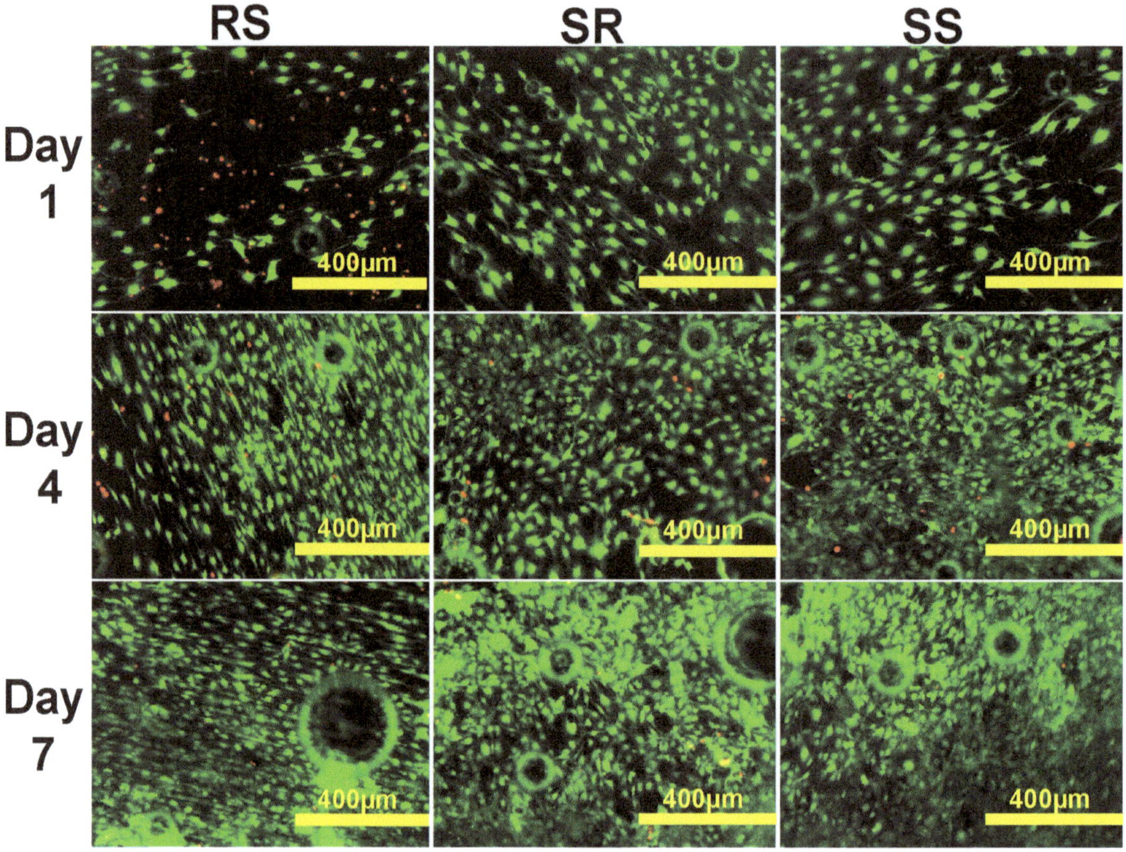

Figure 9. Representative fluorescent images of MC 3T3 cells growing on collagen self-assembled at AZ31 with different surface roughness for 1, 4 and 7 d. Live cells are in green color and dead cells are in red color.

Effect of reaction time

Fig. 4 shows the structures of collagen assembling for different time periods on SS materials at neutral pH. After 4 hours assembling, a cancellous underneath layer was covered by some long collagen ribbons on pure Mg. Similar cancellous structure was found on AZ31, but less thin collagen fibrils could be observed on the top. When the assembly time reached 8 hours, micrometer-wide fibers were present on both pure Mg and AZ31. On pure Mg, a lot of small lamellar sheets were observed between two thick fibers.

Effect of surface roughness

Representative 3-D topographical images of materials after polished by different SiC paper are shown in Fig. 5. The average roughness (R_a) and the corresponding root mean squared roughness (R_q) were also calculated (Fig. 5). Fig. 6 shows the collagen self-assembly on the samples with different surface roughness. On the RS Mg materials, long collagen fibers interwove with each other resulting in a compact woven layer with a few fish-like scales on the top. The woven layer on RS AZ31 was less denser compared with that on RS Mg. Larger and more fish-like scales structures could be seen on the RS AZ31. For both AZ31 and pure Mg with SR, collagen fibers aligned parallelly to each other at some places while randomly intertwined at other places. On SS of both materials, collagen bands were similar as described previously (Fig. 1).

Collagen assembly quantification

We also quantified the amount of collagen on the material surface during its assembly. A standard curve for quantification of collagen content was established (Fig. 7A). The optical absorbance at 540 nm versus collagen content displayed reasonable linearity within the range from 5 to 60 μg/ml. Collagen dynamic adsorption curves on Mg and AZ31 surfaces are shown in Fig. 7B and Fig. 7C. For both AZ31 and pure Mg, less collagen was able to be absorbed onto the SS materials at the initial phase (0.5 h) compared with RS and SR materials. The amount of attached collagen increased as the adsorption time increased and reached equilibrium state after 4 h for all three groups of AZ31. Collagen on the pure Mg with SS kept increasing slowly from 0.5 to 2 h. In comparison, collagen on pure Mg with RS peaked at 2 h and then started to drop. Among all the groups, the highest amount of absorbed collagen was 30.69 ± 1.96 μg on RS Mg at 2 h. Two-way ANOVA analysis revealed that both time and surface roughness had significant effect on collagen adsorption on pure Mg.

Bone cell attachment and proliferation

After Mg and AZ31 with different surface roughness were treated by 200 μg/ml collagen solution for 2 h, the materials were used to test subsequent bone cell attachment (Fig. 8). SEM images showed that both round cells and cells with filopodia were observed on RS AZ31 and pure Mg. On SR materials, most cells were well attached with flattened morphology and a few fibroblast-like cells with webbing could also be seen. In addition, super long

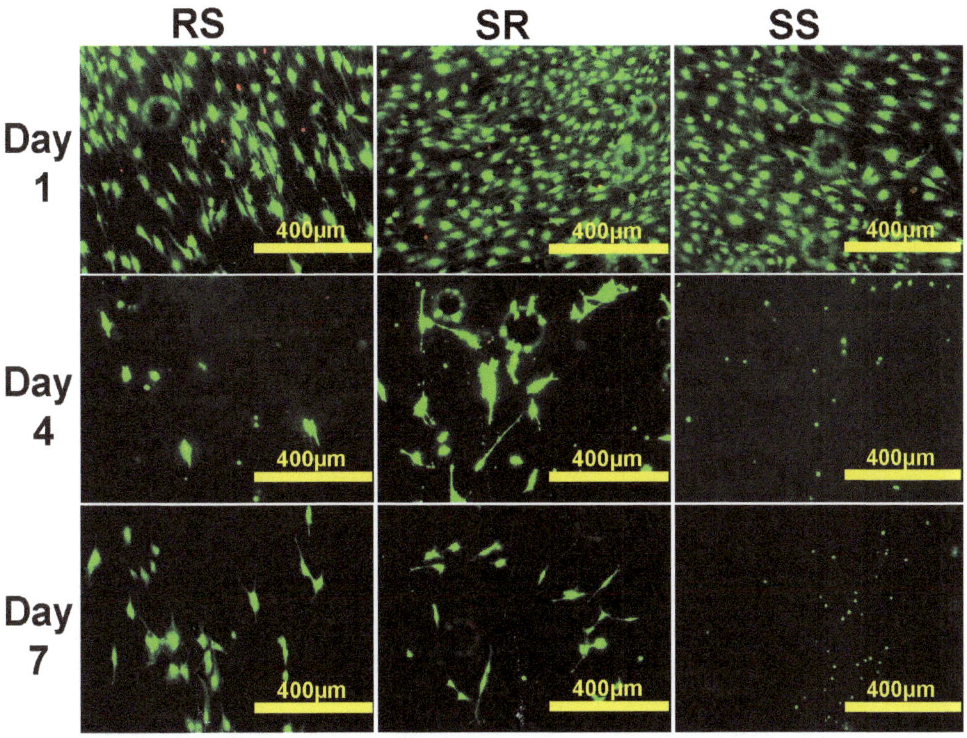

Figure 10. Representative fluorescent image of MC 3T3 cells growing on collagen self-assembled at Mg with different surface roughness for 1, 4 and 7 d. Live cells are in green color and dead cells are in red color.

filopodia from some cells span over a large distance and reached the edge of other cells or an empty area. On the SS materials, most cells were well flattened with very large cell surface area. Cell density on SR and SS was significantly higher than that on RS.

Fluorescent live/dead cell analysis was then performed to examine the bone cell proliferation. MC 3T3 cells on AZ31 and Mg treated with 200 μg/ml collagen solution for 2 h are shown in Fig. 9 and Fig. 10, respectively. Cells displayed healthy morphology in all the groups after one day. Cell densities on SR and SS AZ31 were significantly higher than that on RS AZ31 (Fig. 9). Some dead cells were visible after first day on RS AZ31. Small air bubbles were present in all the materials. After three days' incubation, cell density on all the AZ31 materials increased. On the RS AZ31, cells elongated at the same direction. After seven days, cell density further increased and multiple layers of cells could be seen in all the three groups. Most of the cells on the RS group still aligned in the same direction. Larger hydrogen gas bubbles emerged in all groups. On pure Mg, cell densities significantly decreased at 4th and 7th day. At the first day, cells showed similar uniform elongation on the RS pure Mg. However, normal spreading cells could barely be observed on the SS Mg at 4th and 7th day. Table 1 summarizes the pH value of cell culture media during the culture. In both AZ31 and Mg groups, the materials with RS showed significantly higher pH change than the materials with SR and SS after the first day. Mg^{2+} concentration (Fig. 11) after collagen was incubated with Mg and AZ31 of different surface roughness for 2 h was all around 25 mM. In contrast, Mg^{2+} concentration in the cell culture media was significantly lower than that in the collagen solution for the AZ31 group from 1 day to 7 day.

Table 1. The pH of the cell culture media after incubated with cells on materials.

Material		pH		
		1 day	**4 day**	**7 day**
AZ31	RS	7.93±0.05	7.88±0.03	7.94±0.03
	SR	7.70±0.08	7.92±0.03	7.85±0.05
	SS	7.75±0.08	7.91±0.06	7.76±0.07
Mg	RS	8.30±0.14	7.95±0.03	7.96±0.03
	SR	8.01±0.06	7.92±0.03	7.92±0.03
	SS	8.13±0.06	7.95±0.03	7.87±0.03

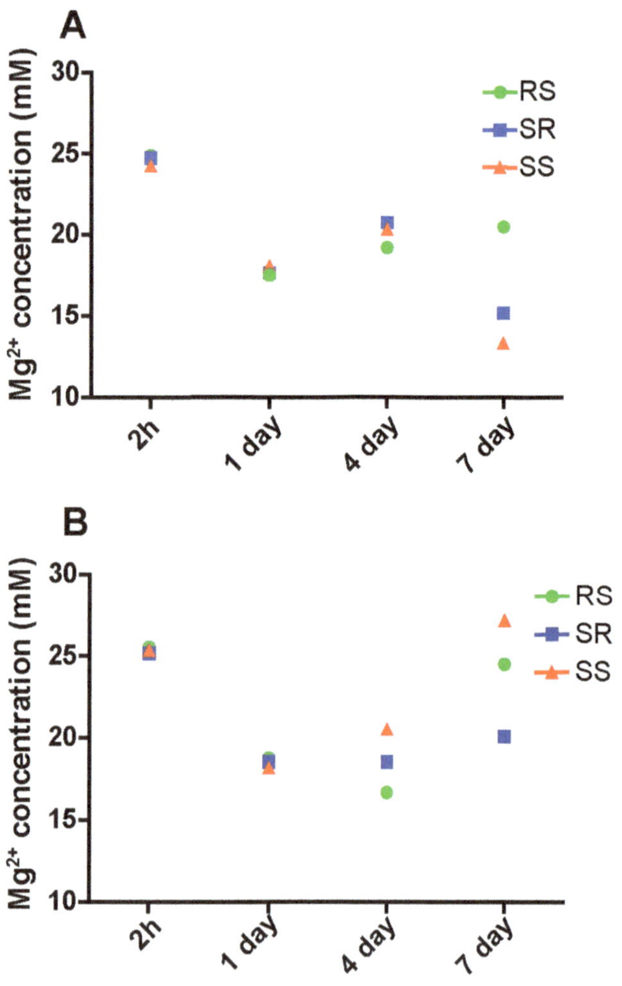

Figure 11. Mg^{2+} concentration after the materials (A-AZ31; B-Mg) were incubated with collagen solution for 2 h and cocultured with cells for 1, 4 and 7 days. Error bars were omitted for clarity purpose.

Discussion

Collagen, the ubiquitous ECM component, is a large family of triple-helical proteins. So far, around 28 types of collagen have been identified. Among them, type I collagen is the most abundant type, which forms the backbone of ECM in a lot of tissues such as bone, dermis, and tendon. 90% of the organic weight of bone is made up of type I collagen [37]. Type I collagen triple helix is composed of heterotrimer of two identical α1(I) chains and one α2(I) chain. Procollagen molecule is synthesized inside cells followed by post-translational modifications and then assemblies into triple helix procollagen with diameter of 1.5 nm and length of 300 nm [37]. Then it is secreted to extracellular space by secretory vesicles and further processed by different proteinases. In vitro, collagen fibrils are formed by self-assembly into cross-striated fibrils with the characteristic D-period of 67 nm [38]. In natural bone tissues, collagen fibrils are the scaffold for biomineralization. It is believed that collagen molecules are secreted as amorphous and non-crystalline forms and then transformed into crystalline forms gradually [39]. Mg-based alloys have promising future for orthopedic applications with respect to their mechanical properties, degradation properties, and biocompatibility. While the exact mechanism of collagen fibril formation on Mg surface in vivo remains unknown, in vitro self-assembly model established in this work provides a simple and alternative way to study how Mg materials interact with collagen molecules.

Collagen fibril formation on mica surface involves the adsorption of collagen molecules, surface diffusion, nucleation of fibrils and fibril elongation [40,41]. A lot of studies have shown that collagen could self-assembly into axially aligned fibrils with D-period similar to native bone tissues [42,43]. However, the assembly of collagen on mica surface could be different from that on Mg-alloy surface due to their distinct surface characteristics and electrostatics. Once in contact with body fluid, the metal elements in Mg materials will be oxidized into metal cations followed by the formation of a layer of metal hydroxide [44]. Metal ion would be released to the fluid during the degradation process and biomacromolecules such as protein, proteoglycan, and glycoprotein can be absorbed to the metal hydroxide layer [44]. It is interesting that for both pure Mg and AZ31 with different surface roughness, Mg ion released to the collagen solution after 2 h incubation didn't show significant difference. This is most likely due to the small total volume of solution (50 μl) and Mg ion was already saturated in the solution. At neutral pH, this metal hydrochloride layer is beneficial for collagen molecule adsorption since collagen molecule is positively charged. The absence of large fibril at low concentration of collagen monomer is most likely caused by the decreased chance for fibril nucleation. The concentration of collagen monomer can also affect the fibril growth rate and single fibrils grow independent from each other until they fuse with adjacent fibrils [40]. In addition, it was shown by Wang et al. that at low concentration collagen monomers form agglomerates in solution containing excessive Mg ions [45]. Similar agglomerates structure was also observed here on pure Mg and AZ31 surface at low collagen concentration. This might be caused by the high Mg^{2+}/collagen ratio and excessive Mg^{2+} could bind to collagen side chain leading to the increase of protein hydrophobicity and the dehydration of collagen [46]. Besides the release of metal ions, pH change accompanying the degradation process is another important factor that could affect collagen assembly. In the absence of other electrolytes, the isoelectric point (pI) of collagen is around 9.3 [47]. When pH approximates pI, the surface charge of collagen monomers is decreased resulting in minimized electrostatic repulsion and better fibril assembly. This is supported by our data (Fig. 3) where collagen fibrils loosely aligned at pH of 7 while they formed a dense layer of sheet at pH 9. As pH increased to 11, negatively charged collagen monomers could inhibit the nucleation of collagen fibril as well as the further attachment to Mg hydroxide layer. With the increase of incubation time to 8 h, small collagen fibrils could merge with adjacent fibrils forming thicker fibers (Fig. 4C and Fig. 4D). It is interesting to see that almost in all experiments spherical particles with different sizes were attached to collagen fibrils regardless of the diameters of collagen fibrils. The shape and size of those particles are very similar to the mineral nucleation reported by Ferreia et al [48]. However, from the EDS elemental analysis (Fig. 2), the particle structures are most likely magnesium compound instead of bone mineral.

It is well documented that implant surface roughness alters osteoblast proliferation, differentiation, and extracellular matrix production [16]. Mendonca et al. showed that rough surface topography can stimulate collagen biosynthesis and accumulation on titanium [17]. Mg materials with RS have relative larger surface area that increases the chance of collagen molecules adsorption. This is probably why the amount of collagen absorbed on the RS and SR materials was significantly higher than that on materials with SS. Also, surface energy could affect collagen adsorption and structural rearrangement. It is noticeable that the amount of absorbed collagen decreased at 8 h on the materials with RS and SR. This phenomenon is most likely caused by severer pitting corrosion on rougher surface compared with smoother surface [49,50]. In addition, surface roughness not only affected the amount of collagen absorbed but also the structure of the fibrils (Fig. 6). The slightly morphological difference of collagen fibrils on Mg and AZ31 is likely caused by the presence of Zn^{2+} and Al^{3+}, the AZ31 degradation products [47]. Therefore, ion release rate, local pH change, hydrogen gas formation, surface energy and surface electrostatic properties can all affect the final fibril structure.

We further studied how those materials affect cell attachment and proliferation. The better cell attachment on the materials with SS (Fig. 8) is consistent with previous studies [51]. On AZ31 material, a lot of dead cells could be observed on the RS materials after the first day (Fig. 9). This is most likely due to the failure of cell attachment or hampered cell attachment on the RS where cells could only anchor themselves at reduced area caused by the existence of the grooves and ridges. The grooves and ridges showed contact guidance effect on cell alignment. It was demonstrated before that the tip of filopodia most likely attaches to the top of the ridges [52]. During cell migration, it would be much easier for cell to move the tip of the adhesion along the ridge than to move the tip of the adhesion perpendicular to the direction of ridges. That may be the reason why cells on the rough surface materials all aligned parallel to the direction of ridges. Cells showed similar proliferation results on AZ31 with different surface roughness indicating that surface roughness and collagen structure won't affect cell proliferation. However, cells didn't show similar proliferation result on pure Mg at 4th day and 7th day. Cell density significantly decreased at the Mg with RS and SR. Healthy spreading cells could hardly be found on the SS of pure Mg materials. At body temperature, melting time for human type I collagen is around several days [53]. Compared with AZ31, the relative faster degradation rate of pure Mg could lead to higher concentration of degradation products and higher pH in solution, which might accelerate dissociation of attached collagen and cause decreased cell density. In addition, the thick collagen ribbon structure doesn't resemble native collagen structure in bone. The collagen fibrils in Fig. 6C and Fig. 6D showed highly similarity with the demineralized circumferential lamellar bone [54]. Ideally, the preferable orthopedic implants should not only be able to stimulate bone cell growth but also to support the assembly of collagen monomer into native fibrils at the bone-implant interface.

This in vitro model was developed to mimic the in vivo interactions between collagen and the Mg implant at the interface. It provided useful information on the molecular mechanism of such an interaction that will influence the fate of the implant. It may also have some limitations. For example, different cell regulations and other protein interactions were neglected. Other types of bone cells and non-collagenous proteins also play important roles in collagen assembly [55]. Therefore, more future studies are needed to address these factors. Additionally, one interesting topic for next step could be to investigate how mineralization happens around the interfaces.

Conclusion

In sum, collagen monomer can form different structures on Mg biomaterial depending on the initial collagen monomer concentration, pH in the solution, assembly time, electrolytes, and the material surface roughness. Only the initial collagen monomer concentration reaches a certain level, can long collagen fibrils form on Mg materials surface. The pH can change the electrostatic properties of collagen and final collagen structure. With the increase of collagen assembly time, small fibrils merge with adjacent fibrils forming thicker fibers. Materials with rough surface show stronger collagen adsorption ability. Initial cell attachment on RS of materials decreased independent of the composition of materials. AZ31 surface roughness and collagen structure did not affect cell proliferation for up to 7 days. Materials with high degradation rate may not change collagen assembly structure but can affect MC 3T3 cell proliferation. More studies on the combined effects of those factors such as pH, material surface roughness, assembly time and electrolytes on collagen assembly, mineralization and bone cell interaction are needed in the future. By controlling those factors, one can not only change the amount of absorbed collagen on Mg material surface but also the collagen fibril structure.

Acknowledgments

We thank Ms. D. Collins, Dr. Z. Xu, and Dr. J. Waterman for experimental trainings.

Author Contributions

Conceived and designed the experiments: NZ DZ. Performed the experiments: NZ. Analyzed the data: NZ DZ. Contributed reagents/materials/analysis tools: DZ. Contributed to the writing of the manuscript: NZ DZ.

References

1. Gu X, Xie X, Li N, Zheng Y, Qin L (2012) In vitro and in vivo studies on a Mg–Sr binary alloy system developed as a new kind of biodegradable metal. Acta Biomater 8: 2360–2374.
2. Li Z, Gu X, Lou S, Zheng Y (2008) The development of binary Mg–Ca alloys for use as biodegradable materials within bone. Biomaterials 29: 1329–1344.
3. Xu L, Yu G, Zhang E, Pan F, Yang K (2007) In vivo corrosion behavior of Mg-Mn-Zn alloy for bone implant application. Journal of Biomedical Materials Research Part A 83: 703–711.
4. Kim W, Kim W (2014) Fabrication of ultrafine-grained Mg-3Al-1Zn magnesium alloy sheets using a continuous high-ratio differential speed rolling technique. Materials Science and Engineering: A 594: 189–192.
5. Hanada K, Matsuzaki K, Huang X, Chino Y (2013) Fabrication of Mg alloy tubes for biodegradable stent application. Materials Science and Engineering: C 33: 4746–4750.
6. Perez P, Onofre E, Cabeza S, Llorente I, del Valle JA, et al. (2013) Corrosion behaviour of Mg-Zn-Y-Mischmetal alloys in phosphate buffer saline solution. Corrosion Science 69: 226–235.
7. Ji DW, Liu CM, Chen ZY, Wang HH, Wang B (2013) Effects of Zn content on microstructures and mechanical properties of as cast Mg-Zn-Y-Zr alloys. Materials Science and Technology 29: 480–486.
8. Bornapour M, Muja N, Shum-Tim D, Cerruti M, Pekguleryuz M (2013) Biocompatibility and biodegradability of Mg-Sr alloys: the formation of Sr-substituted hydroxyapatite. Acta Biomater 9: 5319–5330.
9. Seitz JM, Eifler R, Stahl J, Kietzmann M, Bach FW (2012) Characterization of MgNd2 alloy for potential applications in bioresorbable implantable devices. Acta Biomaterialia 8: 3852–3864.
10. Mao L, Yuan G, Wang S, Niu J, Wu G, et al. (2012) A novel biodegradable Mg–Nd–Zn–Zr alloy with uniform corrosion behavior in artificial plasma. Materials Letters 88: 1–4.
11. Xu S, Oh-Ishi K, Kamado S, Uchida F, Homma T, et al. (2011) High-strength extruded Mg–Al–Ca–Mn alloy. Scripta Materialia 65: 269–272.
12. Staiger MP, Pietak AM, Huadmai J, Dias G (2006) Magnesium and its alloys as orthopedic biomaterials: a review. Biomaterials 27: 1728–1734.
13. Witte F, Kaese V, Haferkamp H, Switzer E, Meyer-Lindenberg A, et al. (2005) In vivo corrosion of four magnesium alloys and the associated bone response. Biomaterials 26: 3557–3563.
14. Choudhary L, Singh Raman R (2012) Magnesium alloys as body implants: Fracture mechanism under dynamic and static loadings in a physiological environment. Acta Biomater 8: 916–923.
15. Zhao N, Zhu D (2013) Application of Mg–based alloys for cardiovascular stents. International Journal of Biomedical Engineering and Technology 12: 382–398.

16. Martin J, Schwartz Z, Hummert T, Schraub D, Simpson J, et al. (1995) Effect of titanium surface roughness on proliferation, differentiation, and protein synthesis of human osteoblast-like cells (MG63). Journal of biomedical materials research 29: 389–401.
17. Mendonça D, Miguez PA, Mendonça G, Yamauchi M, Aragão FJ, et al. (2011) Titanium surface topography affects collagen biosynthesis of adherent cells. Bone 49: 463–472.
18. Nebe J, Lüthen F, Rychly J, Eisenbarth E, Müller M, et al. (2007) Topology-Dependent Cellular Interactions. Metallic Biomaterial Interfaces: 215–248.
19. Anselme K (2000) Osteoblast adhesion on biomaterials. Biomaterials 21: 667–681.
20. Boyan BD, Hummert TW, Dean DD, Schwartz Z (1996) Role of material surfaces in regulating bone and cartilage cell response. Biomaterials 17: 137–146.
21. Nudelman F, Pieterse K, George A, Bomans PH, Friedrich H, et al. (2010) The role of collagen in bone apatite formation in the presence of hydroxyapatite nucleation inhibitors. Nature materials 9: 1004–1009.
22. Ambrose SH (1990) Preparation and Characterization of Bone and Tooth Collagen for Isotopic Analysis. Journal of Archaeological Science 17: 431–451.
23. Bailey LKaAJ (1998) Collagen Cross-Links in Mineralizing Tissues: A Review of Their Chemistry, Function, and Clinical Relevance. Bone 22: 181–187.
24. Franz CM, Muller DJ (2011) Studying collagen self-assembly by time-lapse high-resolution atomic force microscopy. Atomic Force Microscopy in Biomedical Research: Humana Press. 97–107.
25. Stephanopoulos N, Ortony JH, Stupp SI (2013) Self-assembly for the synthesis of functional biomaterials. Acta materialia 61: 912–930.
26. O'Leary LE, Fallas JA, Bakota EL, Kang MK, Hartgerink JD (2011) Multi-hierarchical self-assembly of a collagen mimetic peptide from triple helix to nanofibre and hydrogel. Nature chemistry 3: 821–828.
27. Wang ZL, Yan YH, Wan T, Yang H (2013) Poly(L-lactic acid)/hydroxyapatite/collagen composite coatings on AZ31 magnesium alloy for biomedical application. Proc Inst Mech Eng H 227: 1094–1103.
28. Sader MS, Martins VC, Gomez S, LeGeros RZ, Soares GA (2013) Production and in vitro characterization of 3D porous scaffolds made of magnesium carbonate apatite (MCA)/anionic collagen using a biomimetic approach. Mater Sci Eng C Mater Biol Appl 33: 4188–4196.
29. Ao H, Xie Y, Tan H, Yang S, Li K, et al. (2013) Fabrication and in vitro evaluation of stable collagen/hyaluronic acid biomimetic multilayer on titanium coatings. Journal of The Royal Society Interface 10: 20130070.
30. Fang M, Goldstein EL, Matich EK, Orr BG, Holl MM (2013) Type I collagen self-assembly: the roles of substrate and concentration. Langmuir 29: 2330–2338.
31. Nassif N, Gobeaux Fdr, Seto J, Belamie E, Davidson P, et al. (2010) Self-Assembled Collagen−Apatite Matrix with Bone-like Hierarchy. Chemistry of Materials 22: 3307–3309.
32. Castellani C, Lindtner RA, Hausbrandt P, Tschegg E, Stanzl-Tschegg SE, et al. (2011) Bone–implant interface strength and osseointegration: Biodegradable magnesium alloy versus standard titanium control. Acta Biomater 7: 432–440.
33. Kim HW, Li LH, Lee EJ, Lee SH, Kim HE (2005) Fibrillar assembly and stability of collagen coating on titanium for improved osteoblast responses. Journal of Biomedical Materials Research Part A 75: 629–638.
34. Ao H, Xie Y, Tan H, Yang S, Li K, et al. (2013) Fabrication and in vitro evaluation of stable collagen/hyaluronic acid biomimetic multilayer on titanium coatings. Journal of The Royal Society Interface 10.
35. Chen X, Ergun A, Gevgilili H, Ozkan S, Kalyon DM, et al. (2013) Shell-core bi-layered scaffolds for engineering of vascularized osteon-like structures. Biomaterials 34: 8203–8212.
36. Wu L, Luthringer BJ, Feyerabend F, Schilling AF, Willumeit R (2014) Effects of extracellular magnesium on the differentiation and function of human osteoclasts. Acta Biomaterialia 10: 2843–2854.
37. Gelse K, Pöschl E, Aigner T (2003) Collagens–structure, function, and biosynthesis. Advanced drug delivery reviews 55: 1531–1546.
38. Kadler K, Holmes D, Trotter J, Chapman J (1996) Collagen fibril formation. Biochem J 316: 1–11.
39. Ferreira AM, Gentile P, Chiono V, Ciardelli G (2012) Collagen for bone tissue regeneration. Acta Biomaterialia 8: 3191–3200.
40. Cisneros DA, Hung C, Franz CM, Muller DJ (2006) Observing growth steps of collagen self-assembly by time-lapse high-resolution atomic force microscopy. Journal of structural biology 154: 232–245.
41. Narayanan B, Gilmer GH, Tao J, De Yoreo JJ, Ciobanu CV (2014) Self-assembly of collagen on surfaces: the interplay of collagen-collagen and collagen-substrate interactions. Langmuir 30: 1343–1350.
42. Loo RW, Goh MC (2008) Potassium ion mediated collagen microfibril assembly on mica. Langmuir 24: 13276–13278.
43. Leow WW, Hwang W (2011) Epitaxially guided assembly of collagen layers on mica surfaces. Langmuir 27: 10907–10913.
44. Zheng Y, Gu X, Witte F (2014) Biodegradable metals. Materials Science and Engineering: R: Reports 77: 1–34.
45. Wang L, Guo Y, Li P, Song Y (2014) Anion-Specific Effects on the Assembly of Collagen Layers Mediated by Magnesium Ion on Mica Surface. The Journal of Physical Chemistry B.
46. He L, Cai S, Wu B, Mu C, Zhang G, et al. (2012) Trivalent chromium and aluminum affect the thermostability and conformation of collagen very differently. Journal of inorganic biochemistry 117: 124–130.
47. Zhao N, Workman B, Zhu D (2014) Endothelialization of Novel Magnesium-Rare Earth Alloys with Fluoride and Collagen Coating. International journal of molecular sciences 15: 5263–5276.
48. Ferreira A, González G, González-Paz R, Feijoo J, Lira-Olivares J, et al. (2009) Bone collagen role in piezoelectric mediated remineralization. Acta Microscopica 18: 278–286.
49. Walter R, Kannan MB (2011) Influence of surface roughness on the corrosion behaviour of magnesium alloy. Materials & Design 32: 2350–2354.
50. Walter R, Kannan MB, He Y, Sandham A (2013) Effect of surface roughness on the in vitro degradation behaviour of a biodegradable magnesium-based alloy. Applied Surface Science 279: 343–348.
51. Huang H-H, Ho C-T, Lee T-H, Lee T-L, Liao K-K, et al. (2004) Effect of surface roughness of ground titanium on initial cell adhesion. Biomolecular engineering 21: 93–97.
52. Matschegewski C, Staehlke S, Loeffler R, Lange R, Chai F, et al. (2010) Cell architecture–cell function dependencies on titanium arrays with regular geometry. Biomaterials 31: 5729–5740.
53. Leikina E, Mertts M, Kuznetsova N, Leikin S (2002) Type I collagen is thermally unstable at body temperature. Proceedings of the National Academy of Sciences 99: 1314–1318.
54. Reznikov N, Almany-Magal R, Shahar R, Weiner S (2013) Three-dimensional imaging of collagen fibril organization in rat circumferential lamellar bone using a dual beam electron microscope reveals ordered and disordered sub-lamellar structures. Bone 52: 676–683.
55. Kadler KE, Hill A, Canty-Laird EG (2008) Collagen fibrillogenesis: fibronectin, integrins, and minor collagens as organizers and nucleators. Current opinion in cell biology 20: 495–501.

5

Foreign Body Reaction Associated with PET and PET/ Chitosan Electrospun Nanofibrous Abdominal Meshes

Beatriz Veleirinho[1,2]*, Daniela S. Coelho[3], Paulo F. Dias[3], Marcelo Maraschin[4], Rúbia Pinto[5], Eduardo Cargnin-Ferreira[6], Ana Peixoto[7], José A. Souza[7], Rosa M. Ribeiro-do-Valle[2], José A. Lopes-da-Silva[1]

1 QOPNA Research Unit, Department of Chemistry, University of Aveiro, Aveiro, Portugal, **2** Biotechnology and Biosciences Post-Graduation Program, Federal University of Santa Catarina, Florianópolis, Brazil, **3** Department of Cell Biology, Embryology, and Genetics, Federal University of Santa Catarina, Florianópolis, Brazil, **4** Plant Morphogenesis and Biochemistry Laboratory, Federal University of Santa Catarina, Florianópolis, Brazil, **5** Epagri, Florianopolis, Brazil, **6** Federal Institute of Education, Science, and Technology of Santa Catarina, Garopaba, Brazil, **7** Department of Pediatrics, Federal University of Santa Catarina, Florianópolis, Brazil

Abstract

Electrospun materials have been widely explored for biomedical applications because of their advantageous characteristics, i.e., tridimensional nanofibrous structure with high surface-to-volume ratio, high porosity, and pore interconnectivity. Furthermore, considering the similarities between the nanofiber networks and the extracellular matrix (ECM), as well as the accepted role of changes in ECM for hernia repair, electrospun polymer fiber assemblies have emerged as potential materials for incisional hernia repair. In this work, we describe the application of electrospun non-absorbable mats based on poly(ethylene terephthalate) (PET) in the repair of abdominal defects, comparing the performance of these meshes with that of a commercial polypropylene mesh and a multifilament PET mesh. PET and PET/chitosan electrospun meshes revealed good performance during incisional hernia surgery, post-operative period, and no evidence of intestinal adhesion was found. The electrospun meshes were flexible with high suture retention, showing tensile strengths of 3 MPa and breaking strains of 8–33%. Nevertheless, a significant foreign body reaction (FBR) was observed in animals treated with the nanofibrous materials. Animals implanted with PET and PET/chitosan electrospun meshes (fiber diameter of 0.71 ± 0.28 μm and 3.01 ± 0.72 μm, respectively) showed, respectively, foreign body granuloma formation, averaging 4.2-fold and 7.4-fold greater than the control commercial mesh group (Marlex). Many foreign body giant cells (FBGC) involving nanofiber pieces were also found in the PET and PET/chitosan groups (11.9 and 19.3 times more FBGC than control, respectively). In contrast, no important FBR was observed for PET microfibers (fiber diameter $= 18.9 \pm 0.21$ μm). Therefore, we suggest that the reduced dimension and the high surface-to-volume ratio of the electrospun fibers caused the FBR reaction, pointing out the need for further studies to elucidate the mechanisms underlying interactions between cells/tissues and nanofibrous materials in order to gain a better understanding of the implantation risks associated with nanostructured biomaterials.

Editor: Mário A. Barbosa, Instituto de Engenharia Biomédica, University of Porto, Portugal

Funding: Authors acknowledge Fundação para a Ciência e a Tecnologia (FCT, Portugal), the European Union, QREN, FEDER, COMPETE, for funding the Organic Chemistry Research Unit (QOPNA) (project PEst-C/QUI/UI0062/2013; FCOMP-01-0124-FEDER-037296) and for the PhD grant (SFRH/BD/38881/2007). Authors also acknowledge CNPq (Brazil) for a young talent researcher grant (375205/2012-8) and the financial support from FAPESC (Brazil) (PRONEX 17420/2011/3). The funders had no role in study design, data collection and analysis, decision to publish, or preparation of the manuscript.

Competing Interests: The authors have declared that no competing interests exist.

* E-mail: bveleirinho@ua.pt

Introduction

Electrospinning has attracted the interest of researchers from many fields as a versatile technique to produce nanofibers from synthetic and naturally derived polymers. With typical diameters ranging from 10 nm to a few micrometers, these fibers are usually collected continuously as nonwoven fibrous mats. These mats usually show a tridimensional nanostructure with high surface-to-volume ratio, high porosity, and interconnectivity, and they have demonstrated high potential for biomedical applications, such as tissue engineering scaffolds, vascular grafts, and drug delivery systems [1–3].

In the past decade, our research group has been exploring the potential of electrospinning for different applications [4]. Cell culture studies revealed that hybrid nanofibers of poly(ethylene terephthalate) (PET) and chitosan provide a good substratum for fibroblast adhesion, proliferation, extracellular matrix secretion, and three-dimensional colonization [5], in addition to their interesting surface and mechanical properties [6,7]. The promising results obtained *in vitro* prompted us to test these nondegradable electrospun mats as abdominal meshes for incisional hernia repair.

Incisional hernia is a frequent complication of laparotomy resulting from the decrease of abdominal strength in the injured tissue. The incidence of incisional hernia after abdominal surgery depends on the pattern of the incision performed. For a midline incision, the preferred incision for the upper abdominal surgery, the incidence lies around 10–14% [8–10]. Different patterns of incision, however, such as the transverse incision, yield much lower rates of hernia formation (2%) [10]. In comparison to the traditional hernia repair strategy by primary closure, the implementation of a tension-free repair by using a prosthetic biomaterial, i.e., nonresorbable abdominal mesh, to substitute or reinforce abdominal strength at the damaged area has decreased the recurrence rates significantly [8–12]. Nevertheless, serious complications have also been associated with this procedure, including infection, visceral adhesions to the mesh, seroma, mechanical failure of the mesh, and foreign body reaction [13,14]. Bowel adhesion to the implanted mesh is a major concern as it causes serious complications, including bowel obstruction, enterocutaneous fistula, and chronic pain [15].

Absorbable meshes are also used as scaffolds for abdominal defect correction. In this case, the biodegradable mesh provides support for cell growth and for extracellular matrix secretion, promoting the tissue repair process. A few recent reports have shown successful results in exploring partially degradable or absorbable electrospun mats as abdominal meshes. An electrospun blended fiber mesh prepared from biodegradable poly(ester urethane) urea and poly(lactide-*co*-glycolide), latter loaded with an antibiotic, was shown to provide good mechanical properties, while imparting antibacterial activity and, hence, reducing the risk of infection during application of the composite material to abdominal wall closure [16]. A similar approach used electrospun poly(ester urethane) urea fibers deposited with electrosprayed serum-based culture medium [17] or porcine dermal extracellular matrix digest [18]. When these materials were tested as abdominal wall repair meshes, they were demonstrated to provide adequate mechanical properties and, at the same time, enhanced bioactivity, biocompatibility and cell infiltration, with no herniation, infection, or tissue adhesion. Other examples include the application of electrospun absorbable polycaprolactone scaffolds [19], which were also evaluated for their suitability in hernia repair.

Despite the advantages of absorbable materials, the newly formed tissue has a decreased tensile strength and thereby, re-herniation is a frequent problem after the absorption of the prosthetic material. In this context, electrospun mats of nondegradable polymers have emerged as potential alternative meshes for abdominal defect repair. Because of its unique properties, i.e., unaligned nanofibrous arrangement, microporosity, and high hydrophobicity, the PET electrospun mat has emerged as a potential candidate for abdominal wall repair [6]. In addition to these advantages, PET is a highly biocompatible, biostable, and nondegradable polymer which possesses the mechanical features required for this application. Moreover, the low density (~ 0.091 g/cm^{-3}) and the high malleability of this material may promote enhanced adaptation and, consequently, patient comfort [7]. On the other hand, as a hydrophobic material with small pores, electrospun PET meshes may restrict the integration of parietal conjunctive tissue. McGinty *et al.* have demonstrated that a better incorporation of the mesh in the parietal side reduces the number and the severity of adhesion formations on the visceral side [20]. Hence, with the aim of enhancing the interaction of the mesh with the parietal conjunctive tissue, a hybrid mat of PET/chitosan (PET/C) was developed. Chitosan has shown ideal properties for biomedical applications, including the anti-inflammatory and wound healing effects, which may attenuate the typical symptoms of the post-surgery period and prevent adhesiogenesis [21,22]. Additionally, in the current study, a double-layered mesh (DL), containing one layer of PET (turned to the visceral side) and one layer of PET/C mat (turned to the parietal side), was also developed.

In this paper, we describe the application of three electrospun nonabsorbable mats, including PET, PET/C and DL, in the repair of abdominal defects, comparing the performance of these nanofibrous meshes with that of a commercial polypropylene mesh (Marlex) (control) and a multifilament microfibrous PET mesh. An *in vivo* study with an abdominal hernia Wistar rat model was performed to evaluate the clinical and histological aspects of using these meshes for abdominal hernia repair.

Materials and Methods

Animal experiments were approved by the Animal Ethics Committee of the Federal University of Santa Catarina (PP0406/2009).

Materials

PET pellets and PET woven fabric were kindly supplied by Flexitex (Portugal). Marlex (Intracorp) was purchased from *Cirurgica Passos,* Brazil. Chitosan medium molecular weight (15% acetylation degree) was purchased from Sigma-Aldrich Chemical Company. The molecular weight of the initial chitosan sample (1500 kDa) was reduced to 15 kDa by oxidative depolymerization [23]. All chemicals were of analytical grade and obtained from Sigma-Aldrich Chemical Company.

Mesh Fabrication by Electrospinning

Thirty percent (w/v) PET solution was prepared in a blend of trifluoroacetic acid and dichloromethane [8:2 v/v] by moderate stirring for 2 hours at room temperature. PET/C blend was prepared by adding chitosan (6 wt. %) to the PET 30 wt. % solution and stirring for 3 hours at room conditions.

Electrospinning was performed using a typical experimental setup previously described [6]. The process was conducted at 26 kV of applied voltage, with a flow rate of 0.08 mL/min (V = 20 mL) and a needle tip-to-collector distance of 12 cm.

The double-layer (DL) mesh was fabricated by electrospinning 10 mL of PET solution, followed by 10 mL of PET/C solution. Fibers were collected as a nonwoven fibrous mat on the rotating drum (900 rpm), in air and at room conditions (20±2°C, 45–50% RH), and the mat was dried at 35°C for 24 hours.

Morphological Analysis

The morphology of the fibrous scaffolds was investigated by scanning electron microscopy (SEM). Small sections of the scaffolds were sputter-coated with gold and analyzed using a scanning electron microscope (Hitachi S4100) at an accelerating voltage of 25 kV.

Image processing was performed using ImageJ - 1.37c software (Wayne Rasband, National Institutes of Health, USA). Five random images (1000 X magnification) were obtained for each sample. Fiber diameters were calculated from at least 100 measurements of the sample fibers. SEM images were thresholded, and pore areas were automatically calculated using the "analyze particle" tool of ImageJ ($n > 200$).

Mechanical Properties

Mechanical properties in tension were evaluated using texture analyzer equipment (Model TA HDi, Stable Micro Systems, England) equipped with fixed grips lined with thin rubber on the ends. Test specimens 90 mm long × 10 mm wide were obtained perpendicular to the axis of the collector rotation, and the ends were mounted on the grips using sticky tape. The thickness of the test samples was measured at different locations on each sample using a digital micrometer (Model MDC-25S, Mitutoya Corp., Tokyo, Japan). The initial grip separation was set at 50 mm, and the crosshead speed was 0.5 mm/s. At least eight samples of each mat were tested.

Water Contact Angle (WCA) Measurements

The wettability of scaffolds was assessed by the sessile drop method using an OCA-20 contact angle system (DataPhysics Instruments). A drop of distilled water (1 μL) was automatically dispensed on the scaffold surface, and the WCA and drop life times were calculated using the SCA 20 software. At least 10 measurements were taken for each sample.

Animal Model

Male Wistar rats 3 months of age and weighing 250–300 g were obtained from the Central Biotery of the Federal University of Santa Catarina. Animals were randomly distributed among the 7 treatments (Marlex30, PET30, PET/C, DL, Woven-PET, Marlex90, and PET90), according to Table 1 ($n \geq 8$).

Before biological assay, all meshes were sterilized under UV light for 1 hour of exposure (both faces), immersed in ethanol 70% (v/v) for 10 minutes, and washed with sterile physiologic solution.

A graphical illustration of the surgical procedure is provided (Figure S1). After an intramuscular injection of a mixture of ketamine (90 mg/kg) and xylazine (15 mg/kg) and abdominal shaving, a 5 cm paramedian skin incision was made at the left side, using a sterile scalpel blade. Skin was dissected to expose the underlying abdominal fascia, and a 1.5×1.5 cm defect of anterior abdominal wall was created by the complete resection of abdominal layers. The edges of the meshes (2.0×2.0 cm) were sutured to the remaining muscle of the abdominal wall with interrupted suture and also with simple running suture all over the borders, using 5–0 polypropylene. The skin was closed with intradermal suture with nylon 4–0 monofilament. Animals were allowed to recover from anesthesia, housed in individual cages, and observed daily for evidence of wound complications, such as redness, infection, seroma, abcess, hematoma, or skin dehiscence.

On day 30 and 90 post-surgery, animals were sacrificed in a carbon dioxide chamber, and the presence of intestinal adhesions was analyzed. The abdominal wall was carefully excised well away from the mesh (see Figure S2) to preserve any adherence to the bowel or omentum. After adhesion analysis, tissue was completely excised and collected for histopathological analysis.

Histopathological Analysis

Tissues were excised from the animals and fixed in phosphate buffered formaldehyde solution (4%, pH 7.2, 0.1 M), embedded in paraffin, and sectioned at 4 μm thickness. Giemsa, hematoxylin and eosin (HE), and Garvey's staining were performed [24]. For immunohistochemical analysis, sections were deparaffinized in xylene and rehydrated in a graded ethanol series. Antigen retrieval was performed with 0.05% trypsin and 0.1% calcium chloride (20 minutes, 37°C). Endogenous peroxidase activity was blocked by incubation in a hydrogen peroxide solution. Following, sections were incubated with a monoclonal antibody directed against CD68, clone KP1, dilution 1:100 (Zeta Corporation, CA). Antibody detection was performed using Histofine Simple Stain Max-Po Multi (Nichirei Biosciences, Tokyo, Japan) and 3,3′-diaminobenzidine tetrahydrochloride (Spring Bioscience, CA). Samples were analyzed under a microscope (Nikon Eclipse 50i equipped with a Nikon Digital Sight DS-Fi2). The thickness of the foreign body granuloma was measured and the absolute number of foreign body giant cells (FBGC) per granuloma section was determined by manual counting ($n \geq 8$). A detailed description of morphometric procedures is given in Figure S2.

Statistics

Statistical analysis was carried out using Instat 3.0 software. Results were expressed as the mean ± standard deviation and compared through one-way ANOVA and Tukey-Kramer.

Results

Characterization of Meshes

Figure 1 displays SEM images of PET, PET/C, and DL mats. The electrospun mats showed a typical nonwoven fibrous structure with random fiber orientation and high porosity. Table 2 summarizes some morphological, mechanical, and surface properties of the meshes used in this study. The average diameter of PET fibers was 0.71±0.28 μm, and the average pore area was 9.4 μm^2. Addition of chitosan promoted a substantial increase in fiber diameter and pore area to 3.01±0.72 μm and 89.3 μm^2, respectively. Compared to PET/C, PET mesh showed superior mechanical properties with a higher tensile strength (3.17±0.23 MPa compared to 2.89±0.27 MPa), Young's modulus (120±10 compared to 70±10 MPa), and elongation (32.8±5.7% compared to 8.2±1.3%). Also, a decrease in the hydrophobicity of the mesh was observed by the presence of chitosan (WCA decreased from 133.2±2.9° to 125.2±4.6° with the addition of chitosan).

Surgical Procedure and Post-operative Period

The electrospun mats were evaluated in an incisional hernia experiment with Wistar rats and compared to Marlex, the control. Following the creation of the abdominal defect by resection of 1.5×1.5 cm of the Wistar rats' abdominal muscle, the prosthetic meshes were implanted and fixed through the borders to the remaining muscle (see Figure S1). Suture of electrospun meshes was easily performed without breakage. In fact, electrospun meshes were more suitable and more resistant to suture than

Table 1. Experimental groups of incisional hernia repair.

Group	Mesh chemical composition	Duration of experiment (days)	Mesh thickness (mm)
Marlex30	Polypropylene	30	0.22±0.07
Marlex90	Polypropylene	90	0.22±0.07
PET30	PET	30	0.31±0.02
PET90	PET	90	0.31±0.02
Woven-PET	PET	30	0.49±0.09
PET/C	PET/C 5:1 (w/w)	30	0.52±0.05
DL	PET + PET/C 5:1 (w/w)	30	0.46±0.11

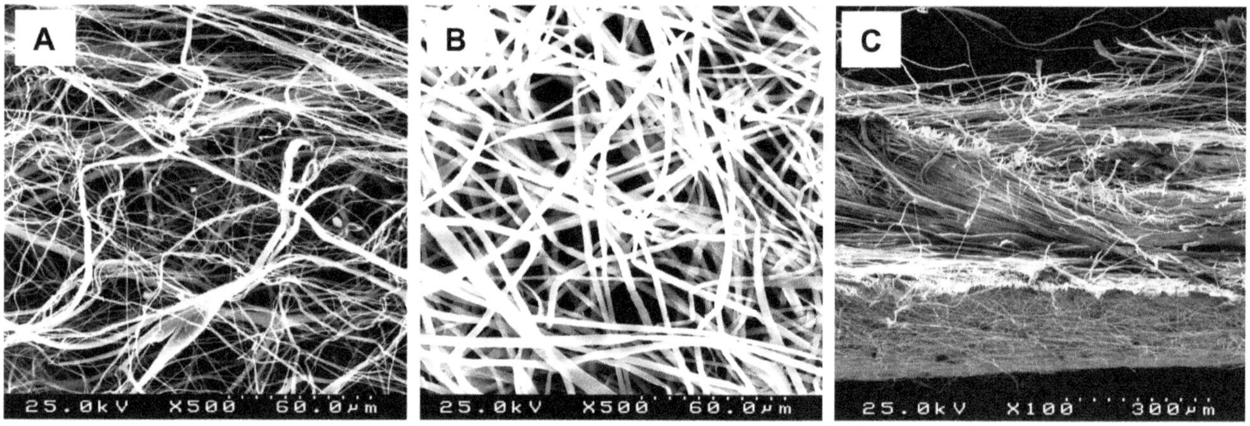

Figure 1. Morphological analysis of the meshes. SEM images of (A) PET, (B) PET/C, and (C) DL meshes. The transversal section of the DL mesh shows PET mat (bottom) and PET/C mat (top). *Bars*: 60 µm (A, B); 300 µm (C).

control, where an extra margin of around 5 mm was used to avoid breaking of filaments.

During the post-operative period, some complications, such as local redness, abscess, or skin dehiscence, were registered. Table 3 displays macroscopic complications found in the post-operative period. No complications were found during the post-operative period for the woven-PET group. In the other experimental groups, local redness, characterized by redness or swelling, was the most common complication, with higher incidence of these complications observed for PET/C and DL groups, occurring in 50% and 75% of the animals, respectively. Redness tended to decrease with time, while abscess and skin dehiscence persisted until the end of the experimental period. Omentum adhesions to the mesh were observed in all animals (see Figure S3), but no visceral adhesions were found in the experimental groups.

Histological Analysis

Histological analysis was performed to evaluate the cellular response to the prosthetic biomaterials. Figure 2 and Figure 3 display representative images of HE-stained sections of animals treated with electrospun meshes and Marlex (low and high magnification, respectively). All animals showed typical nonimmunogenic granulomas (foreign body granulomas) surrounding the mesh structure placed below the abdominal subcutaneous tissue. Representative photomicrographs from Garvey's staining are provided in Figure 4. The foreign body granuloma were mostly composed of macrophages, foreign body giant cells (FBGC), and fibroblasts. Multinucleated cells frequently involved one or more fiber segments. Immunohistochemical analysis confirmed the high density of both macrophages and FBGC (CD68+) in animals treated with the electrospun materials (Figure 5). The mean thickness of the granuloma and the average number of FBGC in the granuloma are plotted in Figure 6A and Figure 6B, respectively. Animals treated with electrospun meshes showed significantly thicker granulomas and a higher number of FBGC compared to Marlex and the woven-PET group. The mean granuloma thickness induced by PET nanofibers was 4-fold higher than control (Marlex) and 10-fold higher than in the woven-PET group. Also a 10-fold increase in the number of FBGC was observed in the PET group compared to control. Hybrid meshes showed even thicker granulomas (1522±277 µm and 1211±547 µm for PET/C and DL, respectively), as well as a large number of FBGC comprising one or more fibrous structures. The woven-PET group produced the weakest inflammatory response with an average granuloma thickness of 87±35 µm and FBGC rarely observed.

Long-term inflammatory response was evaluated for two selected groups (Marlex and electrospun PET) 90 days after mesh implantation. Both groups showed a decrease in inflammation with time. The average granuloma thickness of the electrospun PET group decreased from 959±473 µm to 513±217 µm, and the number of FBGC decreased from 106±30 to 89±12.

Table 2. Fiber diameter, pore area, mechanical properties, and WCA of abdominal meshes.

Group	Fiber diameter (µm)	Pore area (μm^2)	Tensile strength (MPa)	Percentage elongation (%)	Young's modulus (MPa)	WCA (°)
Marlex	177±24	31400	N.A.	N.A.	N.A.	N.A.
PET	0.71±0.28	9.4	3.17±0.23	32.8±5.7	1.2±0.1	133.2±2.9
PET/C	3.01±0.72	89.3	2.89±0.27	8.2±1.3	0.7±0.1	125.2±4.6
Woven-PET	18.9±2.1[a]	N.A.	N.A.	N.A.	N.A.	N.A.
	342±68[b]					

N.A. not available
[a]Average filament diameter.
[b]Average diameter of the multifilament yarn.

Table 3. Occurrence rate (%) of complications during post-operative period (PO) and day euthanized (E).

Mesh	Period	Local redness	Dehiscence	Seroma or abscess	Adhesion
Marlex	PO	37.5	12.5	12.5	N.A.
	E	12.5	12.5	12.5	omentum
PET	PO	37.5	0	12.5	N.A.
	E	25	0	12.5	omentum
PET/C	PO	50	0	0	N.A.
	E	0	0	0	omentum
DL	PO	75	12.5	12.5	N.A.
	E	37.5	12.5	12.5	omentum
Woven- PET	PO	0	0	0	N.A.
	E	0	0	0	omentum

N.A. not available.

Discussion

Several biomaterials have been used as prosthetic meshes for abdominal wall repair, in particular, incisional hernia repair. Among them, nonabsorbable polymers with recognized biocompatibility, such as polypropylene, PET or polytetrafluoroethylene, are the most common. Although a considerable improvement in the recurrence rate has been achieved with these materials in comparison to the traditional suture technique, several problems are still associated with this procedure. Among them, the formation of adhesions between the mesh and the bowel results in several complications, such as chronic pain, bowel obstruction or enterocutaneous fistula. Also, nonresorbable meshes have been associated with a high occurrence of chronic foreign body response and increased risk of infection [25,26]. Improvements have been achieved by using polymer meshes that gradually degrade *in vivo*, promoting improved tissue integration and rapid resorption [19]. However, as scaffolds to support native tissue ingrowth, resorbable synthetic meshes are still limited by their loss of strength. Considering the attractive properties of electrospun polymer meshes, we have developed, for the first time, electrospun mats of nondegradable polymers for the repair of abdominal defects.

Chemical composition, weight, pore size, and filament structure, represent critical parameters employed in surgical mesh design [27]. Apart from preventing adhesion, it is generally accepted that the ideal abdominal mesh should be chemically inert and stable for long periods, promote tissue regeneration to form an adequate barrier against protrusion, cause no immune or inflammatory response, and fulfill the required mechanical needs for the application. The studied PET electrospun mats are nonwoven meshes of nanofibers (average diameter = 0.71 ± 0.28 μm) thought to be ideal for the prevention of visceral adhesion by their highly hydrophobic microporous structure (WCA = $133.2 \pm 2.9°$; average pore area = 9.4 μm^2). Indeed, according to Mathews *et al.*, biomaterials with pores smaller than 75 μm reduce the occurrence of bowel adhesion [28].

With the purpose of manipulating the architecture of PET mesh and also taking advantage of the anti-inflammatory and wound healing effects of chitosan, a hybrid fibrous mat (PET/C) was also developed, with higher fiber diameter and pore area. These morphological differences can have significant effects on cell-biomaterial interactions, as previously demonstrated [5]. The decreased stiffness of the hybrid mesh in comparison to the PET mesh may be attributed to a heterogeneous polymer distribution within fibers as a result of phase separation during the

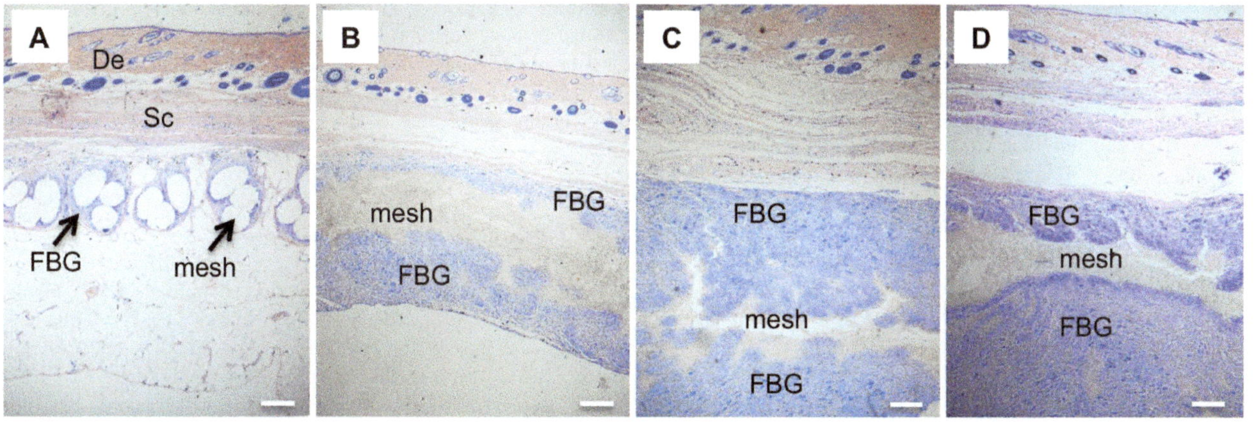

Figure 2. Foreign body granuloma induced by the abdominal meshes. Histological sections evidencing foreign body granuloma (Giemsa staining) of (A) Marlex30, (B) PET30, (C) PET/C, and (D) DL groups. (De) dermis, (Sc) subcutaneous tissue, and (FBG) foreign body granuloma. Bar = 220 μm.

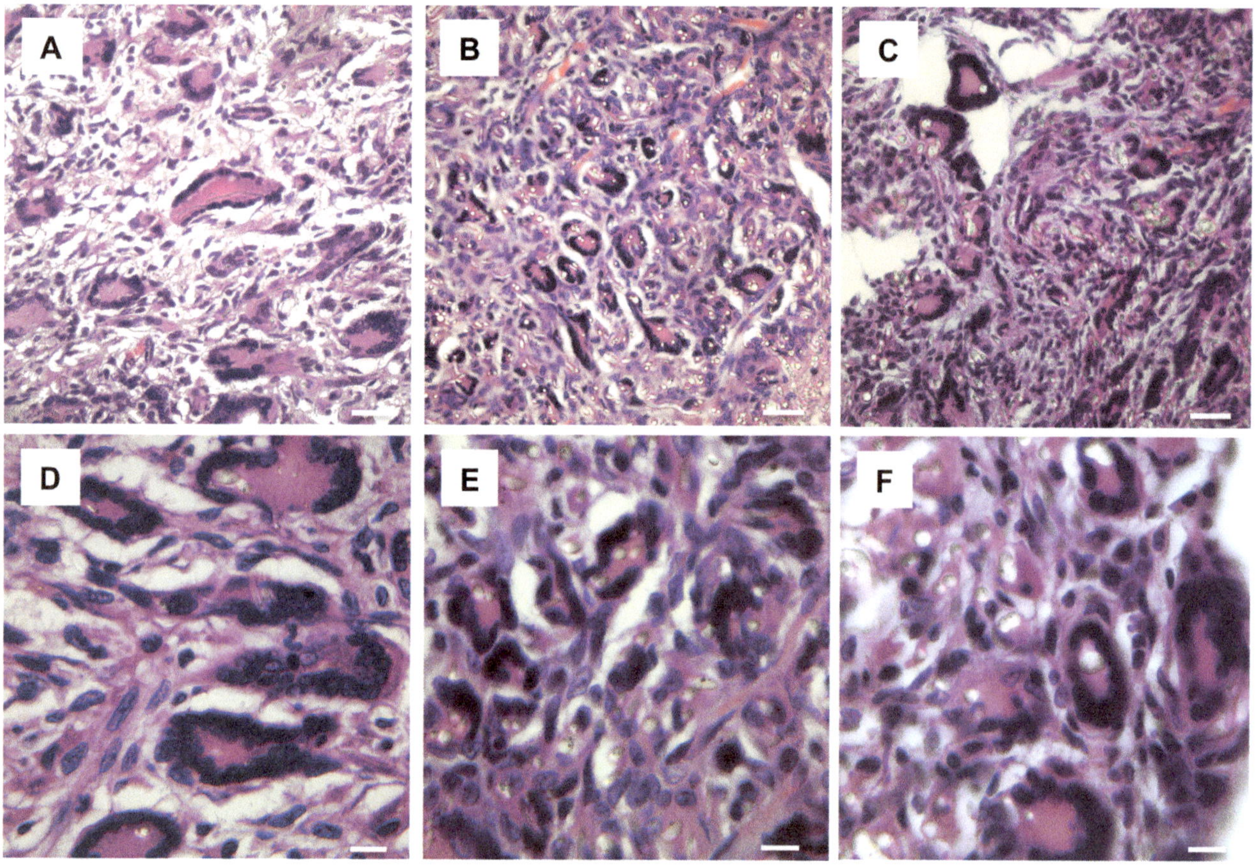

Figure 3. Foreign body giant cells covering nano- and microfibers of electrospun materials. Histological sections (HE staining) evidencing FBGC in (A and D) PET30, (B and E) PET/C, and (C and F) DL groups. *Bars*: 25 μm (A–C); 10 μm (D–F).

electrospinning process [7]. The increased hydrophilic character observed for the PET/C hybrid may be advantageous for mesh integration on the parietal side. A double-layer mesh was also developed, comprising one layer of PET to prevent formation of bowel adhesion and one layer of PET/C nanofibers to stimulate integration of the mesh in the subcutaneous tissue, thus reinforcing the mechanical strength of the prosthetic wall.

In the rat abdominal hernia model, electrospun meshes were demonstrated to be adequate for the surgical procedure, i.e., easy to suture, and as a soft and flexible material, the electrospun meshes adapted well to the abdominal tissues. In contrast, the stiffness of the Marlex material may affect surrounding tissues and cause discomfort. This material was even perceived through the animal's skin by touching. Electrospun meshes, on the other hand, were much softer, as well as more malleable and adaptable, while effectively performing their role of containing visceral components without mesh failure during the experimental period.

In contrast to bowel adhesion that can lead to serious complications, the observed omentum adhesion to the meshes is considered clinically irrelevant [29]. In fact, some authors have even suggested that the interposition of omentum between the prosthetic mesh and viscera is effective in restricting omentum adhesion, both in preclinical and clinical studies [29,30].

Importantly, we observed a significant foreign body reaction associated with the electrospun nanofibrous meshes. This was an unexpected result since the chemical composition of the electrospun mesh is 100% PET, a recognized biocompatible polymer, and many studies have reported the absence of inflammatory response to polymer electrospun fibers by different cell types, both *in vitro* and *in vivo* [31–34]. Numerous studies have reported on the anti-inflammatory and wound healing effects of chitosan; however, in this specific application, chitosan had a negative impact, as an increased inflammatory response was observed in animals treated with chitosan- containing meshes. Similar results were obtained by Barbosa *et al.* [35] when testing the inflammatory response to chitosan scaffolds with different acetylation degree. In fact, the chitosan scaffold with an acetylation degree of 15% induced the formation of a thick granuloma with high infiltration of inflammatory cells, after subcutaneous implantation in mice [35]. Another study also showed that chitosan a scaffold with an acetylation degree of 15% caused a macrophage M1 pro-inflammatory response [36].

The foreign body granulomas were mostly composed of macrophages, FBGC, and fibroblasts. An abundance of FBGC was found in tissues surrounding the electrospun meshes, evidencing a typical foreign body reaction. FBGC were most often found to be enclosing one or more nanofibers as an attempt to isolate the foreign material (*cf.* Figure 3). Extending the experimental time seemed to result in an attenuation of the foreign body reaction; nonetheless, a large FBR persisted 90 days after mesh implantation. Although foreign body reaction has few clinical implications and is usually limited to the close periphery of the implanted material, certain clinical disadvantages are always present as an associated risk condition. Indeed, chronic inflammation and the related proangiogenic process have been assumed

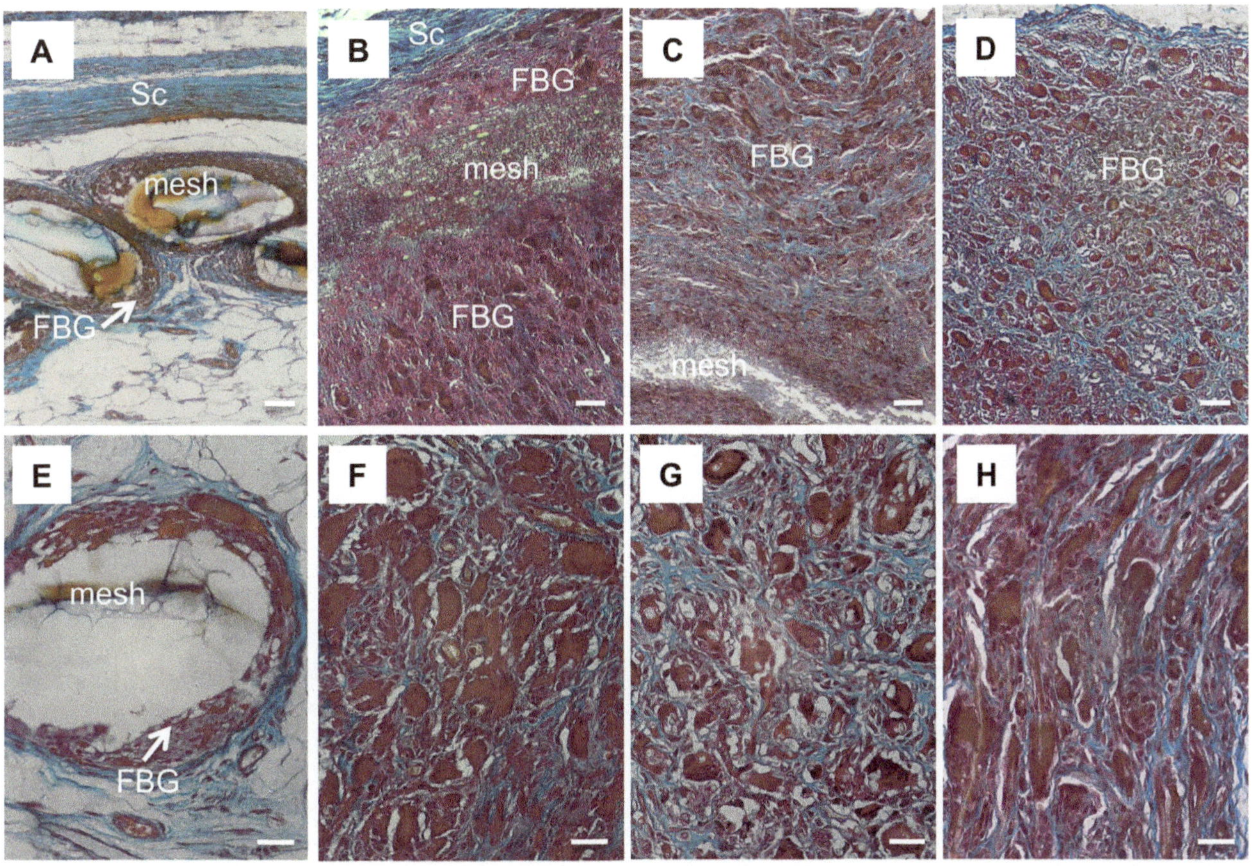

Figure 4. Garvey staining of foreign body granulomas. Histological sections stained with Garvey's staining (nuclei stained black/dark red, cytoplasmic elements stained red and collagen fibers stained light blue). (A,E) Marlex (B,F) PET (C,G) PET/C (D,H) DL. (De) dermis, (Sc) subcutaneous tissue, and (FBG) foreign body granuloma. Bars = 50 μm (A–D); 25 μm (E–H).

to underlie most chronic diseases, including cancer, cardiovascular diseases, and diabetes [37,38].

It is well established that polymer type, mesh construction, fiber size, mesh porosity and contact surface, as well as the specific characteristics of the tissue where the biomaterial is implanted, play important roles in biocompatibility and induced tissue reactions [39,40]. Still, no consensus has been reached with respect to the effects of implanted mesh and the development of inflammation [41–43]. Indeed, cell/tissue-mesh interactions still require further elucidation.

Considering the high biocompatibility of bulk PET and the minimum foreign body reaction found for woven-PET mesh (PET-woven microfibers), we hypothesize that the nanostructure of the electrospun materials underlies the huge foreign body reaction found in animals implanted with electrospun meshes. The reduced diameter of the electrospun fibers and pore size of the meshes, combined with the high surface-to-volume ratio of the electrospun materials, may therefore have important effects on the inflammatory reaction. Among surface properties, the material's ability to adsorb proteins plays a key role, as those proteins, not the material's composition itself, are major contributors of FBR [40]. It is well documented that the high surface-to-volume ratios of electrospun nanofibrous materials contribute to their high protein adsorption capability [44,45] and likely explain the high foreign body reaction observed in electrospun PET meshes. Indeed, higher foreign body reactions are often found for biomaterials, in both in micro- and nanoscale dimensions, with large surface areas [46,47]. Specifically, in abdominal defect repair, Conze *et al.* [48] have verified a pronounced foreign body reaction for a multifilament small-diameter polypropylene mesh. In fact, the diameter of foreign body granuloma 90 days after mesh placement decreased from 106.5 μm to 70.9 μm by increasing the filament diameter from 0.6 mm to 2.5 mm. On the other hand, many reports show a decrease of inflammatory response with the decrease of fiber diameter. Saino *et al.* [49] showed that the decrease of fiber diameter of electrospun polylactic acid fibrous mats reduced *in vitro* macrophage activation and the secretion of proinflammatory molecules. Similarly, Cao *et al.* [50] demonstrated the importance of the nanofibrous scaffold architecture and topography on the *in vivo* and *in vitro* foreign body reaction and showed a decrease of granuloma thickness of subcutaneous implants from 38 μm for a PCL film to ~8 μm and 4 μm for PCL-aligned electrospun nanofibers or nonwoven electrospun nanofibers, respectively.

Other aspects that are thought to have influenced the observed intensive foreign body reaction are related to the extension of the trauma and the specific characteristics of the tissues where the materials had been implanted. Abdominal mesh implantation involves the creation of a 1.5×1.5 cm defect by complete resection of the abdominal wall. This is a severe trauma, considering the relative size of the animal, and, consequently, a large inflammatory response may be induced. Furthermore, specific characteristics of the implantation tissues, such as cell composition and function, vascularization, extracellular matrix composition, and

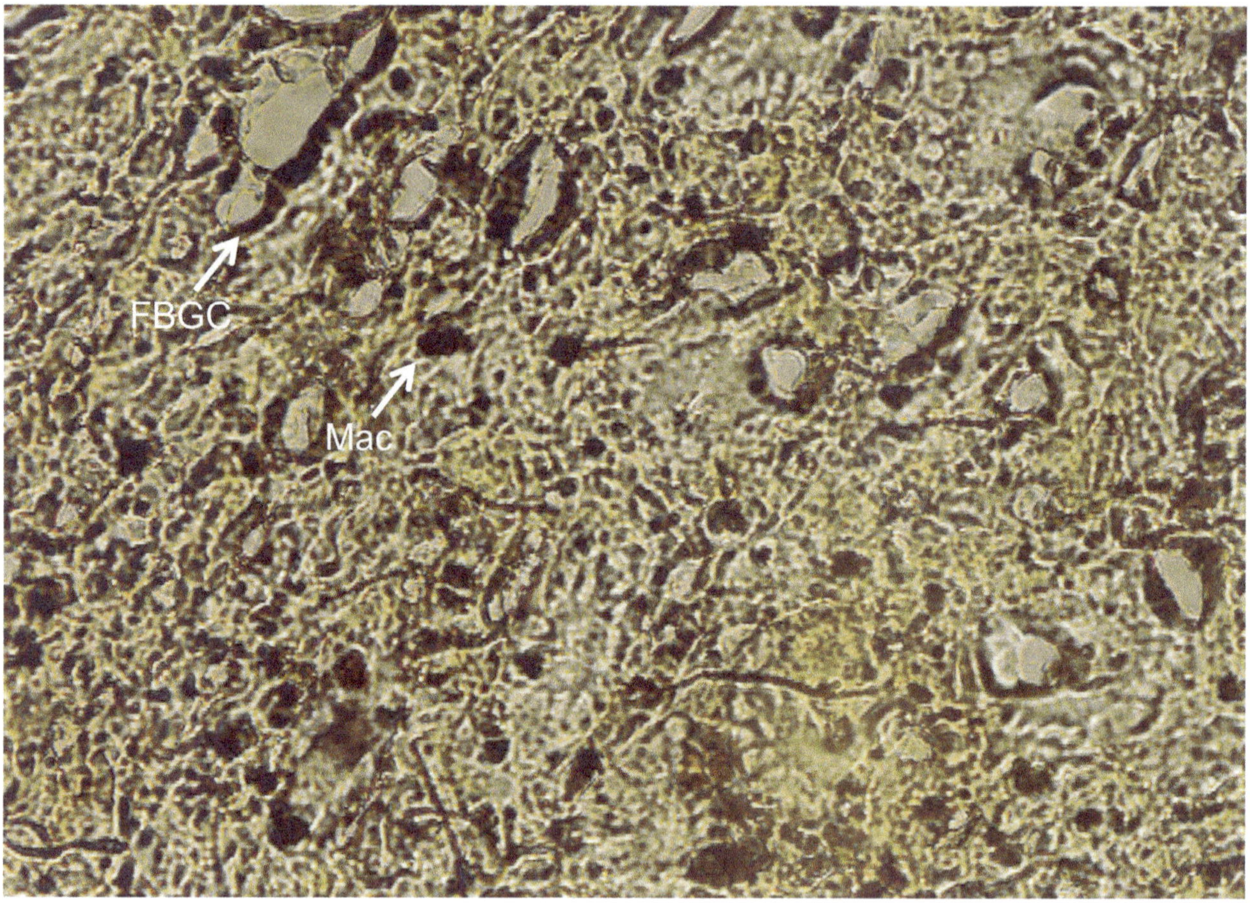

Figure 5. Immunohistochemistry for CD68. Immunohistochemical staining (CD68+) evidencing macrophages (Mac) and foreign body giant cells (FBGC) around a PET electrospun mat implanted in rat as abdominal mesh. Sections were incubated with a monoclonal antibody directed against CD68, clone KP1 (Zeta Corporation, CA), Histofine Simple Stain Max-Po Multi and 3,3′-diaminobenzidine tetrahydrochloride.

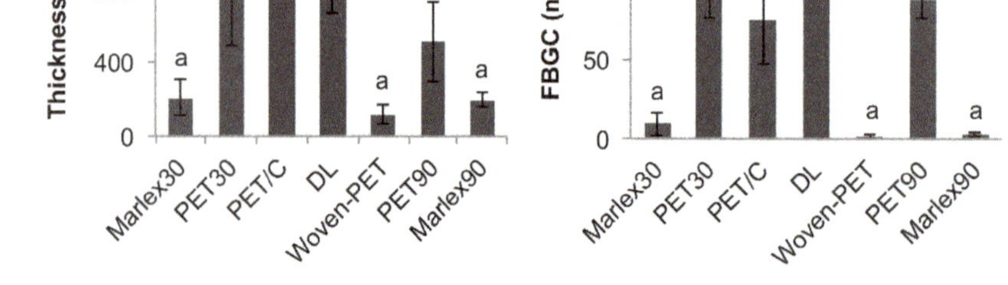

Figure 6. Analysis of the foreign body granuloma induced by abdominal meshes. (A) Granuloma thickness (mean ± standard deviation) and (B) absolute number of FBGC per section of granuloma (mean ± standard deviation) in animals implanted with different abdominal meshes. All experimental groups were compared to each other using Tukey-Kramer multiple comparison tests. For each chart, bars with different letters are significantly different at $P<0.05$.

contact with ascitic fluid, for example, have important effects on foreign body reaction.

Materials are becoming smaller than the basic body unity, i.e. the cell. Despite the enormous progress of the electrospinning technique over the past decade, cellular response and the associated risks involved in the use of nanostructure fibrous biomaterials are still poorly understood [51,52], and further studies are needed to gain more insight.

Conclusions

PET and PET/chitosan electrospun meshes demonstrated good performance during the implantation surgery, adequate mechanical attributes, and no evidence of intestinal adhesion. Nevertheless, a large foreign body reaction was found in animals treated with the electrospun mats. Indeed, the reduced dimension of nanofibers and the high surface-to-volume ratio of electrospun nonwoven materials may induce a high foreign body reaction, depending on the extent and location of the lesion. Refinement may be achieved by the inclusion of biological components on the fiber's surface to enhance bioactivity and biocompatibility, thus increasing the potential of these nondegradable electrospun fiber scaffolds for abdominal wall replacement. Nevertheless, these results demonstrate the need for more studies to elucidate the mechanisms underlying cell/tissue-nanomaterial interactions in order to gain a better understanding of the risks involved in implantation of nanostructured biomaterials.

Acknowledgments

Authors acknowledge Central Laboratory of Electron Microscopy (LCME) and Fluorbeg (UFSC) for microscopic analysis.

Author Contributions

Conceived and designed the experiments: BV JALdS JAS. Performed the experiments: BV DSC RP AP. Analyzed the data: BV MM ECF PFD RMRdV JALdS JAS. Contributed reagents/materials/analysis tools: MM ECF PFD RMRdV JALdS. Wrote the paper: BV JALdS.

References

1. Schofer MD, Roessler PP, Schaefer J, Theisen C, Schlimme S, et al. (2011) Electrospun PLLA nanofiber scaffolds and their use in combination with BMP-2 for reconstruction of bone defects. PLoS ONE 6(9): e25462.
2. Vasita R, Katti DS (2006) Nanofibers and their applications in tissue engineering. Int J Nanomed 1(1): 15–30.
3. Cui W, Zhou Y, Chang J (2010) Electrospun nanofibrous materials for tissue engineering and drug delivery. Sci Technol Adv Mater 11: 014108 (11pp).
4. Veleirinho B, Coelho DS, Dias PF, Maraschin M, Ribeiro-do-Valle RM, et al. (2012) Nanofibrous poly(3-hydroxybutyrate-co-3-hydroxyvalerate)/chitosan scaffolds for skin regeneration. Int J Biol Macromol 51: 343–350.
5. Veleirinho B, Berti FV, Dias PF, Maraschin M, Ribeiro-do-Valle RM, et al. (2013) Manipulation of chemical composition and architecture of non-biodegradable poly(ethylene terephthalate)/chitosan fibrous scaffolds and their effects on L929 cell behavior. Mat Sci Eng C-Biomim 33: 37–46.
6. Veleirinho B, Rei MF, Lopes-da-Silva JA (2008) Solvent and concentration effects on the properties of electrospun poly(ethylene terephthalate) nanofiber mats. J Polym Sci Pol Phys 46(5): 460–471.
7. Lopes-da-Silva JA, Veleirinho B, Delgadillo I (2009) Preparation and characterization of electrospun mats made of pet/chitosan hybrid nanofibers. J Nanosci Nanotechnol 9(6): 3798–3804.
8. Anthony T, Bergen PC, Kim LT, Henderson M, Fahey T, et al. (2000) Factors affecting recurrence following incisional herniorrhaphy. World J Surg 24(1): 95–101.
9. Gecim IE, Kocak S, Ersoz S, Bumin C, Aribal D (1996) Recurrence after incisional hernia repair. Results and risk factors. Surg Today 26(8): 607–609.
10. Halm JA, Lip H, Schmitz PI, Jeekel J (2009) Incisional hernia after upper abdominal surgery: a randomised controlled trial of midline versus transverse incision. Hernia 13: 275–280.
11. Luijendijk RW, Hop WCJ, van den Tol P, de Lange DCD, Braaksma MMJ, et al. (2000) A comparison of suture repair with mesh repair for incisional hernia. New Engl J Med 343(6): 392–398.
12. Kingsnorth A, LeBlanc K (2003) Hernias: inguinal and incisional. Lancet 362: 1561–1571.
13. Carbajo MA, del Olmo JCM, Blanco JI, de la Cuesta C, Toledano M, et al. (1999) Laparoscopic treatment vs open surgery in the solution of major incisional and abdominal wall hernias with mesh. Surg Endosc-Ultras 13(3): 250–252.
14. Robinson TN, Clarke JH, Schoen J, Walsh MD (2005) Major mesh-related complications following hernia repair - Events reported to the Food and Drug Administration. Surg Endosc-Ultras 19(12): 1556–1560.
15. Menzies D, Ellis H (1990) Intestinal-obstruction from adhesions - how big is the problem? Ann Roy Coll Surg 72(1): 60–63.
16. Hong Y, Fujimoto K, Hashizume R, Guan J, Stankus JJ, et al. (2008) Generating elastic, biodegradable polyurethane/poly(lactide-co-glycolide) fibrous sheets with controlled antibiotic release via two-stream electrospinning. Biomacromolecules 9: 1200–1207.
17. Hashizume R, Kazuro L, Fujimoto KL, Hong Y, Amoroso NJ, et al. (2010) Morphological and mechanical characteristics of the reconstructed rat abdominal wall following use of a wet electrospun biodegradable polyurethane elastomer scaffold. Biomaterials 31: 3253–3265.
18. Hong Y, Takanari K, Amoroso NJ, Hashizume R, Brennan-Pierce EP, et al. (2012) An elastomeric patch electrospun from a blended solution of dermal extracellular matrix and biodegradable polyurethane for rat abdominal wall repair. Tissue Eng Pt C-Meth 18: 122–132.
19. Ebersole GC, Buettmann EG, MacEwan MR, Tang ME, Frisella MM, et al. (2012) Development of novel electrospun absorbable polycaprolactone (PCL) scaffolds for hernia repair applications. Surg Endosc 26: 2717–2728.
20. McGinty JJ, Hogle NJ, McCarthy H, Fowler DL (2005) A comparative study of adhesion formation and abdominal wall ingrowth after laparoscopic ventral hernia repair in a porcine model using multiple types of mesh. Surg Endosc 19(6): 786–790.
21. Ueno H, Mori T, Fujinaga T (2001) Topical formulations and wound healing applications of chitosan. Adv Drug Deliver Rev 52(2): 105–115.
22. VandeVord PJ, Matthew HWT, DeSilva SP, Mayton L, Wu B, et al. (2002) Evaluation of the biocompatibility of a chitosan scaffold in mice. J Biomed Mater Res 59(3): 585–590.
23. Tommeraas K, Varum KM, Christensen BE, Smidsrod O (2001) Preparation and characterisation of oligosaccharides produced by nitrous acid depolymerisation of chitosans. Carbohydr Res 333: 137–144.
24. Garvey W (1984) Modified Elastic Tissue-Masson Trichrome Stain. Stain Technol. 59(4): 213–216.
25. Robinson TN, Clarke JH, Schoen J, Walsh MD (2005) Major mesh-related complications following hernia repair: events reported to the Food and Drug Administration. Surg Endosc 19: 1556–1560.
26. Welty G, Klinge U, Klosterhalfen B, Kasperk R, Schumpelick V (2001) Functional impairment and complaints following incisional hernia repair with different polypropylene meshes. Hernia 5: 142–147.
27. Voskerician G, Jin J, White MF, Williams CP, Rosen MJ (2010) Effect of biomaterial design criteria on the performance of surgical meshes for abdominal hernia repair: a pre-clinical evaluation in a chronic rat model. J Mater Sci-Mater M 21(6): 1989–1995.
28. Matthews BD, Pratt BL, Pollinger HS, Backus CL, KercherKW, etal. (2003) Assessment of adhesion formation to intra-abdominal polypropylene mesh and polytetrafluoroethylene mesh. J Surg Res 114(2): 126–132.
29. Karabulut B, Sonmez K, Turkyilmaz Z, Demiroğullari B, Karabulut R, et al. (2006) Omentum prevents intestinal adhesions to mesh graft in abdominal infections and serosal defects. Surg Endosc 20(6): 978–982.
30. Bingener J, Kazantsev GB, Chopra S, Schwesinger WH (2004) Adhesion formation after laparoscopic ventral incisional hernia repair with polypropylene mesh: a study using abdominal ultrasound. JSLS-J Soc Laparoend 8(2): 127–131.
31. Nisbet DR, Rodda AE, Horne MK, Forsythe JS, Finkelstein DI (2009) Neurite infiltration and cellular response to electrospun polycaprolactone scaffolds implanted into the brain. Biomaterials 30: 4573–4580.
32. Pan H, Jiang HL, Kantharia S, Chen WL (2011) A fibroblast/macrophage co-culture model to evaluate the biocompatibility of an electrospun dextran/PLGA

scaffold and its potential to induce inflammatory responses. Biomedical Mat 6(6): 065002.
33. Ni PY, Fu SZ, Fan M, Guo G, Shi S, et al. (2011) Preparation of poly(ethylene glycol)/polylactide hybrid fibrous scaffolds for bone tissue engineering. Int J Nanomed 6: 3065–3075.
34. Bergmeister H, Grasl C, Walter I, Plasenzotti R, Stoiber M, et al. (2011) Electrospun small-diameter polyurethane vascular grafts: Ingrowth and differentiation of vascular-specific host cells. Artificial Organs 36(1): 54–61.
35. Barbosa JN, Amaral IF, Aguas AP, Barbosa MA (2010) Evaluation of the effect of the degree of acetylation on the inflammatory response to 3D porous chitosan scaffolds. J Biomed Mater Res-A 93(1): 20–28.
36. Vasconcelos DP, Fonseca AC, Costa M, Amaral IF, Barbosa MA, et al. (2013) Macrophage polarization following chitosan implantation. Biomaterials 34: 9952–9959.
37. Bartsch H, Nair J (2006) Chronic inflammation and oxidative stress in the genesis and perpetuation of cancer: role of lipid peroxidation, DNA damage, and repair. Langenbeck Arch Surg 391(5): 499–510.
38. Manabe I (2011)Chronic inflammation links cardiovascular, metabolic and renal diseases. Circ J 75(12): 2739–2748.
39. Kamath S, Bhattacharyya D, Padukudru C, Timmons RB, Tang L (2008) Surface chemistry influences implant-mediated host tissue responses. J Biomed Mater Res-A 86(3): 617–626.
40. Hu WJ, Eaton JW, Tang LP (2001) Molecular basis of biomaterial-mediated foreign body reactions. Blood 98(4): 1231–1238.
41. Klosterhalfen B, Klinge U, Schumpelick V, Tietze L (2000) Polymers in hernia repair–common polyester vs. polypropylene surgical meshes. J Mater Sci 35: 4769–4776.
42. Brown CN, Finch JG (2010) Which mesh for hernia repair? Ann R Coll Surg Engl 92(4): 272–278.
43. Binnebösel M, von Trotha KT, Jansen PL, Conze J, Neumann UP, et al. (2011) Biocompatibility of prosthetic meshes in abdominal surgery. Semin Immunopathol 33: 235–243.
44. Shalumon KT, Binulal NS, Deepthy M, Jayakumar R, Manzoor K, et al. (2011) Preparation, characterization and cell attachment studies of electrospun multi-scale poly(caprolactone) fibrous scaffolds for tissue egineering. J Macromol Sci Pure 48(1): 21–30.
45. Leong MF, Chian KS, Mhaisalkar PS, Ong WF, Ratner BD (2009) Effect of electrospun poly(D, L-lactide) fibrous scaffold with nanoporous surface on attachment of porcine esophageal epithelial cells and protein adsorption. J Biomed Mater Res-A 89(4): 1040–1048.
46. Voskerician G, Gingras PH, Anderson JM (2006) Macroporous condensed poly(tetrafluoroethylene). I. In vivo inflammatory response and healing characteristics. J Biomed Mater Res-A 76(2): 234–242.
47. Sanchez VC, Weston P, Yan A, Hurt RH, Kane AB (2011) A 3-dimensional in vitro model of epithelioid granulomas induced by high aspect ratio nanomaterials. Part Fiber Toxicol 8(17): 1–18.
48. Conze J, Rosch R, Klinge U, Weiss C, Anurov M, et al. (2004) Polypropylene in the intra-abdominal position: influence of pore size and surface area. Hernia 8(4): 365–372.
49. Saino E, Focarete ML, Gualandi C, Emanuele E, Cornaglia AI, et al. (2011) Effect of electrospun fiber diameter and alignment on macrophage activation and secretion of proinflammatory cytokines and chemokines. Biomacromolecules 12(5): 1900–1911.
50. Cao H, McHugh K, Chew SY, Anderson JM (2010) The topographical effect of electrospun nanofibrous scaffolds on the in vivo and in vitro foreign body reaction. J Biomed Mater Res-A 93(3): 1151–1159.
51. Soto K, Garza KM, Murr LE (2007) Cytotoxic effects of aggregated nanomaterials. Acta Biomater 3(3): 351–358.
52. Stern ST, McNeil SE (2008) Nanotechnology safety concerns revisited. Toxicol Sci 101(1): 4–21.

6

Increased Infarct Wall Thickness by a Bio-Inert Material Is Insufficient to Prevent Negative Left Ventricular Remodeling after Myocardial Infarction

Aboli A. Rane[1], Joyce S. Chuang[1], Amul Shah[1], Diane P. Hu[1], Nancy D. Dalton[2], Yusu Gu[2], Kirk L. Peterson[2], Jeffrey H. Omens[2], Karen L. Christman[1]*

1 Department of Bioengineering, University of California San Diego, La Jolla, California, United States of America, **2** Department of Medicine, University of California San Diego, San Diego, California, United States of America

Abstract

Background: Several injectable materials have been shown to preserve or improve cardiac function as well as prevent or slow left ventricular (LV) remodeling post-myocardial infarction (MI). However, it is unclear as to whether it is the structural support or the bioactivity of these polymers that lead to beneficial effects. Herein, we examine how passive structural enhancement of the LV wall by an increase in wall thickness affects cardiac function post-MI using a bio-inert, non-degradable synthetic polymer in an effort to better understand the mechanisms by which injectable materials affect LV remodeling.

Methods and Results: Poly(ethylene glycol) (PEG) gels of storage modulus $G' = 0.5 \pm 0.1$ kPa were injected and polymerized *in situ* one week after total occlusion of the left coronary artery in female Sprague Dawley rats. The animals were imaged using magnetic resonance imaging (MRI) at 7 ± 1 day(s) post-MI as a baseline and again post-injection 49 ± 4 days after MI. Infarct wall thickness was statistically increased in PEG gel injected vs. control animals ($p<0.01$). However, animals in the polymer and control groups showed decreases in cardiac function in terms of end diastolic volume, end systolic volume and ejection fraction compared to baseline ($p<0.01$). The cellular response to injection was also similar in both groups.

Conclusion: The results of this study demonstrate that passive structural reinforcement alone was insufficient to prevent post-MI remodeling, suggesting that bioactivity and/or cell infiltration due to degradation of injectable materials are likely playing a key role in the preservation of cardiac function, thus providing a deeper understanding of the influencing properties of biomaterials necessary to prevent post-MI negative remodeling.

Editor: Wei-Chun Chin, University of California Merced, United States of America

Funding: This research was supported in part by the NIH Director's New Innovator Award Program, part of the NIH Roadmap for Medical Research [1-DP2-OD004309-01]. http://commonfund.nih.gov/newinnovator/. A.A.R. would like to thank the National Science Foundation for a Graduate Research Fellowship. The funders had no role in study design, data collection and analysis, decision to publish, or preparation of the manuscript.

Competing Interests: The authors have declared that no competing interests exist.

* E-mail: christman@bioeng.ucsd.edu

Introduction

Progressive left ventricular (LV) remodeling occurs after myocardial infarction (MI), and while remodeling can be beneficial initially, over time it can become maladaptive leading to LV dysfunction [1]. With 5.8 million people suffering from heart failure (HF), treatment of negative LV remodeling after MI, the leading cause of HF, is a pressing clinical need [2]. Coagulative necrosis following MI initiates a cascade of events characterized by myocyte necrosis and their eventual replacement by fibrotic scar tissue. Necrosis of the myocytes triggers an inflammatory response leading to activation of matrix metalloproteinases [3] that cause a breakdown of the extracellular matrix, which results in myocyte slippage and infarct expansion. At the same time fibroblast proliferation occurs, leading to deposition of collagen and the formation of scar tissue. Concurrent TGF-β1 mediated transformation of fibroblast to myofibroblasts leads to deposition of type I and III collagen and scar tissue formation from subsequent collagen cross-linking [1]. Ventricular wall thinning and dilatation also occurs leading to an increase in wall stress, and ultimately resulting in heart failure [1,4,5].

Various injectable biomaterials have been investigated as potential acellular treatments to prevent or reverse this downward spiral of adverse LV remodeling [6]. Collagen [7], fibrin [8], chitosan [9], small intestinal submuscosa (SIS) [10], and a collagen-matrigel blend [11] as well as a few degradable synthetic materials [12,13,14] have shown either an improvement or maintenance of cardiac function in terms of ejection fraction or fractional shortening. Injection of alginate has been shown to cause a decrease in LV diastolic area and systolic area [15,16] while fibrin [17], hyaluronic acid (HA) [18], SIS [10] and a synthetic thermoresponsive polymer [14] reduced infarct size. Injection of biomaterials such as collagen [19], matrigel [19], fibrin [17,19], alginate [20], chitosan [9], self-assembling nanofibers [21], SIS [10], and a degradable thermoresponsive polymer [13] have also increased neovasculature. Though most injectable materials have shown positive effects on cardiac function and LV remodeling, to date, the mechanisms behind this improvement are unknown.

It has been shown that myocardial infarction can lead to an increase in wall stress due to changes in wall curvature and thickness in accordance with Laplace's Law, which states that wall stress is directly proportional to the pressure and radius, and inversely proportional to thickness. Wall thinning due to infarction can increase wall stress, and may contribute to border zone expansion and progression of negative LV remodeling [22]. Hence, it has been hypothesized that injectable materials may affect the passive structural properties of the wall by increasing infarct wall thickness and hence reducing local LV wall stress [6,23]. This decrease in LV wall stress is thought to protect the vulnerable myocardium from stress induced apoptosis and infarct expansion, thus preventing pathological LV remodeling and decline in cardiac function [23]. On the other hand, injectable materials have also been shown to elicit cell infiltration [19,21,24] and neovascularization [19] in the affected area. This leads to the alternate hypothesis that the bioactivity of materials, such as inherent cell adhesion domains or angiogenic degradation products, and/or cell infiltration upon degradation of these materials are playing an important role in prevention of adverse LV remodeling and improvement of cardiac function [6,25,26].

The aim of this study was to begin to understand the mechanisms by which injectable materials preserve cardiac function and prevent negative LV remodeling post-MI by decoupling bioactivity from the structural effects of an injectable polymer. Herein, we utilized an injectable, bio-inert, non-degradable polymer, polyethylene glycol (PEG), along with experimental parameters similar to those employed in previous studies aimed at assessing the effects of injectable biomaterials on LV remodeling and cardiac function. PEG is a well established synthetic polymer that is known to be non-toxic [27] and bio-inert, thus preventing protein and cell adhesion [27,28,29,30]. We demonstrate that passive structural intramyocardial support by itself does not prevent negative LV remodeling or maintain cardiac function, suggesting that other mechanisms such as bioactivity and/or cell infiltration seen with degradable materials likely play a dominant role in the mitigation of LV remodeling and preservation of cardiac function.

Materials and Methods

All experiments in this study were conducted in accordance with the guidelines established by the Institutional Animal Care and Use Committee at the University of California, San Diego and the American Association for Accreditation of Laboratory Animal Care and were approved by the Institutional Animal Care and Use Committee at UCSD (A3033-01). All efforts were made to minimize any pain and suffering felt by the animals.

PEG hydrogel preparation

PEG hydrogels were prepared by mixing solutions of 4-arm polyethylene glycol-amide-succinimidyl glutarate (PEG-ASG) (Sunbio, Anyang City, South Korea, MW 20,000) and trilysine (Sigma-Aldrich, St. Louis, MO) to create PEG hydrogels of desired stiffness at room temperature by chemical crosslinking [31]. Optimal concentrations of PEG-ASG and trilysine were experimentally determined to generate gels of stiffness similar to that of commonly injected biomaterials. Equal volumes of 100 mg/ml PEG-ASG in pH 4.0 phosphate buffer and 0.5–2 mg/ml trilysine in pH 8.2 borate buffer were mixed together to prepare the hydrogels. For *in-vivo* injections of PEG, the solutions were mixed together and injected prior to gelation.

Mechanical testing

Material properties in terms of the storage modulus (G′) and loss modulus (G″) were determined using a parallel plate rheometer (TA Instruments AR-2000) as previously reported [32,33]. Briefly, gels were placed between two parallel plates of the rheometer and compressed to a normal force between 0.2–0.3 N, to avoid slipping. A strain sweep (0.0001 to 10%) was performed at a constant frequency of 1 Hz to confirm that the measurements were within the linear viscoelastic region. 1% constant strain was used for subsequent frequency sweeps (0.01 to 10 Hz) using a logarithmic scale, taking 5 points per decade. All measurements were taken in triplicate and reported as mean ± SEM at a frequency of 1 Hz.

Rat total occlusion model

Left coronary artery total occlusion was performed on female Sprague Dawley rats (225–250 g) under aseptic conditions. Animals were anesthetized using 5% isoflurane, intubated and maintained at 2.5% isoflurane for the surgical procedure. The animals were ventilated using a respirator at 75 breaths/minute. The heart was exposed using a left anterior thoracotomy, and the artery was ligated using a 6-0 silk suture at 1–2 mm below the left atrial appendage as previously reported [34]. The chest was closed and the animals were allowed to recover. ECG was continuously monitored for detection of arrhythmias, and atropine 150 to 200 µl (0.54 mg/ml) was administered I.P. if needed. Buprenorphine was administered I.P. at 5 mg/kg to prevent post-operative pain and Ringer's Lactate (3 cc) was given I.P. to the animals to prevent dehydration.

Echocardiography

Echocardiography was performed 4±1 day(s) post-MI to screen for the presence of an infarct. Animals were anesthetized using 5% isoflurane for 30 seconds and then maintained at 1% isoflurane for the imaging procedure. Parasternal long axis and short axis images were obtained using a Philips, Sonos 5500 system with a 15 MHz transducer. Qualitative assessment of infarction size was rendered at the time of imaging using both planes. Animals with either no MI or very small MI (less than 15% of the LV perimeter in the long axis plane) were excluded from the study prior to injections.

Injection surgery

The animals were randomized 9±1 day(s) after MI, and injected with either 100 µl of PEG hydrogel or saline (control). An incision was made between the fourth and fifth ribs, the anterolateral portion of the heart was exposed, and an injection of polymer or saline was administered using a 27 G needle into the infarct wall. Presence of the injection was verified by temporary discoloration of the tissue. The chest was then closed and the animal was allowed to recover. To prevent post-operative pain in the animals, buprenorphine was administered at 5 mg/kg.

Magnetic resonance imaging

Magnetic Resonance (MR) imaging was used for quantitative assessment of the injected gel on cardiac function. Animals were scanned at 7±1 day(s) post-MI, as a baseline measurement, and again at 49±4 days after MI to assess post-treatment effects on remodeling and cardiac function. A 7T Bruker Horizontal Bore scanner and Bruker single tuned quadrature Transmit/Receive volume coil were used for all measurements. Anatomical cine imaging was performed with an ECG-gated Fast Low Angle Shot (FLASH) sequence as previously reported [35]. MR parameters were set to TR = 7.7 ms, TE = 1.28 ms, flip angle = 15° and 4–6 averages per slice. All slices were 1 mm in thickness, had a field of view of 50 mm×50 mm and were contained in a data matrix of 256×256 pixels. For each slice, 25 frames were taken through the cardiac cycle to capture end diastolic and end systolic images. Animals were anesthetized using 1.5–2% isoflurane and the heart rate, ECG and respiration rate were monitored continuously. The long axis of the heart was identified and 10–12 short axis slices from apex to base were acquired at 1 mm increments. ImageJ was used to trace the endocardial boundary at end diastole (ED) and end systole (ES) on each short axis slice. Simpson's method was used to determine the end diastolic volume (EDV) and end systolic volume (ESV). Ejection fraction (EF) was calculated as [(EDV-ESV)/EDV]×100 similar to previous methods [36].

Finite element modeling

In order to visually represent LV dilatation and wall thickness pre and post-injection, finite element analysis was used to generate geometric models of the LV at ED and ES using prolate spheroidal bi-cubic Hermite finite element surface meshes [35]. Meshes were fit to the endocardial and epicardial boundary points determined from the MRI cine images. The boundary points were fit using a least squares minimization in the λ coordinate direction with bicubic Hermite interpolation. The 40 element mesh was refined to 340 elements and was converted to Cartesian coordinates so that the long axis of the heart aligned with the X- axis and the septum was bisected by the Y-axis.

Histology and immunohistochemisty

Animals were sacrificed 24 hours after post-treatment MR imaging (~49 days post-MI) by an IP injection of Fatal Plus (sodium pentobarbital, 936 mg/kg). Hearts were immediately excised and fresh frozen in Tissue-Tek O.C.T. freezing compound. Hearts were sectioned into 10 μm slices and slides taken every 0.4 mm were stained with hematoxylin and eosin (H&E) to visualize the polymer injection. Images were taken using a Carl Zeiss Observer D.1 and analyzed using AxioVision software. A pathologist qualitatively assessed inflammatory response [37,38,39] to the injection in the infarct, border zone and remote tissue in the H&E stained sections at mid-infarct. Macrophages were identified using an anti-rat CD68 primary antibody [40]. Tissue sections were incubated with 1:50 dilution of mouse anti-rat CD68 and subsequently incubated with horseradish peroxidase (HRP) conjugated goat anti-mouse IgG (1:100 dilution). The HRP was visualized by incubation with diaminobenzidine (DAB) for 10 minutes. Wall thickness was calculated using 3 mid-infarct slides, averaging 5 equally spaced measurements along the infarct wall in each slide. Infarct size was calculated using H&E stained sections by calculating the average of the ratio of the endocardial infarct length to the endocardial circumference and the ratio of the epicardial infarct length to the epicardial circumference over three slides evenly spaced through the infarct as previously shown [41]. To assess arteriole density, immunohistochemistry staining was done on three slides evenly spaced through the polymer region or infarct for each of the PEG and saline injected hearts respectively, according to previously described methods [17,20,24]. Briefly, the sections were fixed with acetone, incubated with anti-smooth muscle actin antibody and then stained with Alexa Fluor 568 anti-mouse antibody. Vessels in the range of 10 μm to 100 μm in the infarct were quantified with Axiovision.

Statistical analysis

Two-tailed Students' t-test was used for infarct wall thickness and arteriole density analyses. Differences between baseline and post-treatment groups with MRI analysis were assessed using a paired two-tailed t-test. All measurements were reported mean ± SEM. Significance was accepted at $p<0.05$.

Results

PEG hydrogel mechanical characterization

To create a PEG hydrogel with desired mechanical properties, the amount of trilysine crosslinker (0.5 mg/ml to 2 mg/ml) was varied. A strain sweep test and a frequency sweep test were performed to generate mechanical spectra of the gels. We sought to test a polymer with similar mechanical properties to commonly injected materials to better delineate between structural effects and bioactivity. A storage modulus (G′) of 0.5±0.1 kPa (trilysine 1 mg/ml) (Figure 1) was selected as it mimicked the mechanical properties of commonly injected polymers [42,43]. As expected the G′ is independent of frequency and G″ (loss modulus) is weakly dependent on frequency [33,44].

LV geometry and cardiac function

In-vivo studies were performed to assess the effect of the bio-inert, non-degradable PEG hydrogel on wall thickness, EF, EDV and ESV. All animals underwent a surgery to induce MI. After screening echocardiography, 17 out of 33 animals were included in the study based on presence of a sizable infarct. One additional animal died during intubation prior to injection. Sixteen animals therefore underwent either polymer (n = 8) or saline (n = 8) injections. 100% of the animals survived the injection surgery; however, 3 animals were removed from functional analysis, 2 animals from the saline group and 1 from the polymer group, due to inconsistent gating during MR imaging. These animals were included in the histological analysis. One animal from the polymer group was excluded from the entire study due to insufficient polymer injection based on histology. In addition, infarct size was not different in the PEG (44.1%±2.8%) and saline (48.0%±2.8%) injected animals (p = 0.336).

Animals injected with the PEG hydrogel showed a significantly thicker infarct wall in comparison to saline injected animals (p = 0.006) as seen via histological analysis (Figure 2). Despite the increase in wall thickness, analysis of the MR images showed a steady decline in cardiac function and expansion of LV volume by comparing values pre- and post-treatment (Figure 3). At 7 weeks there was a significant increase in LV EDV (p = 0.0016) and ESV (p = 0.0008), as well as a decrease in EF (p = 0.006) in the polymer group. As expected in the saline control group, LV EDV

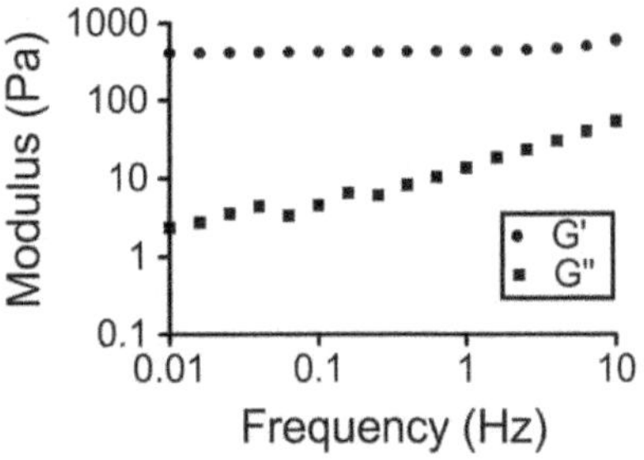

Figure 1. Mechanical spectra. Typical mechanical spectra for a hydrogel formed by crosslinking of 4-arm PEG-ASG (100 mg/ml) and trilysine (1 mg/ml).

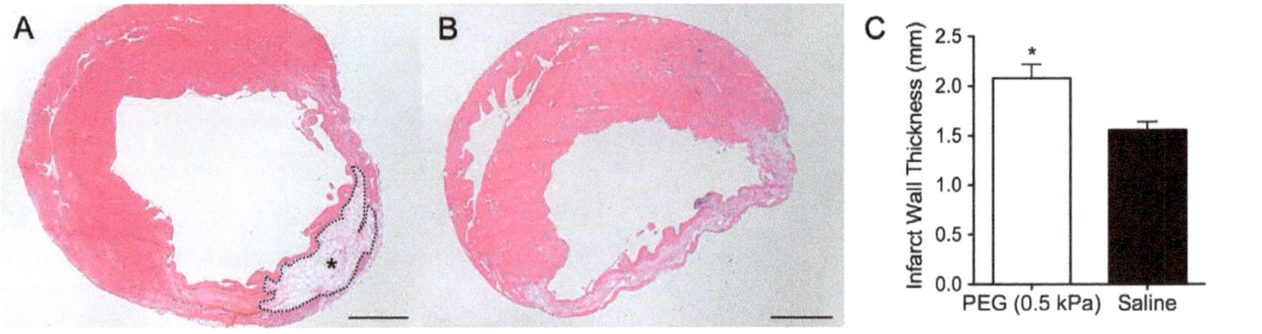

Figure 2. Histological assessment. Histological assessment of polymer and saline injections 7 weeks after MI. Representative slides stained with H&E: (A) PEG hydrogel (G′ = 0.5 kPa) injected in infarct region and (B) saline injected in infarct region. (C) Infarct wall thickness at 7 weeks post-MI. Wall thinning was prevented in the PEG injected group compared to the control. (*$p<0.01$, PEG: n = 7, saline: n = 8) Photomicrographs are taken at 10X and scale bars are 1 mm. PEG area outlined with dotted line and demarcated by an asterisk (*).

(p = 0.000002) and ESV (p = 0.0001) significantly expanded with a corresponding decline in EF (p = 0.005). The heart rate of the animals was continuously monitored during the imaging process and was statistically similar between the PEG and saline injected groups: 278±17 beats/min (PEG pre-injection) and 281±8 beats/min (PEG post-injection); 271±14 beats/min (saline pre-injection) and 285±8 beats/min (saline post-injection). The ECG was carefully monitored during the time of image acquisition. There were no noticeable arrhythmias.

As a 3-D visual representation of the overall effect of polymer injection, finite element models were generated using the endocardial and epicardial boundaries traced from the MR images in ED (Figure 4). The overall change in infarcted LV free wall thickness is seen pre- and post-injection (Figure 4, arrows). The hearts injected with the PEG hydrogel demonstrated a regional increase in wall thickness in the infarcted area while the saline injected hearts showed a thinning of the infarct region at 7 weeks. Dilation of ventricle is clearly depicted by enlargement of the LV lumen at 7 weeks as compared to baseline.

Cellular response to injection

As expected there was a very thin layer of encapsulation around the PEG injection region; however, there were no apparent differences in the inflammatory response (macrophages, fibroblasts, foreign body giant cells) between the polymer and saline injected animals in the infarct, border zone or remote myocardium as expected due to the bio-inert properties of PEG [27,28,29,30] (Figure 5A & 5B). To further confirm this assessment, we examined the tissue for presence of $CD68^+$ macrophages and confirmed that there was no accumulation of macrophages in PEG injected hearts (Figure 5C & 5D). Immunohistochemistry was performed on multiple sections through the hearts to assess the effect of polymer injection on potential neovascularization in the infarct area. There was no significant difference in arteriole density for vessel diameters ranging between 10 μm and 100 μm in the infarct region of the polymer (19±2 arterioles/mm^2) and saline (20±2 arterioles/mm^2) injected hearts, or when vessels were binned as <10 μm, between 10 μm and 25 μm, between 25 μm and 50 μm and between 50 μm and 100 μm, suggesting that polymer injection did not promote increased vessel formation (Figure 6).

Discussion

To date, the exact mechanism by which injectable materials lead to an improvement in cardiac function is unknown. As most injectable biomaterials cause an increase in infarct wall thickness, some have hypothesized that this passive structural enhancement alone may have therapeutic benefit [6,23]. After MI, there is typically a thinning of the myocardial wall and replacement with scar tissue. According to the Law of Laplace, this thinning can cause an increase in wall stress that can cause infarct expansion and apoptosis of the myocytes, which in turn can lead to dilation of LV and subsequent HF [4,23]. Thus, it has been proposed that use of an injectable material to increase wall thickness may lead to a decrease in wall stress and hence prevent adverse LV remodeling [23]. Alternatively, bioactivity or degradation properties of biomaterials may lead to the beneficial effects of these materials in the post-MI setting. The majority of biomaterials that have been examined to date are degradable materials that allow for cell infiltration, as well as biopolymers that have inherent bioactivity. In an effort to begin to understand the mechanisms behind which injectable polymers improve cardiac function, we chose to examine a bio-inert, non-degradable synthetic polymer, PEG, to decouple a material's structural effects from its bioactivity. We chose a compliant gel that mimicked the mechanical properties of commonly injected polymers. Results from this study indicate that injection of PEG leads to an increase in infarct wall thickness, consistent with what has previously been seen with other materials [7,8,14,15]. However, this was insufficient to prevent negative LV remodeling or improve cardiac function as determined through MRI. A strictly structural enhancement of the LV wall may therefore be insufficient for preservation or improvement of cardiac function and/or slowing the progression of negative LV remodeling after MI. These results therefore indicate that other properties of injectable biomaterials may be playing an important role in maintenance or enhancement of function post-MI.

In this study, we show that injection of the non-degradable PEG does not induce post-MI neovascularization, nor does it increase cell infiltration in general. In contrast, many previously injected materials that have shown improvement in cardiac function display increased neovascularization, possibly restoring the blood supply to the affected area and preventing negative LV remodeling [9,19,21]. The increase in new vessel formation could be due to bioactive components of the materials such as angiogenesis inducing degradation products (e.g. Fibrin fragment E [45] and peptide fragment hepIII of collagen IV [46]) Unlike PEG, most of these biomaterials also have integrin binding sites that are important for the influx of cells, such as endothelial cells and myofibroblasts [47]. Huang *et al* have shown that injection of collagen in the infarct induced significant myofibroblast infiltra-

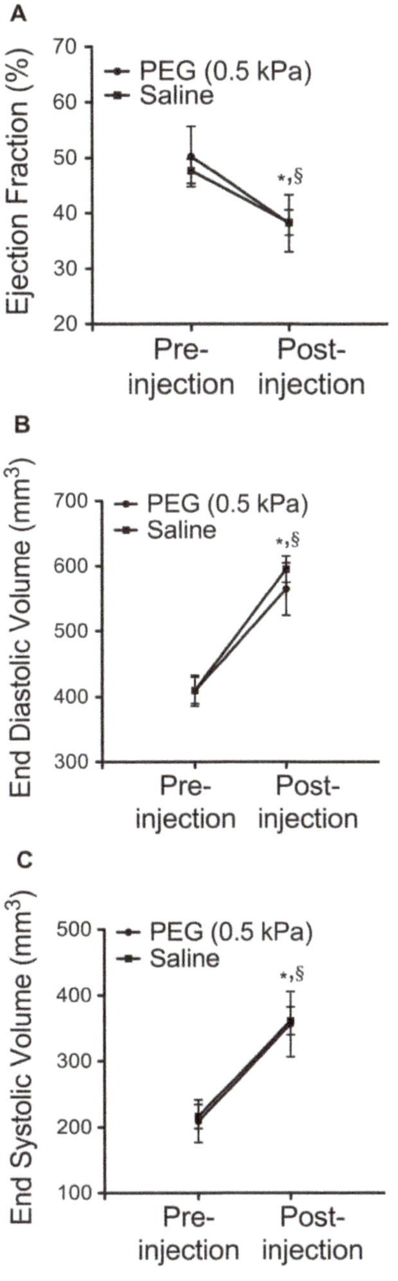

Figure 3. Cardiac function. Comparison of cardiac function baseline (1 week post-MI, pre-injection) and post-treatment (7 weeks post-MI) in PEG (0.5 kPa) injected and saline injected groups. (A) Ejection fraction. (B) End diastolic volume. (C) End systolic volume. Statistical significance determined using paired two-tailed *t*-test between pre and post-injection groups. (*$p<0.01$ PEG compared to baseline, § $p<0.01$ saline compared to baseline, PEG: n = 6, saline: n = 6).

tion. Similarly there was also a trend for increased myofibroblast migration in fibrin and matrigel injected hearts [19]. It has been shown that the contractile nature of myofibroblasts may lead to infarct size reduction and mitigation of infarct expansion, leading to overall improvement in function [48]. In addition to biopolymers, beneficial effects with a few degradable synthetic polymers that allow for cell influx have also been observed [13]. Therefore, cell infiltration due to degradation of injected polymers might be a potential mechanism by which these materials elicit

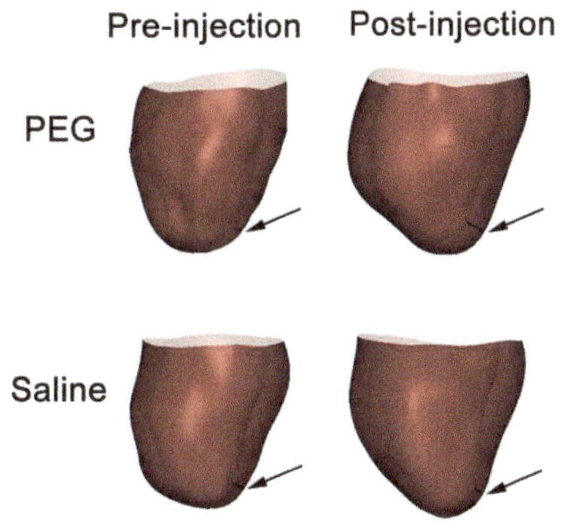

Figure 4. 3-D representation of the LV. Finite element representation of the LV pre and post-injection in both PEG and saline injected groups. Arrow denotes region of infarction. Infarct wall thickening is seen in the PEG injected heart while infarct wall thinning is depicted in the saline injected heart compared to baseline. Also note the dilation of LV lumen pre and post-injection in both groups.

salutatory effects. Though this effect with biopolymers is likely a result of their inherent bioactivity, these materials have also been shown to degrade in the range of 1–6 weeks [9,17,19]. While most of these infiltrating cells cannot restore function to the damaged area, it is possible that the migration of these cells in a timely manner provides beneficial paracrine effects necessary to prevent or lessen infarct scar expansion that is commonly seen post-MI.

While most materials have been degradable, there is one report of a synthetic non-degradable polymer being used for treatment of MI; Dobner *et al* [49], injected a non-degradable PEG in the infarct immediately post-MI and showed that there was temporary retardation of LV remodeling at early time points, but not at later time points. However, the progression of LV remodeling at later time points is aligned with the presented results. It is challenging to make a direct comparison between this study and the presented work as the material properties of the PEG gel were not characterized, and the PEG was injected at a different time point. An additional study has also examined the effects of mechanical properties of materials in an MI setting. Ifkovits *et al* [18], investigated the effect of mechanical modulation of injectable HA hydrogels on LV remodeling and cardiac function in an ovine MI model, finding a smaller infarct area in animals injected with the higher mechanical modulus HA. In that study, the HA spread more interstitially through the infarct, which could have contributed to the differences seen in the results of the presented work and the study performed by Ifkovits *et al.* However, similar to the results presented here, there was a reduction in ejection fraction in all groups compared to baseline. Furthermore, in a study by Garbern *et al*, a pH sensitive, temperature responsive synthetic polymer was injected with or without the addition of bFGF in a rodent infarct model. The control group injected with polymer alone increased infarct wall thickness, yet showed a steady decline in fractional shortening comparable to that of the saline injected control, indicating the polymer injection did not improve cardiac function [50]. While the polymer was designed to degrade, it did not degrade over the course of the study, thus providing additional evidence that degradation may be a key biomaterial parameter for preserving or improving cardiac function post-MI.

In the present study, MRI was the imaging modality of choice. It is well established that MRI is the gold standard for cardiac

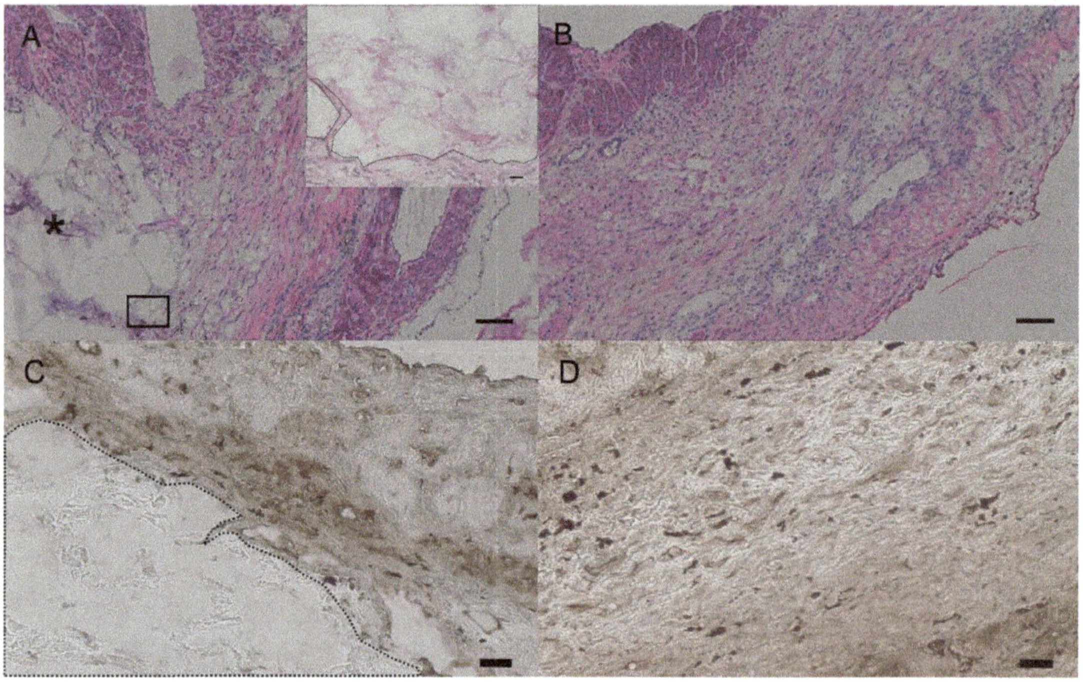

Figure 5. Inflammatory response. Inflammatory response in polymer and saline injected hearts 7 weeks after MI. (A) PEG injected heart (inset shows high magnification image of PEG region showing no cell infiltration in the polymer, scale bar 20 μm) (B) saline injected heart. Scale bars are 100 μm. PEG area is demarcated by an asterisk (*). CD68 stained macrophages in (C) PEG injected heart (D) saline injected heart. PEG area outlined with a dotted line. Scale bars are 20 μm.

imaging [51,52]. Commonly, echocardiography is used for the measurement of cardiac function after injection of a biomaterial. Measurements are made at the level of the papillary muscles that function as an anatomical marker. In the commonly used total occlusion MI model, the anteroapical portion of the heart is most severely affected. The effects of LV dilatation and systolic bulging are thought to be most pronounced in this region. It is possible that by taking measurements at one level the effect of LV dilatation distal to the papillary muscles is overlooked and local effects are being underemphasized. However, using MRI we were able to determine the overall effect of the polymer injection.

As elucidated here, there may be various factors that are potentially contributing to the beneficial effect seen with injectable materials post-MI. In order to better understand the role of biomaterials in MI treatment and compare our findings to previous work, the PEG polymer used in this study was specifically engineered to have properties similar to other previously examined injectable biomaterials. Moreover, we mimicked common study parameters such as the timing of injection and the injection volume. There are however some differences and limitations to our findings. For example, this PEG gels rapidly upon injection limiting the interstitial spread of the material, allowing it to serve as a bulk filler and partial wall replacement. One potential limitation may be the inability of the polymer bolus to distribute through the entire area of the infarct as seen with external LV restraints [53]. This inhomogeneity in spread could lead to regional abnormalities in function as well as allow for infarct expansion in regions where the polymer is not present. Moreover,

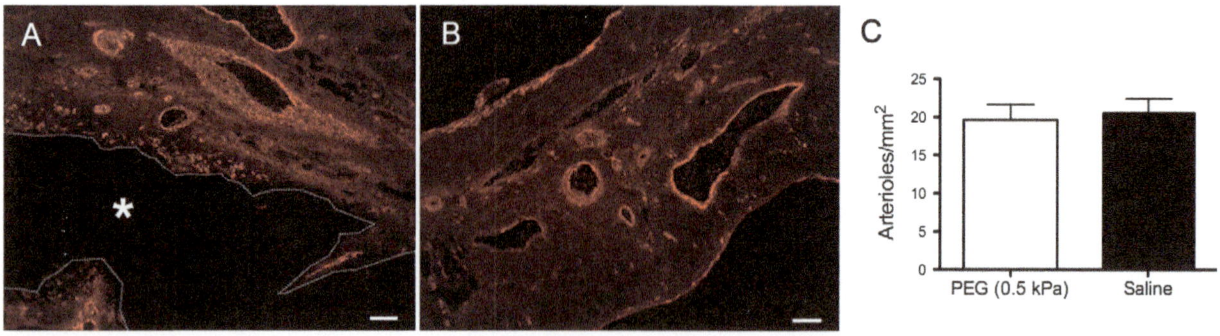

Figure 6. Arteriole density. Arteriole density assessment in polymer and saline injected hearts 7 weeks after MI. (A, B) Representative slides with staining for smooth muscle actin. (A) Arterioles in infarct region of PEG injected animals. PEG area outlined with dotted line and demarcated by an asterisk (*). (B) Arterioles in infarct region of saline injected animals. Scale bars are 100 μm. (C) Arteriole density (vessel diameter between 10 μm and 100 μm) at 7 weeks post-MI. ($p = 0.80$, PEG: $n = 6$, saline: $n = 8$).

there could also still be compensatory expansion of the non-infarcted LV that negates any structural support provided by the material injection. In larger animals, multiple injections can ensure more homogeneous spread of material, although 100% coverage of the wall would still not be achievable. Despite the single injection in the majority of rodent experiments, injection of biomaterials have still improved or preserved cardiac function. In addition to injection spread, timing of injection may also be an important parameter. While some may argue at one week post MI the initial expansion of the ventricle has already occurred, most previous studies injected material in the infarct approximately 1 week post-infarction [7,8,9,15,17,19,54]. Injection at this time point also represents greater clinical translatability due to fear of ventricular rupture at earlier time points. In the literature, volumes of 50 μL to 150 μL have been injected in rat hearts [7,8,9,15,17,19,49,54,55,56,57], hence 100 μL injection volume was selected as it falls in this range and is the volume used most frequently in biomaterial injection studies.

In conclusion, passive structural enhancement by injection of a synthetic, non-degradable polymer into a developing infarct did not improve global cardiac function and was insufficient to alter post-MI remodeling, thereby suggesting that the bioactivity and/or cell infiltration seen with degradable materials may likely play an important role in preserving cardiac function. The results of this study provide additional insight on the important and influencing properties of biomaterials necessary for preventing post-MI negative remodeling. Furthermore, this study highlights the need for further studies that closely examine biomaterial parameters in order to guide the design of enhanced biomaterials therapies for MI.

Acknowledgments

The authors would like to acknowledge the University of California San Diego Center for Functional MRI, and the Small Animal Imaging Core at the Children's Hospital, Los Angeles, for providing functional MRI capabilities. The authors would also like to thank Dr. Nissi Varki for histopathological analysis.

Author Contributions

Conceived and designed the experiments: AAR KLC JHO KLP. Performed the experiments: AAR DPH YG NDD AS. Analyzed the data: AAR JSC AS KLC JHO KLP. Contributed reagents/materials/analysis tools: KLC KLP. Wrote the paper: AAR. Edited manuscript: JSC AS DPH NDD YG KLP JHO KLC.

References

1. Sutton MG, Sharpe N (2000) Left ventricular remodeling after myocardial infarction: pathophysiology and therapy. Circulation 101: 2981–2988.
2. Lloyd-Jones (2010) Heart Disease and Stroke Statistics-2010 Update: A Report From the American Heart Association (vol 121, pg e46, 2010). Circulation 121: E260–E260.
3. Cleutjens JP, Kandala JC, Guarda E, Guntaka RV, Weber KT (1995) Regulation of collagen degradation in the rat myocardium after infarction. Journal of Molecular and Cellular Cardiology 27: 1281–1292.
4. Pfeffer MA (1995) Left ventricular remodeling after acute myocardial infarction. Annu Rev Med 46: 455–466.
5. Whittaker P, Boughner DR, Kloner RA (1991) Role of collagen in acute myocardial infarct expansion. Circulation 84: 2123–2134.
6. Christman KL, Lee RJ (2006) Biomaterials for the treatment of myocardial infarction. J Am Coll Cardiol 48: 907–913.
7. Dai W, Wold LE, Dow JS, Kloner RA (2005) Thickening of the infarcted wall by collagen injection improves left ventricular function in rats: a novel approach to preserve cardiac function after myocardial infarction. J Am Coll Cardiol 46: 714–719.
8. Christman KL, Fok HH, Sievers RE, Fang Q, Lee RJ (2004) Fibrin glue alone and skeletal myoblasts in a fibrin scaffold preserve cardiac function after myocardial infarction. Tissue Eng 10: 403–409.
9. Lu WN, Lü SH, Wang HB, Li DX, Duan CM, et al. (2009) Functional improvement of infarcted heart by co-injection of embryonic stem cells with temperature-responsive chitosan hydrogel. Tissue Eng Pt A 15: 1437–1447.
10. Okada M, Payne TR, Oshima H, Momoi N, Tobita K, et al. (2010) Differential efficacy of gels derived from small intestinal submucosa as an injectable biomaterial for myocardial infarct repair. Biomaterials. pp 7678–7683.
11. Zhang PC, Zhang H, Wang H, Wei YJ, Hu SS (2006) Artificial matrix helps neonatal cardiomyocytes restore injured myocardium in rats. Artif Organs. pp 86–93.
12. Jiang XJ, Wang T, Li XY, Wu DQ, Zheng ZB, et al. (2009) Injection of a novel synthetic hydrogel preserves left ventricle function after myocardial infarction. Journal of biomedical materials research Part A 90: 472–477.
13. Fujimoto KL, Ma Z, Nelson DM, Hashizume R, Guan J, et al. (2009) Synthesis, characterization and therapeutic efficacy of a biodegradable, thermoresponsive hydrogel designed for application in chronic infarcted myocardium. Biomaterials 30: 4357–4368.
14. Wang T, Wu DQ, Jiang XJ, Zhang XZ, Li XY, et al. (2009) Novel thermosensitive hydrogel injection inhibits post-infarct ventricle remodelling. Eur J Heart Fail 11: 14–19.
15. Landa N, Miller L, Feinberg MS, Holbova R, Shachar M, et al. (2008) Effect of injectable alginate implant on cardiac remodeling and function after recent and old infarcts in rat. Circulation 117: 1388–1396.
16. Leor J, Tuvia S, Guetta V, Manczur F, Castel D, et al. (2009) Intracoronary injection of in situ forming alginate hydrogel reverses left ventricular remodeling after myocardial infarction in Swine. J Am Coll Cardiol 54: 1014–1023.
17. Christman KL, Vardanian AJ, Fang Q, Sievers RE, Fok HH, et al. (2004) Injectable fibrin scaffold improves cell transplant survival, reduces infarct expansion, and induces neovasculature formation in ischemic myocardium. J Am Coll Cardiol 44: 654–660.
18. Ifkovits JL, Tous E, Minakawa M, Morita M, Robb JD, et al. (2010) Injectable hydrogel properties influence infarct expansion and extent of postinfarction left ventricular remodeling in an ovine model. Proc Natl Acad Sci USA 107: 11507–11512.
19. Huang NF, Yu J, Sievers R, Li S, Lee RJ (2005) Injectable biopolymers enhance angiogenesis after myocardial infarction. Tissue Eng 11: 1860–1866.
20. Yu J, Gu Y, Du KT, Mihardja S, Sievers RE, et al. (2009) The effect of injected RGD modified alginate on angiogenesis and left ventricular function in a chronic rat infarct model. Biomaterials 30: 751–756.
21. Davis ME, Motion JP, Narmoneva DA, Takahashi T, Hakuno D, et al. (2005) Injectable self-assembling peptide nanofibers create intramyocardial microenvironments for endothelial cells. Circulation 111: 442–450.
22. Jackson BM, Gorman JH, Salgo IS, Moainie SL, Plappert T, et al. (2003) Border zone geometry increases wall stress after myocardial infarction: contrast echocardiographic assessment. Am J Physiol Heart Circ Physiol 284: H475–479.
23. Wall ST, Walker JC, Healy KE, Ratcliffe MB, Guccione JM (2006) Theoretical impact of the injection of material into the myocardium: a finite element model simulation. Circulation 114: 2627–2635.
24. Singelyn JM, DeQuach JA, Seif-Naraghi SB, Littlefield RB, Schup-Magoffin PJ, et al. (2009) Naturally derived myocardial matrix as an injectable scaffold for cardiac tissue engineering. Biomaterials 30: 5409–5416.
25. Singelyn JM, Christman KL (2010) Injectable Materials for the Treatment of Myocardial Infarction and Heart Failure: The Promise of Decellularized Matrices. Journal of cardiovascular translational research 3: 478–486.
26. Nelson DM, Ma Z, Fujimoto KL, Hashizume R, Wagner WR (2011) Intra-myocardial biomaterial injection therapy in the treatment of heart failure: Materials, outcomes and challenges. Acta biomaterialia. pp 1–15.
27. Harris JM (1992) Introduction to Biotechnical and Biomedical Applications of Poly (Ethylene Glycol). In: Harris JM, ed. Poly (Ethylene Glycol) Chemistry. New York & London: Plenum Press. 3 p.
28. Lin-Gibson S, Bencherif S, Cooper JA, Wetzel SJ, Antonucci JM, et al. (2004) Synthesis and characterization of PEG dimethacrylates and their hydrogels. Biomacromolecules. pp 1280–1287.
29. Uhrich KE, Cannizzaro SM, Langer RS, Shakesheff KM (1999) Polymeric systems for controlled drug release. Chem Rev. pp 3181–3198.
30. Ali Khademhosseini JB, Mehmet Toner, Shuichi Takayama (2008) Micro- and Nanoengineering of the Cell Microenvironment, Technologies and Applications (Engineering in Medicine & Biology). Boston and London Artech House.
31. Pathak, Chandrashekhar. "Biocompatible Crosslinked Polymers." Patent 6,566,406. 20 May 2003.
32. Lutolf MP, Hubbell JA (2003) Synthesis and physicochemical characterization of end-linked poly(ethylene glycol)-co-peptide hydrogels formed by Michael-type addition. Biomacromolecules 4: 713–722.
33. Anumolu SS, DeSantis AS, Menjoge AR, Hahn RA, Beloni JA, et al. (2010) Doxycycline loaded poly(ethylene glycol) hydrogels for healing vesicant-induced ocular wounds. Biomaterials 31: 964–974.
34. Iwanaga Y, Hoshijima M, Gu Y, Iwatate M, Dieterle T, et al. (2004) Chronic phospholamban inhibition prevents progressive cardiac dysfunction and pathological remodeling after infarction in rats. J Clin Invest 113: 727–736.

35. Chuang JS, Zemljic-Harpf A, Ross RS, Frank LR, McCulloch AD, et al. (2010) Determination of three-dimensional ventricular strain distributions in gene-targeted mice using tagged MRI. Magn Reson Med 64: 1281–1288.
36. Wang L, Deng J, Tian W, Xiang B, Yang T, et al. (2009) Adipose-derived stem cells are an effective cell candidate for treatment of heart failure: an MR imaging study of rat hearts. Am J Physiol-Heart C. pp H1020–H1031.
37. Allman AJ, McPherson TB, Badylak SF, Merrill LC, Kallakury B, et al. (2001) Xenogeneic extracellular matrix grafts elicit a Th2-restricted immune response. Transplantation. pp 1631–1640.
38. Badylak S, Kokini K, Tullius B, Simmons-Byrd A, Morff R (2002) Morphologic study of small intestinal submucosa as a body wall repair device. J Surg Res. pp 190–202.
39. Valentin JE, Badylak JS, McCabe GP, Badylak SF (2006) Extracellular matrix bioscaffolds for orthopaedic applications - A comparative histologic study. J Bone Joint Surg Am. pp 2673–2686.
40. Badylak SF, Valentin JE, Ravindra AK, McCabe GP, Stewart-Akers AM (2008) Macrophage phenotype as a determinant of biologic scaffold remodeling. Tissue Eng Pt A 14: 1835–1842.
41. Pfeffer JM, Pfeffer MA, Fletcher PJ, Braunwald E (1991) Progressive ventricular remodeling in rat with myocardial infarction. Am J Physiol 260: H1406–1414.
42. Draget KI, Stokke BT, Yuguchi Y, Urakawa H, Kajiwara K (2003) Small-angle X-ray scattering and rheological characterization of alginate gels. 3. Alginic acid gels. Biomacromolecules 4: 1661–1668.
43. Wu DQ, Qiu F, Wang T, Jiang XJ, Zhang XZ, et al. (2009) Toward the development of partially biodegradable and injectable thermoresponsive hydrogels for potential biomedical applications. ACS Appl Mater Interfaces 1: 319–327.
44. Rizzi SC, Hubbell JA (2005) Recombinant protein-co-PEG networks as cell-adhesive and proteolytically degradable hydrogel matrixes. Part I: Development and physicochemical characteristics. Biomacromolecules 6: 1226–1238.
45. Thompson WD, Smith EB, Stirk CM, Marshall FI, Stout AJ, et al. (1992) Angiogenic activity of fibrin degradation products is located in fibrin fragment E. J Pathol 168: 47–53.
46. Mihardja SS, Gao DW, Sievers RE, Fang QZ, Feng JJ, et al. (2010) Targeted In Vivo Extracellular Matrix Formation Promotes Neovascularization in a Rodent Model of Myocardial Infarction. Plos One 5.
47. Nam JO, Kim JE, Jeong HW, Lee SJ, Lee BH, et al. (2003) Identification of the alphavbeta3 integrin-interacting motif of betaig-h3 and its anti-angiogenic effect. J Biol Chem 278: 25902–25909.
48. Jugdutt BI (2003) Ventricular remodeling after infarction and the extracellular collagen matrix: when is enough enough? Circulation 108: 1395–1403.
49. Dobner S, Bezuidenhout D, Govender P, Zilla P, Davies N (2009) A synthetic non-degradable polyethylene glycol hydrogel retards adverse post-infarct left ventricular remodeling. J Card Fail 15: 629–636.
50. Garbern JC, Minami E, Stayton PS, Murry CE (2011) Delivery of basic fibroblast growth factor with a pH-responsive, injectable hydrogel to improve angiogenesis in infarcted myocardium. Biomaterials 32: 2407–2416.
51. Vick GW (2009) The gold standard for noninvasive imaging in coronary heart disease: magnetic resonance imaging. Curr Opin Cardiol 24: 567–579.
52. Karamitsos TD, Francis JM, Myerson S, Selvanayagam JB, Neubauer S (2009) The role of cardiovascular magnetic resonance imaging in heart failure. J Am Coll Cardiol 54: 1407–1424.
53. Blom AS, Mukherjee R, Pilla JJ, Lowry AS, Yarbrough WM, et al. (2005) Cardiac support device modifies left ventricular geometry and myocardial structure after myocardial infarction. Circulation 112: 1274–1283.
54. Hao X, Silva EA, Månsson-Broberg A, Grinnemo KH, Siddiqui AJ, et al. (2007) Angiogenic effects of sequential release of VEGF-A165 and PDGF-BB with alginate hydrogels after myocardial infarction. Cardiovasc Res 75: 178–185.
55. Wang H, Zhang X, Li Y, Ma Y, Zhang Y, et al. (2010) Improved myocardial performance in infarcted rat heart by co-injection of basic fibroblast growth factor with temperature-responsive Chitosan hydrogel. J Heart Lung Transpl. pp 881–887.
56. Hsieh PC, Davis ME, Gannon J, MacGillivray C, Lee RT (2006) Controlled delivery of PDGF-BB for myocardial protection using injectable self-assembling peptide nanofibers. J Clin Invest 116: 237–248.
57. Ryu JH, Kim IK, Cho SW, Cho MC, Hwang KK, et al. (2005) Implantation of bone marrow mononuclear cells using injectable fibrin matrix enhances neovascularization in infarcted myocardium. Biomaterials 26: 319–326.

7

Determining a Clinically Relevant Strategy for Bone Tissue Engineering: An "All-in-One" Study in Nude Mice

Pierre Corre[1,2,5*¶], Christophe Merceron[1,5¶], Caroline Vignes[1,5], Sophie Sourice[1,5], Martial Masson[1,5], Nicolas Durand[1,3,5], Florent Espitalier[1,3,5], Paul Pilet[1,5], Thomas Cordonnier[1,5], Jacques Mercier[2,5], Séverine Remy[4], Ignacio Anegon[4], Pierre Weiss[1,5¶], Jérôme Guicheux[1,5¶]

1 INSERM (Institut National de la Santé et de la Recherche Médicale), UMR (Unité Mixte de Recherche) 791, center for osteoarticular and dental tissue engineering, Université de Nantes, Nantes, France, **2** Centre Hospitalier Universitaire de Nantes, Clinique de Stomatologie et de Chirurgie maxillo-faciale, Nantes, France, **3** Centre Hospitalier Universitaire de Nantes, Clinique d'Oto-Rhino-Laryngologie et de Chirurgie cervico-faciale, Nantes, France, **4** INSERM, UMR 1064, Centre pour la recherche en transplantation et immunologie et Plate-forme Transgenic Rats Nantes, Institut de Transplantation Urologie-Néphrologie (ITUN), Nantes, France, **5** Centre Hospitalier Universitaire de Nantes, Pôle Hospitalo-Universitaire 4, Nantes, France

Abstract

Purpose: Autologous bone grafting (BG) remains the standard reconstruction strategy for large craniofacial defects. Calcium phosphate (CaP) biomaterials, such as biphasic calcium phosphate (BCP), do not yield consistent results when used alone and must then be combined with cells through bone tissue engineering (BTE). In this context, total bone marrow (TBM) and bone marrow-derived mesenchymal stem cells (MSC) are the primary sources of cellular material used with biomaterials. However, several other BTE strategies exist, including the use of growth factors, various scaffolds, and MSC isolated from different tissues. Thus, clinicians might be unsure as to which method offers patients the most benefit. For this reason, the aim of this study was to compare eight clinically relevant BTE methods in an "all-in-one" study.

Methods: We used a transgenic rat strain expressing green fluorescent protein (GFP), from which BG, TBM, and MSC were harvested. Progenitor cells were then mixed with CaP materials and implanted subcutaneously into nude mice. After eight weeks, bone formation was evaluated by histology and scanning electron microscopy, and GFP-expressing cells were tracked with photon fluorescence microscopy.

Results/Conclusions: Bone formation was observed in only four groups. These included CaP materials mixed with BG or TBM, in which abundant *de novo* bone was formed, and BCP mixed with committed cells grown in two- and three-dimensions, which yielded limited bone formation. Fluorescence microscopy revealed that only the TBM and BG groups were positive for GFP expressing-cells, suggesting that these donor cells were still present in the host and contributed to the formation of bone. Since the TBM-based procedure does not require bone harvest or cell culture techniques, but provides abundant *de novo* bone formation, we recommend consideration of this strategy for clinical applications.

Editor: Christophe Egles, Université de Technologie de Compiègne, France

Funding: The authors have no support or funding to report.

Competing Interests: Jérome Guicheux is a PLOS ONE Editorial Board member. This does not alter the authors' adherence to all the PLOS ONE policies on sharing data and materials.

* E-mail: pierre.corre@chu-nantes.fr

¶ These authors contributed equally to this work.

Introduction

Bone reconstruction might be required in patients with craniofacial defects (e.g., craniofacial trauma or malformations) or following resection of oral carcinomas. Autologous bone grafting (BG) is considered the gold standard in bone reconstruction [1]. However, the harvesting procedure requires a second surgical site where potential complications have been reported [2,3]. Moreover, allografts and xenografts are sometimes problematic as they are limited in supply and at risk for cross-contamination [4,5].

Calcium phosphate (CaP) biomaterials, and particularly biphasic calcium phosphate (BCP), have shown clinical efficacy in orthopedic surgery [6,7], avoiding second surgical site morbidity and reducing surgery time compared to autologous grafting [7,8]. However, although CaP biomaterials contribute to bone healing through osteoconduction, they generally lack osteoinductivity for regenerating large bone tissue defects, or in tissues exposed to sources of infection (i.e., oronasal mucosa), because they have little contact with bone. Therefore, clinical applications must be restricted to small bone defects or to regions with significant bone contact [9]. In view of these limitations, surgeons and researchers have focused on developing alternative therapies to BG over the past fifteen years. These mainly included combining osteoprogenitor cells with bone substitutes to improve their osteogenic properties. Thus, extemporaneous mixtures of CaP biomaterials and unprocessed total fresh bone marrow (TBM) have shown osteogenic potential *in vivo*. Indeed, in rat models, this technique successfully induced bone formation in extraosseous sites [10,11] and potentiated bone ingrowth in osseous sites [12,13]. Moreover, the efficacy of CaP biomaterials (i.e., granules, sticks, or blocks) combined with autologous bone marrow or autologous cancellous bone has been shown in clinical studies for defect filling in various orthopedic procedures [14].

More recently, the use of mesenchymal stem cells (MSC) isolated from bone marrow and combined with various scaffolds has emerged as a potential new treatment strategy. Indeed, this bone tissue engineering (BTE) technique showed promising results in small animal models [15,16], as well as in orthotopic and ectopic clinically sized implants [17,18]. However, the efficacy of BTE in humans has not yet been clinically validated [19–21]. Also, to improve the ability of engineered bone to replace autologous BG, new BTE strategies have been developed in recent years, including three-dimensional cell culture in bioreactors [22] and gene therapy. However, preclinical studies aimed at comparing the relative efficiency of these strategies have not been performed to date, and it remains difficult for clinicians to determine which method will offer patients the most benefit. Also, a major concern related to procedures involving MSC is the fate of cells after they enter the host via massive implantation. It remains unknown whether these MSC remain at the grafted site and are specifically responsible for new bone formation, as suggested by some authors [23,24], or if they die early after implantation and release mediators for cell recruitment from the local environment, as suggested in more recent studies [25–27].

Therefore, the aim of the present study was to address some of the uncertainties regarding the clinical use of BTE. Specifically, in an "all-in-one" study, we compared the ectopic bone formation potential of eight well-known bone repair strategies using BCP granules and progenitor cells. Moreover, we assessed if complex BTE strategies were more efficient than *in vivo* BTE procedures, such as TBM combined with CaP biomaterials. In addition, green fluorescent protein (GFP) tracking was used to determine the fate of grafted cells after implantation [28].

Materials and Methods

All procedures involving animals were conducted in accordance with the institutional guidelines of the French Ethical Committee (CEEA.PdL.06): their housing in the Experimental Therapeutic Unit at the Faculty of Medicine of Nantes (France), animal care, the method by which they were anesthetized and sacrificed, and all experimental protocols. All efforts were made to minimize suffering.

The European Community guidelines for the care and use of laboratory animals (DE 86/609/CEE, modified DE 2003/65/CE) have been revised by the new European directive 2010 (DE 2010/63/UE modified 22/09/2010). The translation of the EU regulation into the French one is effective from January 1st 2013. Submission of the project to the new Ethical Committee of the "Pays de la Loire" was not mandatory until January 1st 2013. Nevertheless, all the experiments were conducted prior to January 2013 accordingly to this new regulation.

The Veterinary Service (Center for Preclinical Research and Investigation of the ONIRIS Nantes-Atlantic College of Veterinary Medicine, Food Science and Engineering) approved the senior researcher in charge of the experiments.

Materials

Plastic ware for cell culture was purchased from Corning (Schipol-Rijk, the Netherlands). Sodium L-ascorbate, vitamin D3, dexamethasone, Alizarin Red S, ITS media supplement, 3-isobutyl-1-methylxanthine (IBMX), indomethacin, Oil Red O, and trypan blue were purchased from Sigma-Aldrich (St. Louis, MO). Alpha minimum essential medium (αMEM), phosphate-buffered saline (PBS), penicillin-streptomycin, trypsin-EDTA

Table 1. Gene names and abbreviations, gene bank accession numbers, sequences of primer pairs (Fwd: Forward, Rev: Reverse) and length of PCR products used for real-time RT-PCR analysis.

Gene	Gene Bank Accession Number	Sequence	Base Pairs (bp)
β-Actin	031144.2	Fwd 5′- CCCGCGAGTACAACCTTCT –3′	72
(*Actb*)		Rev 5′- CGTCATCCATGGCGAAC –3′	
Alkaline phosphatase,	013059.1	Fwd 5′- GCACAACATCAAGGACATCG –3′	72
liver/bone/kidney		Rev 5′- TCAGTTCTGTTCTTGGGGTACAT –3′	
(*Alpl*)			
Bone gamma-	013414.1	Fwd 5′- ATAGACTCCGGCGCTACCTC –3′	63
carboxyglutamate (gla)		Rev 5′- CCAGGGGATCTGGGTAGG –3′	
protein (= Osteocalcin)			
(*Bglap*)			
runt-related transcription	053470.2	Fwd 5′- CACAGAGCTATTAAAGTGACAGTGG -3′	86
factor 2		Rev 5′- AACAAACTAGGTTTAGAGTCATCAAGC – 3′	
(*Runx2*)			
Peroxisome proliferator-	013124.2	Fwd 5′- GGTGAAACTCTGGGAGATCCT - 3′	110
activated receptor gamma		Rev 5′- AATGGCATCTCTGTGTCAACC –3′	
(*Pparγ*)			
Leptin	013076.2	Fwd 5′- CCAGGATCAATGACATTTCACA –3′	71
(*Lep*)		Rev 5′- AATGAAGTCCAAACCGGTGA –3′	

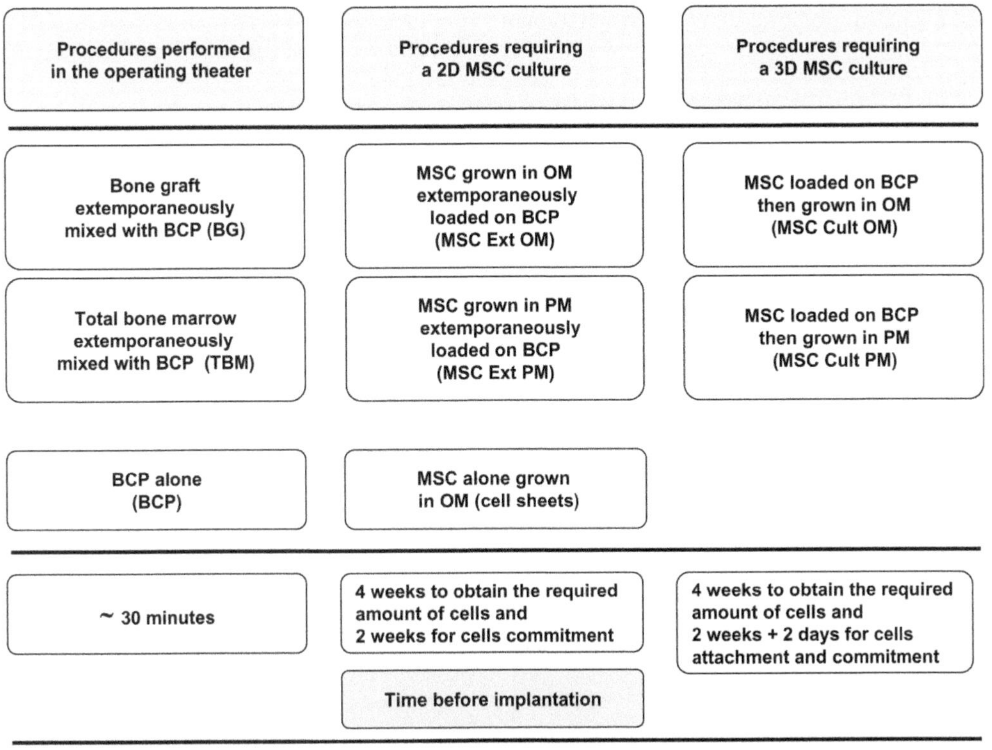

Figure 1. Outline of the protocol illustrating the different procedures and the required time before implantation for each condition. OM: osteogenic medium, PM: proliferative medium, 2D: two-dimensional, 3D: three-dimensional, MSC: mesenchymal stem cells, BCP: biphasic calcium phosphate, Ext: extemporaneously, Cult: cultured.

(0.05%–0.53 mM), L-glutamine, Trizol® reagent, and the superscript III kit were obtained from Invitrogen (Paisley, UK). Fetal calf serum (FCS) was provided by Dominique Dutscher (Brumath, France). β-glycerophosphate was purchased from Calbiochem (Darmstadt, Germany). Brilliant SYBR® Green Master Mix was obtained from Stratagene Europe (Amsterdam Zuidoost, the Netherlands). Polymerase chain reaction (PCR) primers were synthesized by MWG Biotech (Ebersberg, Germany), and Turbo DNase was purchased from Ambion Inc. (distributed by Applied Biosystems; Courtaboeuf, France). All other chemicals were obtained from standard laboratory suppliers and were of the highest purity available.

CaP biomaterials

Biphasic CaP particles made of hydroxyapatite (60%) and beta-tricalcium phosphate (40%) with a size between 500 and 1000 μm (MBCP™) were provided by Biomatlante (Vigneux de Bretagne, France). Tubes containing each 0.07 g of granules were double-packed and sterilized at 121°C for 20 min in an autoclave.

Animals

Twenty-four female nude mice (S/SOPF SWISS; six weeks old) were obtained from a certified breeding center (C. River, l'Arbresle, France) and acclimatized for two weeks to the conditions of the local vivarium (24°C and a 12/12 h light/dark cycle).

In addition, three adult, male, GFP-transgenic rats, previously described [28], were obtained from the transgenic rat facility (INSERM, UMRS 1064, Nantes, France) and specially designated as bone marrow (BM), BG, and MSC donors. Briefly, these GFP-transgenic rats are generated by a lentiviral vector expressing GFP driven by the ubiquitous PGK promoter. These rats strongly express GFP in multiple tissues and in important cell types, especially in mature neurons, in leukocytes subtypes including myeloid and plasmacytoid dendritic cell and in bone marrow where nearly 80% of MSC are GFP-positive.

Bone marrow harvesting

BM was harvested after sacrifice. Animals were anesthetized using in inhaled isoflurane (Forene®, Abott, Rungis, France) and sacrificed by an intracardiac overdose of sodium thiopental (Nesdonal®, Rhône-Merieux, France). Each femur or tibia was flushed with 1 mL of saline. The same rinse (1 mL) was used for all bones (i.e. 6 femurs and 6 tibias) in order to obtain the most concentrated TBM and to minimize the risk of inter-individual variation of BM components. The BM obtained from the 3 donors was immediately and aseptically transferred to heparinized tubes (Venoject II, Terumo Europe, Louvain, Belgium). Then about 300 μL have been used for *in vivo* implantation, corresponding to 30–50 μl per one construct (70 mg of BCP). The remained 700 μL were used for cytological analysis (~200 μL), performed as previously described [29] and for BM culture (~500 μL).

Cancellous bone harvesting

Cancellous bone was harvested with a dental curette from the epiphysis of bone previously cut for BM harvesting. About 30–40 mg of cancellous bone were harvested from each bone (femur or tibia) allowing a combination of ~70 mg of bone roughly crushed with 70 mg BCP granules per construct in a 1:1 weight ratio. The bone was immediately mixed with BCP and implanted subcutaneously in the following 30 minutes.

Table 2. Cells and material combination related to the experimental groups.

Experimental condition	Mass of material per implant (volume)	Density of seeded MSC per mg of material (or per cm^2 of material)	Mass of BG per implant	Volume of TBM per implant
BG	70 mg (~100 µl)	-	~70 mg (~100 µl)	-
BCP	70 mg (~100 µl)	-	-	-
TBM	70 mg (~100 µl)	-	-	~30–50 µl
MSC Ext OM	70 mg (~100 µl)	~$7.5.10^3$ cells/mg (~9.10^4 cells/cm^2)	-	-
MSC Ext PM	70 mg (~100 µl)	~$7.5.10^3$ cells/mg (~9.10^4 cells/cm^2)	-	-
MSC Cult OM	70 mg (~100 µl)	~$7.5.10^3$ cells/mg (~9.10^4 cells/cm^2)	-	-
MSC Cult PM	70 mg (~100 µl)	~$7.5.10^3$ cells/mg (~9.10^4 cells/cm^2)	-	-
Cell sheets	-	~5.10^4 cells/cm^2 of culture dish	-	-

MSC = mesenchymal stem cells, BG = bone graft, TBM = Total bone marrow, MSC Ext OM/MSC Ext PM = Mesenchymal stem cells grown in osteogenic medium/ proliferative medium then extemporaneously loaded onto BCP granules. MSC Cult OM/MSC Cult PM = MSC loaded onto BCP granules then grown in osteogenic/ proliferative medium. BCP = biphasic calcium phosphate alone.

Isolation and expansion of bone marrow-derived mesenchymal stem cells

A portion of the total harvested BM volume was filtered through a 70 µm nylon mesh filter. MSC were then harvested from BM donors and expanded *in vitro* for about six weeks prior to implantation. The BM was seeded in a 75 cm^2 treated polystyrene culture and MSC were isolated based on their adherence capacity after two days. Cells were then grown in proliferative medium, consisting of αMEM supplemented with 1% L-glutamine, 1% penicillin/streptomycin, and 10% FCS, and incubated at 37°C in a humidified atmosphere (95% air, 5% CO_2). The medium was renewed twice weekly until cells were 80–90% confluent. Cells were then detached enzymatically from plastic by an incubation of 3–4 min with 0.25% trypsin/EDTA and counted with a Malassez hemocytometer using trypan blue exclusion dye. To obtain a large number of cells, MSC were further expanded in treated polystyrene culture flasks. All cells used in experiments were between passages 2 and 6.

Osteogenic differentiation

Culture. For *in vitro* osteogenic differentiation of MSC, cells were seeded at a density of 1.10^4 cells/cm^2 in six-well plates and grown in the presence of proliferative or osteogenic medium (OM) for 14 and 28 days as described previously [30]. OM was composed of the proliferative medium supplemented with 10 mM β-glycerophosphate, 50 µM sodium L-ascorbate, and 10 nM vitamin D3. MSC were maintained at 37°C in a humidified atmosphere (5% CO_2 and 95% air), and the media was changed every 2–3 days.

Table 3. Semi-quantitative histological scoring system.

Score	Description
0	No bone
1	1–25% of available pore space filled with bone
2	26–50% of available pore space filled with bone
3	51–75% of available pore space filled with bone
4	76–100% of available pore space filled with bone

Calcium deposition. Calcium deposition was detected at 14 and 28 days by Alizarin Red S staining as described previously [31]. Briefly, MSC were grown as described above and washed with cold PBS followed by staining with 2% Alizarin Red S solution for 2 min. Stained cells were then extensively washed with deionized water to remove any nonspecific precipitates. Stained layers were visualized with phase microscopy using an inverted microscope (Nikon Eclipse TE 2000 E, Badhoevedorp, the Netherlands). Positive red staining indicated the deposition of a calcified matrix on the differentiated cells. Osteogenic differentiation was further characterized through the relative expression of osteogenic marker genes (*Alpl*, *Bglap*, and *Runx2*; see real-time PCR section below).

Adipogenic differentiation

Culture. For *in vitro* adipogenic differentiation of MSC, cells were seeded at a density of 3.10^4 cells/cm^2 in six-well plates and grown in the presence of control or adipogenic medium for 14 days as previously described [30]. Adipogenic medium was composed of the control medium supplemented with 1 µM dexamethasone, 200 µM indomethacin, 0.5 mM IBMX, and ITS (10 µg/mL insulin; 10 µg/mL transferrin; 10 ng/mL sodium selenite). MSC were maintained at 37°C in a humidified atmosphere (5% CO_2 and 95% air), and media was changed every 2–3 days.

Detection of neutral lipid droplets. Adipogenesis of BM-MSC was assessed by Oil Red O staining for lipid droplet detection. BM-MSC were grown as described above, washed with ice-cold PBS, and then fixed with 10% formalin for 5 min (followed by 1 h with fresh formalin at room temperature). The formalin was then discarded, and the wells were rinsed with 60% isopropanol. The wells were left to completely dry, and then 0.35% Oil Red O solution in 60% isopropanol was added for 10 min. Stained cells were extensively washed with deionized water to remove any nonspecific staining. Samples were visualized using a light microscope (Zeiss Axioplan 2, Oberkochen, Germany), and red staining indicated the presence of neutral lipid droplets. Adipogenic differentiation was also characterized through the relative expression of adipogenic marker genes (including *Pparγ* and *Lep*; see real-time PCR section below).

Real-time polymerization chain reaction. For real-time PCR analysis, total RNA was extracted using TRIzol® according to the manufacturer's instructions. After DNase I digestion, RNA was quantified with a UV-spectrophotometer (Nanodrop ND-1000, Labtech, Palaiseau, France) and quality was determined with the Agilent Bioanalyzer 2100 system (Waldbronn, Germany). A total of 500 ng of RNA per sample were reverse transcribed using the superscript III kit in a total volume of 30 μL. Complementary DNA (cDNA) was amplified in a 25 μL (total volume) PCR reaction mix containing 12.5 μL of Brilliant SYBR Green Master Mix® and 30 nM of SYBR green reference dye. The sequences of each primer set are provided in Table 1. The real-time PCR was carried out in a MX3000P real-time PCR system (Stratagene) under the following conditions: 10 min at 95°C followed by 40 successive cycles of 30 s at 95°C, 1 min at 60°C, and 30 s at 72°C. The efficiency and specificity of each primer set were confirmed with standard curves of cycle threshold (Ct) values vs. serial dilution of total RNA and melting profile evaluation. Ct values were normalized to β-actin as a housekeeping gene to control for differences in cDNA quantification. Results were reported as relative expression levels.

Cell culture on biphasic calcium phosphate particles

Cells were seeded on BCP granules and grown for two weeks as previously described [32]. Briefly, to minimize ion release from the BCP particles during cell culture, particles were incubated 48 h in twice-refreshed control medium. Prior to culture, MSC were loaded on the BCP particles at a density of 9.10^4 cells/cm^2 of BCP, corresponding to $7.5.10^3$ cells/mg of BCP as previously described [33]. To favor cell adhesion on ceramic particles, the culture was performed in low attachment 24-well plates. Cells were then grown during 14 days in osteogenic or proliferative medium. The culture media were refreshed every 2–3 days. Osteogenic differentiation was characterized through the relative gene expression of osteogenic marker genes (*Alpl*, *Bglap*, and *Runx2*).

Cell attachment, viability and osteogenic differentiation onto the CaP biomaterials

For the MSC grown on BCP granules in OM and PM, cell attachment was analyzed at day 14 using a methylene blue staining (Certistain, Merck, Darmstadt, Germany) and scanning electron microscopy (SEM). For methylene blue staining, samples were evaluated by light microscopy. For SEM, samples were dehydrated with graded ethanol, fixed with trichlorotrifluoroethane, and then sputtered with a thin layer of gold-palladium (EM Scope, England). Three samples from each group (osteogenic or control medium) were analyzed by SEM (LEO VP1450, Zeiss, Oberkochen, Germany). Cell viability at Day 14 was assessed by tracking GFP-expressing cells on the biomaterial surface using a Nikon Eclipse TE 2000 E microscope (Badhoevedorp, the Netherlands).

Implant preparation

Eight groups of implants were prepared (Fig. 1), thus reproducing eight bone repair procedures that previously demonstrated *in vivo* osteogenic properties in animal models or humans [14,34–39]:

1. BG extemporaneously combined with BCP granules (BG).
2. CaP granules alone (BCP).
3. Total bone marrow graft extemporaneously combined with BCP granules (TBM).
4. MSC differentiated in OM extemporaneously loaded on BCP granules (MSC Ext OM).
5. MSC grown in proliferative medium extemporaneously loaded on BCP (MSC Ext PM).
6. MSC grown on BCP granules in OM (MSC Cult OM).
7. MSC grown on BCP granules in proliferative medium (MSC Cult PM).
8. Osteoprogenitor cell sheet transplantation using hyperconfluent cells (seeded at 5.10^4 cells/cm^2 and grown during 14 days in OM) (Cell sheets).

For extemporaneous mixing procedures (BG, TBM, MSC Ext OM, and MSC Ext PM), cell preparations were allowed to attach to the ceramics for at least 30 min at room temperature prior to implantation. For MSC Ext OM and MSC Ext PM conditions, MSC were loaded on BCP particles at a density of 9.10^4 cells/cm^2 of BCP (~$7.5.10^3$ cells/mg) as previously described [33]. An overview of the combination of cells and scaffold is provided in table 2.

Surgical procedure

All surgical procedures were performed under general anesthesia (4% isoflurane inhalation for induction and 2% for preservation), and the samples were implanted subcutaneously in the backs of 24 nude mice. Briefly, a 1-cm transverse incision was made bilaterally along the dorsum of the back. Blunt dissection was performed to separate the skin from the subcutaneous connective tissues and to form several pockets under the skin into which implants were inserted. Each mouse received two randomly assigned implants (n = 6). Immediate postoperative analgesia was provided through subcutaneous injection of buprenorphine hydrochloride (Buprecare® 0.3 mg/mL, 10 μg/kg, ANIMALCARE, Dunnington, UK), and maintained for 2 days. Implants were removed right after sacrifice via CO_2 overdose, eight weeks after implantation.

Scanning electron microscopy

The explanted specimens were fixed for 24 h in a 4% paraformaldehyde in phosphate-buffered saline, and then dehydrated through a graded series of ethanol treatments. Non-decalcified bone specimens were infiltrated and embedded in glycol-methyl-methacrylate (GMMA) obtained by mixing methyl methacrylate (Prolabo, Paris, France), polyethylene glycol 400 (Prolabo, Paris, France), benzoyl peroxide (Merck, Darmstadt, Germany). Polymerization was started with N-N-dimethyl-aniline (Sigma Aldrich, Saint-Quentin Fallavier, France) at 4°C for 72 h.

Table 4. Cytological analysis of the bone marrow. Results are expressed as a percentage of cells per lineage.

Granulocytic lineages		Erythroblastic	Megacaryocytic	Lymphocyte	Plasmocyte	Monocyte	Blasts
Myeloblastic	Myelocytic	lineages	lineage				
1.5	26	32.5	Physiological	35	0	1	4

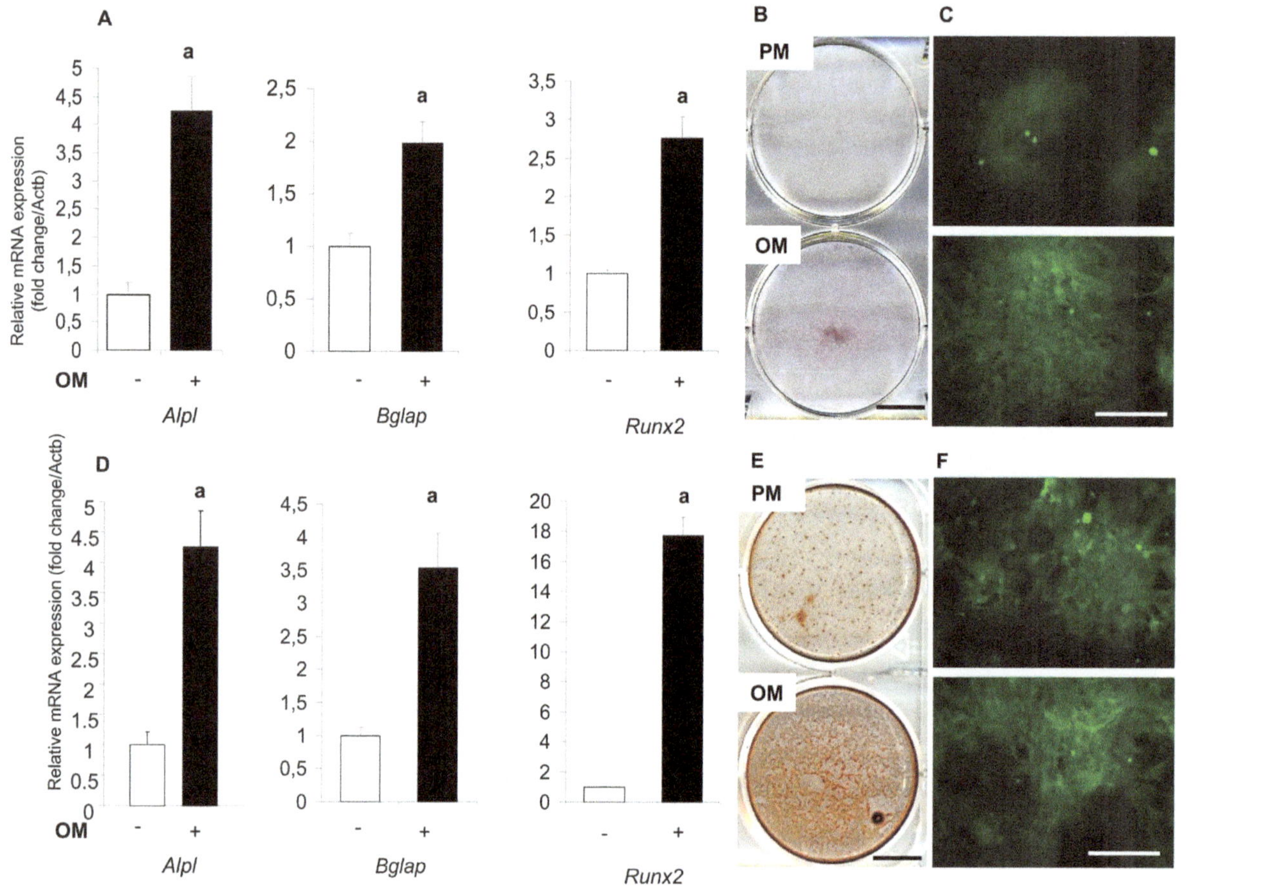

Figure 2. Osteogenic differentiation of GFP-expressing rat bone marrow-derived MSC at 14 and 28 days. Cells were grown for 14 (panels A, B and C) or 28 days (panels D, E and F) in osteogenic medium (OM) or proliferative medium (PM). A and D) Expression of the osteogenic marker genes *Alpl, Bglap*, and *Runx2* was investigated by real-time PCR. Results are expressed as relative expression level compared with the control condition in the absence of OM. a: $p<0.05$. B and E) Calcium deposition was investigated by Alizarin Red staining as described in Materials and Methods. Bar: 1000 μm. C and F) Fluorescence microscopy assessment of GFP-expressing cells was investigated in OM or PM. Bar: 500 μm.

Samples were cut perpendicularly from the implant using a circular diamond saw (Leica, SP1600, Wetzlar, Germany) and then sanded on a Metaserv 2000 (Buehler, Lake Bluff, USA) then gold-palladium-coated on a Desk III (Denton Vacuum, Moorestown, USA). SEM micrographs were taken using backscattered electrons at 15 kV. The surface of the implant was divided into contiguous high-resolution images, and the evaluation was performed with a semiautomatic image analyzer (Leica Quantimeter 500, Cambridge, UK). First, the contours of the defects were traced, then areas of newly formed mineralized bone, BCP granules, and non-mineralized tissues were identified by their grey levels.

Histological assessment

For each sample, serial 5-μm thick sections were cut perpendicularly from the implant using a circular diamond saw (Leica, SP1600, Wetzlar, Germany) and a hard tissue microtome (Leica Polycut SM 2500, Wetzlar, Germany). The sections were stained with Goldner's trichrome and examined with a light microscope (Zeiss, axioplan2, Darmstadt, Germany). Semi-quantitative histological assessment of the amount of bone formed within the pores of the scaffolds was performed on the 6 samples per group as previously described [40]. Three sections, cut approximately 0.75 mm apart depth-wise, were analyzed per sample. A score

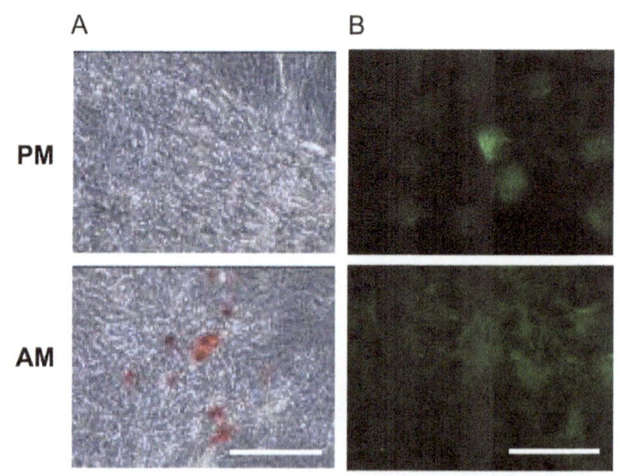

Figure 3. Adipogenic differentiation of GFP-expressing rat bone marrow MSC. Cells were grown in a monolayer for a 14-day period in adipogenic medium (AM) or proliferative medium (PM). A) Adipogenic differentiation was evaluated based on the production of neutral lipid containing vacuoles demonstrated by Oil red O staining. Bar: 500 μm. B) Fluorescence microscopy assessment of GFP-expressing cells was investigated in AM or PM. Bar: 500 μm.

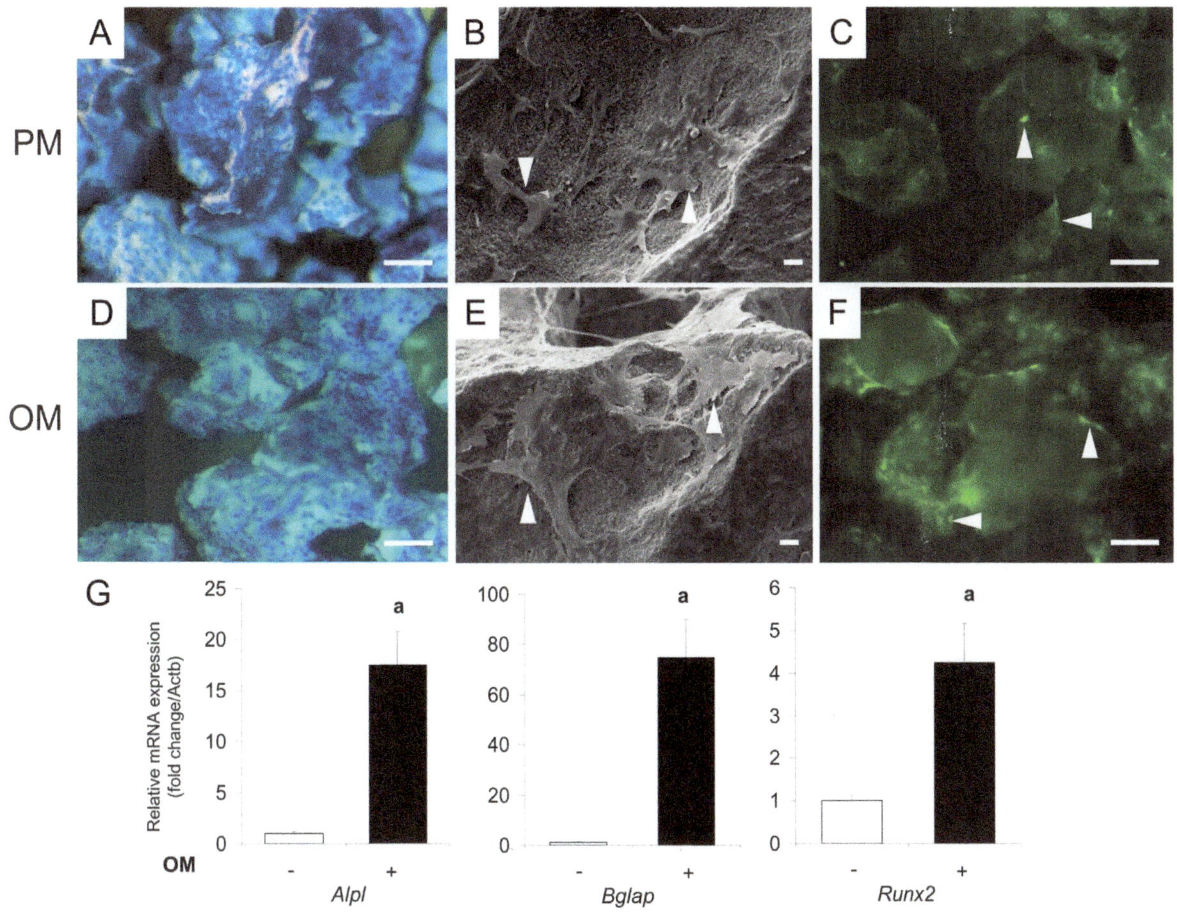

Figure 4. Adhesion and osteogenic differentiation of GFP-expressing transgenic rat bone marrow-derived MSC on BCP granules. At Day 14 in proliferative (PM; A, B and C) and osteogenic (OM; D, E and F) medium. A and D) Cell adhesion was investigated with methylene blue staining. Cells appeared in deep blue. Bar: 100 µm. B and E) Cell morphology was assessed on the surface of biomaterial by SEM after fixation and dehydration (arrow head). Bar: 20 µm. C and F) Microscopic observation of fluorescent GFP-expressing cells on BCP granules (arrow head). Bar: 50 µm. G) Expression of the osteogenic marker genes *Alpl, Bglap*, and *Runx2* was investigated by real-time PCR at the time of implantation (Day 14). Results are expressed as relative expression level compared with the control condition in the absence of OM. a: $p<0.05$.

from 0 to 4 was attributed to each section (Table 3). The mean score per sample was determined. For each section, randomized scoring was performed by two well-versed independent examiners (n = 3 per replicate).

"In vivo" tracking of donor cells by confocal microscopy

GFP expression was first carefully checked in all cell preparations at each of the steps of the study, including after TBM harvest, during cell amplification, and during osteogenic/adipogenic differentiation. Fluorescent cells were examined using a Nikon Eclipse TE 2000 E microscope and images were captured with a digital camera (Cool Snap EZ, Photometrics, Roper Scientific Molecular Devices, Evry, France) using NIS Element Nikon v.3.10. Finally, GFP-expressing cells were tracked *in vivo* in the GMMA-embedded implants. The histological sections were mounted under cover slips with the Prolong Gold Antifade reagent (Molecular Probes, Eugene, OR) and visualized with the same microscope. Image processing was performed using NIKON EZ C1 v.3.60. Wildtype BM from non GFP-expressing adult rats (INSERM, UMRS 1064, Nantes, France) was used as control for "*in vitro*" and "*in vivo*" tracking before and after implantation.

Statistical analysis

Results were expressed as mean ± SD of six samples. Means were then compared using one-way ANOVA followed by a post-hoc test (Fisher's protected least significant difference), and p-values <0.05 were considered statistically significant.

Results

BM myelography

Cytology and myelography showed that the BM contained a normal amount of cells (51.10^6 cells/mm^3). The marrow cell lineages were analyzed (i.e., myeloblasts, myelocytes, proerythroblasts, erythroblasts, megakaryocytes, lymphocytes, plasmocytes, and monocytes). The results (Table 4) confirmed that the harvested BM contained physiologically relevant cell lineages but no blood. GFP-expressing cells were tracked after TBM harvesting. The BM exhibited about 80% of GFP positive cells, while non-GFP wildtype BM did not exhibit any fluorescence (data not shown).

Multidifferentiation potential

A key characteristic of MSC is their ability, with adequate stimuli, to differentiate into multiple lineages, such as osteogenic

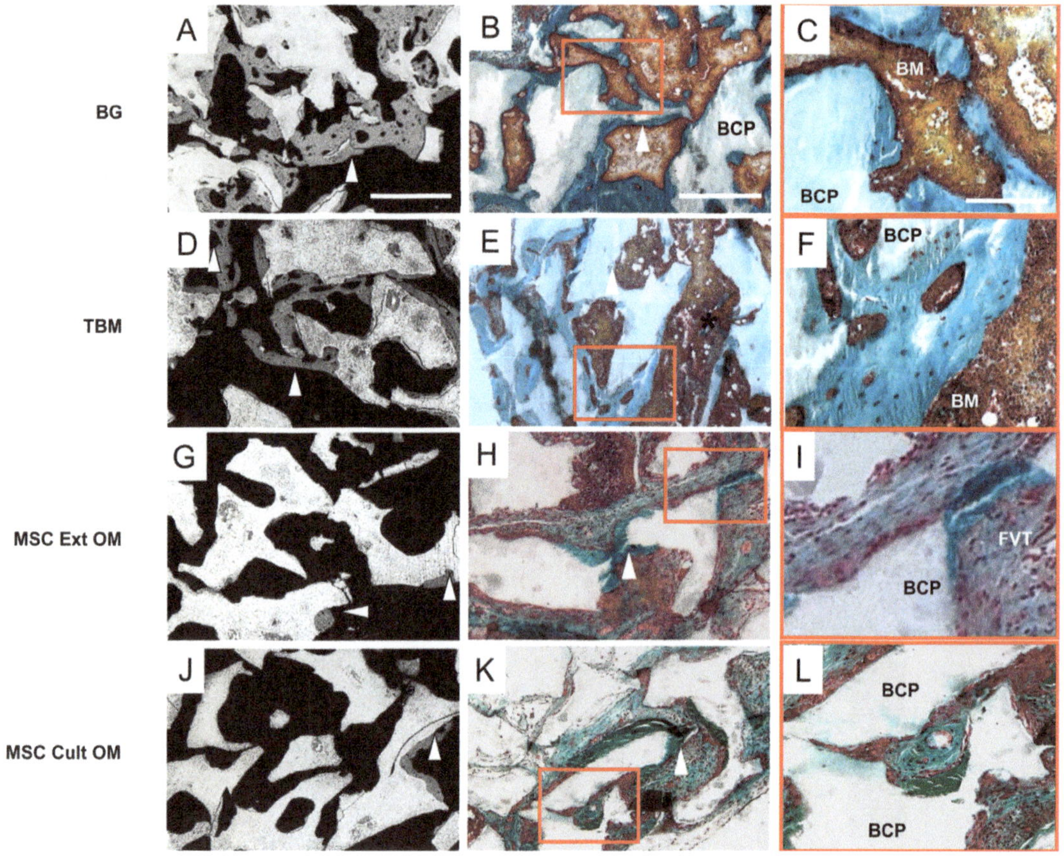

Figure 5. Ectopic bone formation at 8 weeks. Newly formed bone was assessed by SEM in samples embedded in GMMA (A, D, G and J). BCP granules appeared in white, non-calcified tissues in black, and bone in grey (arrow head). Bar: 500 μm. Histological study of newly formed bone and tissues surrounding the material after Goldner trichrome staining as described in Materials and Methods (B, E, H and K). Bone appeared in green (arrow head) BCP: biphasic calcium phosphate. Bar; 500 μm. Magnification of region of interest (red rectangle) showing the surrounding tissues (C,F, I and L). Bone marrow niche were observed only in BG and TBM conditions. In MSC Ext OM and MSC cult OM conditions, fibrovascular tissue was only observed between the BCP granules and bone. BM: bone marrow, FVT: fibrovascular tissue. Bar: 150 μm.

and adipogenic cell lines. For osteogenesis, MSC were grown in a monolayer for a 14 and 28-days period in OM as described in Materials and Methods. Expression levels of osteogenic markers were investigated by real-time PCR at Days 14 and 28. At day 14, the relative expression levels of osteogenic genes (*Alpl, Bglap*, and *Runx2*) for cells grown in OM showed a respective 4, 2, and 2.5 fold increase compared to the non-OM control (Fig. 2A). At day 28, the relative expression levels of osteogenic genes (*Alpl, Bglap*, and *Runx2*) for cells grown in OM showed a respective 4, 3.5, and 18 fold increase compared to the non-OM control (Fig. 2D). Osteogenic differentiation was also evaluated by calcified matrix deposition through Alizarin Red staining. Alizarin Red-positive staining was detected as early as Day 14 and increased until Day 28 when OM was present (Fig. 2B and E). Interestingly, MSC grown in proliferative medium also showed some positive staining at Day 28, but less than the committed cells. GFP expression was preserved throughout the various passages and during cell differentiation (Fig. 2C and F).

For adipogenesis, MSC were grown in a monolayer for a 14-day period in the presence of adipogenic medium as described in Materials and Methods. The expression of the adipogenic markers *Pparγ* and *Lep* was investigated by real-time PCR. Whereas transcripts coding for *Pparg* and *Lep* were detected in adipogenic medium, their expression level remained undetectable in proliferative medium (data not shown). Adipogenic differentiation was also assessed using Oil red O staining through the appearance of neutral lipid containing vacuoles. Oil red O-positive staining was detected at Day 14 for cells grown in the presence of adipogenic medium compared with cells grown in proliferative medium (Fig. 3A). GFP expression was once again preserved throughout the different passages and during cell differentiation (Fig. 3B).

Cell attachment, viability and osteogenic differentiation onto the CaP biomaterials for MSC Cult OM and PM conditions

To assess MSC attachment potential before implantation, methylene blue staining and SEM were performed (Fig. 4). We observed that cells grown in osteogenic or proliferative medium adhered to the surface of BCP particles in similar proportions. Cells were spread over the entire surface of material, but were observed preferentially in the concave areas of the BCP (Fig. 4A and D). When analyzing the BCP surface via SEM, we observed that cells exhibited mainly a branched or spindle morphology (Fig. 4B and E). In addition, the viability of attached GFP-expressing cells was verified using fluorescence microscopy for both the osteogenic and proliferative conditions (Fig. 4C and F). These results also indicated that cells grown in the different conditions expressed GFP before implantation. For osteogenesis, MSC were grown at the surface of BCP particles for a 14 days period in OM as described in Materials and Methods. Expression levels of osteogenic markers were investigated by real-time PCR at Day 14. At Day 14, the relative expression levels of osteogenic genes (*Alpl, Bglap,* and *Runx2*) for cells grown in OM showed a respective 17, 70, and 4 fold increase compared to the non-OM control (Fig. 4G).

Histomorphometric and SEM assessment

The MSC-, BG-, and TBM-loaded ceramics were implanted subcutaneously in nude mice, and revealed inconsistent osteogenic results after eight weeks. Histological observations were consistent with the SEM analysis, and new bone formation was only identified in four groups (BG, TBM, MSC Ext OM, and MSC Cult OM) (Fig. 5 A, D, G and J). In particular, abundant bone formation was seen in BG and TBM implants (Fig. 5 A and D), with woven bone over the ceramic surfaces. In the vicinity of granules, the bone connected individual particles with trabecula (Fig. 5 B and E). Moreover, the bone-lined pores were also filled with accompanying blood vessels. A rich medullar environment, containing hematopoietic progenitors and adipose tissue, was observed between the BCP granules and the bone (Fig. 5 C and F). In contrast, the MSC Ext OM and MSC Cult OM implants showed the least amount of bone formation, which was detected only within the peripheral pores of the implant (Fig. 5 H and K). Mostly loose connective tissue and vascular elements were present in the porous structure of these implants (Fig. 5 I and L). The other implants (granules alone, cells alone and MSC grown in proliferative medium loaded-BCP) failed to form bone. Almost all pores were filled with fibrovascular connective tissue. Conditions in which no bone was detected are not shown.

Assessment of new bone formation

Figure 6 summarizes the results of bone formation analysis using histological score. Bone formation was observed in only four groups (BG, TBM, MSC Ext OM, and MSC Cult OM). Implants with BG or extemporaneously blended TBM with BCP exhibited the highest amount of bone formation. However, implants that required- two or three-dimensional cell culture presented with limited osteogenesis.

Tracking of GFP-labeled cells

To track GFP-expressing cells within the bone forming implants after eight weeks, two contiguous sections were used for each implant. The first was stained with Goldner trichrome to localize the support material, the newly formed bone, the vessels, and the marrow environment (first column of Fig. 7 A, E, I, M and Q). The second section (without any staining) was observed first under transmitted light (not shown) to detect the same elements previously observed with Goldner staining, then under a 535 nm fluorescent light source (second column of Fig. 7 B, F, J, N and R). Next, the same section (without staining) was observed with transmitted light at a higher magnification to detect vessels in newly formed bone or in the peripheral tissue (third column of Fig. 7 C, G, K, O and S). This higher magnification was also used for fluorescence imaging (fourth column of Fig. 7 D, H, L, P and T). Nude mice implanted with non-GFP wildtype BM were used as negative controls (Fig. 7 Q–T). Under fluorescent light, only the groups containing GFP-expressing BG (Fig. 7 B and D), or TBM (Fig. 7 F and H) were positive. Notably, the GFP signal was mainly observed in the marrow environment. A weak signal was also observed within the vessels present in the newly formed bone (Fig. 7 D and H). No signal was detected in osteocytes or osteoblasts. The groups containing GFP-expressing MSC (MSC Ext OM and MSC Cult OM) exhibited a weak signal, mainly in the loose fibrovascular tissue and vessels surrounding the particles. No signal was detected in the vessels or osteoblasts present in the newly formed bone (Fig. 7 J, L, N and P). As expected, non-GFP wild type bone marrow implants did not exhibit any GFP signal in the rich marrow environment, nor in the vessels of the newly formed bone (Fig. 7 R and T).

Discussion

For bone repair, the grafting of fresh whole BM, with or without CaP biomaterials, is simple, safe, inexpensive, and well accepted. In fact, it has been proven to be effective in a wide range of applications including orthopedic surgery for posterolateral spinal fusion [41,42] or tibial fracture healing [43,44]. However, despite studies demonstrating the effectiveness of this technique for bone repair, it is not yet considered a gold standard. Indeed, the results obtained with BM grafting suffer from a lack of reproducibility, which is likely related to the poor number of MSC in BM aspirations [45]. It has been previously observed that human BM aspiration contains approximately 1 stem cell per 1.10^5 nucleated cells in adults [46]. Although the optimal number of MSC required for efficient CaP biomaterial-associated bone formation has not been established and could range between $3.5.10^3$ to up to $7.5.10^4$ cells per mg of material [25,29,47–50], it is generally

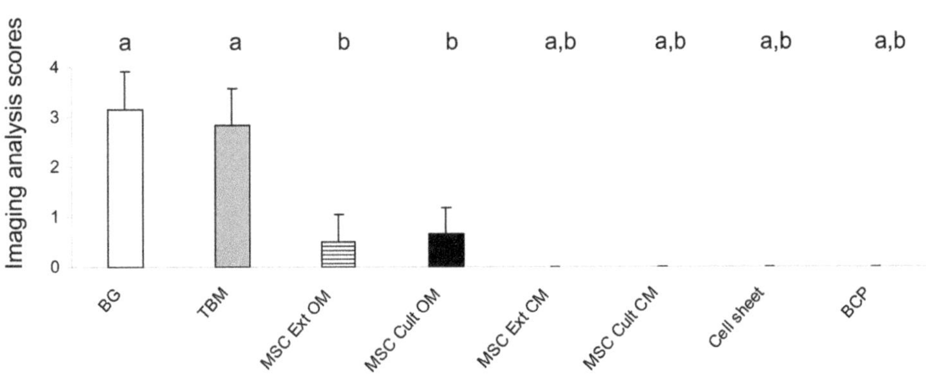

Figure 6. Ectopic bone formation scored at 8 weeks. Data are represented as mean ± S.D. a: $p<0.05$ as compared to MSC Ext OM. b: $p<0.05$ as compared to TBM. BG: bone graft, TBM: Total bone marrow, MSC Ext OM/MSC Ext PM: Mesenchymal stem cells grown in osteogenic medium/proliferative medium then extemporaneously loaded onto BCP granules. MSC Cult OM/MSC Cult PM: MSC loaded onto BCP granules then grown in osteogenic/proliferative medium. BCP: biphasic calcium phosphate alone.

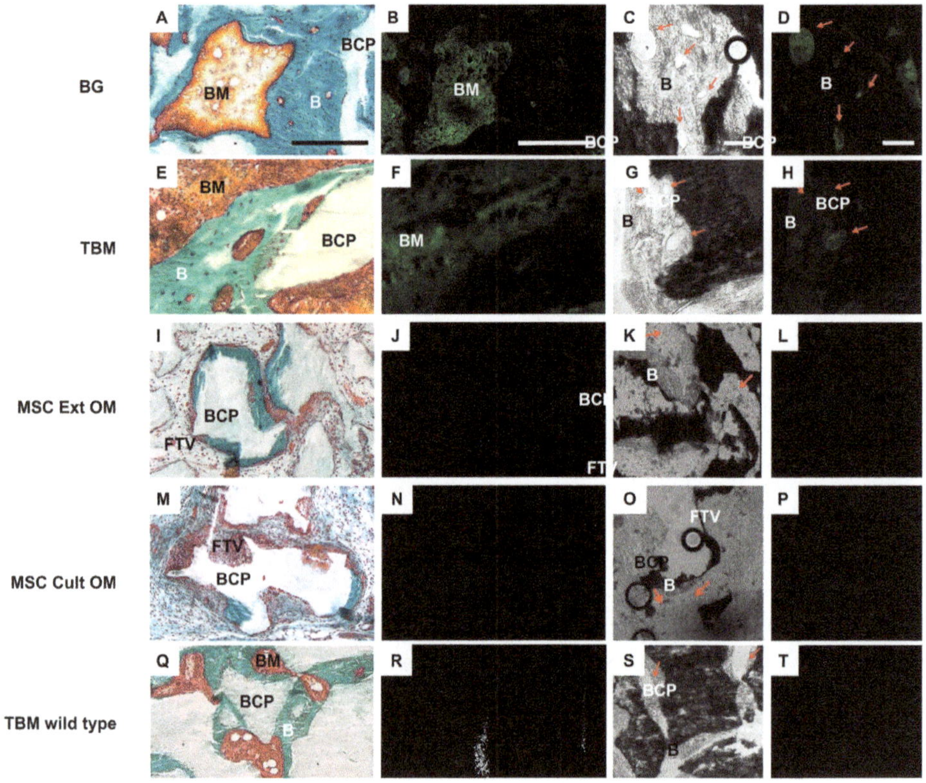

Figure 7. "*In vivo*" tracking of donor cells. Goldner trichrome staining (A, E, I, M and Q). B: bone, BCP: biphasic calcium phosphate, BM: bone marrow, FVT: fibrovascular tissue. Bar: 250 µm. Green fluorescence of GFP retrieved in subcutaneous implants (B, F, J, N and R). Nude mice implanted with non-GFP BM were used as negative controls (TBM wild type). Bar: 250 µm. Transmitted light showing vessels in connective tissues surrounding the BCP granules or in newly formed bone (red arrow) (C, G, K, O and S) Bar: 100 µm. Fluorescent light showing vessels only in TBM and BG groups (red arrow) (D, H, L, P and T) Bar: 100 µm.

believed that increasing the number of grafted MSC will result in better bone formation [33]. This is likely related to the crucial role of MSC in osteoinduction [51].

In this context, it was recently suggested that increasing the number of grafted cells by tissue engineering strategies, which involve *in vitro* culture of MSC on the surface of CaP biomaterials, could make the production of engineered bone constructs possible. Such *in vitro* engineered hybrid constructs have been used in a large number of preclinical animal studies and have provided proof of efficiency [23,50]. This strategy has also been tested in humans for bone regeneration [36,37,52,53], but unfortunately failed to demonstrate a relevant efficiency when compared to BG. Also, it raises strong concerns regarding the regulatory framework governing the use of cell-based therapies and the complexity/cost of cell therapy.

Learning more about the efficacy of the various bone formation methods is necessary for addressing any uncertainties surrounding their clinical use. Thus, we were interested here in determining in an "all-in-one" study the most relevant bone engineering strategy. For this, we compared the effectiveness of eight bone regeneration procedures head-to-head in an ectopic model of osteoinduction. We selected methods that had previously demonstrated *in vivo* osteogenic properties in animal models or humans (i.e., CaP-assisted BG, extemporaneous association of fresh TBM with CaP granules, and several MSC-based bone tissue engineering strategies).

CaP-assisted BG, or bone expansion, was used as a positive control in our studies since it has been shown to promote the formation of well-organized bone in humans [35,54,55]. As a negative control, CaP biomaterials were grafted alone, as they are known to have little to no osteoinductive potential in subcutis ectopic sites [33,56,57]. Also, we examined a condition using cells alone, which consisted of high density committed MSC (known as cell sheets). The *in vivo* osteogenic potential of sheets was previously described [39]. Unprocessed fresh BM was also tested in combination with CaP following a procedure previously described in animal and human studies [14,29]. Finally, several combinations of grown MSC and CaP were evaluated to test parameters influencing MSC-mediated bone formation *in vivo*. For this, we focused on (i) cell commitment towards the osteogenic lineage; (ii) preculture of cells in contact with CaP; and (iii) density of implanted cells [36–38].

The need for cells to be committed towards osteogenic differentiation before implantation is still a large matter of debate. Although in preclinical animal experiments it is unclear whether an osteogenic commitment of MSC is a prerequisite for *in vivo* bone formation [49,57–59], most clinical studies in humans have been performed with precommitted osteogenic cells [37,52,53,60].

To further determine whether osteogenic commitment was required for bone formation, we compared the ability of MSC precommitted or not by an OM [30] to form bone *in vivo*. In our experimental conditions, bone formation was only found when MSC were precommitted into the osteogenic lineage, as evidenced by our *in vitro* data. These results are consistent with those of previous studies [37,52,53,60], and confirm that committed osteogenic MSC are able to promote bone formation.

Among the numerous bone engineering strategies used for human clinical studies, the three-dimensional culture of cells onto the surface of CaP biomaterials before implantation is probably the most widely used [37,53,60]. This culture system allows MSC, committed or not, to adhere to the CaP scaffold and produce early extracellular matrix [47]. The resulting hybrid bone construct is easy to manipulate and demonstrates osteogenic potential in human applications [37,53]. Nonetheless, this complex co-culture process is time consuming, requires many controls, and exhibits safety and cost concerns for human application.

To simplify the procedure, cells may also be expanded in two-dimensional conditions without any CaP scaffold. After proliferation and possibly osteogenic differentiation, cells can then be combined with the scaffold and subsequently implanted. Indeed, this technique has demonstrated efficacy not only in immunocompromised mice and large animals [33,49,58,61], but also in humans [21,36], particularly when MSC were osteogenically differentiated. Consistent with the literature data [62,63], our study revealed that only differentiated MSC, regardless of culture dimensionality, promoted new bone formation at eight weeks. However, the amount of bone formed with MSC was always lower as compared to the group with TBM.

Interestingly, few studies have compared the ability of CaP materials associated with MSC or unprocessed BM to form bone *in vivo*. In large and small animal models of healthy bone critical-sized defects, MSC promoted more bone formation than BM grafting [18,49]. On the contrary, BM grafting was found to be the most efficient strategy to repair critical-sized defects in hypotrophic irradiated bone [29]. These studies seem to be contradictory. However, these data led us to hypothesize that whereas MSC are able to induce bone formation in healthy orthotopic sites [18,49], they might lose osteogenic capabilities in hypotrophic or unfavorable environments (e.g., low nutrients or low oxygen tension) [29,64].

In line with this hypothesis, converging data have recently reported that grafted MSC die early after *in vivo* implantation [25,64]. In fact, it was convincingly demonstrated that grafted human MSC do not survive more than three weeks after subcutaneous implantation in nude mice [25]. In addition, Deschepper *et al.* recently suggested that death of subcutaneously grafted MSC on CaP biomaterials was not only due to hypoxia, but also glucose depletion [64]. Therefore, it seems reasonable to speculate that in our subcutaneous model of implantation, the amount of newly formed bone strongly depended on the survival rate of grafted cells. In order to address this issue, we performed a cell survival tracking experiment using GFP-labeled cells.

Our data revealed that after eight weeks of implantation in the TBM and BG groups, marrow and vessels found in the newly formed bone contained GFP-labeled cells. Even though it was not possible to clearly assign the GFP signal to osteoblasts and osteocytes within the *de novo* bone, these data suggested that grafted cells from TBM and BG groups contributed indirectly to the formation of these tissues. In contrast, when extemporaneous and grown GFP-labeled MSC were implanted, we failed to detect any GFP-labeled cells in the bone, adjacent vessels, or near BCP granules. While we cannot totally rule out the possibility that MSC might still be there but do not express GFP anymore at 8 weeks, it remains unlikely. Indeed, Remy *et al.*, who first described the transgenic GFP-rat strain, previously implanted GFP-neural stem cells in striatum of rats. At 120 days, mature neurons expressed a GFP signal [28]. In our experimental conditions, one can therefore assume that most of the grafted MSC died early after implantation as previously speculated [25,64]. Under these conditions, the residual low-level of bone formation could be a result of remaining MSC releasing some osteotrophic growth factors, which had initially influenced the osteogenic potential of host cells. In addition, MSC were probably not cultured in optimal conditions before implantation and thereby could not support a large bone formation. However, to carry out our comparative study, we had to make a choice between many available bone reparation strategies. Our choice was firstly based on the efficacy of such strategies demonstrating bone formation in animal or human; and secondly on their clinical feasibility with respect to the European regulatory frame. We thus emphasized on cell-based strategies with gradually increasing complexity to provide some realistic data to the clinicians and patients. Though we could have used bone stimulatory factors during the *in vitro* stage, like cytokines, transgenic approach, or bioreactors, but these strategies are currently not transposable in clinic in most of the EU countries and thereby will not be available for clinicians in a near future.

Taken together, these findings suggest that the newly formed bone observed in our study appeared to be a chimera derived both from the grafted and host cells. Similar observations have been previously reported in a model of osteoinduction by Goshima *et al.* using BM cell labeling [65]. Indeed, following subcutis implantation of BM blended CaP biomaterials, they observed that ectopic bone-resident osteoblasts and osteocytes were derived from the grafted cells during the early phase (21 days) and from the host cells in a secondary phase (by 56 to 84 days), with the donor derived bone being gradually replaced by bone derived from the host [66].

Conclusions

The results of this study indicated that more significant bone formation could be obtained with TBM than with several BTE strategies. In addition, they revealed a trophic property of implanted BM, which displayed a capacity to survive and develop eight weeks after implantation. Owing to its simplicity and safety, this method could be performed by surgeons during a single procedure and without additional cost. The variability of BM aspiration quality from donor to donor represents the main limitation of both the TBM grafting and MSC culture [45], but contrarily to the TBM grafting that remains a basic procedure, MSC culture protocols could probably be largely improved in a near future. Nevertheless, in light of increasing regulatory constraints for clinical trials [67] and persistent scientific concerns regarding stem cells in bone regenerative medicine, BM grafting could represent a clinically relevant alternative to BG and BTE. Since ectopic mouse implantation represents a preliminary model in testing clinically relevant bone tissue engineering strategies, further studies are now necessary to assess this method in a bone defect model.

Acknowledgments

We would like to acknowledge Biomatlante SAS (Vigneux de Bretagne) for providing the MBCP™ material. Thanks are due to Dr. S. Wuillem of the Laboratory of Hematology, University Hospital, Nantes, France. The authors also gratefully acknowledge R. Bellouard, A. Bourgine, J. Guiol, R.H. Khonsari, N. Maillard, T. Miramond, M. Perré, S. Portron, E. Rederstorff, and J. Sohier for their helpful assistance.

Author Contributions

Conceived and designed the experiments: PC CM JG PW. Performed the experiments: PC CM ND CV MM SS PP PW TC FE. Analyzed the data: PC CM JG PW. Contributed reagents/materials/analysis tools: SR IA. Wrote the paper: PC CM JG JM.

References

1. Helm GA, Dayoub H, Jane JA Jr (2001) Bone graft substitutes for the promotion of spinal arthrodesis. Neurosurg Focus 10: E4.
2. Burchardt H (1987) Biology of bone transplantation. Orthop Clin North Am 18: 187–196.
3. Rajan GP, Fornaro J, Trentz O, Zellweger R (2006) Cancellous allograft versus autologous bone grafting for repair of comminuted distal radius fractures: a prospective, randomized trial. J Trauma 60: 1322–1329.
4. Al Ruhaimi KA (2001) Bone graft substitutes: a comparative qualitative histologic review of current osteoconductive grafting materials. Int J Oral Maxillofac Implants 16: 105–114.
5. Barboza EP (1999) Clinical and histologic evaluation of the demineralized freeze-dried bone membrane used for ridge augmentation. Int J Periodontics Restorative Dent 19: 601–607.
6. Daculsi G, Passuti N, Martin S, Deudon C, Legeros RZ, et al. (1990) Macroporous calcium phosphate ceramic for long bone surgery in humans and dogs. Clinical and histological study. J Biomed Mater Res 24: 379–396.
7. Passuti N, Daculsi G, Rogez JM, Martin S, Bainvel JV (1989) Macroporous calcium phosphate ceramic performance in human spine fusion. Clin Orthop Relat Res: 169–176.
8. Cavagna R, Daculsi G, Bouler JM (1999) Macroporous calcium phosphate ceramic: a prospective study of 106 cases in lumbar spinal fusion. J Long Term Eff Med Implants 9: 403–412.
9. Toquet J, Rohanizadeh R, Guicheux J, Couillaud S, Passuti N, et al. (1999) Osteogenic potential in vitro of human bone marrow cells cultured on macroporous biphasic calcium phosphate ceramic. J Biomed Mater Res 44: 98–108.
10. Ohgushi H, Goldberg VM, Caplan AI (1989) Heterotopic osteogenesis in porous ceramics induced by marrow cells. J Orthop Res 7: 568–578.
11. Ohgushi H, Okumura M, Yoshikawa T, Inoue K, Senpuku N, et al. (1992) Bone formation process in porous calcium carbonate and hydroxyapatite. J Biomed Mater Res 26: 885–895.
12. Johnson KD, Frierson KE, Keller TS, Cook C, Scheinberg R, et al. (1996) Porous ceramics as bone graft substitutes in long bone defects: a biomechanical, histological, and radiographic analysis. J Orthop Res 14: 351–369.
13. Ohgushi H, Goldberg VM, Caplan AI (1989) Repair of bone defects with marrow cells and porous ceramic. Experiments in rats. Acta Orthop Scand 60: 334–339.
14. Gouin F, Delecrin J, Passuti N, Touchais S, Poirier P, et al. (1995) [Filling of bone defects using biphasic macroporous calcium phosphate ceramic. Apropos of 23 cases]. Rev Chir Orthop Reparatrice Appar Mot 81: 59–65.
15. Bruder SP, Kurth AA, Shea M, Hayes WC, Jaiswal N, et al. (1998) Bone regeneration by implantation of purified, culture-expanded human mesenchymal stem cells. J Orthop Res 16: 155–162.
16. Kruyt MC, Stijns MM, Fedorovich NE, De Bruijn JD, Van Blitterswijk CA, et al. (2004) Genetic marking with the DeltaLNGFR-gene for tracing goat cells in bone tissue engineering. J Orthop Res 22: 697–702.
17. Kruyt MC, Dhert WJ, Yuan H, Wilson CE, van Blitterswijk CA, et al. (2004) Bone tissue engineering in a critical size defect compared to ectopic implantations in the goat. J Orthop Res 22: 544–551.
18. Petite H, Viateau V, Bensaid W, Meunier A, de Pollak C, et al. (2000) Tissue-engineered bone regeneration. Nat Biotechnol 18: 959–963.
19. Cancedda R, Bianchi G, Derubeis A, Quarto R (2003) Cell therapy for bone disease: a review of current status. Stem Cells 21: 610–619.
20. Meijer GJ, de Bruijn JD, Koole R, van Blitterswijk CA (2008) Cell based bone tissue engineering in jaw defects. Biomaterials 29: 3053–3061.
21. Quarto R, Mastrogiacomo M, Cancedda R, Kutepov SM, Mukhachev V, et al. (2001) Repair of large bone defects with the use of autologous bone marrow stromal cells. N Engl J Med 344: 385–386.
22. Sohier J, Corre P, Weiss P, Layrolle P (2010) Hydrogel/calcium phosphate composites require specific properties for three-dimensional culture of human bone mesenchymal cells. Acta Biomater 6: 2932–2939.
23. Cowan CM, Shi YY, Aalami OO, Chou YF, Mari C, et al. (2004) Adipose-derived adult stromal cells heal critical-size mouse calvarial defects. Nat Biotechnol 22: 560–567.
24. Taub PJ, Yau J, Spangler M, Mason JM, Lucas PA (2009) Bioengineering of calvaria with adult stem cells. Plast Reconstr Surg 123: 1178–1185.
25. Boukhechba F, Balaguer T, Bouvet-Gerbettaz S, Michiels JF, Bouler JM, et al. (2011) Fate of bone marrow stromal cells in a syngenic model of bone formation. Tissue Eng Part A 17: 2267–2278.
26. Dumas A, Moreau MF, Gherardi RK, Basle MF, Chappard D (2009) Bone grafts cultured with bone marrow stromal cells for the repair of critical bone defects: an experimental study in mice. J Biomed Mater Res A 90: 1218–1229.
27. Teodelinda M, Michele C, Sebastiano C, Ranieri C, Chiara G (2011) Amniotic liquid derived stem cells as reservoir of secreted angiogenic factors capable of stimulating neo-arteriogenesis in an ischemic model. Biomaterials 32: 3689–3699.
28. Remy S, Tesson L, Usal C, Menoret S, Bonnamain V, et al. (2010) New lines of GFP transgenic rats relevant for regenerative medicine and gene therapy. Transgenic Res 19: 745–763.
29. Espitalier F, Vinatier C, Lerouxel E, Guicheux J, Pilet P, et al. (2009) A comparison between bone reconstruction following the use of mesenchymal stem cells and total bone marrow in association with calcium phosphate scaffold in irradiated bone. Biomaterials 30: 763–769.
30. Merceron C, Portron S, Masson M, Lesoeur J, Fellah BH, et al. (2011) The effect of two- and three-dimensional cell culture on the chondrogenic potential of human adipose-derived mesenchymal stem cells after subcutaneous transplantation with an injectable hydrogel. Cell Transplant 20: 1575–1588.
31. Magne D, Bluteau G, Lopez-Cazaux S, Weiss P, Pilet P, et al. (2004) Development of an odontoblast in vitro model to study dentin mineralization. Connect Tissue Res 45: 101–108.
32. Cordonnier T, Layrolle P, Gaillard J, Langonne A, Sensebe L, et al. (2009) 3D environment on human mesenchymal stem cells differentiation for bone tissue engineering. J Mater Sci Mater Med 21: 981–987.
33. Mankani MH, Kuznetsov SA, Robey PG (2007) Formation of hematopoietic territories and bone by transplanted human bone marrow stromal cells requires a critical cell density. Exp Hematol 35: 995–1004.
34. Corre P, Khonsari RH, Laure B, Cordova Jara L, Bonnet R, et al. (2011) [Medial proximal tibia donor site: contribution to alveolar cleft repair in children]. Rev Stomatol Chir Maxillofac 112: 280–285.
35. Corre P, Khonsari RH, Laure B, Elamrani K, Weiss P, et al. (2012) [Synthetic calcium phosphate ceramics in secondary alveoloplasty]. Rev Stomatol Chir Maxillofac 113: 131–135.
36. Marcacci M, Kon E, Moukhachev V, Lavroukov A, Kutepov S, et al. (2007) Stem cells associated with macroporous bioceramics for long bone repair: 6- to 7-year outcome of a pilot clinical study. Tissue Eng 13: 947–955.
37. Morishita T, Honoki K, Ohgushi H, Kotobuki N, Matsushima A, et al. (2006) Tissue engineering approach to the treatment of bone tumors: three cases of cultured bone grafts derived from patients' mesenchymal stem cells. Artif Organs 30: 115–118.
38. Shayesteh YS, Khojasteh A, Soleimani M, Alikhasi M, Khoshzaban A, et al. (2008) Sinus augmentation using human mesenchymal stem cells loaded into a beta-tricalcium phosphate/hydroxyapatite scaffold. Oral Surg Oral Med Oral Pathol Oral Radiol Endod 106: 203–209.
39. Akahane M, Nakamura A, Ohgushi H, Shigematsu H, Dohi Y, et al. (2008) Osteogenic matrix sheet-cell transplantation using osteoblastic cell sheet resulted in bone formation without scaffold at an ectopic site. J Tissue Eng Regen Med 2: 196–201.
40. Arinzeh TL, Tran T, McAlary J, Daculsi G (2005) A comparative study of biphasic calcium phosphate ceramics for human mesenchymal stem-cell-induced bone formation. Biomaterials 26: 3631–3638.
41. Bansal S, Chauhan V, Sharma S, Maheshwari R, Juyal A, et al. (2009) Evaluation of hydroxyapatite and beta-tricalcium phosphate mixed with bone marrow aspirate as a bone graft substitute for posterolateral spinal fusion. Indian J Orthop 43: 234–239.
42. Moro-Barrero L, Acebal-Cortina G, Suarez-Suarez M, Perez-Redondo J, Murcia-Mazon A, et al. (2007) Radiographic analysis of fusion mass using fresh autologous bone marrow with ceramic composites as an alternative to autologous bone graft. J Spinal Disord Tech 20: 409–415.
43. Khanal GP, Garg M, Singh GK (2004) A prospective randomized trial of percutaneous marrow injection in a series of closed fresh tibial fractures. Int Orthop 28: 167–170.
44. Hernigou P, Mathieu G, Poignard A, Manicom O, Beaujean F, et al. (2006) Percutaneous autologous bone-marrow grafting for nonunions. Surgical technique. J Bone Joint Surg Am 88 Suppl 1 Pt 2: 322–327.
45. Mendes SC, Tibbe JM, Veenhof M, Bakker K, Both S, et al. (2002) Bone tissue-engineered implants using human bone marrow stromal cells: effect of culture conditions and donor age. Tissue Eng 8: 911–920.
46. Muschler GF, Boehm C, Easley K (1997) Aspiration to obtain osteoblast progenitor cells from human bone marrow: the influence of aspiration volume. J Bone Joint Surg Am 79: 1699–1709.
47. Cordonnier T, Langonne A, Corre P, Renaud A, Sensebe L, et al. (2012) Osteoblastic differentiation and potent osteogenicity of three-dimensional hBMSC-BCP particle constructs. J Tissue Eng Regen Med. Jun 11. doi: 10.1002/term.1529.
48. Liu Y, Wang L, Kikuiri T, Akiyama K, Chen C, et al. (2011) Mesenchymal stem cell-based tissue regeneration is governed by recipient T lymphocytes via IFN-gamma and TNF-alpha. Nat Med 17: 1594–1601.
49. Livingston TL, Gordon S, Archambault M, Kadiyala S, McIntosh K, et al. (2003) Mesenchymal stem cells combined with biphasic calcium phosphate ceramics promote bone regeneration. J Mater Sci Mater Med 14: 211–218.
50. van Gaalen SM, de Bruijn JD, Wilson CE, van Blitterswijk CA, Verbout AJ, et al. (2009) Relating cell proliferation to in vivo bone formation in porous Ca/P scaffolds. J Biomed Mater Res A 92: 303–310.
51. Friedenstein AJ (1980) Stromal mechanisms of bone marrow: cloning in vitro and retransplantation in vivo. Haematol Blood Transfus 25: 19–29.
52. Kitoh H, Kitakoji T, Tsuchiya H, Mitsuyama H, Nakamura H, et al. (2004) Transplantation of marrow-derived mesenchymal stem cells and platelet-rich plasma during distraction osteogenesis – a preliminary result of three cases. Bone 35: 892–898.
53. Pradel W, Eckelt U, Lauer G (2006) Bone regeneration after enucleation of mandibular cysts: comparing autogenous grafts from tissue-engineered bone and iliac bone. Oral Surg Oral Med Oral Pathol Oral Radiol Endod 101: 285–290.

54. Guerado E, Fuerstenberg CH (2011) What bone graft substitutes should we use in post-traumatic spinal fusion? Injury 42 Suppl 2: S64–71.
55. Rihn JA, Kirkpatrick K, Albert TJ (2010) Graft options in posterolateral and posterior interbody lumbar fusion. Spine (Phila Pa 1976) 35: 1629–1639.
56. Fellah BH, Gauthier O, Weiss P, Chappard D, Layrolle P (2008) Osteogenicity of biphasic calcium phosphate ceramics and bone autograft in a goat model. Biomaterials 29: 1177–1188.
57. Harris CT, Cooper LF (2004) Comparison of bone graft matrices for human mesenchymal stem cell-directed osteogenesis. J Biomed Mater Res A 68: 747–755.
58. Fang D, Seo BM, Liu Y, Sonoyama W, Yamaza T, et al. (2007) Transplantation of mesenchymal stem cells is an optimal approach for plastic surgery. Stem Cells 25: 1021–1028.
59. Mastrogiacomo M, Scaglione S, Martinetti R, Dolcini L, Beltrame F, et al. (2006) Role of scaffold internal structure on in vivo bone formation in macroporous calcium phosphate bioceramics. Biomaterials 27: 3230–3237.
60. Ohgushi H, Kotobuki N, Funaoka H, Machida H, Hirose M, et al. (2005) Tissue engineered ceramic artificial joint – ex vivo osteogenic differentiation of patient mesenchymal cells on total ankle joints for treatment of osteoarthritis. Biomaterials 26: 4654–4661.
61. Bensaid W, Oudina K, Viateau V, Potier E, Bousson V, et al. (2005) De novo reconstruction of functional bone by tissue engineering in the metatarsal sheep model. Tissue Eng 11: 814–824.
62. Doi K, Kubo T, Hayashi K, Imura K, Akagawa Y (2007) Development of cell-hybrid artificial bone: effect of osteogenic differentiation of bone marrow stromal stem cells on bone formation with newly developed interconnected porous calcium hydroxyapatite. Dent Mater J 26: 162–169.
63. Schubert T, Xhema D, Veriter S, Schubert M, Behets C, et al. (2011) The enhanced performance of bone allografts using osteogenic-differentiated adipose-derived mesenchymal stem cells. Biomaterials 32: 8880–8891.
64. Deschepper M, Oudina K, David B, Myrtil V, Collet C, et al. (2011) Survival and function of mesenchymal stem cells (MSC) depend on glucose to overcome exposure to long-term, severe and continuous hypoxia. J Cell Mol Med 15: 1505–1514.
65. LeDouarin N (1973) A biological cell labeling technique and its use in experimental embryology. Dev Biol 30: 217–222.
66. Goshima J, Goldberg VM, Caplan AI (1991) The origin of bone formed in composite grafts of porous calcium phosphate ceramic loaded with marrow cells. Clin Orthop Relat Res: 274–283.
67. Daculsi G (2011) Association of cells and biomaterials for bone regeneration. ingénierie et recherche biomédicale: 76–79.

8

Extracellular Polymeric Substances (EPS) of Freshwater Biofilms Stabilize and Modify CeO_2 and Ag Nanoparticles

Alexandra Kroll[1]*, Renata Behra[1], Ralf Kaegi[1], Laura Sigg[1,2]

1 Environmental Toxicology, Swiss Federal Institute of Aquatic Science and Technology (Eawag), Dübendorf, Switzerland, **2** Institute of Biogeochemistry and Pollutant Dynamics IBP, Swiss Federal Institute of Technology in Zurich (ETHZ), Zurich, Switzerland

Abstract

Streams are potential receiving compartments for engineered nanoparticles (NP). In streams, NP may remain dispersed or settle to the benthic compartment. Both dispersed and settling NP can accumulate in benthic biofilms called periphyton that are essential to stream ecosystems. Periphytic organisms excrete extracellular polymeric substances (EPS) that interact with any material reaching the biofilms. To understand the interaction of NP with periphyton it is therefore crucial to study the interaction of NP with EPS. We investigated the influence of EPS on the physicochemical properties of selected NP (CeO_2, Ag) under controlled conditions at pH 6, 7.6, 8.6 and light or dark exposure. We extracted EPS from five different periphyton communities, characterized the extracts, and exposed CeO_2 and carbonate-stabilized Ag NP (0.5 and 5 mg/L, both 25 nm primary particle size) and $AgNO_3$ to EPS (10 mg/L) over two weeks. We measured NP size distribution, shape, primary particle size, surface plasmon resonance, and dissolution. All EPS extracts were composed of biopolymers, building blocks of humic substances, low molecular weight (M_r) acids, and small amphiphilic or neutral compounds in varying concentrations. CeO_2 NP were stabilized by EPS independent of pH and light/dark while dissolution increased over time in the dark at pH 6. EPS induced a size increase in Ag NP in the light with decreasing pH and the formation of metallic Ag NP from $AgNO_3$ at the same conditions via EPS-enhanced photoreduction. NP transformation and formation were slower in the extract with the lowest biopolymer and low M_r acid concentrations. Periphytic EPS in combination with naturally varying pH and light/dark conditions influence the properties of the Ag and CeO_2 NP tested and thus the exposure conditions within biofilms. Our results indicate that periphytic organisms may be exposed to a constantly changing mixture of engineered and naturally formed Ag NP and Ag^+.

Editor: Jose Luis Balcazar, Catalan Institute for Water Research (ICRA), Spain

Funding: This work was financially supported by an SNF Ambizione grant (PZ00P2_142533, http://www.snf.ch/E/funding/individuals/ambizione/Pages/default.aspx) to AK. The funders had no role in study design, data collection and analysis, decision to publish, or preparation of the manuscript.

Competing Interests: The authors have declared that no competing interests exist.

* Email: alexandra.kroll@eawag.ch

Introduction

Due to the increased production of engineered nanoparticles (NP), they have become potential environmental contaminants that may be transported in streams. Possible routes of entry have been suggested previously [1] such as spills, treated sewage water, run-off, or rain. Depending on the route of entry, engineered NP may undergo transformation. Sewage treatment for example seems to sulfidize Ag NP depending on their size [2]. Heterogeneity in NP concentration and state of transformation in the environment can thus be expected. Consequently, to be able to estimate the impact of engineered NP on aquatic ecosystems, the interactions of NP with various components of these systems need to be studied.

One potential receiving compartment are biofilms, called periphyton, which colonize submerged surfaces. Periphyton is a taxonomically diverse and dynamic community of photo- and heterotrophic microorganisms, *i.e.* bacteria, algae, and fungi that adapts quickly to environmental changes [3]. It is an essential element of aquatic ecosystems contributing to primary production and release of oxygen [4], influencing the hydrodynamics of the water body [5], and being a habitat for different life stages of uni- and multicellular organisms. Periphyton was shown to be the primary receptor of gold nanorods in a model estuarine system [6] and to accumulate CuO NP when exposed in mesocosms [7]. The accumulation of TiO_2 NP in freshwater biofilms and mono-species algal biofilms in mesocosm experiments positively correlated with the amount of biomass [8]. Sedimentation of micrometer sized particles is influenced by periphyton due to the change in hydrologic conditions near the biofilms [9]. Whether this applies to nano-sized particles has not been investigated, however, engineered NP may occur in micrometer-size agglomerates depending on NP properties, concentration, water chemistry, presence of other particles or colloids, hydrological conditions, solar radiation, and temperature. Relevant NP properties are size and surface charge, which are in turn influenced by NP surface modifications and solubility, as well as ionic strength, presence of bivalent cations, organic matter, and pH [10,11,12,13,14,15]. In particular, these parameters determine whether engineered NP are stabilized, homo- or heteroagglomerate with other particulates, sediment, or remain dispersed.

For the present study, we chose Ag NP and CeO_2 NP as model particles based on usage, the availability of data, and the difference in NP properties. Ag based NP are used in the highest number of NP containing consumer products [16] due to their bactericidal properties [17], which have been attributed to both the NP as well as to Ag^{+} released in aqueous media [18]. Ag^{+} release from Ag NP depends on NP surface modification and the dispersant [19,20]. We selected carbonate stabilized Ag NP which have been characterized regarding stability in artificial and natural stream water as well as toxicity to the model green alga *Chlamydomonas reinhardtii* [10,21]. As the NP were stabilized with carbonate we were able to use the carbonate-bicarbonate buffer to adjust the pH during our experiments, which is the dominant buffer system in streams [22]. CeO_2 based NP have been studied due to Ce^{III}-Ce^{IV} redox-cycling in NP surface atoms, which is associated with the capacity to store and release oxygen [23]. Thus CeO_2 NP are used for example for the oxidative destruction of organic compounds in waste water [24] and have been suggested for biomedical applications involving redox processes [25,26]. Recently, CeO_2 NP have successfully been used to enhance the effect of chemotherapeutics on cancer cells and at the same time protect intact cells *in vitro* [27].

Any material reaching periphyton by settling from the overlaying water will first encounter the extracellular matrix secreted by periphytic organisms. It is made up of hydrated extracellular polymeric substances (EPS) consisting mostly of species-specific polysaccharides and proteins [28,29,30], that mediate adhesion of periphytic organisms to abiotic surfaces and other organisms. Up to 90% of the organic matter associated with biofilms can be EPS [31]. EPS of algal and bacterial monoculture biofilms have been shown in some cases to stabilize Ag NP [32], as well as under other conditions to induce Ag NP agglomeration [33]. EPS have been suggested to trap Ag NP outside bacterial cells possibly lowering their toxicity [33]. Thus, to understand the fate and effects of NP in periphyton, it is essential to study their interaction with the periphytic EPS.

The aim of this study was to investigate the influence of periphytic EPS on physicochemical properties of selected NP over time depending on exposure conditions relevant to the photoheterotrophic community. We exposed CeO_2 and Ag NP to chemically characterized EPS extracted from five independent periphyton communities to account for inter-community variability of EPS. We assessed changes in NP physicochemical characteristics depending on NP material, EPS extract, exposure time and conditions (light, pH).

Materials and Methods

Chemicals

If not stated below, chemicals were purchased from Sigma-Aldrich Switzerland.

Nanoparticles

Carbonate-coated Ag NP with a primary particle size of 25 nm (Table S1) have been described previously [10,21] and were provided by NanoSys GmbH (Wolfhalden, Switzerland) as an aqueous suspension with a nominal concentration of 1 g/L (9.27 mM) Ag. The original suspension was kept in the dark. Unmodified CeO_2 NP with a nominal primary particle size of 25 nm were provided by Nanograde AG (Staefa, Switzerland) (Table S1).

Periphyton colonization

Periphyton was colonized on glass microscope slides (76×26 mm, Thermo Scientific) which were soaked in 0.03 M HNO_3 and washed with nanopure water (18.1 MΩ•cm, Milli-Q) before use and were then placed horizontally in Plexiglas channels (86×10.4×10 cm, polymethyl methacrylate) in a flow-through system fed by water from river Chriesbach (on campus, Dübendorf, Switzerland) [34,35]. An overview of Chriesbach water chemistry is provided in Table S2. A sediment trap (0.51×0.7×2.6 m, residence time ~20 min) was used to remove part of the dispersed particles. The flow rate in the channels was maintained at ~1 cm/s corresponding to a volume flow of 3 L/min. Typical flow rates in the benthic zone of Chriesbach are 1–4 cm/s. Illumination was provided in 12:12 h light/dark cycles by BioSun fluorescent tubes with a radiation similar to the sunlight spectrum (Radium Lampenwerk GmbH, Germany, ML-T8 36W/965/G13B; photon flux during light periods: 100 μE/m^{2}s). Temperature and photosynthetically active radiation (PAR) in the channels were monitored by a HOBO Pendant Temperature/Light Data Loggers (UA-002-64). Oxygen saturation was monitored with a Presens Microx TX3 system and an NTH-PSt1-L2.5-TS-NS40/0.8-YOP-EXT1 oxygen microsensor.

Extraction and characterization of extracellular polymeric substances (EPS) from periphyton

Periphyton was colonized five times independently for three weeks each time in 2011 and 2012. EPS were extracted from periphyton grown on 64 glass slides (Table 1). EPS extracts were analyzed for cell lysis, organic carbon and organic nitrogen size distribution, and protein, calcium, magnesium, and chloride content.

The extraction procedure was adapted from Stewart et al [30]. After 21 days of colonization, the biofilms were scraped off the glass slides with a clean glass slide into 1 mL/slide $NaHCO_3$ (2 mM, pH 7.6) in a 100 mL glass beaker. The buffer contained 1 μg/mL protease inhibitors (1:1:1 Aprotinin (A2132.0010), Leupeptin (L9783), and Pepstatin A (A2205.0010); AppliChem AG). The biomass was resuspended by gentle pipetting and sonication in a water bath (45 kHz 60 W, VWR Ultrasonic Cleaner) for 30 s. Fine sediment and larger biomass was allowed to settle for ~1 min, the supernatant was removed and centrifuged at 1,880·g for 10 min. Biomass was resuspended a second time in 1 mL/slide fresh solution and treated as described above. The resulting biomass pellets were frozen at −80°C and subsequently lyophilized to determine dry weight. All supernatants were sequentially filtered (1 μm glass fiber [VWR], 0.45 μm polypropylene [PALL], and 0.22 μm PES [Millipore] filters). Filters were washed with nanopure water (18.1 MΩ•cm, Milli-Q) prior to use. The procedure resulted in a final volume of about 130 mL per extract. EPS extracts were stored in glass bottles at 4°C (0.02% (w/v) NaN_3). All glassware used for EPS extraction was muffled before use (450°C, 4 h). All extraction steps were performed on ice, the water bath for ultrasound treatment was at room temperature.

To verify that no measurable cell lysis occurred during the extraction procedure, glucose-6-phosphate dehydrogenase (G6P-DH) activity was measured as described previously [30,36]. G6P-DH is an intracellular enzyme that would be released upon cell lysis. At each step of EPS extraction, 20 μL samples were transferred to a 96-well plate in triplicate (Greiner bio-one, black, transparent flat bottom, non-binding, 655906). 180 μL reaction mixture (50 mM Tris Base, 0.15 mM NADP (Merck 481972), 10 mM $MgCl_2$, and 3 mM α-D-glucose-6-phosphate (Merck 346764); pH 8) were added and light absorption of NADPH being formed was measured at 340 nm at 30°C for 30 min. G6P-DH (*Leuconostoc mesenteroides*, Merck, 346774-500U) calibration curves were generated in both water and in the EPS extract to verify that the EPS did not interfere with the detection of G6P-DH activity. G6P-DH activity was below the detection limit in all EPS extracts (limit of detection: 0.00125 U/mL). Lysed periphyton samples served as positive controls as described previously [30].

Table 1. Properties of EPS extracts 1–5 as determined by weighing of dry mass, protein quantification and LC-OCD-OND.

	Extraction Date	Dry mass [g/slide]	DOC/dry mass [mg/g]	cDOC [% DOC]	Protein/dry mass [mg/mg]	DOC/N (biopolymers)	DOC/protein	N/protein	Biopolymers [% cDOC]	HS building blocks [% cDOC]	Low M_r acids [% cDOC]	Neutrals, amphiphilics [% cDOC]
1	05.09.2011	0.06	0.86	94.6	0.07	10.5	14.5	1.4	14.2	15.3	10.3	54.8
2	04.10.2011	0.04	0.71	89.9	0.07	3.9	14.0	3.6	8.8	20.0	6.2	54.9
3	24.10.2011	0.02	1.75	100.3	0.011	5.9	9.3	1.6	15.4	25.3	10.9	48.7
4	11.04.2012	0.03	1.11	89.6	0.20	1.2	5.1	4.3	4.5	21.4	3.2	60.5
5	02.05.2012	0.03	1.14	93.1	0.08	3.3	13.1	4.0	10.3	24.3	13.6	44.9

DOC: dissolved organic carbon, cDOC: chromatographable DOC, N: nitrogen, HS: humic substances, M_r: molecular weight.

Organic carbon and nitrogen size distribution was measured by size-exclusion chromatography – organic carbon detection – organic nitrogen detection (LC-OCD-OND) as described previously [30]. Samples were diluted 1:50 with nanopure water (18.1 MΩ•cm, Milli-Q) right before they were measured. A size exclusion column (250×20 mm, Toyopearl TSK HW-50S) was used to separate EPS compounds. To quantify the carbon background of the extraction protocol, an aliquot of extraction buffer was treated the same way as periphyton suspensions and then assessed by LC-OCD-OND. The mobile phase was phosphate buffer (24 mM, pH 6.6) and the acidification solution was phosphoric acid (60 mM, pH 1.2). The detection limit was 10 µg/L for both OC and ON. Retention times of 35–70 min correspond to 70–0.5 kDa model proteins and 28–0.1 kDa model polysaccharides [30]. The software FIFFIKUS was used to quantify total organic carbon (TOC), dissolved (DOC), and chromatographable DOC compounds (cDOC). The chromatograms obtained from LC-OCD-OND are integrated to determine the amount of biopolymers (high M_r polysaccharides and proteins), humic substances (HS), building blocks of HS, low M_r acids, and amphiphilic/neutral compounds (alcohols, aldehydes, amino acids, and ketones). As an example of the reproducibility of the EPS extraction procedure Figure S4 shows chromatograms of two EPS extracts obtained independently by two individuals on the same day from two different channels after three weeks of colonization.

Total protein in EPS extracts was measured by the Bradford assay [37] using Bradford reagent (Bio-Rad Protein Assay Kit I). Calibration curves were produced with bovine serum albumin (BSA) diluted in equal amounts of EPS extracts to account for any interference of the EPS with protein detection.

Calcium, magnesium, and chloride content of EPS extracts was quantified by ion chromatography (Metrohm 761 Compact IC; with chemical suppression for Cl^- detection). Detection limits were: 5 mg/L Ca^{2+}, 2.5 mg/L Mg^{2+}, and 0.5 mg/L Cl^-.

Exposure of CeO_2 and Ag NP and $AgNO_3$ to EPS

CeO_2 NP, Ag NP, and $AgNO_3$ were dispersed in 20 mL 2 mM $NaHCO_3$ at three different pH values (pH 6, 7.6 or 8.6) with or without EPS (10 mg/L DOC) in light (100 µE/m^{2}s, BioSun fluorescent tubes, see above) or dark resulting in 12 exposure conditions. Samples were prepared in triplicates (n = 3) and were placed in a climate room at 20°C and stirred at 400 rpm (vessels: SNAP-CAP, Huber Switzerland, 8.9227.08; magnetic stir bars, with PTFE coating, 6×20 mm, Huber Switzerland, 13.1120.06). pH was adjusted with HNO_3 and NaOH at the beginning of the experiments and was constant during exposure. Dispersions contained 0.02% (w/v) NaN_3. Final nominal NP concentrations were 5 and 0.5 mg/L for both types of NP and 7.87 and 0.78 mg/L $AgNO_3$ (4.6×10^{-5} and 4.6×10^{-6} mol L^{-1}). Actual concentrations were determined by inductively coupled plasma mass spectrometry (ICP-MS) and used to calculate total and dissolved Ce and Ag during the exposure (see Characterization of NP dispersions below). CeO_2 dispersions were generated from a stock dispersion (50 mg/L). The stock dispersion was made directly before use by adding 20 mL 2 mM $NaHCO_3$, pH 7.6 to 1 mg CeO_2 and stirring for 1 h at daylight conditions. Ag NP were freshly diluted to final concentrations from the original dispersion while stirring. $AgNO_3$ was freshly diluted from a 0.3 g/L stock solution which was kept in the dark.

Experiments were conducted with five different EPS extracts (see Table 1). EPS extracts were used on the day after the extraction. All glassware used was muffled before use (450°C, 4 h). O_2 concentrations in the samples were measured as described above. Average O_2 saturation did not significantly change over time or with sample properties (light/dark, pH, EPS) and were 71.6±1.5% (~6.18 mg/L; Ag NP dispersions), 69.4±3.5% (~5.99 mg/L; $AgNO_3$ suspensions), and 71.1±1.7% (~6.14 mg/L; CeO_2 dispersions).

Characterization of nanoparticle dispersions

NP hydrodynamic diameter was determined by dynamic light scattering (DLS) and nanoparticle tracking analysis (NTA). NP zetapotential was derived from electrophoretic mobility (EPM) measurements. Light absorption of Ag containing dispersions was measured by UV-VIS spectroscopy. Selected samples were analyzed by transmission electron microscopy/energy dispersive X-ray analysis (TEM/EDX) and X-ray absorption spectroscopy (XAS) to determine size, shape, elemental composition and Ag speciation.

The concentration of dissolved Ag and Ce in samples containing 5 mg/L Ag or Ce was determined by ICP-MS.

Samples for NP characterization were taken at 3 h, 24 h, 48 h, 168 h, and 336 h (DLS, NTA, UV-VIS; see below) or 5 min (= "0 h"), 168 h and 336 h (ICP-MS, see below).

Table S4 provides an overview of the sampling scheme.

Dynamic light scattering (DLS) and electrophoretic mobility (EPM)

The hydrodynamic diameter of the NP was derived from DLS measurements and the zetapotential from the EPM, both measured on a Zetasizer Nano ZS (nano ZS, Malvern Instruments). For comparison with data from the literature, the Smoluchowski model was used to calculate the zetapotential from EPM measurements. To compare results within this study we use EPM due to the difference in systems being compared (with and without EPS). Three measurements were taken per sample and the autocorrelation function was analyzed using the cumulant analysis algorithm resulting in a mean size (z-average) and a standard deviation (polydispersity index, PDI). Data sets with a PDI above 0.5 were not taken into account (for settings see supplementary information). DLS measurements performed on aqueous suspension containing EPS did not indicate any particles. Representative count rates for Ag NP, CeO_2 NP and NP formed in $AgNO_3$ are provided in Tables S6–S8.

Nanoparticle tracking analysis (NTA)

NTA (NanoSight LM10 equipped with a LM14 temperature controller, NanoSight Ltd.) was used to determine a number based particle size distribution. Each sample was directly measured three times for 60 s without any further processing (no dilution). All NTA videos were analyzed with the same settings in batch processing mode (for settings see supplementary information). Analyses that resulted in less than 200 tracked particles were not used. Settings were adapted to not detect any EPS particles. Videos were analyzed using the NanoSight NTA 2.2 and 2.3 Analytical Software (NanoSight Ltd.).

UV-VIS absorption

UV–VIS light absorption (190–900 nm) of dispersions and $AgNO_3$ was recorded with a UVIKON 930 spectrophotometer (Kontron Instruments). EPS and $NaHCO_3$ were measured as controls. Ag NP show size- and surface-specific surface Plasmon resonance (SPR) which results in a specific light absorption spectrum [38].

Transmission electron microscopy (TEM)

Selected samples were analyzed by TEM in combination with an energy dispersive x-ray (EDX) analysis system to determine the size and the elemental composition of the particles. For that purpose, a drop of suspension was drawn through a lacy carbon-coated TEM grid (holey Carbon Copper grids, mesh size 200, Plano GmbH, Wetzlar, Germany) with a paper tissue. Samples were analyzed using a TEM (Tecnai F30 STEM, FEI) equipped with an EDX system (EDAX, New Jersey, USA).

X-ray absorption spectroscopy (XAS)

XAS analyses at the Ag K-edge were performed on the Dutch Belgian Beamline (Dubble, BM01B) at the European Synchrotron Radiation Facility (ESRF, Grenoble, France). Frozen samples were measured with a He-Cryostat adjusted to 80 K. Reference samples were prepared as pellets (Ag_2S (Alfa Aesar; 011416)), or as suspensions (Ag NP). XAS data extraction and processing was performed using Athena [39] following standard procedures as described previously [40].

Sample preparation for ICP-MS

NP and $AgNO_3$ dispersions were centrifuged in 2 mL reaction tubes to separate particles (>3.5 nm) and dissolved ions (25°C and 20,800·g for 3 h (Ag, ρ_{Ag} 10.5 g/cm^3) or 4.6 h (CeO_2; ρ_{CeO2} 7.65 g/cm^3). Ultrafiltration was not applicable in this study as we observed varying degrees of sorption of Ag to ultrafiltration membranes (Amicon Ultra-4 Centrifugal Filter Unit with Ultracel-3 membrane, Millipore) depending on pH and presence or absence of EPS. The supernatants (0.6 mL) were transferred to new reaction tubes, acidified (0.65% HNO_3), and stored at 4°C. Non-centrifuged samples for the determination of total NP and $AgNO_3$ concentration were stored in the same manner. All samples were treated by microwave digestion prior to ICP-MS analysis. 0.5 mL of each sample were digested with 4 mL of 65% HNO_3 and 0.5 mL of 30% H_2O_2 in a microwave digestion unit (MLS ultraClave 4; 10 min at 180°C/100 bar, 14 min at 210°C/100 bar) and diluted 1:100 with nanopure water. (18.1 MΩ•cm, Milli-Q). One sample per run contained only HNO_3, H_2O_2, and EPS diluted in $NaHCO_3$ to determine the background concentration of Ce and Ag.

Inductively coupled plasma mass spectrometry (ICP-MS)

The total and dissolved Ag and Ce concentrations in NP and $AgNO_3$ dispersions were measured by ICP-MS (Element 2 High Resolution Sector Field ICP-MS; Thermo Finnigan). The instrument was calibrated with a multi-element mass standard (Merck, 1113550100). The calibration curve for data analysis was made with the calibration standard SCP-33-MS (140-130-321, PlasmaCAL) in the concentration range 0–20 µg/L. A reference with a concentration within the calibration range was measured every 10 samples, the calibration samples were measured every 40 samples. The detection limit of Ce and Ag was 0.01 µg/L. Mean recovery was 88.4±4.4% for total Ce, 96.9±3.7% for total Ag from Ag NP, and 96.5±3.4% for $AgNO_3$.

Data analysis

Calculation of averages, standard deviations, percentages and oneway ANOVA and Tukey posthoc test or unpaired t test (both allowing a significance level $\alpha = 0.05$) were performed with Microsoft Excel 2010 and GraphPad Prism 4 for Windows. Graphs were generated with the same software programs.

Results

Characterization of EPS extracts

EPS were extracted from five different 3-week-old communities between 09/2011 and 05/2012 (Table 1). The average water temperature of the corresponding growth periods ranged from 11.9°C to 18.3°C and is provided in the supplement (Figure S1).

Dry mass ranged from 0.02–0.06 g/slide with the highest mass measured in extract 1 at the beginning of September 2011. DOC concentration in EPS extracts ranged from 0.71–1.75 mg/g dry mass and protein concentrations ranged from 0.007–0.025 mg/mL. The ratios of DOC to organic nitrogen compounds (DOC/N) and to the amount of proteins (DOC/protein) were in the range of 1.2–10.5 and 5.1–14.5, respectively. Extract 4 of April 2012 had the lowest and extract 1 of September 2011 had the highest ratios of DOC to N or protein.

Chromatograms of all EPS extracts obtained by LC-OCD-OND showed four distinct features (Figure S2) which have been reported to correspond to high M_r biopolymers, building blocks of humic substances (HS), low M_r acids, and small amphiphilic/neutral compounds [30,41]. The retention time and shape of the four features varied between the five extracts. Low M_r acids and amphiphilics/neutrals showed less variability in peak retention time and shape than biopolymers and HS building blocks. Extract 4 from April 2012 showed longer retention time in all four groups of components than extracts 1–3 and 5. In particular, the biopolymer signal appeared after 30 min retention time instead of after 25 min as observed in extracts 1–3. Extract 5 also lacked a biopolymer peak between 25–30 min retention time but had a biopolymer concentration in the same range as extracts 1–3. The UV-signal of extract 4 was lower than in the other samples and the peaks were less pronounced as well as shifted to longer retention times (Figure S3).

The percentages of the four groups of EPS components in each extract are listed in Table 1. The concentration of amphiphilic and neutral compounds was typically the highest followed by HS-building blocks, biopolymers, and low M_r acids. Extract 5 was an exception in having a higher fraction of low M_r acids than biopolymers. Extract 4 had the highest amount of neutral/amphiphilic compounds and the lowest percentage of all other components. cDOC/N ratios in the biopolymer fraction ranged from 1.2–10.5 with the lowest ratio observed in extract 4.

Influence of EPS on the stability of CeO_2 NP dispersions

The mean hydrodynamic diameter of CeO_2 NP dispersed in 2 mM $NaHCO_3$ with and without EPS as determined by DLS ranged from ~200 nm to ~600 nm (Figure 1 A, B, Tables S9, S10). In EPS-free samples, CeO_2 NP mean hydrodynamic diameter decreased over time independent of light and pH conditions (from 540 nm [3 h] to 226–376 nm [168 h] in the dark and from 409–553 nm [3 h] to 278–349 nm [168 h] in the light). With all EPS extracts, mean particle sizes were higher in the dark than in the light after 3 h (457–574 nm and 290–330 nm, respectively). Over time, mean sizes decreased slightly in the dark but increased in the light (325–423 nm and 395–591 nm, respectively).

The mean hydrodynamic diameter as determined by NTA was ~200 nm, did not depend on pH or light/dark, or the presence of EPS and did not change over time (Figure 1 C, Tables S9, S10).

The mean EPM of CeO_2 NP incubated in the dark without EPS ranged from −1.68±0.49 μmcm/Vs (pH 6) at 3 h to −2.28±0.24 μmcm/Vs (pH 6) at 168 h (Figure 1 D, Tables S9, S10). The differences between pH 6, 7.6, and 8.6 were not statistically significant. In the presence of EPS in the dark, EPM ranged from −1.91±0.09 μmcm/Vs (pH 6) at 3 h to −2.11±0.11 μmcm/Vs (pH 6) at 168 h. The differences between pH 6, 7.6, and 8.6 were not statistically significant.

In light-treated samples without EPS, EPM ranged from −1.12±0.45 μmcm/Vs (pH 6) at 3 h to −2.46±0.13 μmcm/Vs (pH 6) at 168 h and from −2.0±0.73 μmcm/Vs (pH 6) at 3 h to −1.87±0.18 μmcm/Vs (pH 6) at 168 h in the presence of EPS. After 168 h, EPM were statistically significantly lower without EPS than with EPS. The only statistically significant difference between different pH conditions was between light-exposed samples at pH 6 and pH 7.6/8.6 both in the presence and absence of EPS.

TEM analysis of the CeO_2 suspensions revealed that the NP consisted of 10 −50 nm spherical, cuboidal, and octahedral primary particles that were fused together (Figure 2).

Results for 0.5 mg/L CeO_2 NP dispersions showed the same trends regarding mean hydrodynamic diameter and EPM as dispersions containing 5 mg/L CeO_2 NP (Figure S8, Tables S9, S10).

Release of Ce^{III} from CeO_2 NP in the presence of EPS

At 0 h, dissolved Ce (<0.01 −1 μg Ce/L, or <0.01 −0.03% of total Ce; Figure 3) was close to or below the detection limit. After 168 h and 336 h, all samples contained at least 3 μg/L dissolved Ce (~0.1% of total Ce). At pH 6, the average concentration of dissolved Ce increased to 58.6±14.1 μg/L (~1.6%) in the light and 82.9±17.6 μg/L (~2.4%) in the dark. In EPS containing samples, the concentration of dissolved Ce did not increase significantly at pH 6 as compared to pH 7.6 and 8.6 in the light [5.5±2.9 μg/L (~0.2%)] but increased to 34.0±5.2 μg/L (~1%) in the dark. Analyses after 336 h showed a further increase in dissolved Ce at pH 6 in EPS-free samples in the light and the dark as well as in EPS containing samples in the dark. Overall, the concentration of dissolved Ce at pH 6 was significantly different between light and dark treated samples with EPS but not between EPS-free samples incubated in light or dark.

Influence of EPS on the stability of Ag NP dispersions

According to DLS measurements, the mean hydrodynamic diameter of Ag NP neither changed in EPS-free dispersions over time nor depending on light or pH conditions (Figure 4 A–B; mean diameter: 40±2 nm after 3 h, 40±5 nm after 168 h; Tables S11, S12). The mean hydrodynamic diameter was slightly higher with EPS at 3 h (53±4 nm) but remained unchanged over time in dark treated samples. Incubation in light led to an increase in mean hydrodynamic diameter with decreasing pH within 24 h in the presence of EPS extracts 1–3 and 5. Mean hydrodynamic diameters remained stable until the end of the test period (168 h: pH 6:159±13 nm; pH 7.6:105±34 nm; pH 8.6:66±4 nm). Changes in the mean hydrodynamic diameter occurred later with EPS extract 4 which had the lowest concentration of biopolymers and lowest C/N ratio in the biopolymer fraction. Here, a pH- and light-dependent increase was detectable only between 168 h and 336 h of incubation (Figure 4 B, 336 h values; Tables S11, S12).

NTA data support these observations but showed smaller differences in mean hydrodynamic diameter. The mean hydrodynamic diameter of Ag NP hardly changed in EPS-free dispersions over time (Figure 4 C; 63±15 nm after 3 h, 58±9 nm after 168 h; Tables S11, S12) and slightly decreased in EPS containing dispersions in the dark (81±13 nm after 3 h, 72±8 nm after 168 h). As already described based on DLS data, incubation in light led to an increase in mean size with decreasing pH, however, pH 7.6 and pH 8.6 dispersions were not significantly different

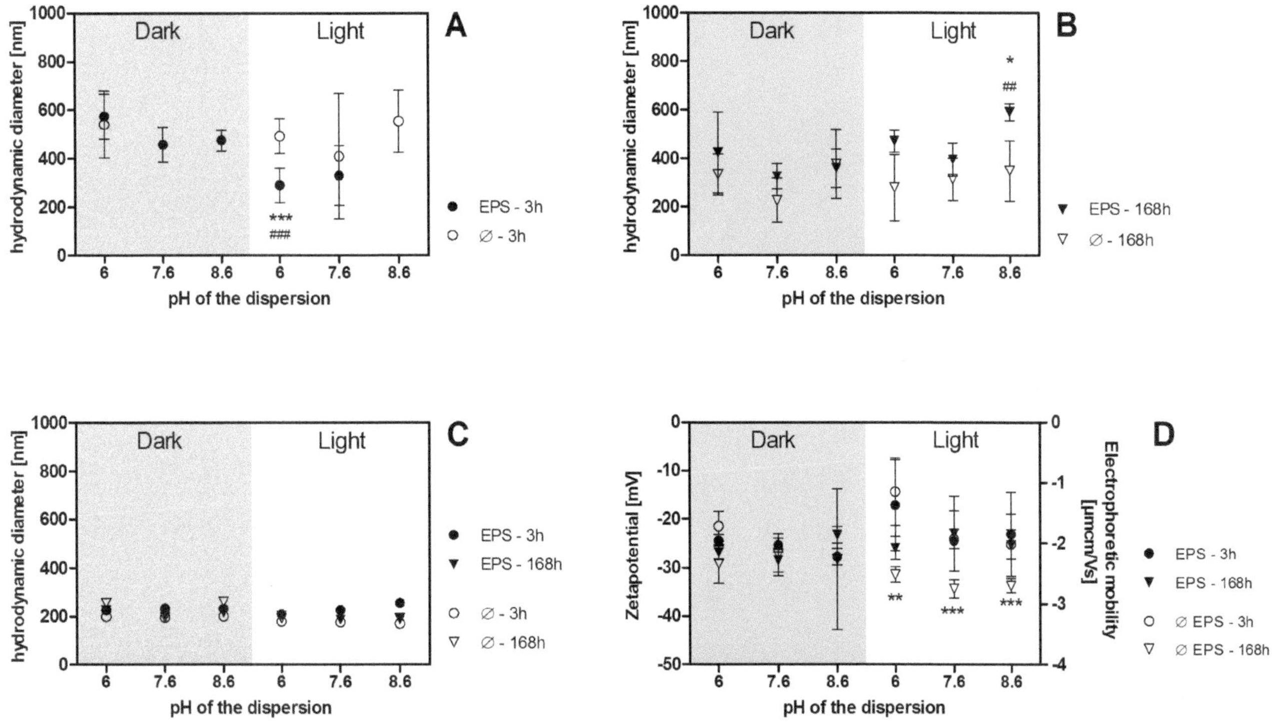

Figure 1. A-D: Mean hydrodynamic diameter of CeO_2 NP with standard deviations in 5 mg CeO_2/L dispersions as determined by DLS (z-average/intensity distribution; **A, B**) and NTA (mean size, **C**) and mean EPM and zetapotential (**D**) with standard deviations determined by electrophoresis/DLS. Samples were incubated in the dark or in light (100 $\mu E/m^2s$). n = 15 from 5 independent experiments with 3 replicates each, each experiment was done with one EPS extract (10 mg/L). 3 h DLS results in panel **A** for CeO_2 NP without EPS at pH 7.6 and 8.6 in the dark and for CeO_2 NP with EPS at pH 8.6 in the light were discarded due to high polydispersity. **A** ** significantly different from light-treated samples without EPS and ### significantly different from dark-treated sample with and without EPS, $p<0.0001$, df = 11 (between groups), F = 15.79. **B** * significantly different from light-treated samples without EPS, ## significantly different from dark-treated samples with and without EPS, p = 0.001, dg = 11 (between groups), F = 6.52. **D** 168 h: ** (pH 6) and *** (pH 7.6 and 8.6) significantly different from light-treated samples without EPS, $p<0.0001$, df = 11 (between groups), F = 11.62.

(mean sizes at 168 h: pH6:144±13 nm; pH 7.6:92±9 nm; pH 8.6:92±9 nm). Again, EPS extract 4 (lowest biopolymer concentration, lowest C/N ratio in the biopolymer fraction) led to a slower change in hydrodynamic diameter with a pH- and light-dependent increase being detectable only between 168 h and 336 h of incubation (Figure 4 C, 336 h values; pH 6:160±22 nm; pH 7.6:96±2 nm; pH 8.6:97±9 nm).

Results for 0.5 mg/L Ag NP dispersions are not displayed as they showed the same trends regarding mean hydrodynamic diameters and EPM as dispersions containing 5 mg/L Ag NP.

The UV-VIS light absorption of Ag NP is indicative of changes in their size and/or surface due to SPR of surface electrons interacting with photons of defined wavelengths. Figure 5 shows light absorption spectra of Ag NP in EPS-free dispersions (A) and EPS-containing dispersions (B). All samples showed strong light absorption around 250 nm. Ag NP containing samples absorbed light around 410–420 nm corresponding to the SPR peak of the primary Ag particles. EPS alone also showed strong light absorption around 250 nm but did not have a second light absorption peak around the SPR signal of Ag NP (Figure 5 B). In the presence of EPS, a second absorption maximum appeared around 650 nm in Ag NP dispersions after 24 h. The change in the absorption spectrum occurred later in the presence of the EPS extract with the lowest biopolymer concentration and lowest C/N ratio in the biopolymer fraction (extract 4; Figure 5 B, 336 h spectrum).

TEM imaging of selected Ag NP samples treated with EPS in the light showed both aggregated NP. In addition, also the size of the primary particles seemed to have increased. However, due to the limited amount of TEM data, this increase remains on a qualitative level. No change of the original spherical shape of the NP was observed (Figure S5A, S5B).

XAS and EDX analyses of selected samples showed that Ag NP treated with EPS were dominantly elemental Ag (Figure S5C, S5D, and S6, Table S5). The LCF fits of the XANES and EXAFS spectra always returned minor fractions (typically 5–10%) of other Ag phases (Ag_2S and Ag_2O). However, the uncertainty of the LCF analysis is in the same range and thus, whether additional Ag phases are present cannot be assessed based on the available dataset.

Dissolved Ag^+ in Ag NP dispersions

The average concentration of dissolved Ag^+ in Ag NP dispersions was ~90 µg/L (~2% of total Ag measured) at the start of the experiments and did not depend on the pH of the dispersion (pH 6, 7.6, or 8.6) or the presence of EPS (Figure 6). After a week of incubation, the concentration of dissolved Ag remained unchanged in EPS-free samples but had increased to 144±21 µg/L (~3.2% of total Ag) in the light and to 123±14 µg/L (~2.7% of total Ag) in the dark. Within the second week of incubation, the concentration of dissolved Ag also increased in EPS-free samples independent of the pH or light/dark conditions (~230 µg/L or ~5% of total Ag). In EPS-containing samples, the

Figure 2. TEM bright field image of CeO_2 NP after 24 h of stirring (5 mg/L, pH 7.6, 2 mM $NaHCO_3$).

dissolved Ag concentration increased to 448±35 µg/L (10% to total Ag) in the light and to 358±16 µg/L (8% to total Ag) in the dark.

Formation of Ag NP from $AgNO_3$

$AgNO_3$ solutions (5 mg/L Ag) with and without EPS (10 mg DOC/L) were analyzed by DLS, NTA, and UV-VIS. An Ag NP

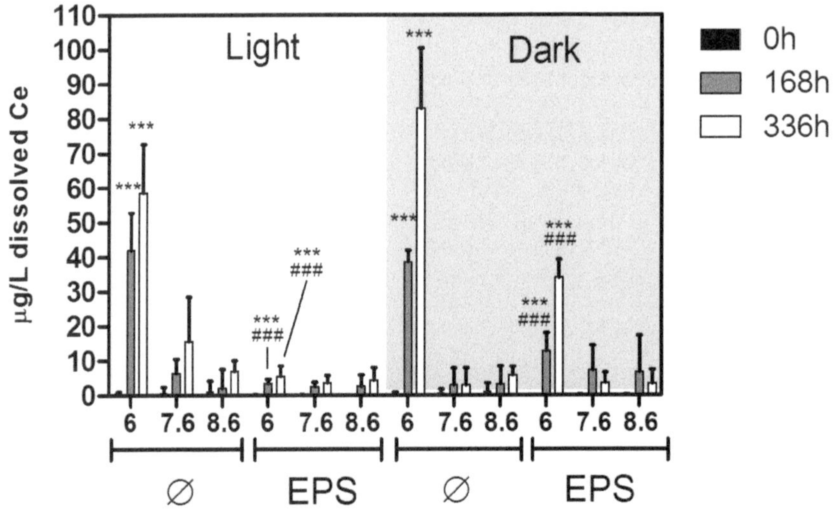

Figure 3. Mean mass concentrations with standard deviations of dissolved Ce in 5 mg/L CeO_2 NP dispersions directly after addition of NP („0 h") and after 168 h and 336 h of incubation as determined by ICP-MS of acid-digested samples. Samples were incubated in the dark or in light (100 µE/m²s). n = 9 in 3 independent experiments, each experiment was done with one EPS extract. *** Significant difference between EPS-free and EPS containing samples; ### significant difference between light and dark treated samples, 168 h: $p<0.0001$, df = 11 (between groups), F = 52.55, 336 h: $p<0.0001$, df = 11 (between groups), F = 25.65.

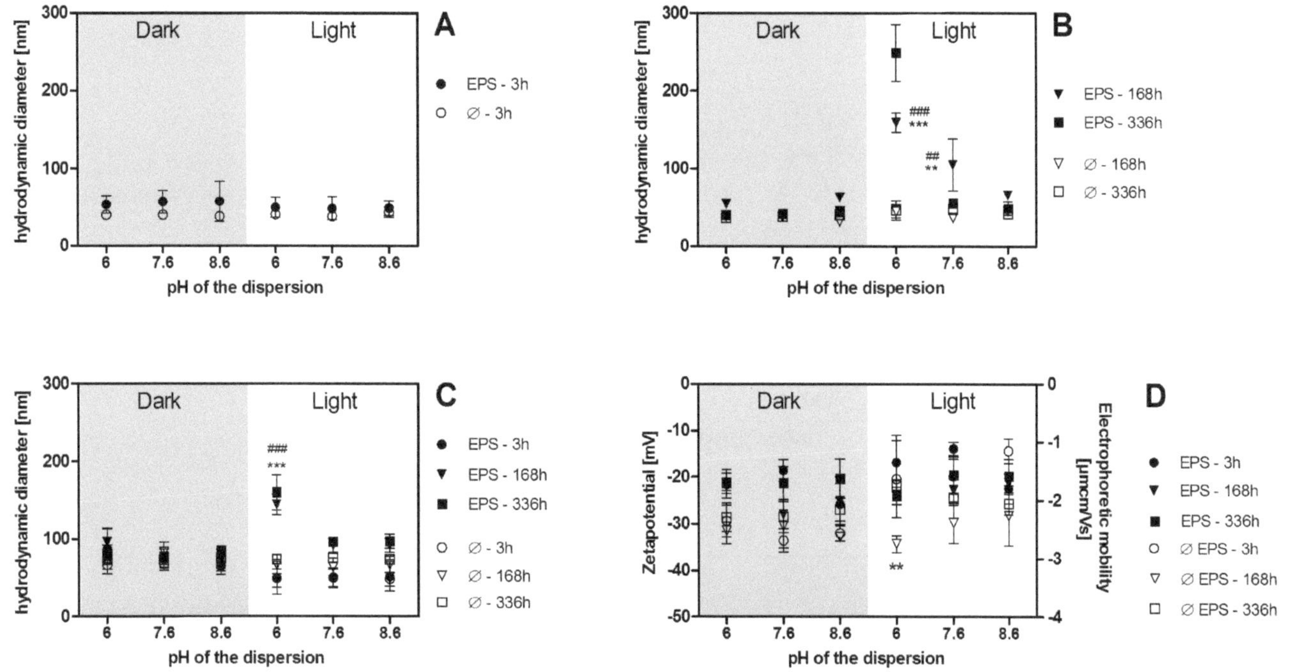

Figure 4. A–D: Mean hydrodynamic diameter of Ag NP with standard deviations in 5 mg Ag NP/L dispersions with and without 10 mg DOC/L EPS as determined by DLS (z-average/intensity distribution; **A, B**) and NTA (mean size, **C**) and mean EPM and zetapotential (**D**) with standard deviations determined by electrophoresis/DLS. 3 h: n = 15, 5 independent experiments (E1–5). 168 h: n = 12, 4 independent experiments (E1–3, 5). 336 h: n = 3, 1 experiment (E4). Each experiment was done with one EPS extract. Samples were incubated in the dark or in light (100 $\mu E/m^2s$). **B** 168 h: *** and ** significantly different from light-treated samples without EPS, ### and ## significantly different from dark-treated samples with and without EPS, $p < 0.0001$, df = 11 (between groups), F = 49.51. **C** 168 h: *** significantly different from light-treated samples without EPS and ### significantly different from dark-treated samples with and without EPS, $p < 0.001$, df = 11 (between groups), F = 77.83. **D** 168 h: * significantly different from light-treated samples with EPS, $p < 0.001$, df = 11 (between groups), F = 4.241.

specific SPR signal was observed under all conditions except in the dark without EPS. In $AgNO_3$ solutions with EPS extracts 1–3 and 5, a SPR signal (400–420 nm) appeared between 0 h and 3 h after the start of incubation in light (Figure 7 B). After 24 h, a shoulder around 600 nm became prominent which had shifted slightly to longer wavelengths after 168 h. The process was delayed in the presence of EPS extract 4 (low biopolymer concentration, low C/N ratio in the biopolymer fraction). It took up to 336 h for the same light absorption spectrum to develop (Figure 7 B). In EPS-free samples, a small SPR maximum around 440 nm was visible after 168 h incubation in light (Figure 7 A). A similar signal was detected in EPS-containing samples in the dark (Figure S7).

Due to a very low or insufficient signal intensity, we could not acquire DLS and NTA data for $AgNO_3$ solutions that were light-treated without EPS and dark-treated with EPS. The mean hydrodynamic diameter of NP detected in $AgNO_3$ solutions with EPS in the light was 86–195 nm (DLS) and 105–185 nm (NTA) (Figure 8 A–B, Tables S13, S14). While there was no significant time dependence, an increase in pH to pH 8.6 resulted in smaller mean hydrodynamic diameters. EPM of the NP formed decreased over time from -1.12 ± 0.23 µmcm/Vs (pH 6) at 3 h to -1.6 ± 0.13 µmcm/Vs (pH 6) at 168 h (Figure 8 C, Tables S13, S14).

Results for 0.5 mg/L Ag in $AgNO_3$ solutions showed the same trends regarding mean sizes and EPM as solutions containing 5 mg/L Ag NO_3 (Tables S13, S14).

Dissolved Ag^+ in $AgNO_3$ solutions

The mean concentration of dissolved Ag in $AgNO_3$ solutions with and without EPS was 4414 ± 201 µg/L directly after the start of the experiments and was equal to the amount of total Ag detected. The concentration of dissolved Ag remained constant over 168 h (4435 ± 304 µg/L) except in EPS-containing samples at pH 6 and 7.6 incubated in light (Table 2). Here, the concentration of dissolved Ag decreased to 2470 ± 280 µg/L (pH 6) and 3575 ± 333 µg/L (pH 7.6), respectively. This corresponds to a decrease by 43% and 17%, respectively, compared to the measured total Ag concentration.

XAS and EDX analyses of selected samples showed that NP formed from $AgNO_3$ with EPS were dominantly metallic Ag (Figure S5E, S5F, and S6, Table S5).

Discussion

EPS characterization

We hypothesized that the interaction of engineered NP with EPS would among others be determined by the composition of EPS which depends on the biofilm community composition. We thus analyzed EPS extracted from periphyton with a focus on inter-community variability of EPS composition. All EPS extracts showed a similar size distribution of DOC indicating an overall constancy in the type of EPS components. Ratios of the individual groups of constituents varied within one order of magnitude indicating a certain community-dependent variability.

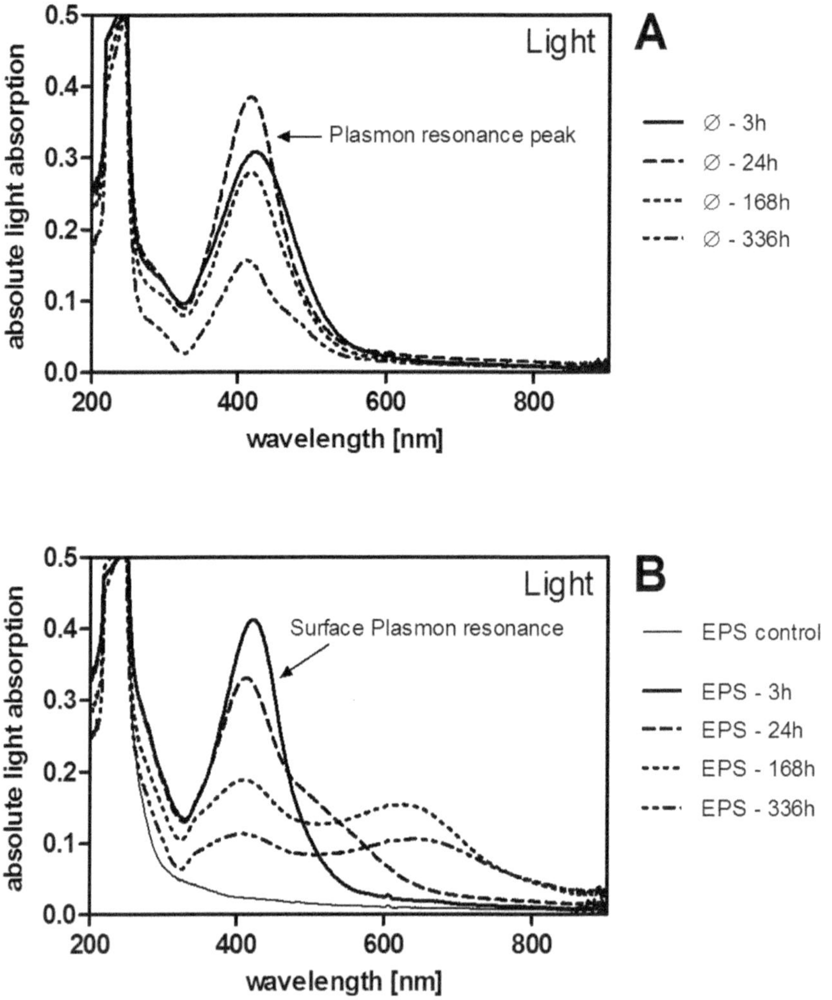

Figure 5. A–B: Spectra of absolute light absorption of 5 mg/L Ag NP as determined by UV-VIS spectroscopy. Samples were incubated in light (100 μE/m^{2}s) at pH 6 without (**A**) or with EPS (**B**, 10 mg DOC/L; 3–168 h spectra obtained with EPS extract 1; 336 h spectrum obtained with EPS extract 4). EPS control (**B**) contained 10 mg DOC/L EPS in 2 mM $NaHCO_3$. Different EPS extracts did not differ in absorption of light >300 nm.

Previously, an overall decrease in EPS C/N ratios has been attributed to an increase of protein [42]. Here, the overall ratio of cDOC to N and the amount of measured proteins did not correlate indicating a sample specific composition of nitrogen containing organic compounds. The biopolymer fraction of EPS mostly contains polysaccharides and proteins. A decrease in the C/N ratio in the biopolymer fraction is thus indicative of the ratio of polysaccharides to proteins [30]. Consequently, the concentration of polysaccharides is higher in EPS extracts with a higher C/N ratio in the biopolymer fraction and a higher percentage of biopolymers. Of the five EPS extracts studied here, extract 4 is thus expected to contain the lowest concentration of polysaccharides. The overall similarity as well as the difference in protein and polysaccharide concentration of the five extracts corroborate the results on the influence of EPS on CeO_2 and Ag NP and $AgNO_3$ as discussed below.

None of the samples showed a UV signal typical of humic substances at 45 min retention time (Figure S3) indicating that no HS were present [41]. This fact is important to distinguish the interaction of EPS extracts with NP in the present study from reports on the interaction of HS on NP stability [43,44].

Comparison of DLS and NTA results

DLS and NTA analyses resulted in different mean hydrodynamic diameters which is expected due to the different signals used to derive the particle sizes. The DLS derived z-average represents an intensity based mean size distribution while the mean diameter derived from NTA is based on the number of particle tracks acquired. In polydisperse samples, DLS derived average hydrodynamic diameters are expected to be larger than those derived from NTA due to the strong size dependence of the light scattering, especially for particles <100 nm. This probably explains the smaller average diameters obtained by NTA for CeO_2 dispersions compared to those obtained by DLS measurements.

However, mean diameters of AgNP dispersions below < 100 nm determined by NTA were larger than those determined by DLS, whereas those >100 nm were larger according to DLS. Tables S9, S11, and S13 list mean and mode of the size distribution determined by NTA next to z-averages determined by DLS. The difference in mean and mode is indicative of the polydispersity and non-normal size-distribution of the samples. We used NTA mean values to compare to DLS z-averages in the manuscript. The less polydisperse the samples are, the closer are z-

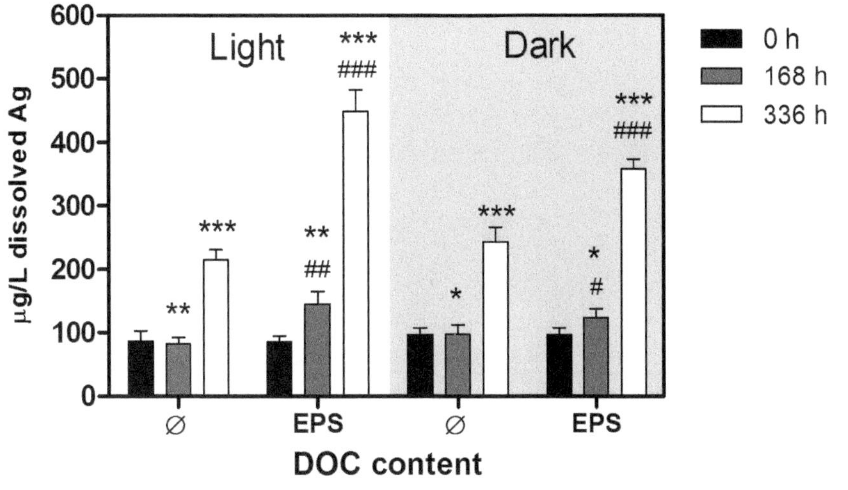

Figure 6. Mean mass concentrations with standard deviations of dissolved Ag in 5 mg/L Ag NP dispersions directly after addition of NP („0 h") and after 168 h and 336 h of incubation as determined by ICP-MS of acid-digested samples. Samples were incubated in the dark or in light (100 µE/m²s). n = 9 from 3 independent experiments, each experiment was done with one EPS extract (10 mg DOC/L). * Significant difference between EPS-free and EPS containing samples; # significant difference between light and dark treated samples, 168 h: $p = 0.0012$, $df = 3$ (between groups), $F = 5.774$, 336 h: $p < 0.0001$, $df = 3$ (between groups), $F = 66.56$.

average (DLS) and mean diameter (NTA). Consequently, the higher the polydispersity the stronger are the results influenced by the different biases of the methods [45].

Stability and dissolution of CeO_2 NP

Electron microscopy investigations revealed that primary particles (25 nm, Figure 2) were mostly aggregated as typically observed for particles synthesized by flame pyrolysis.

Average hydrodynamic diameters determined by DLS (~200–600 nm) and mean sizes measured by NTA (~200 nm) indicated that the aggregates detected by TEM remained stable over time independent of the experimental condition (pH, light/dark). As the primary particles were strongly sintered together, EPS did not disperse the aggregates into single primary particles and the average particle size remained unchanged. This observation is supported by the fact that the total Ce recovered from CeO_2 NP dispersions as determined by ICP-MS remained stable over time. The amount of suspended CeO_2 NP was thus constant.

The decreasing EPM within one week of light exposure in the presence of $NaHCO_3$ as with EPS also indicates a stabilization of CeO_2 NP. The decrease was not influenced by pH. According to speciation data modeled with Visual MINTEQ software, the major carbonate species were H_2CO_3* (aq) and HCO_3^- (68% and 32%) at pH 6 and HCO_3^- (95%/97%) at pH 7.6/8 (see table S3). These calculations suggest comparable effects of different carbonate species and negatively charged EPS components on the surface charge of CeO_2 NP. The negative EPM observed in this study are in line with the expected negative charge of EPS components at the pH conditions applied here. The negative charge is due to polysaccharide residues such as glucose, xylose, fucose, galactose, and uronic acids [29,46] and the presence of low M_r acids [41]. Stabilization of CeO_2 NP by artificial mixes of natural organic matter (NOM) has been reported previously [47]. As in our study which used a natural mix of NOM, mean zetapotentials (derived from EPM by the same model) were −21.8 mV to −55.8 mV corresponding to the overall negative charge of the organic molecules used.

We observed an increase in dissolved Ce over time. This may be explained by a redox-cycling of Ce atoms between Ce^{III} and Ce^{IV} on the surface of CeO_2 NP [23]. In contrast to Ce^{III}, Ce^{IV} is insoluble in aqueous environments [48]. Thus, the dissolved Ce was either already present as Ce^{III} and directly released from the NP surfaces or was released following reduction of Ce^{IV} to Ce^{III}. The amount of dissolved Ce was significantly larger at pH 6 than at pH 7.6 and 8.6. A higher Ce^{III}/Ce^{IV} ratio at lower pH is expected according to the pH-dependence of the reduction reaction [49]. An increase of Ce^{III} with a decrease in pH from 8.6 to 6 was found in perchlorate solution [49] and in natural sea water in the dark where the oxidation rate of Ce^{III} to Ce^{IV} increased between pH 7 to 7.5 but did not further increase between pH 7.5 and 8.6 [50]. Furthermore, the higher concentration of H^+ at pH 6 may lead to a competition between H^+ and Ce^{III} on the surface of CeO_2 NP and thus a more efficient desorption of Ce^{III}.

We found that the amount of dissolved Ce decreased in the presence of EPS. Light further reduced the concentration of dissolved Ce in the presence EPS while there was no significant difference between light/dark treated samples without EPS. The amount of free Ce^{III} at pH 6 was thus decreased in the presence of EPS as opposed to $NaHCO_3$ only. According to the above discussed hypothesis, this may be explained by a lower Ce^{III}/Ce^{IV} ratio following oxidation of Ce^{III} to Ce^{IV}. This is possibly due to two parallel processes. On the one hand, Ce^{III} ions on CeO_2 surfaces are potential sites for catalysis and photooxidation and -reduction [51]. Thus, Ce^{III} ions on the CeO_2 NP studied here may have reduced EPS components enhanced by photocatalysis which reduced the amount of Ce^{III} and thus the concentration of dissolved Ce. On the other hand, EPS components may have formed ROS in the light which oxidized Ce^{III} resulting in a lower amount of dissolved Ce^{III}. These processes would be in line with the observation that artificial mixes of NOM alone decreased the fraction of Ce^{III} in CeO_2 NP at neutral pH [52,53] and induced ROS in the light which then oxidized metal ions [54].

Stability, dissolution and *de novo* formation of Ag NP

Ag NP were in the same size range when dispersed in $NaHCO_3$ as reported previously (40–50 nm (DLS), pH 6–10 [10]) and remained stable independent of pH in the range of pH 6 to

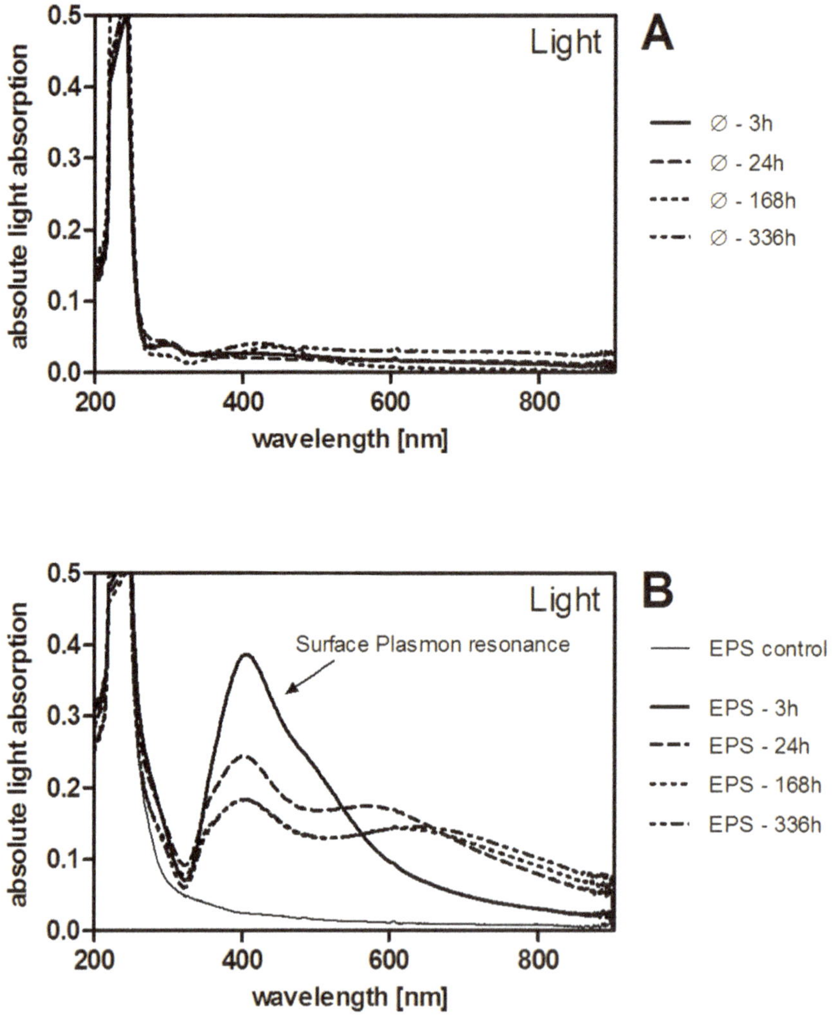

Figure 7. A–B: Spectra of light absorption of $AgNO_3$ solutions (5 mg/L Ag) as determined by UV-VIS spectroscopy. Samples were incubated in light (100 μE/m^{2}s) at pH 6 without (**A**) or with EPS (**B**, 10 mg DOC/L; 3 h –168 h spectra obtained with EPS extract 1; 336 h spectrum obtained with EPS extract 4). EPS control (**B**) contained 10 mg DOC/L EPS in 2 mM $NaHCO_3$. Different EPS extracts did not differ in the absorption of light >300 nm.

pH 8.6 and light/dark conditions. Similarly, EPS did not impact the stability of Ag NP dispersions in the dark. However, when EPS-containing Ag NP dispersions were exposed to light, Ag NP mean diameter as determined by DLS and NTA first increased in size with decreasing pH and then remained stable. In the presence of EPS, EPM were similar to those of EPS-treated CeO_2 NP dispersions indicating an interaction of EPS components with Ag NP despite the negative charge of both Ag NP and EPS. Previously, PVP-stabilized Ag NP were found to aggregate in the presence of EPS extracted from a single bacterial strain [33], while other Ag NP (unknown surface modification) were stabilized by bacterial EPS [55]. In both cases, pH and light/dark dependency of the processes were not assessed. Nevertheless, our results and the existing reports indicate that Ag NP surface modifications will influence the effect of different types of EPS on NP stability.

SPR spectra indicate that the increase in Ag NP average diameter in our study (pH 6 with EPS in light) was both due to the formation of stable NP agglomerates and an increase in individual NP diameter. In particular, SPR spectra showed a decrease in light absorption at the initial peak position along with a second maximum at higher wavelengths. The observed change in light absorbance of Ag NP dispersions results from agglomeration as well as an increase in individual particle size [56]. This observation corresponds to the increase in mean hydrodynamic diameter detected by DLS and NTA. TEM images of selected samples qualitatively support these data.

An increase in particle diameter will change the diffusion of Ag NP in periphyton as has been shown for heterotrophic biofilms [57] and thus the exposure of periphytic organisms.

As dominantly metallic Ag was detected by XAS and EDX in NP dispersions incubated with EPS at pH 6 in the light, we hypothesized that the underlying mechanism was pH-dependent and that EPS-enhanced photoreduction of Ag^+ on Ag NP surfaces was occurring. Previous studies have reported on interaction of Ag^+ with NOM [58] and Ag NP formation from Ag^+ in the presence of NOM (humic acid [44], humic and fulvic acids [43]) and whole cell extracts (marine algae [59], soy bean [60]), thus the formation of Ag NP from extracellular substances can be expected. Photoreduction of Ag^+ to metallic Ag in particular was observed in the presence of NOM [61] and without an additional reducing agent (reviewed by [62]). Formation of Ag_2S is not expected under

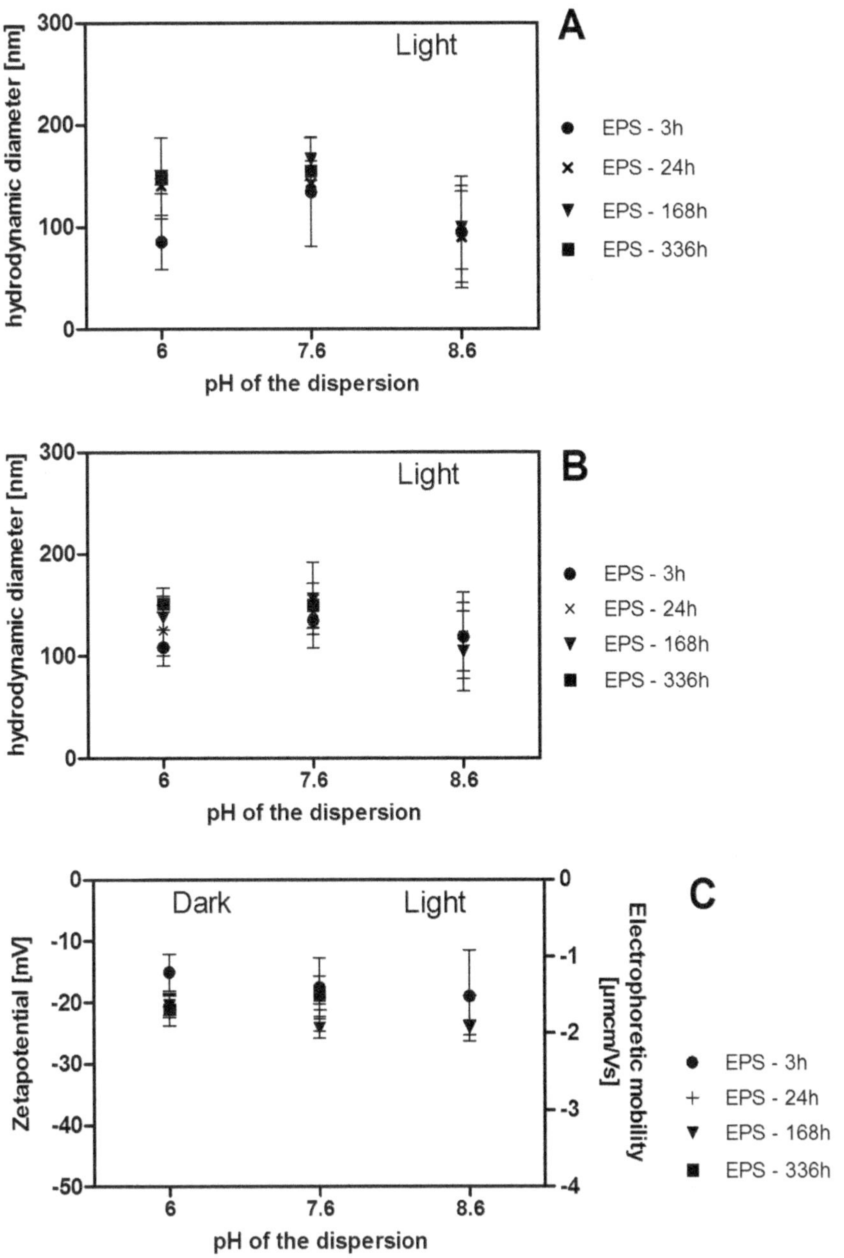

Figure 8. A–C: Mean hydrodynamic diameter of NP formed from $AgNO_3$ (5 mg/L dispersions) with standard deviations as determined DLS (intensity distribution (A) and NTA (B). Mean EPM and zetapotential with standard deviations determined by electrophoresis/DLS (C). Samples were incubated in light (100 μE/m^{2}s). 3 h: n = 15, 5 independent experiments (EPS extracts 1–5, 10 mg DOC/L). 168 h: n = 12, 4 independent experiments (EPS extracts 1–3, 5). 336 h: n = 3, 1 experiment (EPS extract 4).

these conditions, because all experiments were carried out in the presence of oxygen and there were no obvious sources of sulfide.

Experiments assessing the independent and combined effects of EPS, light and pH with $AgNO_3$ as a source of Ag^+ confirmed this hypothesis. We observed an Ag NP specific SPR signal under all conditions except in the dark without EPS and detected NP by DLS and NTA in $AgNO_3$ solutions with EPS in the light. However, the efficiency of NP formation was low in the light without EPS and in the dark with EPS resulting in a NP concentration too low for analysis by DLS and NTA. These results indicate that light and EPS alone reduce Ag^+ to metallic Ag inducing Ag NP formation and that a combination of both increases the reduction rate. This behavior is in contrast to the system CeO_2– EPS, where EPS seem to act as oxidizing agent. Effects of EPS to Ag^+/Ag°-NP and to $Ce^{IV}O_2$/Ce^{III} are in line with the different redox potential of these redox couples. Despite the size increase of individual Ag particles in the presence of EPS and light, the concentration of Ag^+ increases over time similar to EPS-free samples in the light/dark and to EPS-containing samples in the dark. Consequently, Ag NP must be oxidized to release Ag^+ parallel to photoreduction. Under supersaturated conditions, colloidal systems stabilize at a system-dependent size distribution steered by a process called Ostwald-ripening. The process is based on the reduction of energy in the system. Small crystals dissolve

Table 2. Average concentrations of total and dissolved Ag with standard deviations (SD) in $AgNO_3$ dispersions as determined by ICP-MS of acid-digested samples incubated in light (100 μE/m²s) with (+) or without (−) EPS (10 mg DOC/L) for 168 h.

		[Ag] μg/L			
EPS	**pH**	**Total Ag**	**SD**	**Dissolved Ag**	**SD**
−	6	4344	±279	4430	±150
−	7.6	4234	±70	4580	±162
−	8.6	4265	±113	4593	±78
+	6	4314	±173	2470	±280
+	7.6	4259	±44	3575	±333
+	8.6	4280	±42	4241	±129

n = 24 from 3 independent experiments.

driven by their high surface tension, while large crystals grow due to supersaturation of the solution and less surface tension to counteract growth. This process has been described for Ag NP synthesis [63]. In our system, the concentration of Ag^+ increases over time at all experimental conditions while the size-distribution only shifts to larger NP sizes in the presence of EPS and light in a pH-dependent manner. Reduction of Ag^+ to Ag° is only possible if a reductant is present (here the natural EPS components) and is more efficient under light conditions. Under these conditions the formation of Ag° at the surface of the larger Ag NP then leads to increased size.

The difference in the rate of Ag NP size increase and Ag NP formation between EPS extract 4 and 1–3/5 must be due to the different composition of EPS as there was no difference in dissolved Ag^+ in the presence of the different extracts. In a previous study, the formation of Ag nanoplates by algae extracts was explained by proteins reducing Ag^+ and controlling the shape of the nanomaterial with carboxyl groups and hydroxyl groups potentially being responsible for Ag^+ reduction [64]. However, polysaccharides which contain hydroxyl groups as well as hemiacetal ends are expected to be more reactive for reduction of Ag^+ [65]. We found that the extract that induced a slower increase in size and formation of Ag NP had the highest protein content, less high M_r biopolymers and low M_r acids and the lowest concentration of polysaccharides in the biopolymer fraction. This suggests that acidic groups of high M_r polysaccharides and/or low M_r acids in the EPS extracts were mainly involved in the reduction of Ag^+.

As the total Ag concentration remained stable in $AgNO_3$ solutions in which Ag NP formation had occurred, the decrease in dissolved Ag can be explained by the formation of Ag NP. As DLS and NTA cannot be used to quantify particle number, we used a combination of the NP diameter range determined by DLS and NTA and the decrease in dissolved Ag to estimate the number of particles formed. Based on an average NP diameter of 100 nm – 200 nm, roughly 3.6×10^{11}–4.5×10^{10} NP/L were formed at pH 6 and roughly 1.8×10^{11} -2.2×10^{10} NP/L were formed at pH 7.6. As samples incubated at pH 8.6 produced a DLS signal and contained NP according to NTA, a decrease in dissolved Ag^+ could be expected. However, the average DLS signal counts (Table S7) indicate only a small amount of Ag NP at pH 8.6. The corresponding decrease in dissolved concentration may thus fall in the range of the standard deviation of the ICP-MS measurements. In summary, the formation of Ag NP was most efficient at pH 6. Furthermore, the size increase of Ag NP was largest at pH 6. Consequently, the reduction of Ag^+ to Ag was most efficient at the lowest pH studied.

The reactive components of EPS that probably contributed to EPS-enhanced photoreduction (M_r biopolymers, especially polysaccharides, and low M_r organic acids) are expected to be more protonated at pH 6 compared to pH 7.6/8.6 resulting into a less negative charge. The more efficient reduction of Ag^+ and growth of Ag NP at lower pH seems counterintuitive. Less negative charge on relevant functional groups of EPS components at pH 6 may have resulted in a weaker interaction with Ag^+, and thus a less efficient detachment of Ag^+ ions from Ag NP surfaces, and a more efficient photoreduction and formation of Ag° at the NP surface. In the case of Ag NP formation from $AgNO_3$, a weaker interaction of Ag^+ with EPS components may also have facilitated the reduction to elemental Ag and Ag NP formation. Furthermore, the carbonate coating of the Ag NP is less negatively charged at lower pH, so that interactions with the EPS may be facilitated. At the same time, the interaction of EPS components, Ag NP and Ag^+ is probably more dynamic at pH 6 as both reduction and oxidation of Ag^+ will occur simultaneously, so that released Ag+ may be reduced again and lead to growth of the Ag NP. A pH-dependent process with an inverse relationship as described here was found for humic and fulvic acids which induced Ag NP formation from Ag^+ with an increase in pH (6.1 to 9) accelerating the formation of NP [66]. On the other hand, the release of Ag^+ from citrate-stabilized Ag NP in the presence of humic acids increased with increasing pH [67]. The processes of AgNP formation and growth are thus dependent on the pH dependence of the functional groups of EPS, of the surface coating of AgNP and of the reduction reactions.

Environmental relevance

In this study we present results from basic research on the interaction of selected engineered NP with EPS under controlled conditions. pH, light intensity, and the buffer system were selected according to their relevance to freshwater periphyton. The chemical make-up of stream water, natural colloids, and DOC other than EPS were not included to be able to analyze the specific effects of EPS.

Despite differences in relative chemical composition, the five EPS extracts studied had the same effect on CeO_2 NP stability and release of Ce^{III} as well as on photoreduction of Ag^+ resulting in pH-dependent Ag NP size increase and NP formation from $AgNO_3$. Merely the kinetics of Ag^+ reduction were slower in the presence of the EPS extract that probably contained the lowest concentration of polysaccharides as part of the biopolymer fraction and the lowest concentration of low M_r acids. These results indicate that EPS contains reactive biomolecules that may be decisive regarding the fate of Ag NP and Ag^+ in periphyton.

The similarity of newly formed and engineered Ag NP observed in this study demonstrates that it will be challenging to distinguish between both particle types in environmental samples. Knowing the origin of Ag NP is crucial to understanding sources of Ag species in freshwater systems, to tracking engineered Ag NP and to understanding potentially variable effects of Ag in the environment.

Conclusions

CeO_2 NP were stabilized by EPS extracted from periphyton and dissolved to a certain extent with decreasing pH and with time. Ag NP were likewise stabilized by EPS and increased in size with decreasing pH by EPS-enhanced photoreduction. EPS also enhanced the formation of Ag NP from $AgNO_3$ with a decrease in pH increasing the efficiency of *de novo* NP formation. The composition of the EPS extracts regarding biopolymers and possibly low M_r acids determined the kinetics of the photoreduction. The parameters tested in this study, namely photon flux, pH, and EPS change with light irradiation, the photosynthetic and respiratory activity of periphytic organisms, and with species composition. We conclude that local conditions in periphyton will influence the properties of the tested types of nanomaterials and thus the exposure of periphytic organisms.

Supporting Information

Figure S1 Water temperature in the channel used for periphyton colonization measured by a HOBO Pendant Temperature/Light Data Logger. Blue numbers indicate the EPS extract produced at the end of the respective period. Blue lines indicate the mean water temperature during the respective period.

Figure S2 Organic carbon (OC) signal chromatograms of EPS extracts 1–5 as determined by LC-OCD-OND. Samples were diluted 1:50 with nanopure water (Ω 18, Milli-Q).

Figure S3 UV signal (254 nm) chromatograms of EPS extracts 1–5 as determined by LC-OCD-OND. Samples were diluted 1:50 with nanopure water (Ω 18, Milli-Q).

Figure S4 OC chromatograms of two extracts obtained independently by two individuals on the same day from periphyton from 32 slides from two different channels after three weeks of colonization.

Figure S5 ***A*** TEM bright field image of Ag NP, ***B*** TEM bright field image of Ag NP exposed to EPS for 24 h (pH 6, light), ***C*** High angular annular dark field (HAADF) image and ***D*** Energy dispersive x-ray (EDX) spectrum of Ag NP exposed to EPS for 24 h (pH 6, light); ***E*** HAADF image and ***F*** EDX spectrum of $AgNO_3$ exposed to EPS for 24 h (pH6, light).

Figure S6 XANES(left) and EXAFS (right) spectra of two experimental samples (red) and the reference materials (black) used for the linear combination fitting (Table S5).

Figure S7 Spectra of light absorption of $AgNO_3$ solutions (5 mg/L Ag) as determined by UV-VIS spectroscopy. Samples were incubated in the dark at pH 6 with EPS (10 mg DOC/L; spectra obtained with EPS extract 1) EPS control contained 10 mg DOC/L EPS in 2 mM $NaHCO_3$. Different EPS extracts did not differ in the absorption of light >300 nm.

Figure S8 Representative DLS spectra of CeO_2 NPs in 2 mM $NaHCO_3$ without EPS after 168 h of incubation in light. A: 5 mg Ce/L CeO_2 NP, B: 0.5 mg Ce/L CeO_2 NP.

Table S1 NP properties.

Table S2 Chriesbach water chemistry.

Table S3 Speciation of the components of Ag NP dispersions and $AgNO_3$ solutions according to Visual MINTEQ modelling software. NO_3^- was 99.95% dissolved as NO_3^- and 0.05% $NaNO_3$ (aq) in all cases.

Table S4 Sampling scheme for NP characterization. 1 DLS, 2 NTA, 3 UV-VIS, 4 ICP-MS$^{\$}$. L : light, D : dark; $^{\$}$ only performed for 5 mg/L dispersions; * only for EPS extract 4 and corresponding controls.

Table S5 Fractions of Ag(0), Ag_2O and Ag_2S derived from LCF results of XANES and EXAFS spectra from experimental samples. The XANES spectra were fitted from 24490 to 26000 eV and the EXAFS from 3–10k. The individual fractions were constrained to range between 0 and 1 and the sum of the fitted fractions was left unconstrained (XANES) or was constraint to 1 (EXAFS).

Table S6 Average derived DLS count rates (kilocounts per second, kps) with standard deviations of Ag NP dispersions in 2 mM $NaHCO_3$ without EPS after 168 h of incubation in light.

Table S7 Average derived DLS count rates (kilocounts per second, kps) with standard deviations and estimated NP concentration in samples containing EPS and $AgNO_3$ (5 mg Ag/L) incubated at pH 6, 7.6, or 8.6 in the light for 168 h.

Table S8 Average derived DLS count rates (kilocounts per second, kps) with standard deviations of CeO_2 NP dispersions in 2 mM $NaHCO_3$ without EPS after 168 h of incubation in light.

Table S9 Z-averages (DLS), polydispersity (PDI), mode and mean diameters (NTA), zetapotential, and EPM of CeO_2 NP dispersions dependent on pH, light/dark, EPS content, and time. Each value is derived from three measurements of three replicates (3x 3n). Blue: data used to calculate mean values represented in Figures 1 A–D.

Table S10 Standard deviations of Z-averages (DLS), polydispersity (PDI), mode and mean diameters (NTA), zetapotential, and EPM of CeO_2 NP dispersions dependent on pH, light/dark, EPS content, and time.

Table S11 Mean Z-averages (DLS), polydispersity (PDI), mode and mean diameters (NTA), zetapotential, and EPM of Ag NP dispersions dependent on pH, light/dark, EPS content, and time. Each value is derived from three measurements of three replicates (3x 3n). Blue: data used to calculate mean values represented in Figures 4 A–D.

Table S12 Standard deviations of Z-averages (DLS), polydispersity (PDI), mode and mean diameters (NTA), zetapotential, and EPM of Ag NP dispersions dependent on pH, light/dark, EPS content, and time.

Table S13 Z-averages (DLS), polydispersity (PDI), mode and mean diameters (NTA), zetapotential, and EPM of NP formed in $AgNO_3$ solutions dependent on pH, light/dark, EPS content, and time. ns: no signal. Each value is derived from three measurements of three replicates (3x 3n). blue: data used to calculate mean values represented in Figures 8 A–C.

Table S14 Standard deviations of Z-averages (DLS), polydispersity (PDI), mode and mean diameters (NTA), zetapotential, and EPM of NP formed in $AgNO_3$ solutions dependent on pH, light/dark, EPS content, and time.

Text S1 Settings for DLS and NTA measurements.

Acknowledgments

We would like to thank Brian Sinnet for help with the sample preparation for TEM and XAS analyses, Andreas Voegelin for supporting the XAS analyses, Theodora Stewart and Carmen Gil for help with the periphyton facility, David Kistler and Adrian Ammann for maintaining the ICP-MS setup, and Jacqueline Traber for maintaining the LC-OCD setup. The Swiss Norwegian Beamline at the European Synchrotron Radiation Facility (SNBL, ESRF, Grenoble, France) are acknowledged for the allocation of beam time. We thank Sergey Nikitenko and Dipanjan Banerjee (both ESRF) for support at the Dutch Belgian Beamline (DUBBLE) at ESRF.

Author Contributions

Conceived and designed the experiments: AK RB LS. Performed the experiments: AK RK. Analyzed the data: AK RK. Contributed reagents/materials/analysis tools: AK RK. Wrote the paper: AK RK RB LS.

References

1. Navarro E, Baun A, Behra R, Hartmann N, Filser J, et al. (2008) Environmental behavior and ecotoxicity of engineered nanoparticles to algae, plants, and fungi. Ecotoxicology 17: 372–386.
2. Kaegi R, Voegelin A, Ort C, Sinnet B, Thalmann B, et al. (2013) Fate and transformation of silver nanoparticles in urban wastewater systems. Water Research 47: 3866–3877.
3. Lowe RL, LaLiberte GD (2007) Chapter 16 - Benthic Stream Algae: Distribution and Structure. Methods in Stream Ecology (Second Edition). San Diego: Academic Press. pp. 327–339.
4. Lamberti GA (1996) The Role of Periphyton in Benthic Food Webs. In: Stevenson RJ, Bothwell, M L., & Lowe, R L., editor. Algal ecology: Freshwater benthic ecosystems. San Diego: Academic Press.
5. Nikora VI, Goring DG, Biggs BJF (1998) A Simple Model of Stream Periphyton-Flow Interactions. Oikos 81: 607–611.
6. Ferry JL, Craig P, Hexel C, Sisco P, Frey R, et al. (2009) Transfer of gold nanoparticles from the water column to the estuarine food web. Nat Nano 4: 441–444.
7. Lawrence JR, Dynes JJ, Korber DR, Swerhone GDW, LeppardD GG, et al. (2012) Monitoring the fate of copper nanoparticles in river biofilms using scanning transmission X-ray microscopy (STXM). Chemical Geology 329: 18–25.
8. Kulacki KJ, Cardinale BJ, Keller AA, Bier R, Dickson H (2012) How do stream organisms respond to, and influence, the concentration of titanium dioxide nanoparticles? A mesocosm study with algae and herbivores. Environ Toxicol Chem 31: 2414–2422.
9. Salant NL (2011) Sticky business': The influence of streambed periphyton on particle deposition and infiltration. Geomorphology 126: 350–363.
10. Piccapietra F, Sigg L, Behra R (2012) Colloidal Stability of Carbonate-Coated Silver Nanoparticles in Synthetic and Natural Freshwater. Environmental Science & Technology 46: 818–825.
11. Rodea-Palomares I, Boltes K, Fernández-Piñas F, Leganés F, García-Calvo E, et al. (2011) Physicochemical characterization and ecotoxicological assessment of CeO2 nanoparticles using two aquatic microorganisms. Toxicological Sciences 119: 135–145.
12. Rogers NJ, Franklin NM, Apte SC, Batley GE, Angel BM, et al. (2010) Physico-chemical behaviour and algal toxicity of nanoparticulate CeO(2) in freshwater. Environmental Chemistry 7: 50–60.
13. Schulze C, Kroll A, Lehr CM, Schäfer UF, Becker K, et al. (2008) Not ready to use - Overcoming pitfalls when dispersing nanoparticles in physiological media. Nanotoxicology 2: 51–61.
14. Veriansyah B, Chun M-S, Kim J (2011) Surface-modified cerium oxide nanoparticles synthesized continuously in supercritical methanol: Study of dispersion stability in ethylene glycol medium. Chemical Engineering Journal 168: 1346–1351.
15. Jiang XJ, Tong MP, Li HY, Yang K (2010) Deposition kinetics of zinc oxide nanoparticles on natural organic matter coated silica surfaces. Journal of Colloid and Interface Science 350: 427–434.
16. Klaine SJ, Alvarez PJ, Batley GE, Fernandes TF, Handy RD, et al. (2008) Nanomaterials in the environment: behavior, fate, bioavailability, and effects. Environ Toxicol Chem 27: 1825–1851.
17. Maynard AD, Aitken RJ, Butz T, Colvin V, Donaldson K, et al. (2006) Safe handling of nanotechnology. Nature 444: 267–269.
18. Morones JR, Elechiguerra JL, Camacho A, Holt K, Kouri JB, et al. (2005) The bactericidal effect of silver nanoparticles. Nanotechnology 16: 2346–2353.
19. Kent RD, Vikesland PJ (2011) Controlled Evaluation of Silver Nanoparticle Dissolution Using Atomic Force Microscopy. Environmental Science & Technology 46: 6977–6984.
20. Odzak N, Kistler D, Behra R, Sigg L (2014) Dissolution of metal and metal oxide nanoparticles in aqueous media Environ Poll: 132–138.
21. Navarro E, Piccapietra F, Wagner B, Marconi F, Kaegi R, et al. (2008) Toxicity of silver nanoparticles to Chlamydomonas reinhardtii. Environ Sci Technol 42: 8959–8964.
22. Allan JD CM (2007) The Bicarbonate Buffer System. In: Allan JD CM, editor. Sream Ecology. 2nd Edition ed. Dordrecht, The Netherlands: Springer. pp. 67.
23. Imagawa H, Suda A, Yamamura K, Sun S (2011) Monodisperse CeO2 Nanoparticles and Their Oxygen Storage and Release Properties. The Journal of Physical Chemistry C 115: 1740–1745.
24. Binnemans K (2006) Chapter 229 Applications of tetravalent cerium compounds. In: Karl A. Gschneidner J-CGB, Vitalij KP, editors. Handbook on the Physics and Chemistry of Rare Earths: Elsevier. pp. 281–392.
25. Kim T, Hyeon T (2014) Applications of inorganic nanoparticles as therapeutic agents. Nanotechnology 25.
26. Shcherbakov AB, Usatenko AV, Ivanov V (2009) Structure-sensitive properties and biomedical applications of nanodispersed cerium dioxide. Russian chemical reviews 78: 855–871.
27. Sack M, Alili L, Karaman E, Das S, Gupta A, et al. (2014) Combination of Conventional Chemotherapeutics with Redox-Active Cerium Oxide Nanoparticles-A Novel Aspect in Cancer Therapy. Molecular Cancer Therapeutics 13: 1740–1749.
28. Flemming HC, Wingender J (2010) The biofilm matrix. Nat Rev Microbiol 8: 623–633.
29. Bellinger BJ, Gretz MR, Domozych DS, Kiemle SN, Hagerthey SE (2010) Composition of extracellular polymeric substances from periphyton assemblages in the florida everglades. Journal of Phycology 46: 484–496.
30. Stewart T, Traber J, Kroll A, Behra R, Sigg L (2013) Characterization of extracellular polymeric substances (EPS) from periphyton using liquid chromatography-organic carbon detection–organic nitrogen detection (LC-OCD-OND). Environmental Science and Pollution Research 20: 3214–3223.
31. Flemming H-C, Wingender J, Griebe T, Mayer C (2000) Physico-chemical properties of biofilms. Biofilms: Recent Advances in Their Study and Control: 19–34.
32. Khan SS, Mukherjee A, Chandrasekaran N (2011) Impact of exopolysaccharides on the stability of silver nanoparticles in water. Water Res 45: 5184–5190.
33. Joshi N, Ngwenya BT, French CE (2012) Enhanced resistance to nanoparticle toxicity is conferred by overproduction of extracellular polymeric substances. Journal of Hazardous Materials 241–242: 363–370.
34. Navarro E, Robinson CT, Wagner B, Behra R (2007) Influence of ultraviolet radiation on UVR-absorbing compounds in freshwater algal biofilms and Scenedesmus vacuolatus cultures. Journal of Toxicology and Environmental Health - Part A: Current Issues 70: 760–767.

35. Bradac P, Navarro E, Odzak N, Behra R, Sigg L (2009) Kinetics of cadmium accumulation in periphyton under freshwater conditions. Environmental Toxicology and Chemistry 28: 2108–2116.
36. Esposito S, Guerriero G, Vona V, Di Martino Rigano V, Carfagna S, et al. (2006) Glucose-6P dehydrogenase in Chlorella sorokiniana (211/8k): an enzyme with unusual characteristics. Planta 223: 796–804.
37. Bradford MM (1976) A rapid and sensitive method for the quantitation of microgram quantities of protein utilizing the principle of protein-dye binding. Anal Biochem 72: 248–254.
38. Evanoff DD Jr, Chumanov G (2005) Synthesis and optical properties of silver nanoparticles and arrays. Chemphyschem 6: 1221–1231.
39. Ravel B, Newville M. (2005) ATHENA, ARTEMIS, HEPHAESTUS: data analysis for X-ray absorption spectroscopy using IFEFFIT. Journal of Synchrotron Radiation 12: 537–541.
40. Kaegi R, Voegelin A, Sinnet B, Zuleeg S, Hagendorfer H, et al. (2011) Behavior of Metallic Silver Nanoparticles in a Pilot Wastewater Treatment Plant. Environmental Science & Technology 45: 3902–3908.
41. Huber SA, Balz A, Abert M, Pronk W (2011) Characterisation of aquatic humic and non-humic matter with size-exclusion chromatography – organic carbon detection – organic nitrogen detection (LC-OCD-OND). Water Research 45: 879–885.
42. Souza FD, Garg A, Bhosle NB (2005) Seasonal variation in the chemical composition and carbohydrate signature compounds of biofilm. Aquatic Microbial Ecology 41: 199–207.
43. Akaighe N, MacCuspie RI, Navarro DA, Aga DS, Banerjee S, et al. (2011) Humic acid-induced silver nanoparticle formation under environmentally relevant conditions. Environmental Science and Technology 45: 3895–3901.
44. Dubas ST, Pimpan V (2008) Humic acid assisted synthesis of silver nanoparticles and its application to herbicide detection. Materials Letters 62: 2661–2663.
45. Baalousha M, Lead JR (2013) Nanoparticle dispersity in toxicology. Nat Nanotechnol 8: 308–309.
46. De Brouwer JFC, Stal LJ (2002) Daily fluctuations of exopolymers in cultures of the benthic diatoms cylindrotheca closterium and nitzschia SP. (Bacillariophyceae)1. Journal of Phycology 38: 464–472.
47. Quik JT, Lynch I, Van Hoecke K, Miermans CJ, De Schamphelaere KA, et al. (2010) Effect of natural organic matter on cerium dioxide nanoparticles settling in model fresh water. Chemosphere 81: 711–715.
48. Elderfield H, Whitfield M, Burton JD, Bacon MP, Liss PS (1988) The Oceanic Chemistry of the Rare-Earth Elements [and Discussion]. Philosophical Transactions of the Royal Society of London Series A, Mathematical and Physical Sciences 325: 105–126.
49. Hayes SA, Yu P, O'Keefe TJ, O'Keefe MJ, Stoffer JO (2002) The Phase Stability of Cerium Species in Aqueous Systems: I. E-pH Diagram for the Ce - HClO 4 - H 2 O System. Journal of The Electrochemical Society 149: C623–C630.
50. Moffett JW (1994) The relationship between cerium and manganese oxidation in the marine environment. Limnology and Oceanography 39: 1309–1318.
51. Campbell CT, Peden CHF (2005) Oxygen vacancies and catalysis on ceria surfaces. Science 309: 713–714.
52. Baalousha M, Le Coustumer P, Jones I, Lead JR (2010) Characterisation of structural and surface speciation of representative commercially available cerium oxide nanoparticles. Environmental Chemistry 7: 377–385.
53. Thill A, Zeyons O, Spalla O, Chauvat F, Rose J, et al. (2006) Cytotoxicity of CeO2 nanoparticles for Escherichia coli. Physico-chemical insight of the cytotoxicity mechanism. Environmental Science and Technology 40: 6151–6156.
54. Buschmann J, Canonica S, Sigg L (2005) Photoinduced oxidation of antimony(III) in the presence of humic acid. Environmental Science and Technology 39: 5335–5341.
55. Khan SS, Mukherjee A, Chandrasekaran N (2011) Impact of exopolysaccharides on the stability of silver nanoparticles in water. Water Research 45: 5184–5190.
56. Kelly KL, Coronado E, Zhao LL, Schatz GC (2002) The Optical Properties of Metal Nanoparticles: The Influence of Size, Shape, and Dielectric Environment. The Journal of Physical Chemistry B 107: 668–677.
57. Peulen T-O, Wilkinson KJ (2011) Diffusion of Nanoparticles in a Biofilm. Environmental Science & Technology 45: 3367–3373.
58. Wirth SM, Lowry GV, Tilton RD (2012) Natural Organic Matter Alters Biofilm Tolerance to Silver Nanoparticles and Dissolved Silver. Environmental Science & Technology 46: 12687–12696.
59. Prasad TNVKV, Kambala VSR, Naidu R (2013) Phyconanotechnology: Synthesis of silver nanoparticles using brown marine algae Cystophora moniliformis and their characterisation. Journal of Applied Phycology 25: 177–182.
60. Vivekanandhan S, Misra M, Mohanty AK (2009) Biological Synthesis of Silver Nanoparticles Using Glycine max (Soybean) Leaf Extract: An Investigation on Different Soybean Varieties. Journal of Nanoscience and Nanotechnology 9: 6828–6833.
61. Glover RD, Miller JM, Hutchison JE (2011) Generation of metal nanoparticles from silver and copper objects: Nanoparticle dynamics on surfaces and potential sources of nanoparticles in the environment. ACS Nano 5: 8950–8957.
62. Sharma VK, Yngard RA, Lin Y (2009) Silver nanoparticles: Green synthesis and their antimicrobial activities. Advances in Colloid and Interface Science 145: 83–96.
63. Lin XZ, Teng X, Yang H (2003) Direct Synthesis of Narrowly Dispersed Silver Nanoparticles Using a Single-Source Precursor. Langmuir 19: 10081–10085.
64. Xie J, Lee JY, Wang DIC, Ting YP (2007) Silver Nanoplates: From Biological to Biomimetic Synthesis. ACS Nano 1: 429–439.
65. Park Y, Hong YN, Weyers A, Kim YS, Linhardt RJ (2011) Polysaccharides and phytochemicals: a natural reservoir for the green synthesis of gold and silver nanoparticles. IET Nanobiotechnol 5: 69–78.
66. Adegboyega NF, Sharma VK, Siskova K, Zbořil R, Sohn M, et al. (2013) Interactions of aqueous Ag+ with fulvic acids: Mechanisms of silver nanoparticle formation and investigation of stability. Environmental Science and Technology 47: 757–764.
67. Fabrega J, Fawcett SR, Renshaw JC, Lead JR (2009) Silver nanoparticle impact on bacterial growth: Effect of pH, concentration, and organic matter. Environmental Science and Technology 43: 7285–7290.

9

New Therapy of Skin Repair Combining Adipose-Derived Mesenchymal Stem Cells with Sodium Carboxymethylcellulose Scaffold in a Pre-Clinical Rat Model

Cristiano Rodrigues[1], Adriano M. de Assis[1,4], Dinara J. Moura[2], Graziele Halmenschlager[3], Jenifer Saffi[2], Léder Leal Xavier[5], Marilda da Cruz Fernandes[6], Márcia Rosângela Wink[1]*

1 Laboratórios de Pesquisa em Biologia Celular, Departamento de Ciências Básicas da Saúde, Universidade Federal de Ciências da Saúde de Porto Alegre, Porto Alegre/RS, Brazil, **2** Genética Toxicológica, Departamento de Ciências Básicas da Saúde, Universidade Federal de Ciências da Saúde de Porto Alegre, Porto Alegre/RS, Brazil, **3** Estresse Oxidativo e Poluição Atmosférica, Departamento de Ciências Básicas da Saúde, Universidade Federal de Ciências da Saúde de Porto Alegre, Porto Alegre/RS, Brazil, **4** Programa de Pós-Graduação em Ciências Biológicas: Bioquímica, Departamento de Bioquímica, Universidade Federal do Rio Grande do Sul, Porto Alegre/RS, Brazil, **5** Laboratório de Biologia Celular e Tecidual, Departamento de Ciências Morfofisiológicas, Pontifícia Universidade Católica do Rio Grande do Sul, Porto Alegre/RS, Brazil, **6** Laboratório de Pesquisa em Patologia, Departamento de Ciências Básicas da Saúde, Universidade Federal de Ciências da Saúde de Porto Alegre, Porto Alegre/RS, Brazil

Abstract

Lesions with great loss of skin and extensive burns are usually treated with heterologous skin grafts, which may lead rejection. Cell therapy with mesenchymal stem cells is arising as a new proposal to accelerate the healing process. We tested a new therapy consisting of sodium carboxymethylcellulose (CMC) as a biomaterial, in combination with adipose-derived stem cells (ADSCs), to treat skin lesions in an *in vivo* rat model. This biomaterial did not affect membrane viability and induced a small and transient genotoxicity, only at the highest concentration tested (40 mg/mL). In a rat wound model, CMC at 10 mg/mL associated with ADSCs increased the rate of cell proliferation of the granulation tissue and epithelium thickness when compared to untreated lesions (Sham), but did not increase collagen fibers nor alter the overall speed of wound closure. Taken together, the results show that the CMC is capable to allow the growth of ADSCs and is safe for this biological application up to the concentration of 20 mg/mL. These findings suggest that CMC is a promising biomaterial to be used in cell therapy.

Editor: Gobardhan Das, University of KwaZulu-Natal, South Africa

Funding: C. Rodrigues was the recipient of a Master fellowship from CAPES-REUNI program (Coordenação de aperfeiçoamento de Pessoal de Nível Superior); M.R. Wink, L.L. Xavier, and J. Saffi are recipients of research fellowships from CNPq (Conselho Nacional de Desenvolvimento Científico e Tecnológico). This study was supported by PRONEX and PRONEM - FAPERGS (Fundação de Amparo à Pesquisa do Estado do Rio Grande do Sul) and PROBITEC-CAPES. The funders had no role in study design, data collection and analysis, decision to publish, or preparation of the manuscript.

Competing Interests: The authors have declared that no competing interests exist.

* E-mail: mwink@ufcspa.edu.br

Introduction

The skin is the largest organ of the human body and its integrity allows body functions such as elimination of toxins, resistance to mechanical stress, prevention of water loss, participation in immune responses and protection against external agents [1,2,3]. However, many factors and systemic conditions may compromise the dermal function, such as, diabetes mellitus, trauma, inflammatory diseases, metabolic abnormalities, bleeding disorders, smoking, immunosuppression, malnutrition and obesity [1]. Therefore, promotion of skin lesion healing, regardless of the nature of the causative agent, is of utmost importance [4].

After injury to skin, a new epithelium has to form in order to close the wound and restore the barrier function. This requires the proliferation and directional migration of keratinocytes, fibroblasts and mesenchymal stem cells (MSCs) to wound edges [5,6]. MSCs are essential for skin homeostasis and repair of the lesions, by promoting cell differentiation, immune-modulation, secretion of growth factors to drive re-epithelialization and neovascularization and modulating resident stem cells [7,8]. Both, proliferation and migration of these cells are controlled by extracellular hormones, e.g., glucocorticoids and paracrine factors [5,6], which provides attractive opportunities for therapeutic interventions.

In the last decade, studies have indicated that human adipose tissue contains populations of MSCs, called adipose-derived stem cells (ADSCs) [9]. In contrast to stem cells obtained from bone marrow (BM-MSCs), whose access can be painful for the patient, requiring general or spinal anesthesia and most often resulting in low numbers of MSCs after processing [10], ADSCs can be collected from adipose tissue, as the waste material after liposuction in aesthetic surgeries [11]. It is estimated that 1 mL of lipoaspirate contains at least 7.3×10^3 fibroblast colony forming (CFU-F), corresponding to 0.5–5% of the total population of stromal vascular fraction (SVF) (4.04 ± 3.26 on average). Therefore, the SVF adipose tissue contains a quantity of CFU-F, 400 times higher than the bone marrow, where the frequency of CFU-F is estimated to be about 1 per 10^5 cells [12].

Thus, the possibility of deriving ADSCs from adult adipose tissues, the relatively easy protocols for their isolation and expansion in culture, and the fact that they are immunologically well tolerated, allowing the transplant from one individual to another of the same species, makes them attractive candidates for new studies in translational clinical research [9].

Besides the availability and abundance of ADSCs in the body, several steps remain to be solved before these cells become effectively and safely used as an established therapy. One of the obstacles is to find the correct substances or molecules capable to offer the correct environment to these cells, allowing their safe deployment in tissues or lesions. The long-term success of these therapies requires biocompatibility of the biomaterial that hosts the tissue while retaining the functionality of the cells within the implanted tissue [13]. Sodium carboxymethylcellulose (CMC) is synthesized by the hydration of the cellulose with sodium hydroxide and from an alkaline pulp catalyzed reaction with chloroacetic acid [13]. This synthetic polymer has been shown to be suitable for the protection and treatment of skin wounds, including partial thickness burns [14]. It has a strong ability to absorb and transport fluids and protect from bacterial exposure of the external environment [14]. These attributes make CMC suitable for use as a polymer matrix for wound healing, although its toxicity was not yet fully evaluated [15]. Over the years, CMC has been used as a single material with or without combination with drugs and co-excipients in partial thickness wounds, diabetic ulcers, lesions of the foot and deep dermal filler [15].

Considering that ADSCs have a great potential for clinical use in tissue repair and regeneration, but its efficacy depends of biomaterials that allow the proper tissue-engineering products, the aim of this study was to test the applicability of CMC as a scaffold for the ADSCs, to treat dermal lesions in a pre-clinical model.

Materials and Methods

Ethical Aspects

The protocols used in this study were approved by the Ethics Committee on Animal Use (CEUA) of Universidade Federal de Ciências da Saúde de Porto Alegre (UFCSPA), under the number **10-009**, following the resolutions of the the CONCEA (Conselho Nacional de Controle de Experimentação Animal). The NIH "Guide for the Care and Use of Laboratory Animals" (NIH publication N° 80–23, revised 1996) was followed in all experiments. All surgery was performed under ketamine and xylazine anesthesia and all efforts were made to minimize the animals suffering.

Chemicals

Dulbecco's modified Eagle medium low glucose (DMEM), Ca^{2+} and Mg^{2+} free Hank's balanced salt solution containing 10 mM sodium HEPES (HB-CMF-HBSS), collagenase Type I, trypsin–EDTA, penicillin/streptomycin and carboxymethylcellulose sodium (medium viscosity) were obtained from Sigma Chemical Co. (St. Louis, M.O., USA); fetal bovine serum (FBS) from Cultilab (São Paulo, S.P., Brazil); trypan blue (TB) was purchase from GIBCO (Grand Island, N.Y., USA). The measurement of the lactate dehydrogenase activity was determined by commercial kit (Labtest, Lagoa Santa, M.G., BR). Low-melting point agarose and agarose were obtained from Invitrogen (Carlsbad, C.A., USA).

Extraction and Cultivation of ADSCs

The ADSCs were extracted from visceral adipose tissue of Wistar rats (6–8 weeks), as previously described [16]. Briefly, the abdominal fat tissue from 2 rats was transferred to a petri dish, washed with Hanks solution and dissociated mechanically with a scalpel. The fat fragments were transferred to a Falcon tube containing 3 mg/mL collagenase Type I solubilized in DMEM medium without FBS and incubated in a water bath (37°C) for 45 minutes (being vortexed every 15 minutes). After this time twice the volume of DMEM supplemented with 10% FBS was added, to inactivate the collagenase. Cells were centrifuged for 10 minutes at 1000 rpm. The supernatant was discarded and cells were resuspended in a known volume of DMEM. Cell count was performed microscopically, and cell viability was assessed by trypan blue exclusion. 5×10^6 cells/cm^2 were seeded in six-well plates and cultured at 37°C in an incubator with 5% CO_2 in the presence of 3 mL DMEM low glucose medium supplemented with FCS 10%, penicillin (100 U/mL), streptomycin (100 mg/mL). After 24 hours of incubation, half of the culture medium was removed and replaced with fresh medium. At the end of 72 hours, the culture medium was removed completely and the non-adherent fraction was subtracted. The attached fraction was maintained until confluence, generally the fourth passage, when cells were harvested to be used in experiments.

Osteogenic and Adipogenic Differentiation of ADSCs

Cell differentiation was performed according the technique already described [17]. Osteogenic differentiation was induced by culturing ADSCs for up to 8 weeks in DMEM supplemented with 10^{-8} M dexamethasone, 5 µg/mL ascorbic acid 2-phosphate and 10 mM β-glycerophosphate. In order to observe calcium deposition, cultures were washed once with PBS, fixed with 4% paraformaldehyde in PBS for 15–30 minutes, and stained for 5 minutes with Alizarin Red S stain at pH 4.2. Excess of stain was removed by several washes with distilled water. To induce adipogenic differentiation, ADSCs were cultured for up to 8 weeks in DMEM supplemented with 10^{-8} M dexamethasone, 2.5 µg/mL insulin, 100 µM indomethacin and, in some experiments, 3.5 µM rosiglitazone. Adipocytes were easily discerned from the undifferentiated cells by phase-contrast microscopy. To further confirm their identity, cells were fixed with 4% paraformaldehyde in PBS for 1 hour at RT and stained with Oil Red O solution (three volumes of 3.75% Oil Red O in isopropanol plus two volumes of distilled water).

Immunophenotyping Analysis

The ADSCs were characterized to confirm the presence or absence of MSC surface markers using flow cytometry. The cells were trypsinized, washed twice with PBS, centrifuged and incubated for 25 minutes at 4°C with the antibodies described in the Table 1. Excess of antibody was removed by washing with PBS. The cells were analyzed using a FACScalibur cytometer equipped with 488 nm argon laser (Becton Dickinson, San Diego, CA) and with the CellQuest software. Gating was set by using unstained cells and at least 10,000 events were collected [18].

Scaffold Preparation

Sterile sodium carboxymethylcellulose (CMC) salt of medium viscosity was diluted in laminar flow with the aid of magnetic stirrer and glass rod in low glucose DMEM medium, previously supplemented with 3.7 g of sodium bicarbonate and 2.5 g of HEPES per liter and 10% FBS. The CMC was diluted to achieve the concentrations of 10 mg/mL, 20 mg/mL and 40 mg/mL and was used to test different viscosities and behavior of cells in culture.

Table 1. List of Antibodies.

Primary antibody	Host	Dilution	Supplier	Cat. N°
CD45 - PE	Mouse	1:1000	Invitrogen	MR6404
CD29 - PE	Mouse	1:1000	Invitrogen	A14886
CD90.1 - PerCP Cy5.5	Mouse	1:1000	Invitrogen	A14798
CD11b*	Mouse	1:1000	Invitrogen, USA	MR6200

*The unconjugated antibody CD11b was incubated with the secondary Anti-Mouse IgG – FITC antibody from Sigma Chemical Co. (St. Louis, M.O., USA) for 25 minutes

Lactate Dehydrogenase Activity Assay

MSCs on the 5th passage were plated with 3 mL of DMEM without phenol red in each well of 6-well plates (in triplicate), and allowed to growth until reaching a 70% confluence. Then, cells were treated with CMC at the concentrations of 10, 20 and 40 mg/mL. As a negative control we used cells cultured with standard culture medium, and as a positive control, cells lysed with Triton X-100 1%. After 24 hours or five days of growth in contact with CMC, the LDH cytotoxicity assay was performed following the manufacture protocol. The LDH activity was quantified at absorbance of 500 nm by a multiwell spectrophotometer (SpectraMax Plus Microplate Spectrophotometer, Molecular Devices, US). LDH activity in each cell supernatant was expressed by comparing it to the total enzyme activity in cells lysed with 1% of Triton X-100.

The Alkaline Comet Assay

The alkaline comet assay was performed as described in literature with minor changes [19,20]. After treatment of the cells with the biomaterial, cells were washed with ice-cold PBS, trypsinized with 100 μL trypsin (0.15%) and re-suspended in complete medium. Next, 20 μL of cell suspension (~10^6 cells/mL) was suspended in 0.75% low-melting point agarose and immediately spread onto a glass microscope slide pre-coated with a layer of 1% normal melting point agarose. The slides were incubated in ice-cold lysis solution (2.5 M NaCl, 10 mM Tris, 100 mM EDTA, 1% Triton X-100 and 10% DMSO, pH 10.0) at 4°C for a minimum of 1 hour to remove cellular proteins. The slides were then placed in a horizontal electrophoresis box containing a freshly prepared alkaline buffer (300 mM of NaOH and 1 mM of EDTA, pH ~13.0) at 4°C for 20 minutes to allow DNA unwinding. DNA electrophoresis was performed with a 300 mA and 25 V (0.90 V/cm) electric current for 20 minutes. The slides were then neutralized (0.4 M of Tris, pH 7.5), stained with silver nitrate as described in literature and analyzed using an optical microscope [21]. Images of 100 randomly selected cells (50 cells from each of two replicate slides) were analyzed for each test of different substance concentration. Cells were scored visually into five classes, according to tail size (from undamaged - 0, to maximally damaged - 4), and a damage index (DI) value was assigned to each comet according to its class. Visual scoring of comets is a valid evaluation method determined by international guidelines and recommendations for the comet assay [22]. The DI is an arbitrary score calculated for cells in different damage classes that were visually scored by measuring the DNA migration length and the amount of DNA in the tail. The DI ranged from 0 (no tail: 100 cells × 0) to 400 (with maximum migration: 100 cells × 4).

Micronucleus Assay

The micronucleus assay was performed according to Bonacker et al. (2004) with minor modifications. After treatment, cultures were washed twice with the media, and cytochalasin B (Cyt-B) was added at a final concentration of 2 μg/mL. Cultures were harvested 36 hours after Cyt-B addition. Cells were trypsinized and the cellular suspension was centrifuged at 2000 rpm for 5 minutes. Cells were then resuspended in 75 mM KCl solution and maintained at 4°C for 3 minutes (mild hypotonic treatment). Subsequently, the cells were centrifuged and a methanol/acetic acid (3:1) solution was gently added. This fixation step was repeated twice and cells were then resuspended in a small volume of methanol/acetic acid and dropped onto clean slides. The fixed cells were hydrolyzed in HCl and stained according to the Feulgen method [23]. Slides were mounted and coded prior to analysis. Micronuclei were counted in 1000 binucleated cells (BNCs) with well-preserved cytoplasms. The identification of micronuclei was carried out according to described in literature [24].

Skin Wound Model and Tissue Preparation

We used a total of 16 adult male Wistar rats (8 weeks), to avoid the influence of hormonal changes that could cause interferences in the healing process. The dorsal region was chosen to avoid animal access to their own wound and animals were housed one per box to prevent cross-access to the lesions. The animals were anesthetized with an intraperitoneal injection of ketamine and xylazine (90 mg/kg and 10 mg/kg, respectively), followed by shaving the dorsal region [25]. Then, with a punch, the animals were subjected to four circular lesions, measuring 7 mm of diameter, two upper and two lower, comprising dermis and epidermis. Immediately afterwards, the animals received topical treatment in the following order: Injury upper left (lesion 1): 100 μL of CMC 20 mg/mL+26×10^6 of ADSCs; injury upper right (lesion 2): 100 μL of CMC 10 mg/mL+ 26×10^6 of ADSCs; injury bottom left (lesion 3): 100 μL of CMC 20 mg/mL and injury bottom right (lesion 4): Sham, untreated. After the procedure, to ensure that the biomaterial stays housed in the lesion, the wounds were bandaged with polyurethane from 3 M (St. Paul, M.N., USA), permeable to oxygen, but acting as a barrier to liquids and bacteria. This bandage normally fell off naturally in the first or second day after surgery. After topical treatment, to analyze the wound healing, the animals were randomly organized in 4 groups, with 4 rats each: Group 1, the animals were euthanized four days after surgery; group 2, eight days after surgery; group 3, twelve days after surgery and group 4 were euthanized sixteen days after surgery. Animals were euthanized in CO_2 chamber and the lesions were removed with a scalpel, keeping 2 mm of intact skin around. Tissues were stored in phosphate buffered formaldehyde 4% for subsequent histological analysis.

Determination of Lesion Area

Lesions were photographed every day, next to one scale, using Sony Cyber Shot camera (Full HD 1080, 7.2 mega pixels, DSC-W120) and the area of lesion was measured using Image Pro Plus 6.1 software (IPP 6.1) (Media Cybernetics, CA, USA), after calibration.

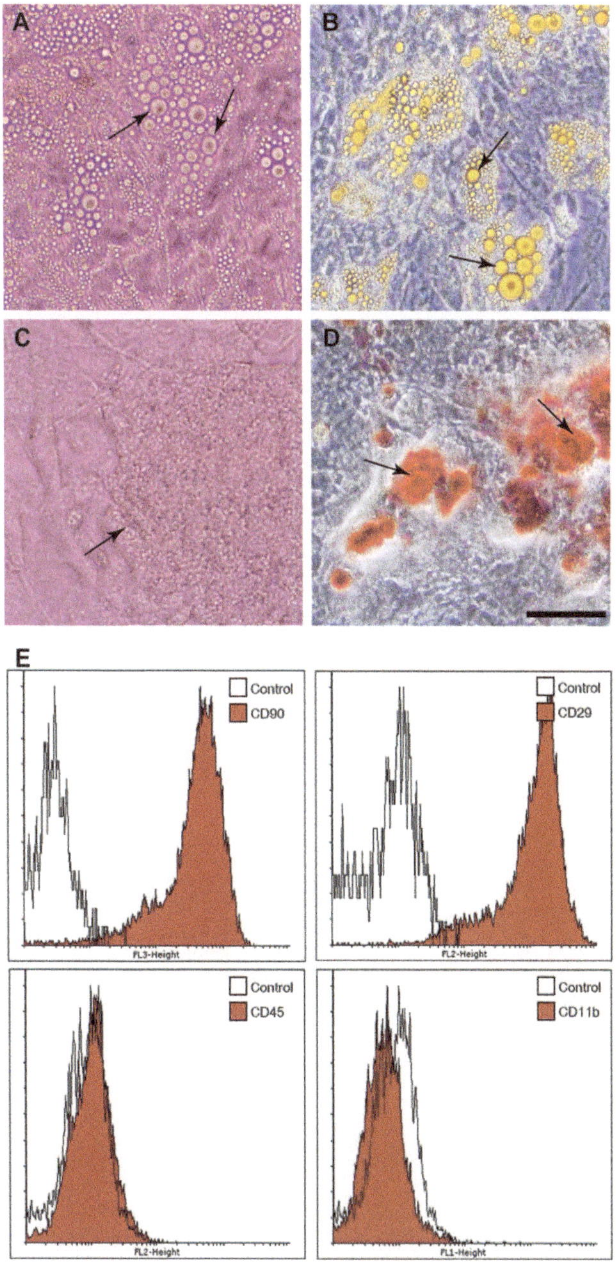

Figure 1. Analysis of the stem cell character of ADSCs extracted from rats. The differentiation of ADSCs is shown in light microscopy at 200X magnification. A. Cells differentiated into adipocytes showing the lipid vacuoles unstained. B. Cells differentiated into adipocytes showing the lipid vacuoles stained with Oil Red. C. Cells differentiated into osteoblasts showing the bone matrix in phase contrast. D. Cells differentiated into osteoblasts showing the bone matrix stained by Alizarin Red. A–D: Scale bar = 100 µm. E. Immunophenotyping analysis of surface marker expression in ADCSs. Flow cytometry histograms show the expression (shaded) of selected molecules (CD90, CD29, CD11b and CD45) by ADSC populations compared with controls (unshaded peaks).

Immunostaining of Cytokeratin

The tissue was included in paraffin being sectioned into slices of 4 µM and de-paraffinized. After, the slides were placed in sodium citrate pH 6.0, heated at 92°C for 40 minutes, with drawn to cool to room temperature, for 20 minutes and then passed into distilled water for antigen retrieval. The endogenous peroxidase was blocked by placing the slides in a mixture of 95 mL of methanol plus 5 mL of hydrogen peroxide 30 V for 5 minutes, protected from light. The tissues were washed twice with distilled water and then with PBS for 5 minutes, under stirring. To block the unspecific bindings, the samples were incubating with a 1% BSA, for 30 minutes. Subsequently, each slide was incubated with a volume of 50 mL of PBS plus a primary antibody at a dilution of 1:1000. The incubation occurred for 30 minutes at room temperature, and after at 4°C overnight. The next day, the slides were washed in PBS twice for 5 minutes and incubated with the secondary antibody (DAKO kit ADVANCE HPR), applied on each slide for 40 minutes at room temperature. Samples were washed with PBS twice for 5 minutes and incubated with tertiary antibody of the same kit. Revelation was performed with chromogen substrate (DAB), dehydration twice for 5 minutes, with xylene and assembled with Entellan. Reaction product of immunohistochemistry for cytokeratin was measured semi-quantitatively, by densitometry, using a Olympus BX-50 microscope, optical lens (10X/0.30 Ph1-UplanFI), coupled to a Motican 2500 camera and Image Pro Plus Software 6.1 (IPP 6.1) (Media Cybernetics, C.A., USA). The background correction and background staining subtraction were performed in accordance with a previously described protocol [26]. To measure optical density, at least two sections were analyzed in each case. Four images were obtained per section and four areas of interest (AOIs) were laid over the cytoplasmatic portions of epithelium in each digitized image. The epithelium thickness was measured in the same images, using an IPP 6.1 tool. At least four measurements were done in each image.

Epithelium Thickness

Slides with hematoxylin-eosin staining were also prepared in order to measure the epithelium thickness. This measurement was performed using Image Pro Plus 6.1 software (IPP 6.1) (Media Cybernetics, C.A., USA), after adjustment of calibration blade. For acquisition was used BX-50 Olympus microscope with optical lens (10X/0.30 Ph1-UplanFI), coupled to a camera Motican 2500. For each wound of each animal, two images were made, and four measurements for each image.

Collagen Fibers Analysis

After the preparation of histological slides, we used the staining of picrosirius, in order to examine whether the treatment used for healing was able to change the number of collagen fibers [27]. The images were taken with optical lens Olympus BX-50 microscope (10X/0.30 Ph1-UplanFI), using polarizing filter to highlight the collagen Type I and Type III on a black background. For each lesion of each animal, three fields of six sections were captured, using microscope images software DP2-BSW ver. 2.2 (Build 6212). The analysis was performed by the Image Pro Plus 6.1 software, proceeding with the marking of pixels according to their birefringence pattern (greenish/yellow-greenish or orange, orange-reddish). Quantification was performed on the percentage histogram of the software itself, totaling 16 points through injury.

Statistics

The toxicity experiments were repeated three times independently. The results were expressed as mean ± SD. Data were analyzed by one-way variance (ANOVA) for the toxicological analysis and two-way for histological parameters. The means were compared by Tukey's test for multiple comparisons, with $p<0.05$ considered statistically significant. Statistical analysis was performed using Prism 4.0 software (GraphPad, La Jolla, CA).

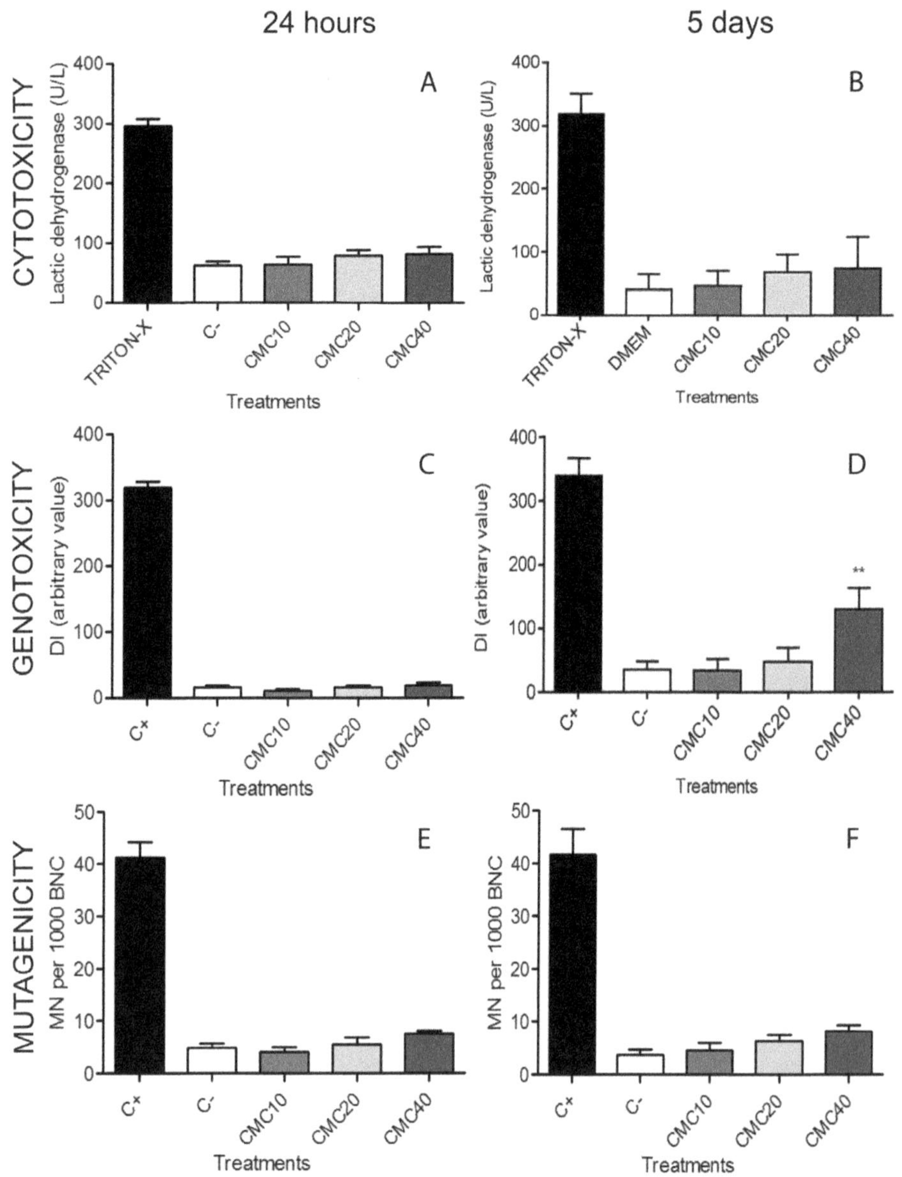

Figure 2. Determination of CMC toxicity on ADSCs cells. The cytotoxicity of CMC (A and B) was evaluated by measuring the LDH enzymatic activity on supernatant of ADSCs. The results are expressed in U/L and were analyzed by one-way ANOVA, followed by Tukey's multiple comparisons test. The values were considered significantly different from controls, when $P<0.05$. The potential genotoxicity of CMC (C and D) was analyzed by alkaline comet assay, as described in material and methods. Results are expressed as mean of damage index (DI) ± standard deviation of three independent experiments performed in triplicate. There was statistical significance when comparing CMC in the concentration of 40 mg/mL with the negative control (**$P<0.01$/One-way ANOVA Tukey's multiple comparison test). The potential of mutagenicity of CMC (E and F) was evaluated by the *in vitro* cytokinesis-block micronucleus method. The quantification was performed by counting the micronuclei in 1000 binucleated cells (BNCs) with well-preserved cytoplasms. The results were analyzed by one-way ANOVA, followed by Tukey's multiple comparisons test. The values were considered significantly different from controls, when $P<0.05$. All toxicity tests were performed in short and long periods of exposition of ADSCs to biomaterial, 24 hours (A, C and E) and 5 days (B, D and F), respectively. CMC: Sodium carboxymethylcellulose; ADSC: Adipose-derived stem cell. MN: micronucleus.

Results

Confirmation of the ADSC Stem Cell Character

Adherent cells, derived from samples of adipose tissue of rats, showed elongated fusiform morphology with a central nuclei. Cells could be maintained in culture for at least 18 passages. The multipotency of the populations studied was demonstrated by incubating the cells in media promoting differentiation into the osteogenic or adipogenic lineage. ADSCs differentiated into osteoblasts or adipocytes, as evidenced by calcium-rich mineralized matrix deposit or by the presence of intracellular lipid droplets, respectively (Figure 1A–D). The analysis of phenotype of ADSCs used in this study, by flow cytometry, showed that these cells express CD90 (glycosylphosphatidylinositol-anchored glycoprotein) and CD29 (Integrin β1 chain) and were negative for hematopoietic markers, CD11b and CD45 (Figure 1E). According to the literature, the capacity of differentiation and self-replication,

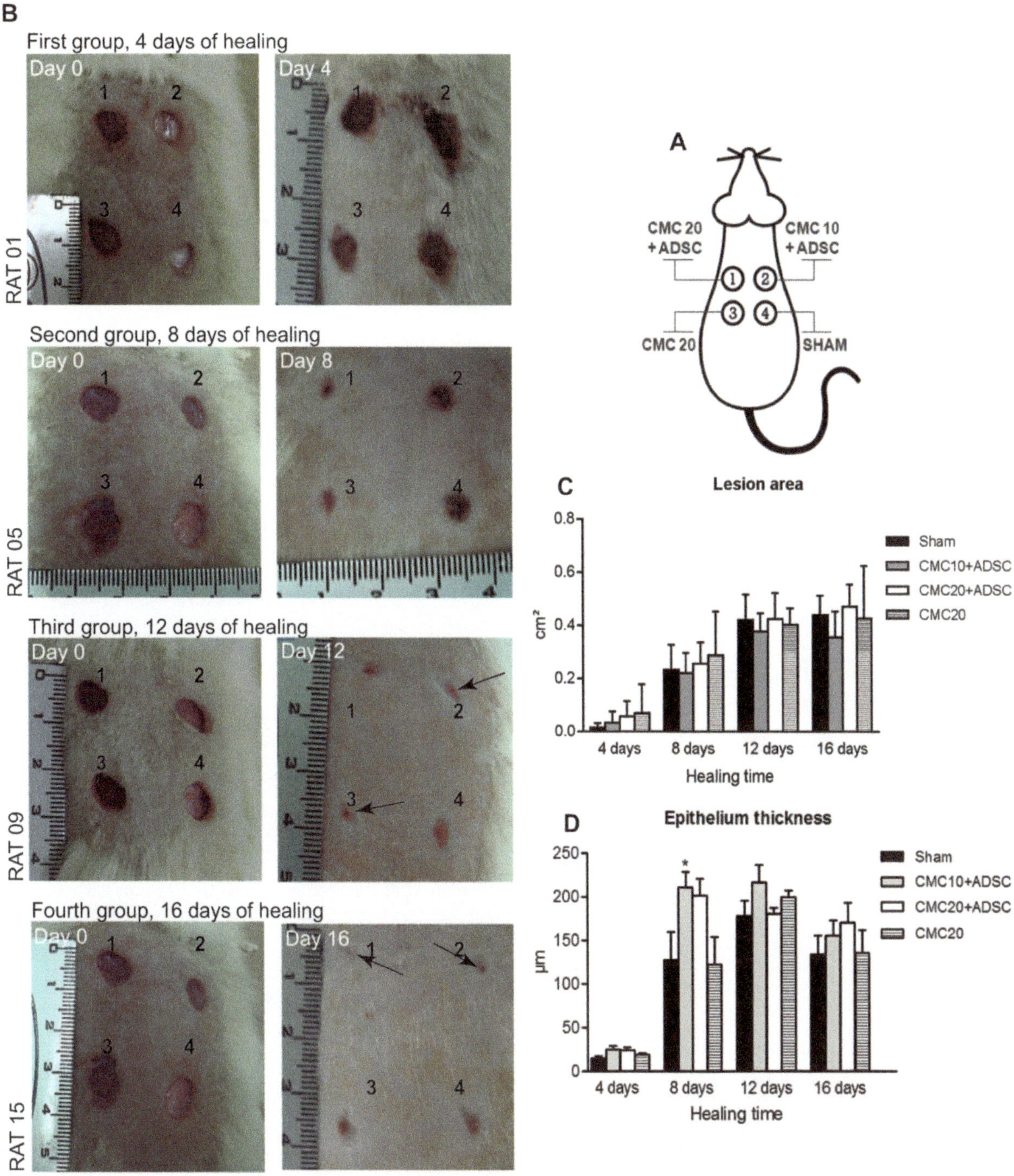

Figure 3. Progression of wound healing. A. Wound model. The scheme shows the four back injuries in a Wistar rat model, comprising the dermis and epidermis layers. The treatments were, in the upper left injury (1): 100 μL of CMC 20 mg/mL+26×10^6 of ADSCs; upper right (2): 100 μL CMC 10 mg/mL+26×10^6 of ADSCs; bottom left (3): 100 μL of CMC 20 mg/mL; bottom right (4): Sham. B. The figure is showing the image of one animal, representative of each group, from day 0 to 16, where the closure of wounds was followed and documented daily. Photographs in the left column were taken in day zero, immediately after surgery. In the right column, in the euthanasia day. C. Evaluation of the area of lesions. The graphic shows, in cm^2, the difference between the size of the initial lesion and the size after different days of healing. There was no difference among the groups. D. Determination of epithelium thickness. Values are expressed in micrometers (μM) and demonstrate a significant increase in the epithelium thickness in the group of animals treated with 10 mg/mL of CMC + ADSCs (lesion 2), with eight days of healing, when compared with control group (injury 4, Sham). (*$P<0.05$, two-way ANOVA test). CMC: Sodium carboxymethylcellulose; ADSC: Adipose-derived stem cell; Sham: Without treatment.

as well as adherence on plastic, are parameters acceptable to the confirmation of the identity of MSCs [16,17,18].

CMC Toxicity

In order to evaluate whether the CMC could damage the cell membrane, we measured the activity of LDH in the supernatant of

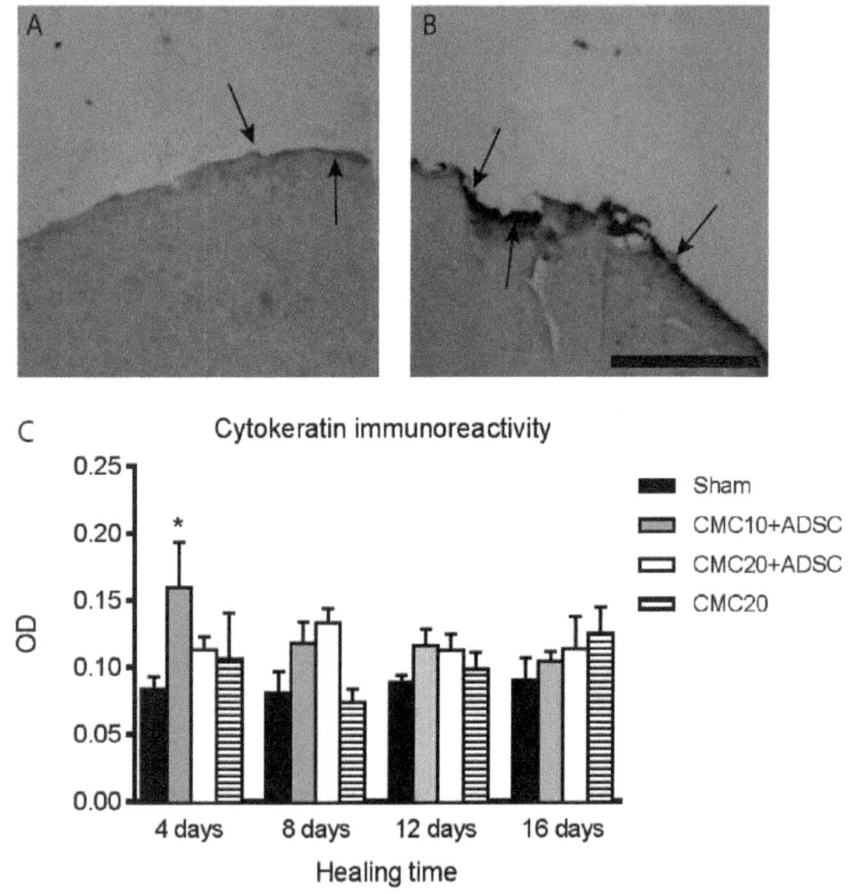

Figure 4. Cytokeratin immunoreactivity in the lesions. A and B. Show representative photomicrographs of the cytokeratin expression in the granulation tissue at the lesion site. The arrows indicate the points where the expression was higher in the lesions: A. Control lesion (Sham); and B. Treated with CMC 10 mg/mL+26×10^6 ADSCs, with four days of healing. Optical microscopy (400x). C. Quantification of cytokeratin expression. The animal groups were separated by healing time and the different treatments are represented in the legend at right. Analyzes demonstrated statistically significant increase in the immunoreactivity in the first group of animals which had four days of healing, (treated with CMC 10 mg/mL + ADSCs) when compared to Sham. Results are expressed as mean of optical density (OD) (*$P<0.05$, two-away ANOVA test). CMC: Sodium carboxymethylcellulose; ADSC: Adipose-derived stem cell; Sham: Without treatment. A–B: Scale bar = 10 μm.

treated cells for 24 hours or 5 days (Figure 2A and 2B). None of the CMC concentrations tested on ADSCs promoted a significant increase in LDH release. Cells that were cultured for as long as five days in contact with the biomaterial, remained adhered to the plate, similarly to cells grown in normal medium, despite the increased density of the culture medium, as a consequence of the presence of CMC. Cells replicated until reaching 100% of confluence.

The induction of single-strand, double-strand and alkali-labile breaks in ADSCs cells was analyzed using the alkaline comet assay (Figure 2C and 2D). The DNA damage index induced in ADSCs that were grown for 5 days in the presence of CMC was increased at the concentration of 40 mg/mL, when compared to the negative control (DMEM) (Figure 2D). Importantly, this concentration of CMC did not alter the proportions of micronuclei in short (24 hours) or long (5 days) periods (Figures 2E and 2F), suggesting that the small damage observed with the comet assay did not lead to a permanent damage of the cells.

Healing Parameters

The evaluation of the healing process of lesions treated with CMC in the presence or absence of ADSCs was determined by different histological parameters. The observation of lesions during the sixteen days of treatment showed that there was no visible suppurative inflammation. Also, none of the animals became sick or died during the treatment. The representative images of lesions of one animal in each group, at baseline (day zero) and end of healing (day of euthanasia) can be seen in Figure 3B. There was no statistical significance when comparing the sizes of the area of the lesions, treated with CMC, when compared with control lesions (Sham) (Figure 3C). The analysis of histological sections from animals that after treatment had 4 days of healing (first group of animals), showed an epithelium not completely formed to be measured. Among the groups, there was a statistical significant increase, in the second group of animals (8 days in the wound healing treated with CMC 10 mg/mL + ADSCs), when compared with lesions in the control group (Sham) (Figure 3D). The animals euthanized on day 4 (first group) showed a higher expression of cytokeratin with the treatment of CMC 10 mg/mL + ADSCs, when compared with the lesion in negative controls (Sham) (Figures 4 A–C).

The collagen fibers production increased progressively in the lesions during the sixteen days of treatment. There was a slight increase in collagen with the treatment of CMC 20 mg/mL associated with ADSCs, during the healing process. However the

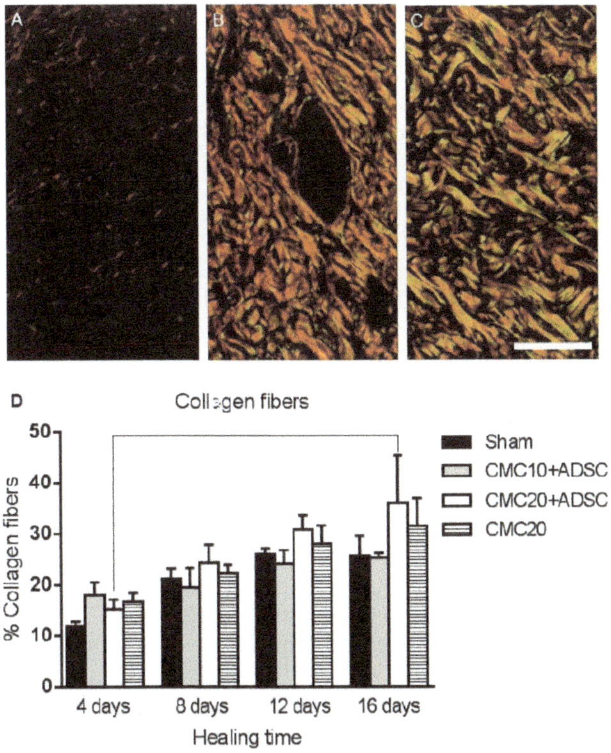

Figure 5. Evaluation of collagen fibers from scar tissues by picrosirius staining. Representative photomicrographs from scar tissues that were subjected to polarization microscopy for visualization of collagen fibers: A. Representative image from the first group with four days of healing. B. From the fourth group with sixteen days of healing, both having received treatment CMC 20 mg/mL+26×10^6 of ADSCs. C. Shows the collagen fibers in a biopsy of normal skin, not subjected to injury. Optical microscopy. Picrosirius staining (200x). D. The statistical analysis showed significance only in the progression of fibers, when were compared the groups of animals with 4 and 16 days of healing, with the treatment of CMC 20 mg/mL associated with ADSCs. There was no statistical difference in the amount of collagen fibers in any group treated, when compared with control lesions (Sham). CMC: Sodium carboxymethylcellulose; ADSCs: Adipose-derived stem cells; Sham: Without treatment. A–C: Scale bar = 50 μm.

quantification of microphotographs stained by picrosirius technique, showed that there was no statistical difference in the amount of collagen fibers in any treated groups, when compared with Sham (Figure 5D).

Discussion

CMC is a low cost biomaterial that has been used as dressings for wounds, with the purpose of protection and as an excipient for some drugs [28]. It has been tested as a biomaterial for growing cells [29,30], however, there is no data evaluating its potential risks and the safe concentrations to be used in therapies. Thus, the first concern in our study was to evaluate the range of safe CMC concentrations to use as a scaffold in cell therapy. For this purpose we performed cytotoxic, genotoxic and mutagenic assays, before testing it in the injured area of an animal. The results showed that there was no significant toxicity on cells cultured with CMC for five days, considering that concentrations of 10 to 40 mg/mL did not cause a significant membrane rupture.

The alkaline comet and micronucleus have been the most commonly used assays for the routine screening of potential genotoxic agents [22]. While the alkaline version of the comet assay [31] detects primary and repairable DNA damage, the micronucleus test detects DNA lesions after their fixation into chromosome mutations. Our results have shown that only at the highest concentration of CMC tested (40 mg/mL), after five days the ADSCs presented increase in DNA damage. However, this damage seems to be repaired, considering that the micronucleus assay did not detect chromosomal damage in ADSCs with the CMC concentrations from 10–40 mg/mL, even with five days of treatment. These results suggest that CMC should be used up to 20 mg/mL.

The epithelization process is an important event for wound healing, ensuring that the new epithelium is formed [32]. In this study, the epithelization rates were improved by CMC containing ADSCs. The cell differentiation and secretion of paracrine factors have been reported to justify the mechanisms by which MSCs contributed to repair damaged tissues [33]. Different molecules are involved in signaling pathways of ADSCs, such as, basic FGF (fibroblast growth factors), VEGF (vascular endothelial growth factor) and PDGF-A (platelet-derived growth factor) [12]. The early development of granulation tissue suggests that the biological events that characterize the dynamic healing are occurring more rapidly, in the group that received treatment of CMC 10 mg/mL associated with ADSCs. An increase in cytokeratin expression, between treated and untreated lesions, suggests a highly active biochemical and bioactive epithelial response, indicative of a migratory epithelium and migratory keratinocyte phenotype, common to the cutaneous wound healing response [34].

MSCs can modulate the host immune responses by changing macrophage, inducing regulatory T cells and inhibiting TH17 formation by both, direct cell-cell contact and indirect mechanisms [35]. This immunomodulatory property could be indirectly related to the epithelization process, since activated macrophages release expressive amounts of FGF [36], a cytokine involved in epithelial proliferation process [37]. A significant increase in cytokeratin expression was observed only in the group of animals, where the healing process occurred by four days. Considering that the therapeutic intervention occurred only once, after the surgery, these results could explain why the animals whose time of healing was 8, 12 and 16 days did not present a significant increase.

The primary route of wound healing in human beings is the formation of granulation tissue and re-epithelialization. Granulation tissue is essential, since it is formed on the surface of the lesion, protecting, nourishing and supporting fibroblasts, new capillaries, and cells from the immune system [38]. The histological slices of lesions stained with HE from animals in the first group (animals with four days of healing), showed a mix of cells, consisting mainly in fibroblasts and leukocytes, which characterize the granulation tissue in its first stage. In the second group, (animals with eight days of healing) the new epithelium observed inside of the wound, was clearly different from the first group, presenting the stratified epithelium. This group also presented an increased epithelium thickness in the lesions that received a concentration of 10 mg/mL of CMC associated with the ADSCs. Interestingly, this result is in accordance with the cytokeratin expression being higher in the group where the healing time was 4 days, with the same treatment. This suggests that during the first 4 days there was an increase in the epithelial cell proliferation, which in the next four days originated a new epithelium layer, capable of being measured and significantly different from the Sham group. Considering that four days of healing were not enough to form a new epithelium, the eight days healing group was the best for this analysis, because the other groups with longer healing time (12 and 16 days), already presented the new epithelium completely formed, independently of treatments.

Although there was a significant increase in the proliferation of epithelial cells in the lesions, with four days of healing, and increased epithelium thickness, in lesions with eight days of healing, when treated with CMC 10 mg/mL plus ADSC, there was no significant difference in the measurements in the injury area when compared to Sham. This improvement of healing at the histological level may be important for large wounds, which would be the ones in which this kind of therapy would be used in the future.

Collagen plays a key role in the wound healing process. The control of its production is a complex process involving a fine balance between its synthesis and degradation [39]. Our analysis by picrosirus stain showed that the treatments did not increase the number of these fibers in comparison with controls (Sham), even in the groups where the epithelium thickness and cytokeratin expression were improved by the treatment. This could be important, because in the healing process, when there is an overproduction of collagen, a fibrosis injury occurs, leading to loss of tissue function and leaving a compromised aesthetic aspect [40,41]. It is possible that ADSCs regulate the synthesis of collagen in the lesion, by releasing anti-fibrogenic molecules [39]. Taken together, the results suggest that CMC in association with the ADSCs contributed to wound healing in the *in vivo* model studied. Despite promising, this new therapy will need further studies to better investigate the basic mechanisms behind this process, and thus be adapted to clinical applications.

Finally, the results presented here demonstrate an important application for CMC as a biomaterial that presents the properties required for ADSCs growth at concentrations of 20 mg/mL or below. In addition, we showed that CMC at 40 mg/mL has some genotoxic effects, but which do not translate into severe chromosomal damage leading to micronuclei. This can open new avenues for the exploration of the potential pharmaceutical uses of CMC in therapies.

Acknowledgments

The authors would like to thank Terezinha Stein and Rosalva Thereza Meurer (Laboratório de Pesquisa em Patologia, UFCSPA) for excellent technical assistance with histological analysis and Dr. Guido Lenz (Laboratório de sinalização e plasticidade celular, UFRGS) for critical editing of the manuscript.

Author Contributions

Conceived and designed the experiments: CR AMA DJM MCF MRW. Performed the experiments: CR AMA DJM. Analyzed the data: CR DJM GH LLX MCF MRW. Contributed reagents/materials/analysis tools: JS MCF MRW. Wrote the paper: CR MRW.

References

1. Chen M, Przyborowski M, Berthiaume F (2009) Stem cells for skin tissue engineering and wound healing. Crit Rev Biomed Eng 37: 399–421.
2. Simpson CL, Patel DM, Green KJ (2011) Deconstructing the skin: cytoarchitectural determinants of epidermal morphogenesis. Nat Rev Mol Cell Biol 12: 565–580.
3. Wendt H, Hillmer A, Reimers K, Kuhbier JW, Schafer-Nolte F, et al. (2011) Artificial skin-culturing of different skin cell lines for generating an artificial skin substitute on cross-weaved spider silk fibres. PLoS ONE 6: e21833.
4. Neuman MG, Nanau RM, Oruna L, Coto G (2011) In vitro anti-inflammatory effects of hyaluronic acid in ethanol-induced damage in skin cells. J Pharm Pharm Sci 14: 425–437.
5. Kirschner N, Houdek P, Fromm M, Moll I, Brandner JM (2010) Tight junctions form a barrier in human epidermis. Eur J Cell Biol 89: 839–842.
6. Balbino CA, Pereira LM, Curi R (2005) Mechanisms involved in wound healing: a revision. Rev Bras Cienc Farm 41: 27–51.
7. Balaji S, Keswani SG, Crombleholme TM (2012) The Role of Mesenchymal Stem Cells in the Regenerative Wound Healing Phenotype. Advances in wound care 1: 159–165.
8. Fuchs E, Horsley V (2008) More than one way to skin. Genes Dev 22: 976–985.
9. Zuk PA, Zhu M, Ashjian P, De Ugarte DA, Huang JI, et al. (2002) Human adipose tissue is a source of multipotent stem cells. Mol Biol Cell 13: 4279–4295.
10. Zuk PA, Zhu M, Mizuno H, Huang J, Futrell JW, et al. (2001) Multilineage cells from human adipose tissue: implications for cell-based therapies. Tissue Eng 7: 211–228.
11. Sterodimas A, de Faria J, Nicaretta B, Pitanguy I (2009) Tissue engineering with adipose-derived stem cells (ADSCs): current and future applications. J Plast Reconstr Aesthet Surg 63: 1886–1892.
12. Peroni D, Scambi I, Pasini A, Lisi V, Bifari F, et al. (2008) Stem molecular signature of adipose-derived stromal cells. Exp Cell Res 314: 603–615.
13. Sudhakar Y, Kuotsu K, Bandyopadhyay AK (2006) Buccal bioadhesive drug delivery–a promising option for orally less efficient drugs. J Control Release 114: 15–40.
14. Shimamura MK, Kamata K, Yao A, Deguchi T (2005) Scattering functions of knotted ring polymers. Phys Rev E Stat Nonlin Soft Matter Phys 72: 041804.
15. Ramlia NA, Wonga TW (2011) Sodium carboxymethylcellulose scaffolds and their physicochemical effects on partial thickness wound healing. Int J Pharm 403: 73–82.
16. Meirelles L S, Nardi NB (2009) Methodology, biology and clinical applications of mesenchymal stem cells. Front Biosci 14: 4281–4298.
17. Phinney DG, Kopen G, Isaacson RL, Prockop DJ (1999) Plastic adherent stromal cells from the bone marrow of commonly used strains of inbred mice: variations in yield, growth, and differentiation. J Cell Biochem 72: 570–585.
18. Gutiérrez-Fernández M, Rodríguez-Frutos B, Ramos-Cejudo J, Vallejo-Cremades TM, Fuentes B, et al. (2013) Effects of intravenous administration of allogenic bone marrow- and adipose tissue-derived mesenchymal stem cells on functional recovery and brain repair markers in experimental ischemic stroke. Stem Cell Res Ther. 4: 11.
19. Speit G, Hartmann A (1999) The comet assay (single-cell gel test). A sensitive genotoxicity test for the detection of DNA damage and repair. Methods Mol Biol 113: 203–212.
20. Tice RR, Strauss GH (1995) The single cell gel electrophoresis/comet assay: a potential tool for detecting radiation-induced DNA damage in humans. Stem Cells 13: 207–214.
21. Nadin SB, Vargas-Roig LM, Ciocca DR (2001) A silver staining method for single-cell gel assay. J Histochem Cytochem 49: 1183–1186.
22. Hartmann A, Speit G (1997) The contribution of cytotoxicity to DNA-effects in the single cell gel test (comet assay). Toxicology Letters 90: 183–188.
23. Thomas P, Holland N, Bolognesi C, Kirsch-Volders M, Bonassi S, et al. (2009) Buccal micronucleus cytome assay. Nat Protoc 4: 825–837.
24. Fenech M (2000) The in vitro micronucleus technique. Mutat Res. 455: 81–95.
25. Kim CH, Lee JH, Won JH, Cho MK (2011) Mesenchymal stem cells improve wound healing in vivo via early activation of matrix metalloproteinase-9 and vascular endothelial growth factor. J Korean Med Sci 26: 726–733.
26. Xavier LL, Viola GG, Ferraz AC, Da Cunha C, Deonizio JM, et al. (2005) A simple and fast densitometric method for the analysis of tyrosine hydroxylase immunoreactivity in the substantia nigra pars compacta and in the ventral tegmental area. Brain Res Brain Res Protoc 16: 58–64.
27. de Souza DB, Silva D, Cortez CM, Costa WS, Sampaio FJ (2012) Effects of chronic stress on penile corpus cavernosum of rats. J Androl 33: 735–739.
28. van Zuijlen PP, Vloemans JF, van Trier AJ, Suijker MH, van Unen E, et al. (2001) Dermal substitution in acute burns and reconstructive surgery: a subjective and objective long-term follow-up. Plast Reconstr Surg 108: 1938–1946.
29. Lipp C, Kirker K, Agostinho A, James J, Stewart P (2010) Testing wound dressings using an in vitro wound model. J Wound Care 19: 220–226.
30. Liuyun J, Yubao L, Chengdong X (2009) Preparation and biological properties of a novel composite scaffold of nano-hydroxyapatite/chitosan/carboxymethyl cellulose for bone tissue engineering. J Biomed Sci 16: 65.
31. Singh NP, McCoy MT, Tice RR, Schneider EL (1988) A simple technique for quantitation of low levels of DNA damage in individual cells. Exp Cell Res 175: 184–191.
32. Albuquerque-Júnior RLC, Barreto ALS, Pires JA, Reis FP, Ribeiro MAG, et al. (2009) Effect of bovine type-I collagen-based films containing red propolis on dermal wound healing in rodent model. Int J Morphol 27: 1105–1110.
33. Hocking AM, Gibran NS (2010) Mesenchymal stem cells: paracrine signaling and differentiation during cutaneous wound repair. Exp Cell Res 316: 2213–2219.
34. Wong P, Coulombe PA (2003) Loss of keratin 6 (K6) proteins reveals a function for intermediate filaments during wound repair. J Cell Biol 163: 327–337.
35. Khosrotehrani K (2013) Mesenchymal stem cell therapy in skin: why and what for? Exp Dermatol 22: 307–310.
36. Logan A, Frautschy SA, Gonzalez AM, Baird A (1992) A time course for the focal elevation of synthesis of basic fibroblast growth factor and one of its high-affinity receptors (flg) following a localized cortical brain injury. J Neurosci 12: 3828–3837.

37. Werner S, Grose R (2003) Regulation of wound healing by growth factors and cytokines. Physiol Rev 83: 835–870.
38. Harding KG, Moore K, Phillips TJ (2005) Wound chronicity and fibroblast senescence - implications for treatment. Int Wound J 2: 364–368.
39. Motawi TM, Atta HM, Sadik NA, Azzam M (2013) The therapeutic effects of bone marrow-derived mesenchymal stem cells and simvastatin in a rat model of liver fibrosis. Cell Biochem Biophys 13: 9698–9701.
40. Zhou S, Salisbury J, Preedy VR, Emery PW (2013) Increased collagen synthesis rate during wound healing in muscle. PLoS One 8: e58324.
41. Johnson A, DiPietro LA (2013) Apoptosis and angiogenesis: an evolving mechanism for fibrosis. FASEB J 27: 3893–3901.

Histological Evaluation of the Biocompatibility of Polyurea Crosslinked Silica Aerogel Implants in a Rat Model: A Pilot Study

Firouzeh Sabri[1]*, John D. Boughter Jr[2], David Gerth[3], Omar Skalli[4], Thien-Chuong N. Phung[1], George-Rudolph M. Tamula[1], Nicholas Leventis[5]

1 Department of Physics, University of Memphis, Memphis, Tennessee, United States of America, 2 Department of Anatomy and Neurobiology, UTHSC, Memphis, Tennessee, United States of America, 3 Department of Otolaryngology Head and Neck Surgery, UTHSC, Memphis, Tennessee, United States of America, 4 Department of Biological Sciences, University of Memphis, Memphis, Tennessee, United States of America, 5 Department of Chemistry, Missouri University of Science and Technology, Rolla, Missouri, United States of America

Abstract

Background: Aerogels are a versatile group of nanostructured/nanoporous materials with physical and chemical properties that can be adjusted to suit the application of interest. In terms of biomedical applications, aerogels are particularly suitable for implants such as membranes, tissue growth scaffolds, and nerve regeneration and guidance inserts. The mesoporous nature of aerogels can also be used for diffusion based release of drugs that are loaded during the drying stage of the material. From the variety of aerogels polyurea crosslinked silica aerogels have the most potential for future biomedical applications and are explored here.

Methodology: This study assessed the short and long term biocompatibility of polyurea crosslinked silica aerogel implants in a Sprague-Dawley rat model. Implants were inserted at two different locations a) subcutaneously (SC), at the dorsum and b) intramuscularly (IM), between the gluteus maximus and biceps femoris of the left hind extremity. Nearby muscle and other internal organs were evaluated histologically for inflammation, tissue damage, fibrosis and movement (travel) of implant.

Conclusion/Significance: In general polyurea crosslinked silica aerogel (PCSA) was well tolerated as a subcutaneous and an intramuscular implant in the Sprague-Dawley rat with a maximum incubation time of twenty months. In some cases a thin fibrous capsule surrounded the aerogel implant and was interpreted as a normal response to foreign material. No noticeable toxicity was found in the tissues surrounding the implants nor in distant organs. Comparison was made with control rats without any implants inserted, and animals with suture material present. No obvious or noticeable changes were sustained by the implants at either location. Careful necropsy and tissue histology showed age-related changes only. An effective sterilization technique for PCSA implants as well as staining and sectioning protocol has been established. These studies further support the notion that silica-based aerogels could be useful as biomaterials.

Editor: Jeongmin Hong, Florida International University, United States of America

Funding: The first author would like to thank the FedEx Institute of Technology, University of Memphis, for finanical support. The funders had no role in study design, data collection and analysis, decision to publish, or preparation of the manuscript.

Competing Interests: The authors have declared that no competing interests exist.

* E-mail: fsabri@memphis.edu

Introduction

Porous biocompatible materials have received particular attention in recent years for a broad range of applications. From filters and prostheses to scaffolds for tissue engineering, porous biomaterials have been under constant development and improvement for biological and biomedical applications [1,2,3,4,5,6,7]. Complications such as poor mechanical performances, batch-to-batch purity variations, and large pore sizes (μm –range) have limited the extent of use of naturally occurring biomaterials [8] even though they most closely simulate the native cellular environment.

Large pore sizes restrict the use of this class of porous materials to whole cell penetration and tissue infiltration applications only, making them unsuitable for drug delivery and protein loading applications where the physical size of the entities of interest are on the order of 10 s of nanometers rather than microns [9,10]. Studies have also shown that cellular uptake of drugs is more efficient at the nm scale level [11] emphasizing the need for nm-scale porous materials and membranes as ideal tools for drug delivery. The cell size of porous biomaterials also plays an important role in the formation of scar tissue and fibrosis where major effort is invested to minimize these formations. There is evidence to suggest that closely spaced nanometer-sized pores prevent formation of extensive fibrous connective and scar tissue and in fact promotes superior tissue integration [12,13]. Additionally, from a mechanical behavior point of view stress concentration greatly depends on pore size and is proven to be significantly less for materials with smaller pore diameters such as mesoporous materials (pore diameter<50 nm) [14,15].

Aerogels are nanostructured, open-mesoporous (pore diameter<50 nm) ultra low-density lightweight materials with a high surface-to-volume ratio [16] and tunable surface and bulk properties allowing for control over potentially key parameters such as surface wettability, density, opacity, pore size and shape to name a few [17]. Of particular interest to biological applications are the polyurea crosslinked silica aerogels (PCSA) where significant mechanical strength has been accomplished by covalent crosslinking of the skeletal nanoparticles with polyurea without significant compromise of the porosity and the low bulk density of the native material [18,19]. The biocompatibility of silica particles has been widely explored and has shown a great deal of promise and compatibility with living matter [20,21,22,23]. Drug delivery by porous silica has also been explored and offers great promise [24,25]. At the cellular level, recent *in vitro* studies on PCSA has shown good biocompatibility of this material [26,27,28] and the ability to manipulate the growth of neurons on the PCSA surface with the aid of a laminin layer [26].

The versatility of sol-gel method for material preparation allows for doping and pigmenting techniques to be incorporated at the synthesis stage of aerogels with relative ease [29,30]. Previous investigations of pigmenting PCSA demonstrated a stable and non-leaching chemistry while also strengthening the mechanically properties and retaining the porosity of the aerogels [31]. This can be of significant interest for biomedical applications since the majority of polymeric implants used today are translucent or transparent. Post-surgical identification of a colored implant may be easier.

In this work, we report the first *in vivo* assessment of the biocompatibility of both clear and pigmented PCSA implants in a rat model sterilized following an ethylene oxide (EtO) sterilization protocol. The presence of the implants did not result in any morbidity. In addition, histological evaluation of the tissues surrounding the implants as well as distant organ sites did not demonstrate inflammation or substantial tissue damage. The results of the PCSA implant study were compared to three control groups: 1) Suture material present (peripheral nerve transection followed by suture repair), 2) sham surgery, and 3) no surgery. Results presented in this study demonstrate the biocompatibility of PCSA thereby supporting the notion that these materials could be useful for biomedical applications.

Materials and Methods

Preparation of clear and pigment-doped aerogels

Pigmented and clear polyurea crosslinked silica aerogels were synthesized according to our previously described formulation [30]. The aerogel implants were cut and roughly shaped from the bulk material using a diamond tipped abrasive disk mounted in a rotary tool. The samples were then polished by hand to a final size of $5 \times 2 \times 2$ mm^3 for the back muscle study and $5 \times 2 \times 2$ mm^3 for the deep muscle study. The final stage of implant preparation was exposure to a small vibratory tumbler containing aluminum oxide grinding media (approximately 200 mesh), and processed for about 2 days. The surface contact angle of these aerogels was measured to be around 45° and a density of 0.4 g/cm^3 for clear and 0.5 g/cm^3 for pigmented samples was measured. All implants were sterilized for 24 hrs by a standard ethylene oxide sterilization process in an Amprolene system (Anderson Products) prior to surgery.

Surgical procedure(s)

All surgical procedures were performed on male or female Sprague-Dawley rats weighing 200–300 gr. Rats were anesthetized 15 min prior to surgery via an initial intramuscular injection of Telazol (0.03–0.05 ml at 0.3–0.5 mg/kg), followed by isoflurane inhalation at 1–2%. During the post-surgery recovery period animals were not restrained and were allowed to continue with their normal grooming routine. They were fed a routine diet and kept under close observation for signs of infection or abnormal behavior. Rats were euthanized post implant recovery by an overdose of carbon dioxide. This study was approved by the Animal Care and Use Committee at the University of Memphis.

Surgical group 1: Subcutaneous Implant insertion (n = 4)

An area on the back, approximately 3 cm^2, was shaved and a 1 cm incision was made with a scalpel. One pigmented and one clear implant was inserted subcutaneously in the back of rats (total of 8 implants). Samples were inserted and immobilized without the use of any sutures, adhesives, or staples. Figure 1 shows clear (Figure 1a) and pigmented (Figure 1b) SC PCSA implants after incubation period, prior to extraction.

Surgical group 2: Intramuscular implant insertion (n = 2)

The left hind limb of rats were abducted and shaved. Under a dissecting microscope, a transverse incision was made through the skin of the limb halfway between the iliac crest and the femur's articulation with the tibia. A self retracting retainer was replaced, and dissection was carried down between the biceps femoris and gluteus muscle until the sciatic nerve was identified. The implant

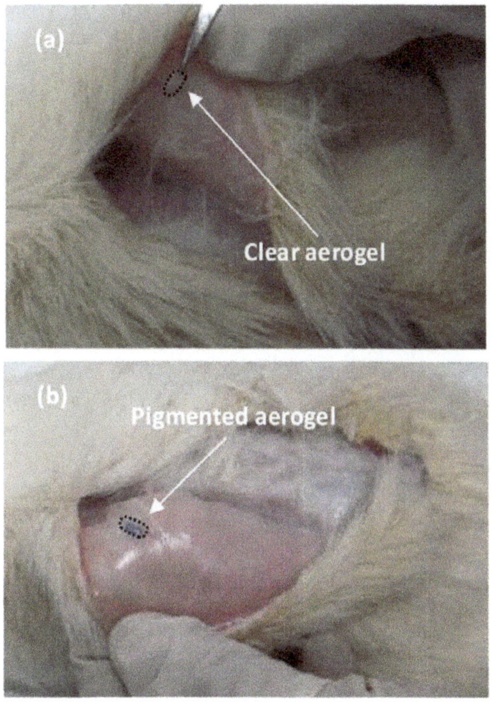

Figure 1. Subcutaneous implant retrieval (surgery site 1). Optical images of (a) clear and (b) pigmented aerogel implants being retrieved after twenty months of subcutaneous incubation in Sprague-Dawley rats. Arrows indicate the embedded implants. A mild fibrosis can be observed for both pigmented and clear PCSA implants. The pigmented aerogel samples were much easier to identify compared to the clear implants.

was positioned between the muscle and the sciatic nerve such that the nerve was in direct contact with the surface of the implant. No adhesive or sutures were used for immobilizing the aerogel implant. The skin was then closed with staples and the animal was then allowed to emerge from anesthesia.

Surgical group 3: Suture repair (n = 2)

The sciatic nerve branches of rats were exposed and transected sharply as described above. The nerve segments were then coapted using standard epineural suture technique with interrupted 9-0 nylon.

Surgical group 4: Sham surgery (n = 1)

The sciatic nerve of rats was exposed and severed again sharply, similar to the method used for suture repair but after severance, nerve endings were abandoned and no coaptation was attempted.

Implant and tissue retrieval

At time of sacrifice each animal was weighed and its weight recorded. All animals under investigation in this study continued to gain weight steadily throughout the implant incubation period at the same rate as the control groups. At appropriate recovery time points previous incisions were reopened and the aerogel implant was identified and removed with a cuff of the surrounding tissue intact for further analysis. At two weeks, four of the subcutaneously implanted clear and pigmented PCSA samples (surgical group 1) were removed for analysis. At twenty months four more subcutaneously inserted clear and pigmented PCSA samples (surgical group 1) were removed along with vital organs. Finally, at seven months time point PCSA implants and vital organs were removed from surgical group 2. For the sake of comparison tissue samples from the surgery sites as well as vital organs were removed from surgical groups 2, 3, and 4 at the same time point. Post surgery, animals were anesthetized with inhaled agent and sacrificed at each time point with an overdose of carbon dioxide. All tissue samples were transferred to jars containing 10% formaldehyde.

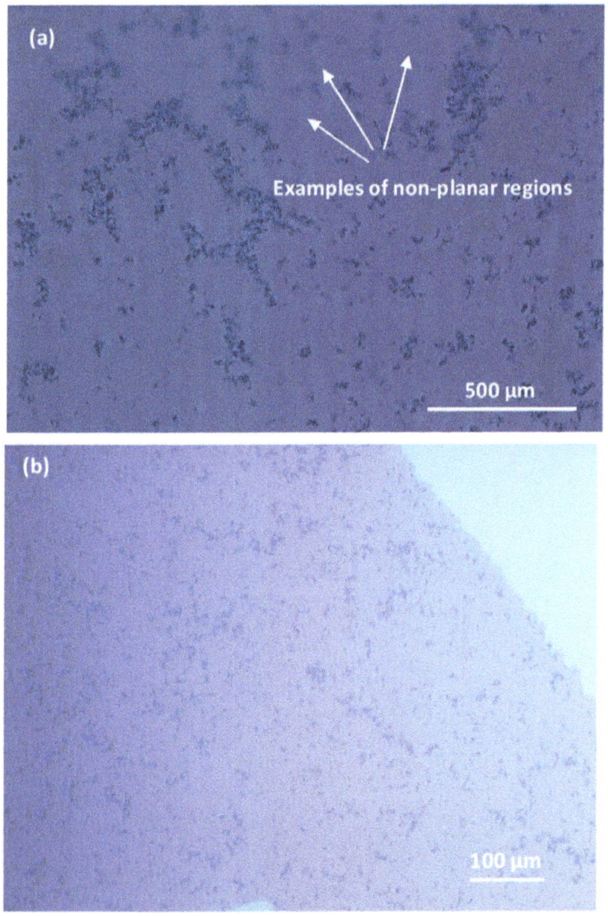

Figure 2. Stain uptake by PCSA sections. H&E stain uptake by 5 μm thick sections of (a) clear and (b) pigmented aerogel samples. The out-of-focus regions of the image (indicated by arrows) reflect the three dimensional nature of the aerogel material. While the pores cannot be visualized by a light microscope, the darker regions are interpreted as openings to the three dimensional porous structure. Images were taken with an upright Nikon Eclipse 800 microscope. Striations noticed in (a) are attributed to sectioning artifacts.

Histological staining and examination

Immediately after dissection, tissues containing implants as well as tissues from distant organs were fixed for at least 18 hrs in 10% formaldehyde. Tissue segments obtained from organs without implant were embedded in paraffin and the sections obtained from these paraffin blocs were stained by hematoxylin eosin (H&E) following routine procedures.

Initial experiments revealed that adhesion of sectioned PCSA aerogel implants to glass slides was challenging. Two protocols were tested to overcome this difficulty. First, implant containing tissue samples were embedded in polymethylmethacrylate (PMMA) and the sections obtained from these blocs were mounted onto plastic slides for staining. This resulted in satisfactory adhesion of tissue/aerogel samples onto the slides but the refractive index of the plastic hindered optimal light microscopy observations. In the second protocol, the implant containing tissue sections were embedded in paraffin following standard procedures. Sections of 5 μm thickness were obtained from the paraffin blocks and various functionalized glass slide surfaces were tested with respect to their ability to bind the implant present in the tissue sections. Slides coated with silane (Electron Microscopy Sciences), poly-L-lysine (Electron Microscopy Sciences), or super-frost coating (Thermo Fisher Scientific) were unsatisfactory as the aerogel implants detached readily early during the staining process. In contrast, the PCSA implant material adhered through the whole staining and mounting procedure on glass slides coated

Table 1. Summary of short and long term effect of subcutaneous aerogel implant on local tissue.

Incubation Period Response	**Two weeks**		**Twenty months**	
	Clear	**Pigmented**	**Clear**	**Pigmented**
Inflammation	Mild	Mild	Mild	Mild
Fibrosis	Mild	Mild	Mild	Mild
Infection	None	None	None	None
Implant travel	None	None	None	None
Age of rat at time of extraction	~3 months		~2 years	

Slides from both control and implant-containing rats were reviewed blindly by pathologist, specifically looking for signs of inflammation, infection, fibrosis, and implant travel.

Table 2. Summary of the effect of intramuscular aerogel implant on local tissue.

Seven month period				
Procedure / **Response**	**Sham**	**Suture**	**Implant-clear**	**Implant-pigmented**
Inflammation	Mild	Mild	Mild	Mild
Fibrosis	Mild	Mild	Mild	Mild
Infection	None	None	None	None
Age of rat at time of extraction	9–10 months old			

Slides from both control and implant-containing rats were reviewed blindly by pathologist, specifically looking for signs of inflammation, infection, and fibrosis.

with egg albumin (Newcomers Supply). Sections containing the implant were stained with H&E. Double blinded tissue section evaluations were performed by a pathologist who, in the case of sections without implants, was unaware of whether or not the tissue examined was obtained from an animal with or without implants. Figure 2 shows the effect of the staining protocol on the clear (Figure 2a) and pigmented (Figure 2b) control PCSA sections. It can be seen that the PCSA alone does uptake the stain to a certain extent. Striations seen in the image are considered to be staining artifacts and do not reflect PCSA surface morphology. At times, sectioning the paraffin-embedded tissue-implant samples was difficult due to the hardness of the PCSA implant and several attempts had to be made in order to create a uniform section.

Results and Discussion

Implant incubation times were chosen based on the three contact duration categories that biomaterials and medical devices are recommended for testing namely 1) Limited (<24 hrs), 2) Prolonged (>24 hrs and <30 days) and 3) Permanent (>30 days) [32]. Aerogel-based biomaterials are expected to serve under classifications 2 and 3 and incubation periods were chosen accordingly

Both clear and pigmented PCSA samples were implanted a) subcutaneously (SC) at the dorsum and b) intramuscularly (IM) between the gluteus maximus and biceps femoris of the left hind extremity in Sprague–Dawley rats. Subcutaneous implants were removed at two weeks and twenty month time points while IM implants were extracted after seven months of incubation.

Implant Condition

The physical appearance and size of the recovered implants was not altered as a result of short or long term *in vivo* incubation. Macroscopic and microscopic (optical) evaluation of implants showed no signs of erosion, surface deterioration, or fragmenting. Additionally, the implants inserted at the different locations did not appear to have traveled from their original location even though no immobilization (sutures, adhesives, etc) was used. This

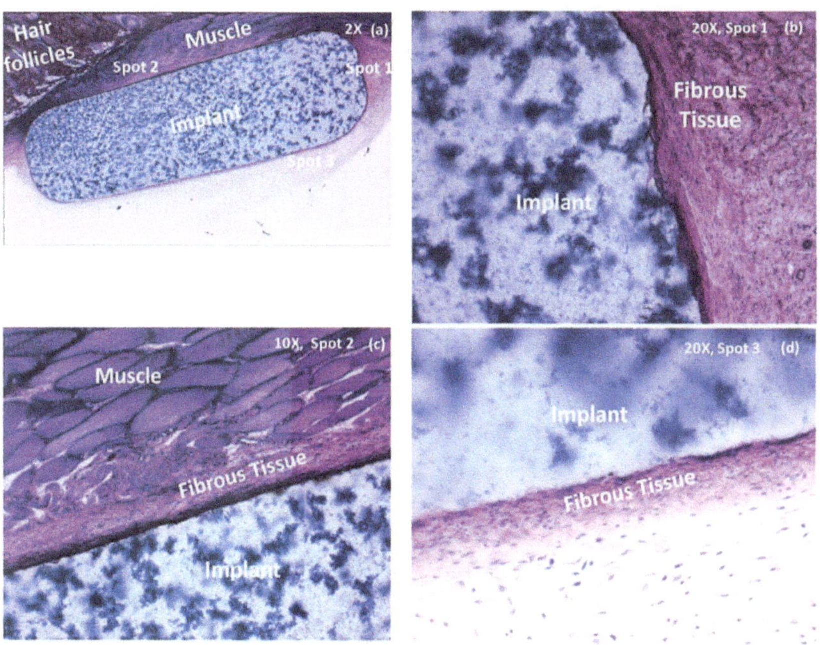

Figure 3. Histological evaluation of the short-term effect of subcutaneous PCSA implants on nearby tissue. Histology of pigmented aerogel implant extracted after two weeks and stained with methylene blue/basic fucshin. (a) All of the aerogel implant at X 2 magnification, (b) X 20 magnification for spot 1, (c) X 20 magnification of spot 2 and (d) X 10 magnifications of spot 3. A mild fibrosis is observed but no inflammation. Images were taken with an Olympus BX51 microscope.

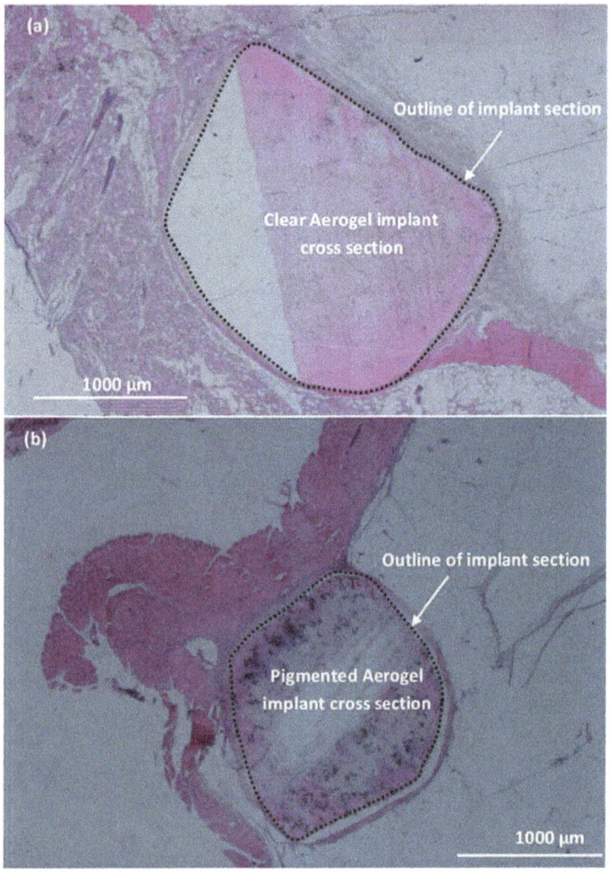

Figure 4. Histological evaluation of the long-term effect of subcutaneous PCSA implants on nearby tissue. Histology of (a) clear and (b) pigmented aerogel implant extracted after seventeen months of subcutaneous incubation in Sprague-Dawley rats and stained with H&E. A mild fibrosis is observed but no inflammation. The dotted line outlines the boundary between PCSA and nearby tissue.

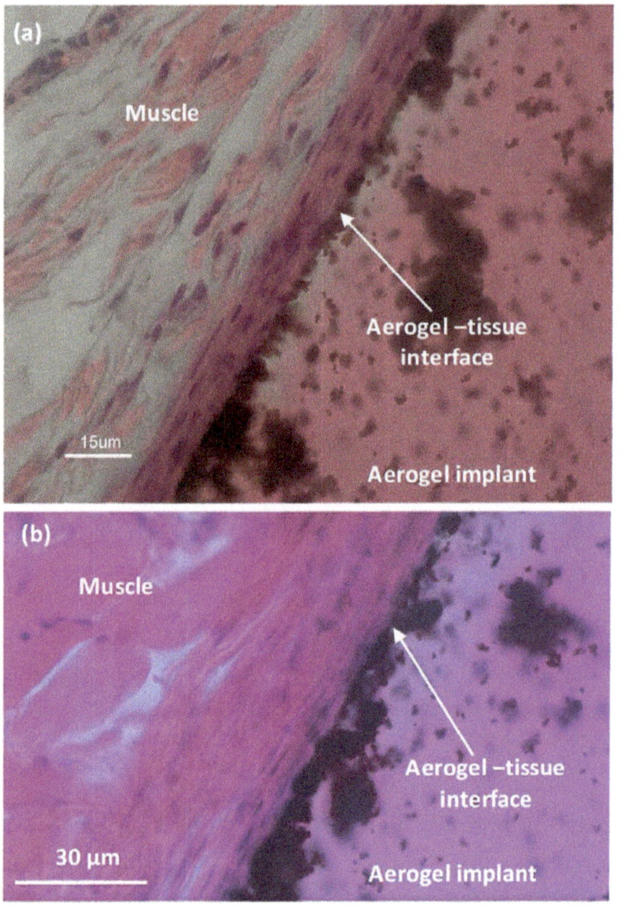

Figure 5. High magnification images of aerogel-muscle interface, intramuscular implantation. Histology of (a) clear and (b) pigmented aerogel implants extracted after seven months of IM incubation. At the interface between aerogel and muscle a mild fibrosis is observed but no inflammation. Darker regions seen on the aerogel side are associated with the three dimensional, no-planar structure of aerogel. Images were taken with a Nikon Eclipse 800 microscope.

is particularly surprising in the case of the implants inserted in the hind leg, since the animal had continuous motion and grooming ability and its motion was not restricted during the recovery period. We attribute this to the low density and light weight nature of the aerogel implant. It is likely that the nanoporous/mesoporous and three dimensional nature of the surface of the aerogel implants created anchoring sites between the nearby tissue and the surface of the PCSA implant.

Inflammation

In general, irritation and inflammation of the tissue that is in direct contact with implants is a concern [33]. In this case however, no significant inflammation was observed for any of the time points, at the interface of the aerogel implants and the nearby tissue for both SC and IM surgery types. The amount of inflammation (according to pathologists report) was categorized as mild in all cases such that the amount of inflammation observed for PCSA implants was comparable to the amount of inflammation observed for biocompatible materials particularly surgical steel and nylon sutures (Tables 1 and 2). This is likely due to the nanoporous/mesoporous nature of the aerogel implant surface that while it is highly non-uniform at the nanometer scale, to the large muscle cells nearby it presents itself smooth and as a result causes minimum irritation and inflammation. A mild inflammation is expected for all foreign materials including biomaterials and is recognized as part of the body's foreign body response [34]. The response seen in this study to the PCSA implants is identified as a normal reaction.

Fibrosis

In general, the biocompatibility of novel biomaterials with tissue is evaluated based on the *in vivo* inflammatory responses as well as the fibrosis formed around the implant [35]. Fibrous capsule formation is a well-established reaction to implanted biomaterials and is recognized as the end stage of the foreign body reaction [34]. In this study a mild fibrosis and capsule formation was observed for almost all of aerogel implant-tissue interfaces. Figure 3 shows the SC PCSA-tissue interface behavior for a pigmented sample. In Figure 3a the whole implant is imaged while Figures 3b, 3c, and 3d show high magnification images of interface and the fibrous layer formed around the implant for spots 1, 2, and 3 respectively as identified on the image (Figure 3a).

In Figure 4 example cross sectional images of the SC implant insertion after twenty months of incubation are shown for both clear (Figure 4a) and pigmented (Figure 4b) PCSA samples. The dotted line in each case outlines the boundary between the aerogel

Table 3. Summary of long term effect of intramuscular and subcutaneous aerogel implants on organs.

Procedure Organ	Sham surgery	Suture surgery	Control (no surgery)	Pigmented/clear aerogel implant-SC	Pigmented/clear aerogel implant-IM
Spleen	Unremarkable	Unremarkable	Unremarkable	Undistinguishable	Undistinguishable
Lung	Unremarkable	Unremarkable	Unremarkable	Undistinguishable	Undistinguishable
Heart	Unremarkable	Unremarkable	Unremarkable	Undistinguishable	Undistinguishable
Kidney	Unremarkable	Unremarkable	Unremarkable	Undistinguishable	Undistinguishable
Intestine	Unremarkable	Unremarkable	Unremarkable	NA	NA

and the nearby tissue. Some staining of the PCSA section can be seen in each case. The thickness of the fibrous layer for the SC surgery, around the clear and pigmented aerogel implants between two weeks (Figure 3) and twenty months (sample images shown in Figure 4) time points did not show a significant difference. In all cases native cellular tissue had grown up to the implant surfaces. It is not clear from the tests performed here if any tissue had grown into the aerogel pores and interstices. From the histological evaluations it was concluded that the fibrosis around the pigmented implants wasn't significantly greater than fibrosis around clear implants. Table 1 summarizes these results.

Next, the fibrous layer formed around biocompatible materials such as surgical sutures and surgical tools (images not shown) were compared with the fibrous layer formed around PCSA implants inserted intramuscularly. High magnification images of the interface of clear (Figure 5a) and pigmented (Figure 5b) PCSA with nearby tissue extracted after seven months of IM incubation was studied and results are summarized in Table 2. Fibrosis observed in all cases was classified as "mild" and was in agreement with the amount of fibrosis observed at the surgery site for suture and sham surgery procedures. Again, cellular tissue had grown at least up to the implant and perhaps growth has also occurred into the structure of the aerogels although, no specific tests were performed to confirm this.

Infection

No infection was observed in any of the implant or control conditions, at any of the time points, suggesting that ethylene oxide sterilisation is an effective method for sterilising PCSA implants.

Toxicity

Toxicity of novel biomaterials is often evaluated by monitoring the response of vital organs distant to the implant. In particular organs such as spleen, lung, heart, kidney, and intestine must be evaluated histopathologicaly [36] for any signs of abnormal responses. In Table 3 the long term effect of aerogels' presence in the system of Sprague-Dawley rats has been summarized. The evaluation of organs removed from animals with 1) suture surgery, 2) no surgery (control), and 3) sham surgery was classified as unremarkable. The organs removed from animals carrying an aerogel implant of some kind were indistinguishable from those previously mentioned. This indicates that in the initial study, no noticeable systemic reaction or immune response has been triggered by the presence of the aerogel implants.

Summary and Conclusion

This study involves the utilization of *in vivo* tests to determine the general biocompatibility of PCSA as a biomaterial. Double blinded reviews by pathologists showed no statistical difference between tissue samples collected from different surgical groups suggesting tolerance and biocompatibility of PCSA. Early *in vivo* assessment of tissue compatibility presented here can be used to influence the design criteria of future PCSA-based medical devices.

Ethylene oxide sterilization has proven to be an effective method for sterilizing aerogel implants. The *in vivo* studies performed here on a small group of Sprague-Dawley rats have demonstrated biocompatibility of polyurea crosslinked silica aerogels over a maximum of twenty month incubation period. Fibrosis observed was at the normal level that would be observed with any foreign object including sutures. Although this work was performed on a small group of animals, it provides the basis for continuation of the study on a larger group of animals and shows promise for PCSA as a biomaterial. The nanometer scale roughness of the aerogel surface seems to play a significant role in the limited range of motion and travel of the implant from the surgery site therefore, eliminating the need for sutures. The results of our *in vivo* experiments suggest the need for further careful study to better understand the fundamental processes involved in the interaction between this specific type of material and the living body.

Acknowledgments

The authors would like to acknowledge Ms. Renada Scott at the Integrated Microscopy Center (IMC), the University of Memphis for histology and sectioning work as well as University of Alabama at Birmingham, Center for Metabolic Bone Disease–Histomorphometry and Molecular Analysis Core Laboratory. Authors would also like to thank Dr Anand Kulkarni and his team at UTHSC- Department of Pathology for pathology evaluations. Finally we are grateful to Dr Karyl Buddington for all surgical procedures conducted at UoM.

Author Contributions

Conceived and designed the experiments: FS JDB. Performed the experiments: TCP GRT DG. Analyzed the data: FS JDB OS. Contributed reagents/materials/analysis tools: NL. Wrote the paper: FS JDB DG OS NL.

References

1. Power M, Hosticka B, Black E, Daitch C, Norris P (2001) Aerogels as biosensors: viral particle detection by bacteria immobilized on large pore aerogel. Journal of Non-Crystalline Solids 285:303–308.
2. Ren L, Tsuru K, Hayakawa S, Osaka A (2002) Novel approach to fabricate porous gelatin-siloxane hybrids for bone tissue engineering. Biomaterials 23:4765–4773.

3. Chang MC, Ko CC, Douglas WH (2003) Preparation of hydroxyapatite-gelatin nanocomposite. Biomaterials 24:2853–2862.
4. Lyons L, Cauble M, Cole JA, Sabri F (1999) Investigation of biological adhesives and polyurea-crosslinked silica-based aerogels. 76th Annual Meeting of the Southeastern Section of APS Vol. 54, No. 16
5. Karageorgiou V, Kaplan D (2005) Porosity of 3D biomaterial scaffolds and osteogenesis. Biomaterials 26:5474–5491.
6. Thomas V, Dean DR, Vohra YK (2006) Nanostructured Biomaterials for Regenerative Medicine. Current Nanoscience 2:155–177.
7. Yang S, Leong KF, Du Z, Chua CK (2001) The Design of Scaffolds for Use in Tissue Engineering. Part I. Traditional Factors. Tissue Engineering 7:679–689.
8. Kim BS, Mooney DJ (1998) Development of biocompatible synthetic extracellular matrices for tissue engineering. TIBTECH 16: 224–229.
9. Jahanshahi M, Babaei Z (2008) Protein nanoparticle: A unique system as drug delivery. Vehicles African Journal of Biotechnology 7:4926–4934.
10. Hughes GA (2005) Nanostructure-mediated drug delivery Nanomedicine: Nanotechnology, Biology, and Medicine 1:22–30.
11. Liu H, Liu T, Li L, Hao N, Tan L, et al. (2012)Size dependent cellular uptake, in vivo fate and light–heat conversion efficiency of gold nanoshells on silica nanorattles. Nanoscale 4: 3523–3529.
12. Slack M, Sandhu JS, Staskin DR, Grant RC (2006) In Vivo Comparison of Suburethral Sling Materials. International Urogynecology Journal 17:106–110.
13. Klinge U, Klosterhalfen B, Birkenhauer V, Junge K, Conze J, et al. (2002) Impact of Polymer Pore Size on the Interface Scar Formation In a Rat Model. Journal of Surgical Research 103: 208–214.
14. Boccaccini AR (1998) Influence of stress concentrations on the mechanical property-porosity correlation in porous materials. Journal of Materials Science Letters 17: 1273–1275.
15. Liu DM (1997) Influence of porosity and pore size on the compressive strength of porous hydroxyapatite ceramic. Ceramics International 23: 135–139
16. Leventis N, Lu H (2011) Polymer Crosslinked Aerogels. In: Aegerter M, Leventis N, Koebel M, editors. Aerogels Handbook - Advances in Sol-Gel Derived Materials and Technologies. Springer: New York. pp 251–285.
17. Hrubesh LW (1998) Aerogel Applications. Journal of non-Crystalline Solids 225:335–342.
18. Leventis N, Sotiriou-Leventis C, Zhang G, Rawashdeh AMM (2002) Nanoengineering strong silica aerogels. Nano Letters 2: 957–960.
19. Leventis N (2007) Three dimensional core-shell superstructures: Mechanically strong aerogels. Acc Chem Res 40: 874–884.
20. Roy I, Ohulchanskyy TY, Bharali DJ, Pudavar HE, Mistretta RA, et al. (2005) Optical tracking of organically modified silica nanoparticles as DNA carriers: A nonviral, nanomedicine approach for gene delivery. Proceedings of the National Academy of Sciences of the United States of America 102: 279–284.
21. Giri S, Trewyn BG, Lin VSY (2007) Mesoporous silica nanomaterial-based biotechnological and biomedical delivery systems. Nanomedicine 2: 99–111.
22. Giri S, Trewyn BG, Stellmaker MP, Lin VSY (2005) Stimuli-Responsive Controlled-Release Delivery System Based on Mesoporous Silica Nanorods Capped with Magnetic Nanoparticles. Angewandte Chemie International Edition, 44: 5038–5044.
23. Smirnova I, Suttiruengwong S, Arlt W (2004) Feasibility study of hydrophilic and hydrophobic silica aerogels as drug delivery systems. Journal of Non-Crystalline Solids 350:54–60.
24. Lai CY, Trewyn BG, Jeftinija DM, Jeftinija K, Xu S, et al. (2003) Mesoporous Silica Nanosphere-Based Carrier System with Chemically Removable CdS Nanoparticle Caps for Stimuli-Responsive Controlled Release of Neurotransmitters and Drug Molecules. Journal of the American Chemical Society 125: 4451–4459.
25. Torney F, Trewyn BG, Lin VSY, Wang K (2007) Mesoporous silica nanoparticles deliver DNA and chemicals into plants. Nature Nanotechnology 2:295–300.
26. Sabri F, Cole JA, Scarbrough MC, Leventis N (2012) Investigation of Polyurea-Crosslinked Silica Aerogels as a Neuronal Scaffold: A Pilot Study. PLoS ONE 7(3): e33242. doi:10.1371/journal.pone.0033242
27. Sabri F, Cole JA, Scarbrough MC, Leventis N (2011) Investigation of crosslinked silica Aerogels for implant applications. In Proceedings of the Third Annual ORNL Biomedical Science and Engineering Conference: 15–17 March 2011; Knoxville. Catalog number CFP1147G-ART.
28. Yin W, Venkitachalam SM, Jarrett E, Staggs S, Leventis N, et al. (2010) Biocompatibility of surfactant-templated polyurea–nanoencapsulated macroporous silica aerogels with plasma platelets and endothelial cells. Journal of Biomedical Materials Research Part A 92:1431–1439.
29. Dunn B, Miller JM, Dave BC, Valentine JS, Zink JI (1998) Strategies for encapsulating biomolecules in sol gel matrices. Acta Materialia 46:737–741.
30. Schneider M, Baiker A (1997) Titania-based aerogels. Catalysis Today 35:339–365.
31. Sabri F, Leventis N, Hoskins J, Schuerger AC, Sinden-Redding M, et al. (2011) Spectroscopic evaluation of polyurea crosslinked aerogels, as a substitute for RTV-based chromatic calibration targets for spacecraft. Advances in Space Research 47: 419–427.
32. Schuh JCL (2008) Medical Device Regulations and Testing for Toxicologic Pathologists. Toxicologic Pathology 36:63–69.
33. Anderson JA (2001) Biological Responses to Materials. Annu. Rev. Mater. Res. 31:81–110.
34. Doherty P (2009) Inflammation, Carcinogenicity and Hypersensitivity. In: Roger Narayan editor. Biomedical Materials. Springer: New York. pp.201–214.
35. Spector M, Cease C, Tong-Li X (1989) The local tissue response to Biomaterials. Crit. Rev. Biocompatibility 5:269–295.
36. Park JC, Lee DH, Suh H (1999) Preclinical Evaluation of Prototype Products. Yonsei Medical Journal 40:530–535.

11

Development of Implant/Interconnected Porous Hydroxyapatite Complex as New Concept Graft Material

Kazuya Doi[1]*, Hiroshi Oue[1], Koji Morita[1], Shiho Kajihara[1], Takayasu Kubo[1], Katsunori Koretake[1], Vittoria Perrotti[2], Giovanna Iezzi[2], Adriano Piattelli[2], Yasumasa Akagawa[1]

1 Department of Advanced Prosthodontics, Hiroshima University Graduate School of Biomedical Sciences, Hiroshima, Japan, 2 Department of Medical, Oral and Biotechnological Sciences, Chieti-Pescara University, Chieti, Italy

Abstract

Background: Dental implant has been successfully used to replace missing teeth. However, in some clinical situations, implant placement may be difficult because of a large bone defect. We designed novel complex biomaterial to simultaneously restore bone and place implant. This complex was incorporated implant into interconnected porous calcium hydroxyapatite (IP-CHA). We then tested this Implant/IP-CHA complex and evaluated its effect on subsequent bone regeneration and implant stability *in vivo*.

Methodology/Principal Findings: A cylinder-type IP-CHA was used in this study. After forming inside of the cylinder, an implant was placed inside to fabricate the Implant/IP-CHA complex. This complex was then placed into the prepared bone socket in the femur of four beagle-Labrador hybrid dogs. As a control, implants were placed directly into the femur without any bone substrate. Bone sockets were allowed to heal for 2, 3 and 6 months and implant stability quotients (ISQ) were measured. Finally, tissue blocks containing the Implant/IP-CHA complexes were harvested. Specimens were processed for histology and stained with toluidine blue and bone implant contact (BIC) was measured. The ISQs of complex groups was 77.8±2.9 in the 6-month, 72.0±5.7 in the 3-month and 47.4±11.0 in the 2-month. There was no significant difference between the 3- or 6-month complex groups and implant control groups. In the 2-month group, connective tissue, including capillary angiogenesis, was predominant around the implants, although newly formed bone could also be observed. While, in the 3 and 6-month groups, newly formed bone could be seen in contact to most of the implant surface. The BICs of complex groups was 2.18±3.77 in the 2-month, 44.03±29.58 in the 3-month, and 51.23±8.25 in the 6-month. Significant difference was detected between the 2 and 6-month.

Conclusions/Significance: Within the results of this study, the IP-CHA/implant complex might be able to achieve both bone reconstruction and implant stability.

Editor: Ryan K. Roeder, University of Notre Dame, United States of America

Funding: This study was supported in part by a Grant-in Aid for Scientific Research (No24390435) from the Ministry of Education, Culture, Sports, Science and Technology, Japan and Brain circulation program to develop new leaders for international dental education course through international collaborative dental research. The funders had no role in study design, data collection and analysis, decision to publish, or preparation of the manuscript.

Competing Interests: The authors have declared that no competing interests exist.

* E-mail: kazuya17@hiroshima-u.ac.jp

Background

The use of dental implants has become widespread as prosthetic therapy for patients with missing teeth. The success of an implant relies on the presence of adequate bone quantity and quality at the placement site because the implant needs to undergo "osseointegration" [1]. However, in some clinical situations, the condition of bone is not always optimal to allow the implant to be stably integrated into the placement site. In such cases, bone augmentation, including guided bone regeneration, has received considerable attention [2,3,4,5].

In a large bone defect is caused by injury or tumor, implant treatment becomes more difficult. Moreover, if the functional implant site becomes infected, leading to peri-implantitis, the circumference of the affected bone in contact with the implant is resorbed, resulting in enlargement of the surrounding area owing to an inflammatory reaction. In such cases, the implant and circumference of the affected bone require surgical removal, resulting in a large bone defect, which allows implant treatment only after bone reconstruction using some bone grafts, however, bone augmentation could be a tricky procedure, healing period may be very long, multiple surgeries are usually required, with a consequent increased risk of morbidity for patients due to the multiple step approach. In particular, when implants are not placed simultaneously to bone augmentation procedures a further decrease in the bone gain should be expected. Therefore, growing need to develop a novel approach to simultaneously restore bone tissue and place implant exists. In our knowledge, there is no reports evaluated such approach. Recently, interconnected porous calcium hydroxyapatite (IP-CHA) ceramics have been introduced for use as scaffold for bone regeneration [6], and are now widely used in the clinical field [7,8,9,10]. IP-CHA has a systematic arrangement of spherical uniform pores with interconnections and has been reported to be successfully used as bone-grafting material in the field of tissue engineering. Also, there are some reports that biomaterials of porous structure had applied for implant graft material [11,12]. Previous studies conducted in our laboratory demonstrated that both of cell-hybrid artificial bone where bone marrow stromal calls were injected to IP-CHA and polyphosphate adsorbed IP-CHA could enhance bone formation [13,14,15,16].

Basing on the specific characteristic of the IP-CHA, we designed a novel original complex to be used as graft material (Fig. 1A). Specifically, the complex consists of dental implant fixture surrounded by a hollow block of IP-CHA, which is thought to enhance osteoconduction and improve implant stability. The hypothesis is that the complex replaces lost bone tissue, supports new bone formation and allows a simultaneous implant placement with a shortening of the time and surgical steps. To the best of our knowledge, the application of such a complex has hitherto not been reported. This complex is expected to be useful in the fields of dentistry and orthopedics as a novel, simplified approach to achieve bone rehabilitation. Therefore, the aim of the present study was to test the implant/IP-CHA complex developed and to assess its effects on subsequent bone regeneration and implant stability in an animal model.

Materials and Methods

Materials

The hydroxyapatite (HA) used in this study was custom-fabricated into a hollow cylinder of IP-CHA (outer diameter: 6 mm, inner diameter: 3 mm, height: 8.5 mm)(NEOBONE®; Covalent Materials, Tokyo, Japan) with 75% porosity, and a mean porous diameter of 150 μm, and all pores were connected by interconnected pores of 40 μm diameter (Fig. 2). The IP-CHA structure was manufactured using the "form-gel" technique [6]. The implants used were commercially available pure titanium implants (diameter: 3.75 mm, length: 8.5 mm; Brånemark System® Mk III, Nobel Biocare, Sweden).

Fabrication of the Implant/IP-CHA Complex

Fabrication of the complex was performed as follows: first, a screw thread was prepared at the inner surface of a hollow cylinder of IP-CHA using a special electric engine (Nobel Biocare Japan Inc, Tokyo, Japan) with serial cutting drills and a screw tap (Nobel Biocare Japan Inc), and then, countersinking was performed at upper section. After preparation of the IP-CHA cylinder, the titanium implant was incorporated into it to fabricate the complex (Fig. 3). This procedure was performed in accordance to with the Brånemark® system manual.

Scanning Electron Microscopy Image of the Prepared Cylindrical IP-CHA

After the screw tap and countersink preparation, before incorporation of the implant, one of the hollow IP-CHA cylinders was bisected, and the inner portion was observed by scanning electron microscopy (SEM, JMS-7300, Nihon Denshi Oyo, Tokyo, Japan).

Observation of the Implant/IP-CHA Complex

One of the Implant/IP-CHA complex was dehydrated using increasing concentrations of ethanol, cleared with styrene monomer, and then embedded in light-polymerized polyester resin (Technovit 7200VLC; Kulzer, Wehrheim, Germany). To achieve complete polymerization of the resin block, photopolymerization equipment was used (BS5000; EXAKT APARATEBAU, Hamburg, Germany). The resin block was cut using a diamond saw system at the center of the Implant/IP-CHA complex, and the cross-section was observed.

Animal Experiments

The animal research protocol was in accordance with the current version of the Japan Law on the Protection of Animals. This study was approved by the Research Facilities Committee for Laboratory Animal Science at the Hiroshima University School of Medicine, Hiroshima, Japan. All surgery was performed under general anesthesia, and all efforts were made to minimize suffering during experimental period.

The study design of the animal procedure is shown in Fig. 4. Three male Beagle-Labrador hybrid dogs weighing 20–25 kg and aged 20–24 months were fed in their cages for 1 month to allow them to acclimatize to the environment. The Implant/IP-CHA complexes were placed into prepared bone sockets (diameter:

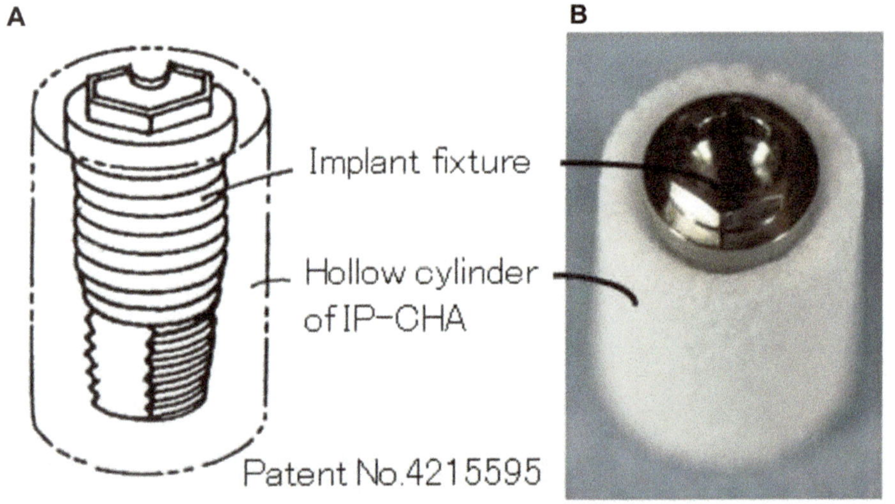

Figure 1. Implant/interconnected porous calcium hydroxyapatite (IP-CHA) Complex. (A) Outline (B) fabricated complex, showing the assembled graft consisting of the IP-CHA and implant fixture.

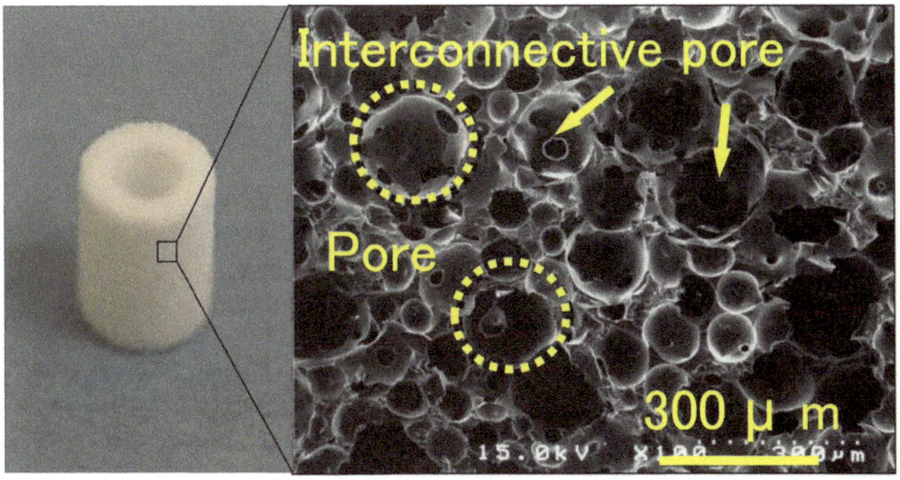

Figure 2. IP-CHA fabricated of a hollow cylinder. (A) Photograph of the prepared IP-CHA (B) A SEM image of the surface. IP-CHA has a systematic arrangement of uniform pores, all of which are connected by a network of smaller interconnected pores.

6 mm, height: 10 mm) in the center of the left femur of each of the 3 dogs (Fig. 5). Additionally, titanium implants without IP-CHA were placed beside the Implant/IP-CHA complex as a control. All surgical procedures were performed under general anesthesia with sodium pentobarbital (40 mg/kg) and local infiltration anesthesia with 2% lidocaine and 1:80,000 noradrenaline. The animals were sacrificed under deep anesthesia and perfused with 10% neutral formalin through the aorta, and the femurs were dissected. Samples were retrieved at different time points (2, 3, and 6 months) and processed to obtain histological sections. Finally, the Implant/IP-CHA complex at 2, 3, and 6 months and the control Implants were set in position (n = 3). The placement design is shown in Table 1.

Histological Evaluation

The specimens were immediately fixed in 10% buffered formalin and processed to obtain thin ground sections. Tissue blocks containing implants and Implant/IP-CHA complex were dehydrated using ascending concentrations of ethanol, cleared with styrene monomer, and then embedded in light-polymerized polyester resin (Technovit 7200VLC, Kulzer). To achieve complete polymerization of the resin block, photo polymerization equipment was used (BS5000; EXAKT APARATEBAU). After polymerization, the specimens were sectioned with a high-precision diamond disc to produce a 200-μm-thick cross section. The undecalcified specimens were ground to approximately 70-μm-thick sections (MG5000; EXAKT APPARATEBAU), which were

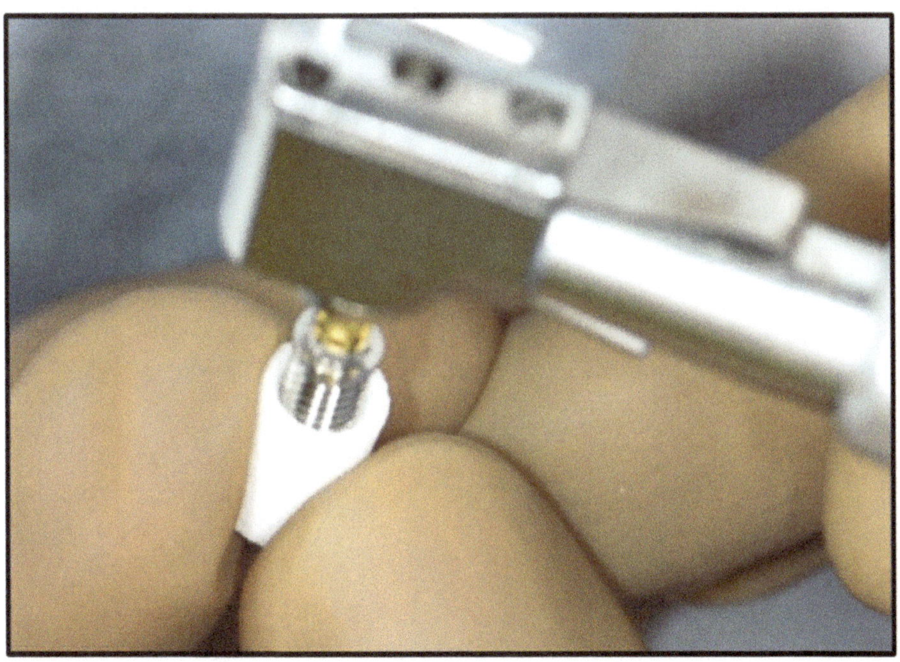

Figure 3. In corporation of a titanium implant into the prepared IP-CHA. After preparation of the IP-CHA, commercially available implants were incorporated into the IP-CHA to fabricate the graft complex. This process was performed under sterile condition.

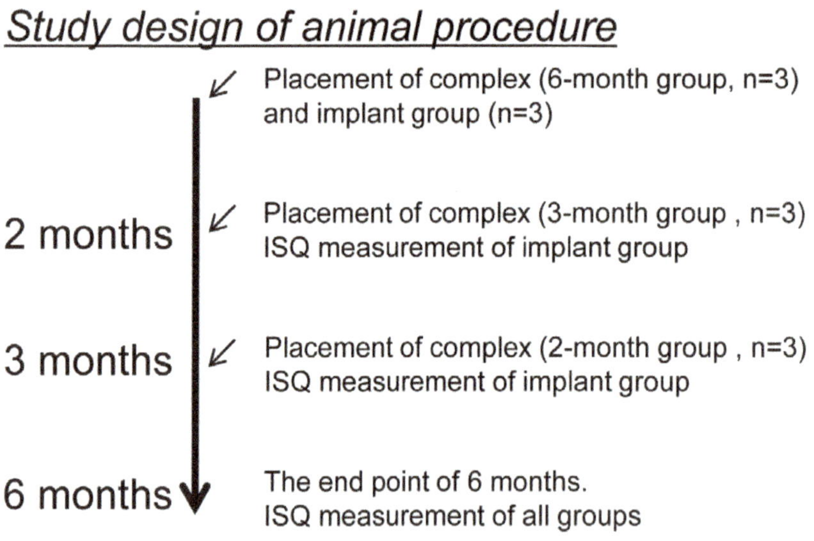

Figure 4. Design of the animal experiment. Three animals were used in this study. Both control implants and test complexes were placed and evaluated over 6 months period. The ISQ of the control implant group was measured continually and the ISQs of the complex groups are measured at the end point.

stained with toluidine blue. A light microscope was used for histological examination of the Implant/IP-CHA complex specimens.

Measurement of Implant Stability Quotient

The implant stability quotient (ISQ) of implants was measured repeatedly at 2, 3 and 6 months. ISQs of Implant/IP-CHA complexes were measured at 6 months ISQ was assessed using Ostell ISQ^{TM} (Ostell Mentaor, Göteborg, Sweden) as resonance frequency analyzer (Fig. 6). After 2, 3, and 6 months placement, the ISQs of all samples was measured.

Measurement of Bone Implant Contact (BIC) Ratio

The ratio of BIC was measured as the ratio of contact length of newly formed bone in total length from the bottom of cortical bone detected to the top of implant shoulder part [17]. This measurement was performed in the cortical bone area. BIC were measured using NIH Image J (National Institutes of Health, Bethesda, MD).

Statistical Analysis

The data obtained were expressed as means +/− standard deviations.

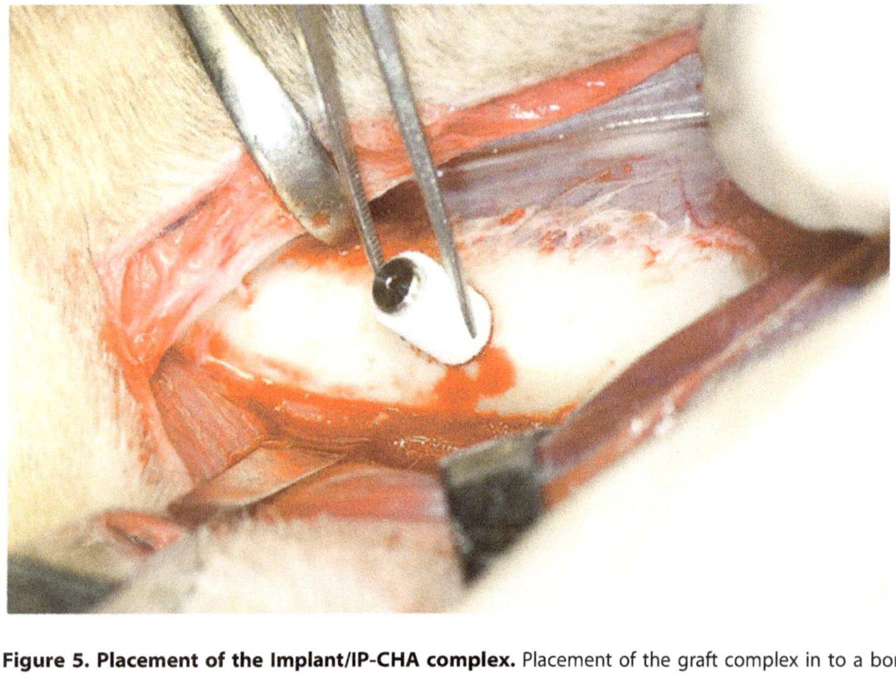

Figure 5. Placement of the Implant/IP-CHA complex. Placement of the graft complex in to a bone sockets in the femur.

Table 1. Placement part of implants and Implant/IP-CHA complexes.

	Right side		Left side	
Animal A	Implant	Complex:6M	Complex:3M	Complex:2M
Animal B	Implant	Complex:3M	Complex:2M	Complex:6M
Animal C	Implant	Complex:2M	Complex:3M	Complex:6M

The values obtained were statistically analyzed using one-way analysis of variance and Tukey's HSD test for multiple comparisons, with significance level set at 5%.

Results

SEM Image of Formed IP-CHA

Fig. 7 shows an SEM image of the inner surface of the IP-CHA cylinder. The shape of the formed screw-thread is evident. After preparation of the screw-thread, the characteristic structure of IP-CHA could still be seen, and the interconnected pores remained intact.

Observation of the Implant/IP-CHA Complex

Fig. 8 shows the central part of the Implant/IP-CHA complex. The implant screw threads and formed IP-CHA were almost fully integrated, and there was very little space between the screw threads and IP-CHA.

Histological Evaluation

At 2-month group of Implant/IP-CHA complex, newly formed bone mainly colonized the pores of the IP-CHA located in the vicinity of pre-existing cortical bone. In the central part of the IP-CHA, only in few pores, new bone tissue could be observed. At the interface between IP-CHA and implant no bone tissue, but only loose connective tissue was detected. In the coronal and middle portions of the specimens cortical bone were evident, while in the apical part marrow spaces. Newly formed bone could easily be distinguished by pre-existing bone as it appeared more intensely stained (Fig. 9A).

At 3-month group of Implant/IP-CHA complex, the pores of the IP-CHA located in the vicinity of pre-existing cortical bone and the ones in the central part of the biomaterial were filled by bone tissue. Bone could also be observed at the interface between the biomaterial and the implant mainly in the coronal and middle portion of specimens, while in the apical part very small and scarce trabeculae were present. The different intensities of the staining are less markedly distinguishable, owing to the process of maturation of the regenerated bone tissue (Fig. 9B).

At 6-month group of Implant/IP-CHA complex, the Implant/IP-CHA complex seemed to be integrated. Bone could be observed in the coronal, middle and apical portions of the specimens and especially in the coronal part many areas of contact between bone and implant surface were evident. Regenerated bone tissue could not be distinguished by the pre-existing one (Fig. 9C).

Measurement of ISQ

Fig. 10 shows the ISQs of the Implant/IP-CHA complex and control implants, measured using an Ostell^{TM} as the RFA. The ISQs of the Implant/IP-CHA complex was 47.4 ± 11.4 at 2 months, 72.0 ± 5.68 at 3 months, and 77.8 ± 2.88 at 6 months. The ISQs of the control Implant groups were 77.2 ± 2.57 at 2 months, 78.4 ± 1.94 at 3 months, and 81.0 ± 3.28 at 6 months. The differences decreased over time.

At 2 months, the ISQ of the Implant/IP-CHA complex was significantly lower than that of the control Implant. On the other hand, at 3 and 6 months, the Implant/IP-CHA complex ISQs were not significantly different from those of the control Implant group. The ISQs of Implant/IP-CHA complexes were significantly higher at 3 and 6 months than at 2 months (Table 2).

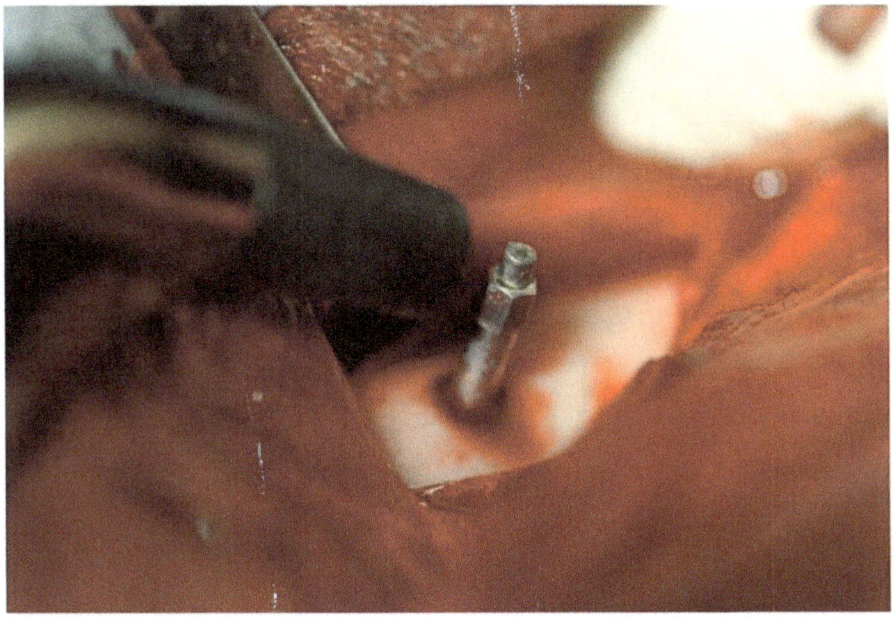

Figure 6. Measurement of ISQ. Stability was measured in the implant control group and the complex groups using an Ostell ISQ^{TM} as RFA.

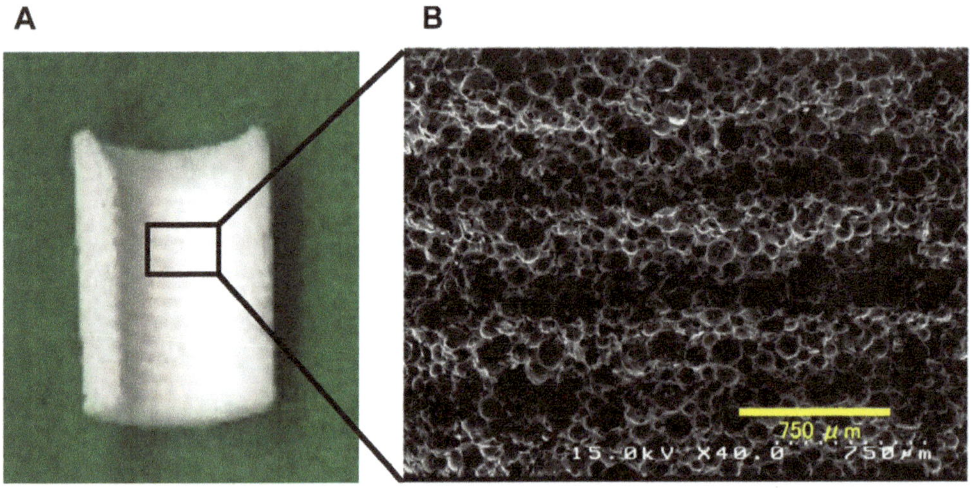

Figure 7. Inner surface of the formed IP-CHA. (A) Photograph of the inner. (B) SEM image of the formed IP-CHA. The shape of the formed screw-thread is evident. After preparation, the characteristic structure was not destroyed, and the interconnected pores remained intact.

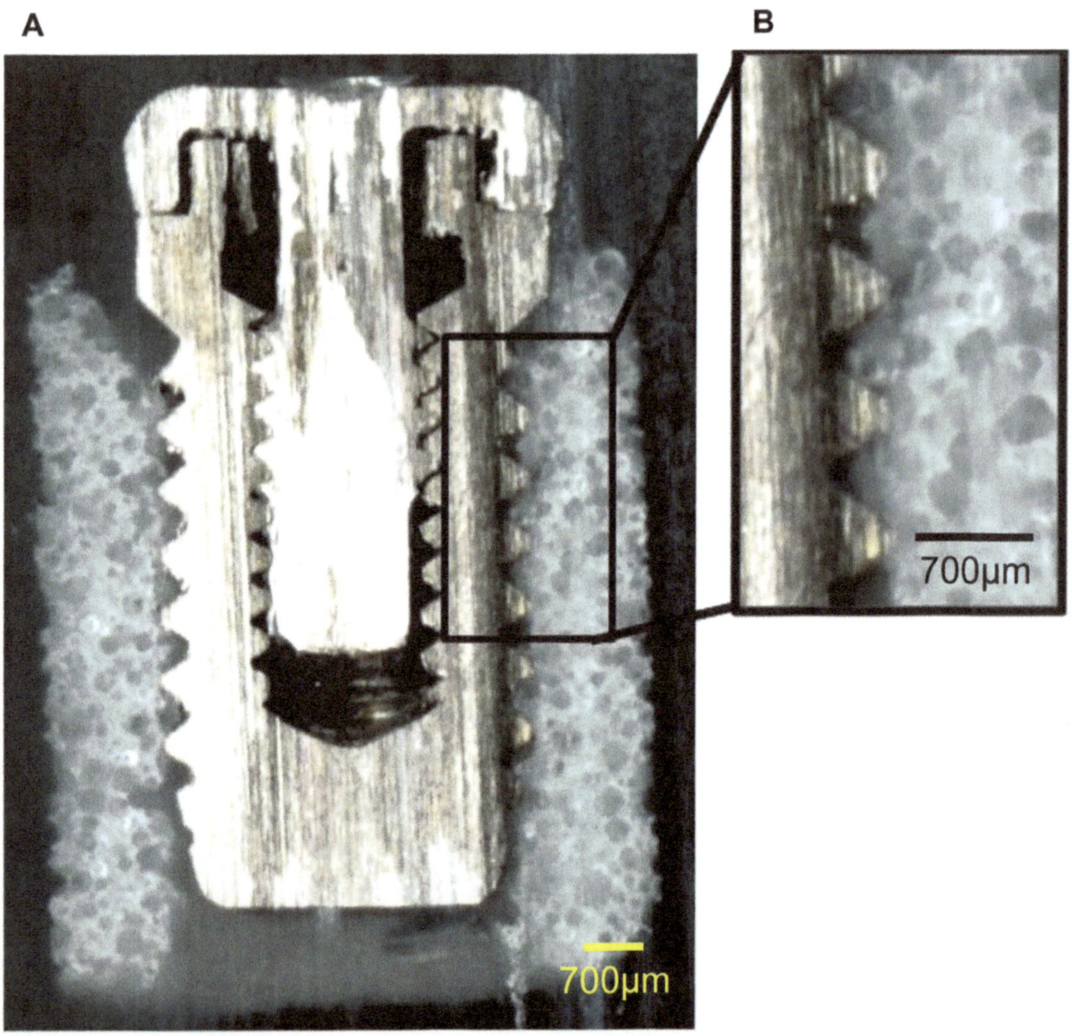

Figure 8. Cross-sectional shape of the Implant/IP-CHA complex. Observation of the cross section of the complex, showing that the implant screw threads and the formed IP-CHA are almost fully3 integrated and there is very little space between the implant threads and the formed IP-CHA.

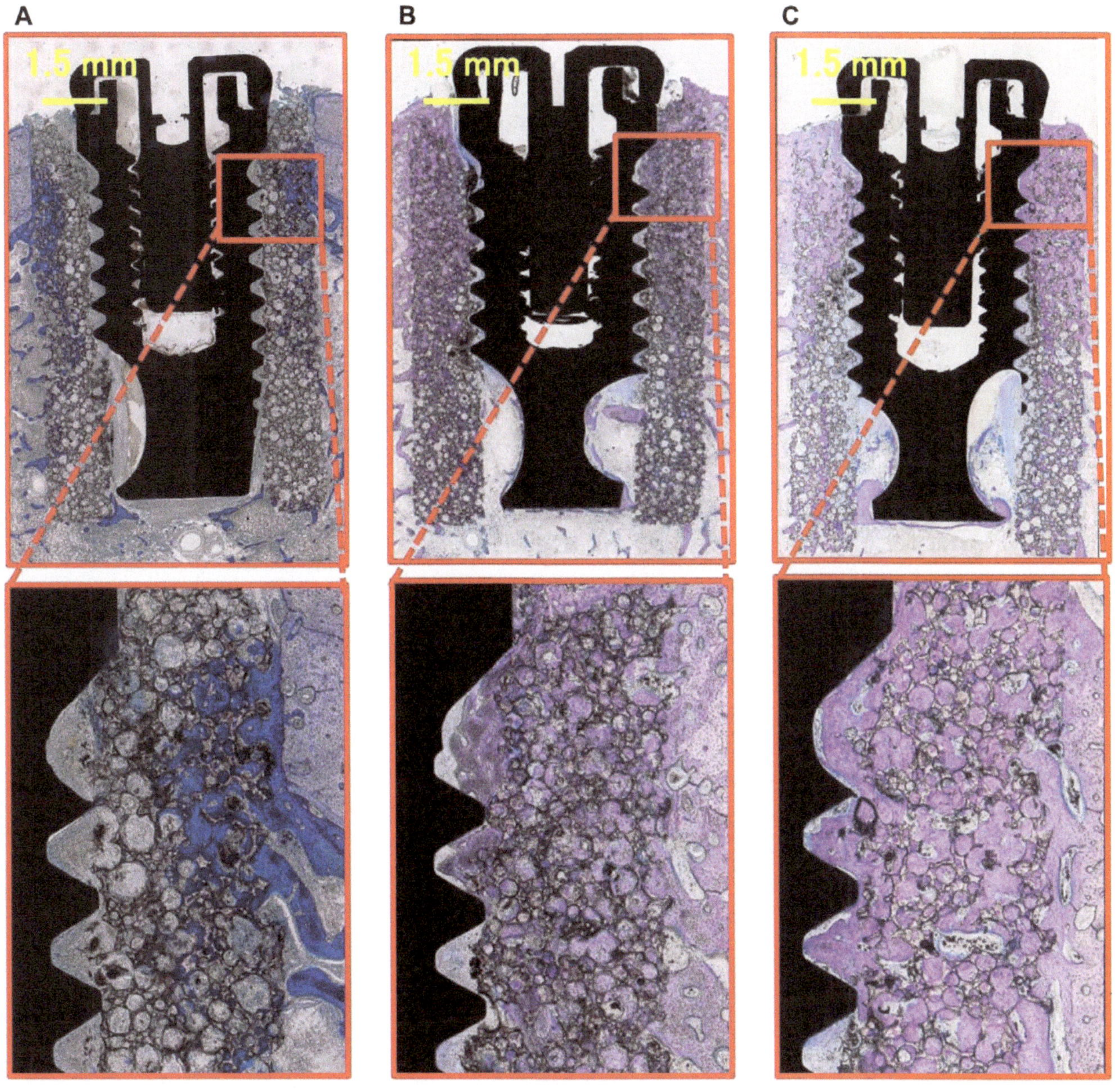

Figure 9. Histology of the Implant/IP-CHA complex groups. (A) Specimen of the 2-month complex, showing newly formed bone, connective tissue and bone marrow detected in the IP-CHA part of the cortical area. At the implant interface, no newly formed bone is observed and the apace is occupied mainly by connective tissue and bone marrow. Therefore, osseointegration is not achieved. (B) Specimen of the 3-month complex, newly formed bone is detected not only in the IP-CHA part but also at the implant interface. Newly formed bone is in contact with the implant surface, and osseointegration is achieved. (C) Specimen of the 6-month complex, showing similar bone formation to the 3-month complex and clear increase in new bone formation at the implant interface.

Measurement of BIC

The BICs of the Implant/IP-CHA complexes were 2.18±3.77 at 2 months, 44.04±29.58 at 3 months, and 51.02±8.25 at 6 months. The BICs of Implant/IP-CHA complexes were significantly higher at 6 months than at 2 months (Table 3).

Discussion

In this study, the Implant/IP-CHA complex achieved both bone regeneration and implant stability.

The complex developed in this study could be built in two ways: incorporating the implant into the shaped IP-CHA, or burning apatite and implant simultaneously. The latter method, however, has some disadvantages. It is difficult to shape arbitrarily apatite, and titanium surface is formed oxide film because of high temperature burning. Therefore, it is considered that the former method is the most suitable to fabricate Implant/IP-CHA complex. The IP-CHA can be prefabricated into optimal forms and shapes using drilling, cutting or computer-aided design/manufacturing system [18]. Thus, IP-CHA can easily be formed into the desired shape using a cutting instrument. The shape of the

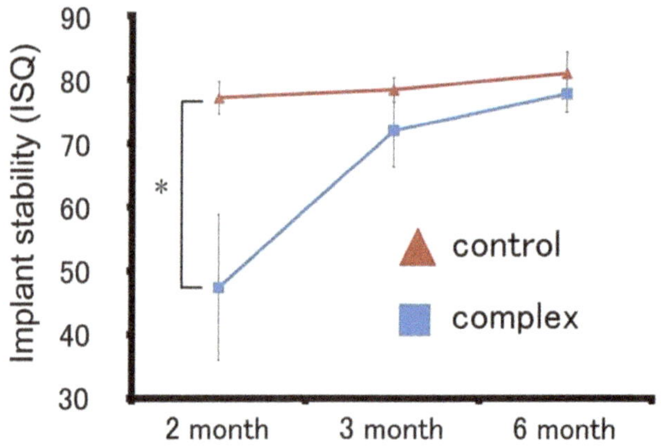

Figure 10. Mean ISQ of the Implant group and the Implant/IP-CHA complex groups. Asterisks indicate a significant difference with respect to the 2-month Implant as control group and the 2-month Implant/IP-CHA complex group (p<0.05).

screw-thread was evident in the SEM image of the formed IP-CHA (Fig. 4). Additionally, observation of the fabricated implant/IP-CHA complex showed that the area between the implant and the formed IP-CHA was mostly integrated (Fig. 5). Moreover, the distance between the threads of the implant used in this study was set at approximately 0.7 mm. At this size, it can be formed into an optimal shape by a screw instrument with same shape. These results indicate that the implant/IP-CHA complex may not only be applicable to this implant system but also to other implant systems used in the fields of dentistry or orthopedics. The requirements for a biomaterial used in bone regeneration are biocompatibility, good osteoconduction, mechanical strength, and availability of space for tissue and blood vessel regeneration. Moreover, the ability to use tissue engineering techniques, such as a creating scaffold for stem cells, or incorporating several growth factors is required. In this study, we used IP-CHA as a complex structure owing to its specific characteristics. Porous HA without interconnected pores has previously been used in situations where bone regeneration is needed [19,20,21]. While it possesses biocompatibility and promotes good bone formation, its osteoconduction is limited to its surface because of the lack of adequate connections between each pore [22]. It is widely accepted that pore sizes more than 10 μm in diameterare required to permit osteoconduction because the size of the nucleus in most mammalian cells is more than 10 μm [23]. In a clinical case, it was reported that bone ingrowth into porous HA without interconnected pores penetrated less than 300 μm from the HA surface [24]. Using such HA for an implant and HA complex would limit bone formation to the HA surface, and osseointegration would barely be achieved. In contrast, the IP-CHA has a systematic arrangement of uniform pores and almost all pores are onterconnected. The average diameter of the interconnection between pores is 40 μm, allowing efficient migration of bone-producing cells from pore-to-pore and colonization by blood vessels, which are essential for new bone formation.

In our previous study, we reported that IP-CHA was a potentially useful biomaterial for use as a scaffold; it has been shown to integrate bone marrow stromal cells and growth factors, and exhibited good osteoconduction in an animal model [13,14,15,16]. With regard to histology, while bone formation clearly occurred at the IP-CHA implantation area in all Implant/IP-CHA complex groups, bone formation at the implant surface

Table 2. Comparison of Mean ISQ for Implant/IP-CHA complex groups.

	Mean ISQ (SD)	Turkey test
Complex: 2M	47.4 (11.4)	
Complex: 3M	72.0 (5.68)	P=0.0017*
Complex: 6M	77.8 (2.88)	P=0.0064**

*between Complex: 2M and Complex: 3M.
**between Complex: 2M and Complex: 6M.
SD, standard deviation.

was only detected at 3 and 6 months. On the other hand, at 2 months, fibrous connective tissue or bone marrow was predominant at the implant surface. Considering the lengthy healing period, it may take more than 3 months to achieve osteoconduction throughout the implant surface. The IP-CHA of the complex has initial compressive mechanical strength equivalent to that of cancellous bone because it has a porosity rate of approximately 75%. While it is difficult to support an implant with this strength, mechanical strength is increased by the development of newly formed bone within the IP-CHA. In one study, the compressive strength of implanted IP-CHA was reported to increase steadily with bone ingrowth into the pores [6]. In this study, to evaluate osseointegration of Implant/IP-CHA, implant stability and bone implant contact ratio were measured. Measuring implant stability is considered to be an important method for evaluating the success of an implant [25]. The RFA technique has been shown to be an effective method of measuring implant stability [26,27,28]. This system measures ISQ using RFA, which measures emitting frequency via a vibration transducer attached to the abutment or fixture [29,30].

The ISQ is given as a number from 1 to 100, where 1 is the lowest and 100 the highest degree of implant stability. ISQs of successfully stabilized implants are reported to range from 57 to 82 [27] and due to bone deflection, ISQ values as 90 or more are uncommon in clinical situation.

Additionally, histomorphological studies suggest that the bone-to-implant ratio and RFA values are related [31]. We measured the implant stability of all groups at 2, 3, and 6 months (Fig. 10). The ISQs of control Implant groups were found to be 70 or more, and significant differences were not observed in control implant groups. This is likely because the control Implants were placed in sufficient bone and initial fixation was obtained. The ISQ is known to increase in proportion to the stiffness of the bone-implant interface with that of the surrounding bone [32,33]. Furthermore, no significant differences were observed between the ISQs of the control group and Implant/IP-CHA complex at 3 and 6 months.

Table 3. Comparison of BIC for Implant/IP-CHA complex groups.

	BIC Mean (SD)	Turkey test
Complex: 2M	2.18 (3.77)	
Complex: 3M	44.04(29.58)	
Complex: 6M	51.02 (8.25)	P=0.0356*

*between Complex: 2M and Complex: 6M.
SD, standard deviation.

However, at 2 months, the ISQ of the Implant/IP-CHA complex was lower than that of the control Implant group or the 3- and 6-month complex groups. A significant correlation between the ISQ and bone cortical thickness was reported [34]. However, regarding ISQ associated with bone contact, there is a controversy. Abrahamsson et al reported there was no correlation of between BICs and ISQs [35], while, Blanco et al showed a there was positive correlation between the increase of ISQs and BICs [36].

In our result of BIC measurements of Implant/IP-CHA complex significant difference was observed between at 2 months and 6 months. This result seems to correlate between BICs and ISQs, however, the BIC at 3 month of Implant/IP-CHA complex was not significantly at 2month. It was due to that the BIC at 3 month had large standard deviation which suggested that bone contact formation was unstable.

On the basis of our histological evaluation, we consider that Implant/IP-CHA complex had stabilized and exhibited osteoconduction into the complex surrounding the implant at 6 months. A healing period of 2 months is not enough for osteoconduction to lead to adequate implant stability for the Implant/IP-CHA complex. The results indicated that implant stability was achieved using the Implant/IP-CHA after a 6-month healing period.

It is the authors' intention to conduct further studies with experimental underloaded condition or mechanical properties to investigate detailed bone aspects in clinical model.

Conclusions

Our results indicate that placed Implant/IP-CHA complex could achieve osteoconduction of newly formed bone and osseointegration at the implant interface. We concluded that the Implant/IP-CHA complex could be expected to achieve bone reconstruction and implant stability. The results of this study may also contribute to provision of increased knowledge of the development of more predictable bone graft materials.

Author Contributions

Conceived and designed the experiments: KD HO YA. Performed the experiments: KD HO TK. Analyzed the data: KK GI AP. Contributed reagents/materials/analysis tools: KD HO KM SK. Wrote the paper: KD VP YA.

References

1. Brånemark P, Zarb G, Albrektsson T (1985) Tissue-integrated prostheses. Chicago: Quitessence Publishing: 11–43.
2. Becker W, Becker BE (1990) Guided tissue regeneration for implants placed into extraction sockets and for implant dehiscences: surgical techniques and case report. Int J Periodont Rest Dent 10: 376–391.
3. Hermann JS, Buser D (1996) Guided bone regeneration for dental implants. Curr Opin Periodontol 3: 168–177.
4. Kalk WW, Raghoebar GM, Jansma J Boering G (1996) Morbidity from iliac crest bone harvesting. J Oral Maxillofac Surg 54: 1424–1429.
5. Chiapasco M, Abati S, Romeo E, Chiapasco M, Abati S, et al. (1999) Clinical outcome of autogenous bone blocks or guided bone regeneration with e-PTFE membranes for the reconstruction of narrow edentulous ridges. Clin Oral Impl Res 10: 278–288.
6. Tamai N, Myoui A, Tomita T, Nakase T, Tanaka J, et al. (2002) Novel hydroxyapatite ceramics with an interconnective porous structure exhibit superior osteoconduction in vivo. J Biomed Mater Res. 59: 110–117.
7. Deie M, Ochi M, Adachi N, Nishimori M, Yokota K (2008) Artificial bone grafting [calcium hydroxyapatite ceramic with an interconnected porous structure (IP-CHA)] and core decompression for spontaneous osteonecrosis of the femoral condyle in the knee. Knee Surg Sports Traumatol Arthrosc. 16: 753–758.
8. Kuriyama K, Hashimoto J, Murase T, Fujii M, Nampei A, et al. (2009) Treatment of juxta-articular intraosseous cystic lesions in rheumatoid arthritis patients with interconnected porous calcium hydroxyapatite ceramic. Mod Rheumatol. 2009 19: 180–186.
9. Tamai N, Myoui A, Kudawara I, Ueda T, Yoshikawa H (2010) Novel fully interconnected porous hydroxyapatite ceramic in surgical treatment of benign bone tumor. J Orthop Sci 15: 560–568.
10. Shigeishi H, Takechi M, Nishimura M, Takamoto M, Minami M, et al. (2012) Clinical evaluation of novel interconnected porous hydroxyapatite ceramics (IP-CHA) in a maxillary sinus floor augmentation procedure. Dent Mater J 31: 54–60.
11. Bobyn JD, Stackpool GJ, Hacking SA, Tanzer M, Krygier JJ (1999) Characteristics of bone ingrowth and interface mechanics of a new porous tantalum biomaterial. J Bone Joint Surg Br 81: 907–914.
12. Tsao AK, Roberson JR, Christie MJ, Dore DD, Heck DA, et al. (2005) Biomechanical and Clinical Evaluations of a Porous Tantalum Implant for the Treatment of Early-Stage Osteonecrosis. J Bone Joint Surg Am 87: 22–27.
13. Doi K, Kubo T, Hayashi K, Imura K, Akagawa Y (2007) Development of cell-hybrid artificial bone: effect of osteogenic differentiation of bone marrow stromal stem cells on bone formation with newly developed interconnected porous calcium hydroxyapatite. Dent Mater J 26: 162–169.
14. Yuan Q, Kubo T, Doi K, Morita K, Takeshita R, et al. (2009) Effect of combined application of bFGF and inorganic polyphosphate on bioactivities of osteoblasts and initial bone regeneration. Acta Biomater 5: 1716–1724.
15. Morita K, Doi K, Kubo T, Takeshita R, Kato S, et al. (2010) Enhanced initial bone regeneration with inorganic polyphosphate-adsorbed hydroxyapatite. Acta Biomater 6: 2808–2815.
16. Kubo T, Doi K, Hayashi K, Morita K, Matsuura A, et al. (2011) Comparative evaluation of bone regeneration using spherical and irregularly shaped granules of interconnected porous hydroxylapatite. A beagle dog study. J Prosthodont Res 55: 104–109.
17. Hayashi K, Kubo T, Doi K, Tabata Y, Akagawa Y (2007) Development of new drug delivery system for implant bone augmentation using a basic fibroblast growth factor-gelatin hydrogel complex. Dent Mater J 26: 170–177.
18. Yoshikawa H, Tamai N, Murase T, Myoui (2009) A Interconnected porous hydroxyapatite ceramics for bone tissue engineering J. R. Soc. Interface 6: 341–348.
19. Ayers RA, Wolford LM, Bateman TA, Ferguson VL, Simske J (1999) Quantification of bone ingrowth into porous block hydroxyapatite in humans. J Biomed Mater Res 47: 54–59.
20. Cordioli G, Mazzocco C, Schepers E, Brugnolo E, Majzoub Z (2001) Maxillary sinus floor augmentation using bioactive glass granules and autogenous bone with simultaneous implant placement. Clinical and histological findings. Clin Oral Implants Res 12: 270–278.
21. Woodard JR, Hilldore AJ, Lan SK, Park CJ, Morgan AW, et al. (2007) The mechanical properties and osteoconductivity of hydroxyapatite bone scaffolds with multi-scale porosity. Biomaterials 28: 45–54.
22. Ayer RA, Simske SJ. Nunes CR, Wolford LM (1998) Long-term bone ingrowth and residual microhardness of porous block hydroxyapatite implants in humans. J. Oral Maxillofac Surg 56: 1297–1301.
23. Steinkamp JA, Hansen KM, Crissman HA (1976) Flow microfluorometric and light-scatter measurement of nuclear and cytoplasmic size in mammalian cells. J Histochem Cytochem 24: 292–297.
24. Anderson HC (1995) Molecular biology of matrix vesicles. Clin Orthop Relat Res314: 266–280.
25. Zarb GA, Schmitt A (1991) Osseointegration and the edentulous predicament. The 10-year-old Toronto study. Br Dent J 70: 439–444.
26. Meredith N, Alleyne D, Cawley P (1996) Quantitative determination of the stability of the implant-tissue interface using resonance frequency analysis. Clin Oral Implants Res 7: 261–267.
27. Friberg B, Sennerby L, Linden B, Gröndahl K, Lekholm U (1999) Stability measurements of one-stage Brånemark implants during healing in mandibles. A clinical resonance frequency analysis study. Int J Oral Maxillofac Surg 28: 266–272.
28. Glauser R, Sennerby L, Meredith N, Rée A, Lundgren A, et al. (2004) Resonance frequency analysis of implants subjected to immediate or early functional occlusal loading. Successful vs. failing implants. Clin Oral Implants Res 15: 428–434.
29. De Sants E, Botticelli D, Pantani F, Pereira FP, Beolchini M, et al. (2011) Bone regeneration at implants placed into extraction sockets of maxillary incisors in dogs. Clin Oral Implants Res 22, 430–437.
30. Isoda K, Ayukawa Y, Tsukiyama Y, Sogo M, Matsushita Y, et al. (2011) Relationship between the bone density estimated by cone-beam computed tomography and the primary stability of dental implants. Clin Oral Implants Res 23, 832–836.
31. Huang HM, Pan LC, Lee SY, Chiu CL, Fan KH, et al. (2000) Assessing the implant/bone interface by using natural frequency analysis. Oral Surg Oral Med Oral Pathol Oral Radiol Endod 90: 285–291.
32. Hui E, Chow J, Li D, Liu J, Wat P, et al. (2001) Immediate provisional for single-tooth implant replacement with Brånemark system: preliminary report Clin Implant Dent Relat Res 3: 79–86.
33. Tawse-Smith A, Perio C, Payne AG, Kumara R, Thomson WM (2001) One-stage operative procedure using two different implant systems: a prospective study on implant overdentures in the edentulous mandible. Clin Implant Dent Relat Res 3: 185–193.

34. Miyamoto I, Tsuboi Y, Wada E, Suwa H, Iizuka T (2005) Influence of cortical bone thickness and implant length on implant stability at the time of surgery-clinical, prospective, biomechanical, and imaging study. Bone 37: 776–880.
35. Abrahamsson I, Linder E, Lang NP (2009) Implant stability in relation to osseointegration: an experimental study in the Labrador dog. Clin Oral Implants Res. 20: 313–318.
36. Blanco J, Alvarez E, Muñoz F, Liñares A, Cantalapiedra A (2011) Influence on early osseointegration of dental implants installed with two different drilling protocols: a histomorphometric study in rabbit. Clin Oral Implants Res. 21: 92–99.

12

Ectopic Osteoid and Bone Formation by Three Calcium-Phosphate Ceramics in Rats, Rabbits and Dogs

Liao Wang[1], Bi Zhang[2], Chongyun Bao[1]*, Pamela Habibovic[3], Jing Hu[1], Xingdong Zhang[4]

1 State Key Laboratory of Oral Diseases, West China Hospital of Stomatology, Sichuan University, Chengdu, China, **2** The Affiliated Hospital of Stomatology, Chongqing Medical University, Chongqing, China, **3** Department of Tissue Regeneration, MIRA-Institute for Biomedical Technology and Technical Medicine, University of Twente, Enschede, the Netherlands, **4** National Engineering Research Center for Biomaterials, Sichuan University, Chengdu, China

Abstract

Calcium phosphate ceramics with specific physicochemical properties have been shown to induce *de novo* bone formation upon ectopic implantation in a number of animal models. In this study we explored the influence of physicochemical properties as well as the animal species on material-induced ectopic bone formation. Three bioceramics were used for the study: phase-pure hydroxyapatite (HA) sintered at 1200°C and two biphasic calcium phosphate (BCP) ceramics, consisting of 60 wt.% HA and 40 wt.% TCP (β-Tricalcium phosphate), sintered at either 1100°C or 1200°C. 108 samples of each ceramic were intramuscularly implanted in dogs, rabbits, and rats for 6, 12, and 24 weeks respectively. Histological and histomorphometrical analyses illustrated that ectopic bone and/or osteoid tissue formation was most pronounced in BCP sintered at 1100°C and most limited in HA, independent of the animal model. Concerning the effect of animal species, ectopic bone formation reproducibly occurred in dogs, while in rabbits and rats, new tissue formation was mainly limited to osteoid. The results of this study confirmed that the incidence and the extent of material-induced bone formation are related to both the physicochemical properties of calcium phosphate ceramics and the animal model.

Editor: João Costa-Rodrigues, Faculdade de Medicina Dentária, Universidade do Porto, Portugal

Funding: This work was supported by the National Natural Science Foundation of China (Grant No. 81371181 and 81171005). The funders had no role in study design, data collection and analysis, decision to publish, or preparation of the manuscript.

Competing Interests: The authors have declared that no competing interests exist.

* Email: cybao9933@scu.edu.cn

Introduction

Bone defects caused by tumor resection, trauma, infection and congenital anomalies remain an important challenge in the field of orthopaedic and craniofacial surgery. Clinical treatment of such defects involves the use of natural grafts, like autograft and allograft, or synthetics [1,2]. Autograft is still considered the gold standard for bone substitution, owning to the absence of immunogenic reaction, excellent osteogenicity, osteoinductivity and osteoconductivity. However, shortage of donor supply, persistence of pain, nerve damage, fracture, infection and cosmetic disability at the donor site remain important drawbacks. The use of allograft, where availability is less of a problem, suffers from clinical risks of disease transmission and immunogenic reaction. Another major drawback of allograft is its inconsistent quality from both a biological and a mechanical perspective [1,3].

Tissue engineering is defined as an interdisciplinary field that applies the principles of engineering and life sciences toward the development of biological substitutes that restore, maintain, or improve tissue function or a whole organ [4]. Along with stem cell therapy that is currently being used for the treatment of various diseases and disorders [5–8], tissue engineering approaches have been developed and tested for various clinical applications. In orthopedics and cranio-maxillo-facial surgery, tissue engineering is considered a promising method to replace the autograft and allograft and it developed rapidly in the last two decades as a result of progress in cell biology and material science research [9–11]. In tissue engineering, constructs are built consisting of degradable biomaterials that serve as scaffold and transplanted (stem) cells that adhere, proliferate and differentiate on the scaffold, and, in some cases, additional growth factors to control cell fate [12,13]. Independent of the cell sources used in bone tissue engineered constructs, one always faces the challenges of limited cell availability, and in the case of allogeneic cells, risk of infection, immunogenic reaction and disease transmission [12]. Furthermore, the complex procedure of cell harvesting, expansion and maintenance makes the tissue engineering time-, money- and labor-consuming [3]. Cell survival and vascularization are similarly challenging issues, especially when attempting to create large tissue constructs, despite developments of various *in vitro* bioreactors [14–17]. In 2004, Warnke et al. reported a novel method for repairing a defect in human mandible by *in vivo* tissue engineering, where the patient's body served as a natural bioreactor. By combining ceramic particles, recombinant human Bone Morphogenetic Protein 7 (rh-BMP-7) and autologous bone marrow cells inside a titanium mesh, a sufficiently large piece of bone tissue was successfully grown inside the *latissimus dorsi* muscle of the patient prior to transplantation to repair the mandible defect [17,18].

Calcium phosphates ceramics are extensively used as natural bone graft substitutes owning to their similarity to the mineral phase of natural bone, absence of immunogenic reactions, excellent biocompatibility and osteoconductive potential [19]. In the past five decades, a large number of publications have also shown osteoinduction by a range of calcium phosphate biomaterials, demonstrated by *de novo* bone formation upon implantation in ectopic sites, such as in the muscle of baboon [20–22], goat [23], sheep [19,24–26], dog [21,27–30], rabbit [21,31,32], rat [28] and mouse [33,34] and in the subcutis of dog [28,35], goat [36] and mouse [32]. The osteoinductive properties of calcium phosphate ceramics put forward a new idea for *in vivo* bone tissue engineering in which bone graft could be constructed *in vivo* by implanting osteoinductive calcium phosphate ceramics, without the addition of growth factors and/or stem cells in the muscle or subcutis of patient for several weeks, followed by the transplantation of the formed tissue to repair the bone defect. By doing so, the step in which cells from the patient are harvested becomes unnecessary. To demonstrate the feasibility of this method, our group has implanted an osteoinductive calcium phosphate ceramic in the femoral muscle of dogs for 8 weeks, followed by transplantation to a bone defect created unilaterally in the dog mandible to simultaneously support a dental implant. 8 weeks after transplantation, the graft fused well with the host bone, and a satisfactory osseointegration was achieved between the implant and the graft [37]. It has furthermore been shown that osteoinductive ceramics have the potential to complete the process of regeneration of large, critical-sized bone defects as successfully as autograft can [24], suggesting that even the step of intramuscular implantation prior to transplantation to bone defect may not be needed.

Osteoinduction of a great number of materials has been studied in the past five decades, using different animal species, however the exact parameters influencing and mechanisms governing ectopic bone formation are still not fully understood. It is suggested that the important parameters for successful osteoinduction are (1) chemical composition of the material and related dissolution properties, which create a microenvironment rich in calcium and phosphate ions; (2) geometry (i.e. porosity, interconnectivity and pore size), which aid nutrition and oxygen supply and recruitment of stem cells; (3) surface topography and micropores, which determines the specific surface area and affects the osteogenic differentiation of undifferentiated cells either directly (adhesion mediated differentiation) or indirectly (via adsorption/precipitation of relevant endogenous proteins); (4) the species and genetic background of animal models; and (5) implantation site [21,23,24,38–42].

Previous studies have demonstrated that osteoinduction by biomaterials predominantly occurs in large animals [19–36,39,43], while in small animals the incidence of ectopic bone formation is relatively low [32–34]. The reason for these differences between animal species is still unclear. Experiments in large animal models are associated with issues of limited availability, high cost and ethical concerns, which is why studies in large animals usually involve relatively small sample size, limiting their reliability. Testing osteoinductivity of biomaterials in small animals, such as mice and rats, would not only allow studies with a large sample size, but it would also enable more in-depth immunohistochemical analyses which are readily available for studies in small animals. Such in-depth analyses could provide more insight in biological mechanisms governing bone induction by synthetic biomaterials.

The aim of this study was to determine material parameters that are essential for osteoinduction to occur. In addition, this study is aimed at investigating influence of species on material-induced ectopic bone formation, as an attempt to find most suitable model for this type of studies.

Materials and Methods

Preparation and characterization of the ceramics

Raw materials as well as final calcium phosphate ceramic scaffolds were provided by the National Engineering Research Center for Biomaterials (Chengdu, Sichuan, China). Briefly, the ceramics were prepared in house using wet precipitation and H_2O_2 foaming method, respectively, as previously described [37]. Hydroxyapatite (HA) was sintered at 1200°C (HA1200) and biphasic calcium phosphate (BCP), consisting of 60 wt.% HA and 40 wt.% β-tricalcium phosphate (TCP) was sintered at either 1100°C (BCP1100) or 1200°C (BCP1200) for 3 hours. Ceramic cylinders with dimensions φ5 mm×8 mm, φ4 mm×6 mm and

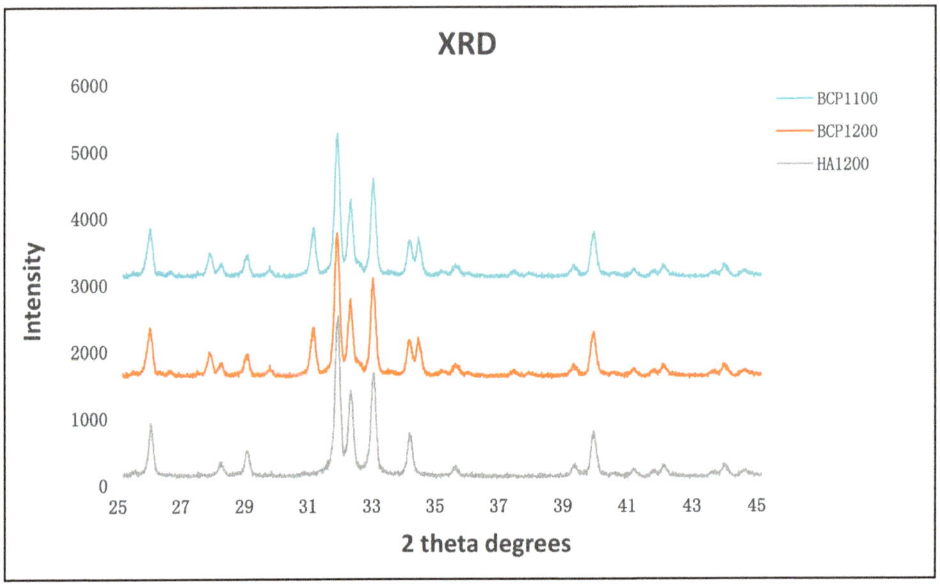

Figure 1. XRD patterns of the three ceramics. The patterns of BCP1100 and BCP1200 demonstrated biphasic nature of the ceramics consisting of HA and β-TCP. No apparent differences were found as a result of difference in sintering temperature. The pattern of HA1200 was typical of phase-pure hydroxyapatite.

φ2 mm×3 mm intended for implantation in dogs, rabbits and rats, respectively were cut and cleaned in ultrasonic baths. The size of implants varied per animal model as to adapt to the size of the implantation site. All the samples were sterilized by autoclaving before implantation. The chemical composition of the ceramics was analyzed by X-ray diffraction (XRD, Miniflex, Rigaku, Japan). The ultrastructure and the surface microstructure of the ceramics were investigated using scanning electron microscopy (XL30 ESEM FEG, Philips, The Netherlands). For quantification of the average macropore size and macroporosity, ceramic blocks (n = 8) were embedded in methylmethacrylate (MMA), and allowed to polymerize, after which 10–20 μm thick sections were cut using a diamond saw (Leica, Germany). Semi-automated measurements were performed using a light microscope (Nikon, Japan, Objective×10), connected to a KS400 imaging system (Carl Zeiss, Germany).

Animal experiments

The experimental protocol was approved by the Animal Care and Use Committee of Sichuan University. To determine the minimum sample size needed to obtain statistical differences between animal species and three types of materials, the sample sizes were estimated by applying the equation for comparative studies [44]. Surgery was performed on 27 mongrel dogs (10–15 kg, male), 36 New Zealand white rabbits (2.5–3.0 kg, male) and 54 S.D. rats (200–250 g, male) at the Experimental Animal Centre of Sichuan University. After anesthesia of the animals with 3% sodium pentobarbital (Sigma Chemical, St. Louis, MO, USA), the operation sites were shaved and the skin was sterilized with iodine. A longitudinal incision (10 cm, 5 cm and 2 cm in dogs, rabbits and rats, respectively) was created and the muscles were exposed by blunt separation. The muscle incisions were made by scalpel and the separated intramuscular pockets of appropriate size, with a distance of approximately 1 cm from one another were created by blunt separation. Each pocket received a single ceramic implant, after which the wound was closed in layers by using silk suture. After surgery, penicillin was administered intramuscularly for 3 consecutive days to prevent infection. In total 108 samples of each ceramic was implanted in each of the three species: four cylinders (φ5 mm×8 mm) of each ceramic type were implanted in the dorsal muscle of each dog, three cylinders (φ4 mm×6 mm) were implanted in the dorsal muscle of each rabbit, and two cylinders (φ2 mm×3 mm) of each ceramic were implanted in the femoral muscle of each rat. Animals were sacrificed by an overdose of sodium pentobarbital 6, 12 or 24 weeks post-implantation in a randomized manner. These survival times were selected based on earlier studies on ectopic bone formation by synthetic biomaterials [34].

Retrieval of implants, histology, and histomorphometry

At explantation, the samples were collected with surrounding tissues and fixed in 4% buffered formaldehyde solution. The fixed samples were then decalcified in 10% ethylene diamine tetraacetic acid (EDTA) for 15 days and washed with phosphate buffer solution (PBS), dehydrated with gradient of ethanol solutions at 70%, 80%, 90%, 95% and 100%×2, and then embedded in paraffin. Continuous 5 μm sections were made parallel to the long axis of the implants and then stained with hematoxylin and eosin

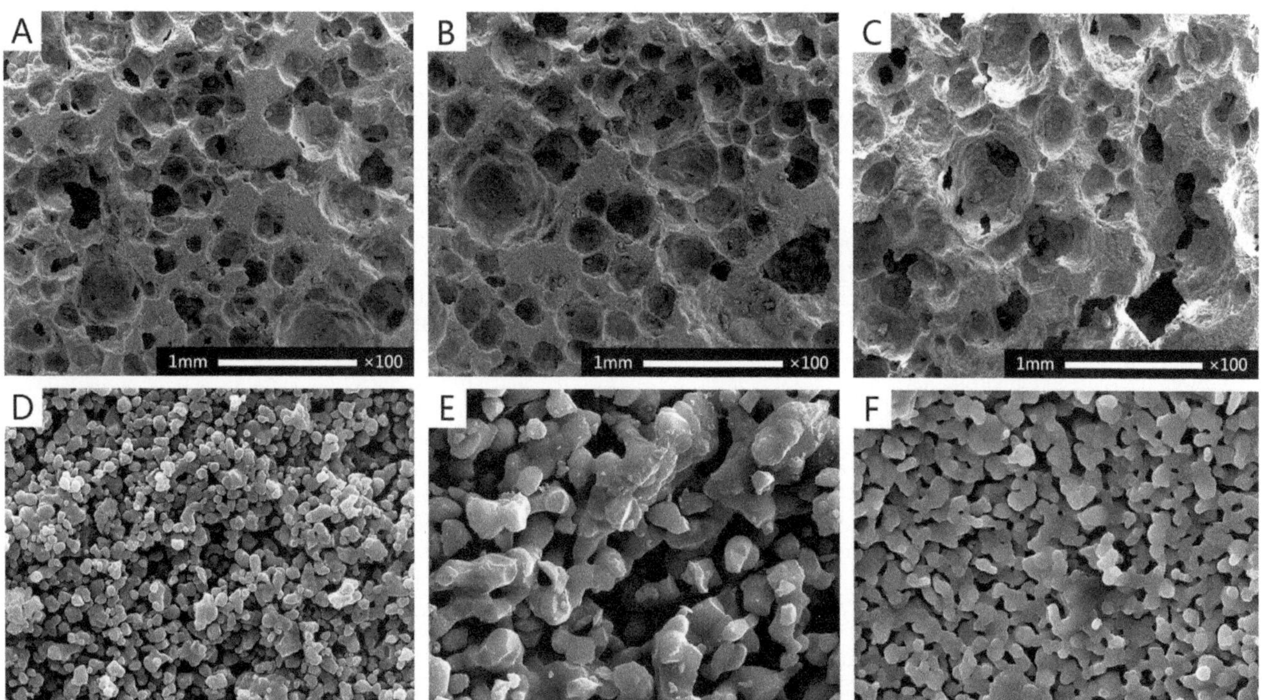

Figure 2. SEM micrographs showing macrostructure (A–C) and surface microstructure (D–F) of BCP1100 (A,D), BCP1200 (B,E) and HA1200 (C,F). All ceramics were porous with similar macroporous structure (A–C). The surface of the BCP1100 (D) ceramic exhibited smaller grains and a larger number of micropores than the chemically identical BCP1200, that was sintered at 1200°C (E). Grain size and number of micropores in HA1200 were between those of BCP110 and BCP1200 (F), which is in accordance with specific weight measurements in Table 1. Scale bar = 1 mm for A–C and 10 μm for D–F.

Table 1. Physico-chemical characterization of biphasic calcium phosphate sintered at 1100°C (BCP1100), of biphasic calcium phosphate sintered at 1200°C (BCP1200) and of phase-pure hydroxyapatite sintered at 1200°C (HA1200).

	BCP1100	*BCP1200*	*HA1200*
Weight ratio HA/β-TCP	60/40	60/40	100/0
Macropore size (μm)	380±50	400±50	400±50
Overall porosity (%)	70.8	65.2	66.9
Specific weight (g/cm^3)	0.970	1.21	1.10

(HE) (for all time points) or Masson's trichrome staining (for 12 weeks, which is representative for other time points) for histological analysis. Occurrence of osteoid tissue formation was assessed by three experienced pathologists separately without knowledge of the source of the specimens.

Histomorphometry was performed on 25 randomly selected fields at a magnification 10× (objective) of a single section taken from the middle of each implant, by using a light microscope (Nikon, Japan) coupled to a KS400 imaging system (Carl Zeiss, Germany). The macropore area and the area of bone or osteoid tissue were semi-automatically measured and the percentage of the induced bone and osteoid tissue inside the macropores (b% and o% respectively) were calculated. While b% was determined in all samples in which ectopic bone formation was induced, o% was determined in samples where osteoid, but no bone formation was found. The reason for this is the fact that, while pathologist can easily distinct between osteoid and bone, it is very difficult to draw the line between the area where the osteoid stops and bone tissue starts, making a reliable histomorphometry impossible to perform. Therefore, in samples with both osteoid and bone, the total area was used to calculate %b, whereas only the samples without bone were used for calculating %o. This makes results on bone and osteoid incidence and %b and %o complementary.

Statistical analysis

Light micrograph images were analyzed with IPP (Image-Pro-Plus) software, and statistical analysis was performed on histomorphometry data using SPSS19.0. Pearson's chi-square test was conducted to compare the overall constituent ratios of bone and osteoid tissue incidence in the three groups and differences with $p<0.05$ were considered significant. The adjusted test of level (α') was used to compare the constituent ratios of each two groups and the differences with $p<0.017$ were considered significant (Partitions of chi-square method) with $\alpha' = 2\alpha/k(k-1)$ $k=3$, $\alpha=0.05$.

Results

Properties of the ceramics

The X-ray diffraction analysis (**Fig. 1**) of BCP1100 and BCP1200 demonstrated biphasic nature of the ceramics and no apparent differences were found as a result of difference in sintering temperature. The pattern of HA1200 was typical of phase-pure HA. The SEM analysis (**Fig. 2**) showed that all ceramics were porous. The pore size of macropores in BCP1100, BCP1200 and HA1200 were 380±50 μm, 400±50 μm and 400±50 μm respectively. However at the temperature of 1100°C the surface of the BCP ceramics was rougher with more micropores between the ceramic grains as compared to the ceramic sintered at 1200°C (**Fig. 2**). Characterization of the three ceramics is summarized in Table 1.

Histological and histomorphometrical analyses

There were no surgical complications and all samples were retrieved. At retrieval, all implants were surrounded by well-vascularized muscle tissue. No signs of an inflammatory tissue response directly related to the implants were observed. **Table 2** and **figure 3** provide an overview of the osteoid and bone incidence, as well as percentage of bone and osteoid tissue in the pores after 6, 12 and 24 weeks of implantation in the three animal species.

In rats, no bone formation was observed in any of the three ceramics, independent of the implantation time. At 6 weeks, no osteoid was observed in any of the three ceramics (**Fig. 4A–C**), and similarly, HA1200 did not induce osteoid formation at the later time points either (**Fig. 5C or Fig. 6C**). In HA1200, the tissue in the pores was fibrous in nature, only containing fibroblast-like cells (**Fig. 4C, 5C and 6C**). After 12 weeks of implantation, a thin band of osteoid with some compacted cuboidal cells adjacent to pore walls was observed in 9 out of 36 samples of BCP1100 and in 2 out of 36 samples of BCP1200. The percentage of osteoid tissue formation was relatively low (0.96±0.31% for BCP1100 and 0.83±0.16% for BCP1200 respectively). Only the difference between HA1200 and the other two ceramics was statistically significant ($p<0.017$). At 24 weeks, incidence of osteoid formation was 7/36 (0.40±0.23%) and 5/36 (0.39±0.17%) in BCP1100 and BCP1200, respectively, a difference that was not statistically significant, while there was a statistical difference between BCP1100 and HA1200 (**Fig. 3A** and **Table 2**).

Intramuscular implantation of the three ceramics in rabbits did not lead to a pronounced ectopic bone formation. Limited amount of bony tissue and bone trabeculae (1.2±0.8%) was only observed in three BCP1100 samples from the 12-week group, all coming from one animal (**Fig. 5D** and **Table 2**). Osteoid formation was observed in the three ceramics at all time points, although the incidence was strongly dependent on the material. In HA1200, only few samples showed osteoid formation (1/36, 1/36 and 2/36 after 6, 12 and 24 weeks respectively), and the pores of the ceramic were predominantly occupied by fibrous tissue containing fibroblasts, neutrophils, lymphocytes and plasma cells (**Fig. 4F, 5F** and **6F**). Osteoid incidence was 4/36, 26/36 and 13/36 for BCP1100 and 2/36, 13/36 and 2/36 for BCP1200 after 6, 12 and 24 weeks respectively (**Fig. 3B, 4D–E, 5D–E, 6D–E** and **Table 2**). In terms of osteoid incidence, no statistically significant differences were observed among the three ceramics after 6 weeks of implantation. BCP1100 showed a higher osteoid incidence than BCP1200 and HA1200 at both 12 and 24 weeks ($p<0.017$).

Similar to bone incidence, in rabbits, the area percentage of ceramic pores occupied by osteoid tissue was highest in BCP1100, increasing from 0.40±0.27% at week 6 to 1.57±0.51% at 24 weeks, respectively. In BCP1200 and HA1200, the %o remained below 2% at all time points, and considering the very low osteoid

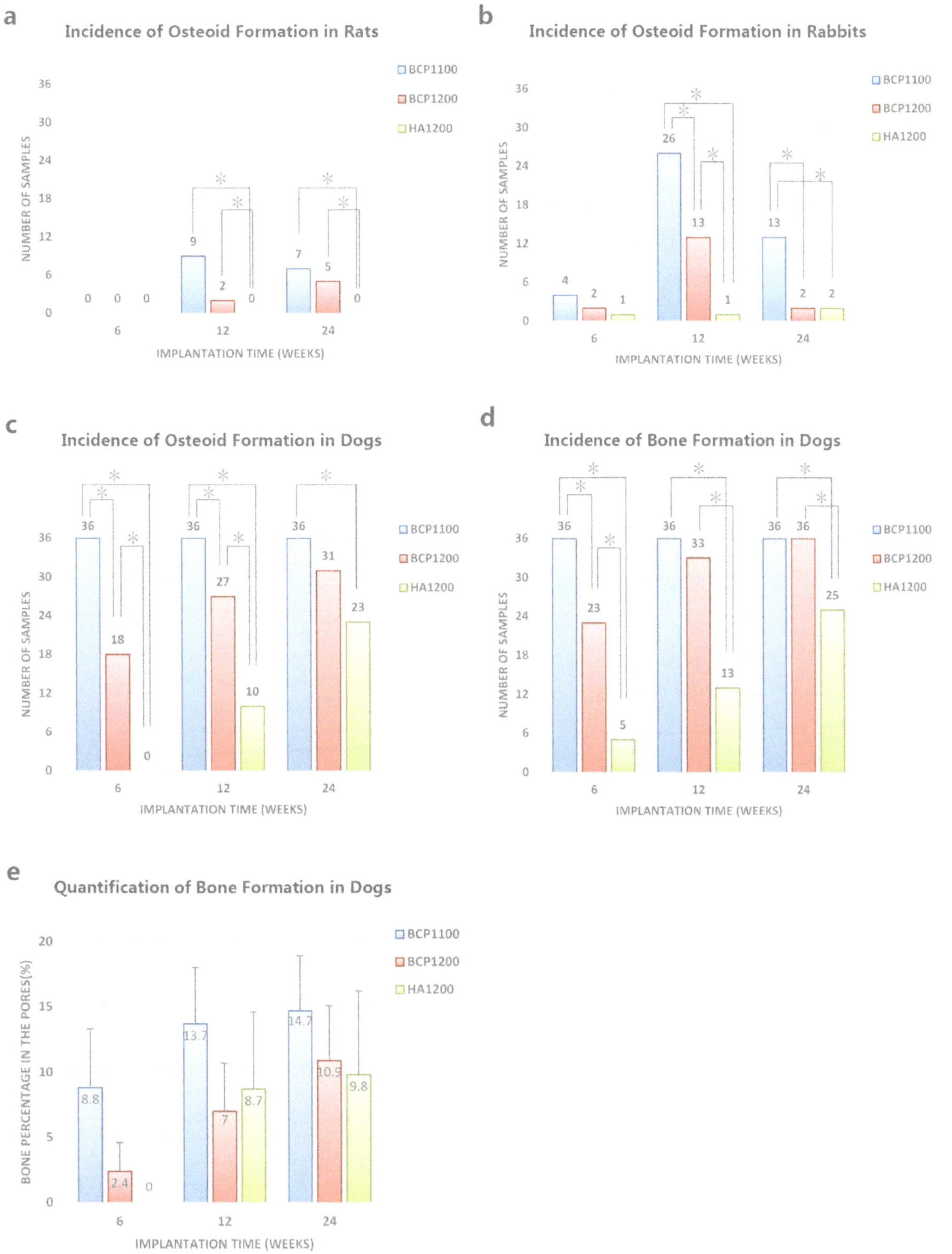

Figure 3. Incidence of bone and osteoid tissue formation in different animals (A–D) and percentage of bone area in pores of dogs (E) at 6, 12 and 24 after implantation. Osteoid formation in rats was not observed at 6 weeks in any of the ceramics, while at 12 and 24 weeks, only BCP ceramics showed signs of osteoid formation (A). In rabbits, all ceramics showed ectopic osteoid formation at all time points, with a highest incidence at 12 weeks. At all time points, BCP1100 showed a higher osteoid incidence than BCP1200, which was again higher than osteoid incidence in HA1200 (B). In dogs, osteoid formation was observed in all ceramics at all time points, except in HA1200 at 6 weeks. Like in rabbits, a decreasing osteoid incidence was observed from BCP11 to BCP1200 to HA1200 (C). All ceramic showed ectopic bone formation in dogs at all time points (D) following a trend that was similar to that of osteoid incidence (C). Quantification of % of ectopic bone formation in available pore area showed that, in general, BCP1100 showed higher values than BCP1200 and HA1200; however, these differences were not statistically significant (E).

Table 2. A summary of tissue response to the three types of ceramics in different animals.

Species		Rats			Rabbits			Dogs		
Intervals (weeks)		6	12	24	6	12	24	6	12	24
BCP1100	Incidence of bone	0/36	0/36	0/36	0/36	3/36	0/36	36/36	36/36	36/36
	Percentage of bone in available pore area (b%)	0	0	0	0	1.2±0.8	0	8.8±4.5	13.7±4.3	14.7±4.2
	Incidence of osteoid	0/36	9/36	7/36	4/36	26/36	13/36	36/36	36/36	36/36
	Percentage of osteoid tissue in available pore area (o%)	0	0.96±0.31	0.40±0.23	0.40±0.27	1.22±0.26	1.57±0.51	N/A	N/A	N/A
BCP1200	Incidence of bone	0/36	0/36	0/36	0/36	0/36	0/36	18/36	27/36	31/36
	Bone area in available pore area (b%)	0	0	0	0	0	0	2.4±2.2	7.0±3.7	10.9±4.2
	Incidence of osteoid	0/36	2/36	5/36	2/36	13/36	2/36	23/36	33/36	36/36
	Percentage of osteoid tissue in available pore area (o%)	0	0.83±0.16	0.39±0.17	2.21±0.20	1.02±0.46	1.62±0.34	0.56±0.24	1.15±0.19	1.42±0.26
HA1200	Incidence of bone	0/36	0/36	0/36	0/36	0/36	0/36	0/36	10/36	13/36
	Bone area in available pore area (b%)	0	0	0	0	0	0	0	8.5±5.9	8.8±6.4
	Incidence of osteoid	0/36	0/36	0/36	1/36	1/36	2/36	5/36	13/36	25/36
	Percentage of osteoid tissue in available pore area (o%)	0	0	0	0.65±0.08	1.34±0.11	0.70±0.18	1.43±0.38	1.22±0.41	1.51±0.46

incidence, a statistical comparison of histomorphometrical data was not sensible (**Table 2**).

In dogs, the incidence and the amount of osteoid and ectopic bone were higher for all the ceramics as compared to tissue formation found in rats and rabbits (**Fig. 3C–D, 4G–I, 5G–I, 6G–I** and **Table 2**). In the pores of all BCP1100 samples, formation of *de novo* bone adjacent to the material surface was observed as early as at 6 weeks, and bone incidence 36/36 remained at the later time points. With time, the bone appeared more trabecular, with lacunae containing osteocytes (**Fig. 4G–6G and Fig. 7A**). In BCP1200, pores were mainly filled with star-shaped cells or thin, elongated cells at 6 weeks and bone trabeculae were observed in a half of the samples (**Fig. 4H**). After 12 weeks, bone trabeculae were observed in the pores of the ceramic in most samples (27/36), with a layer of osteoblasts aligning bone (**Fig. 5H and Fig. 7B**). At week 24, bone incidence further increased to 31/36 and maturation of the trabeculae and the formation of new blood vessels, along with appearance of myeloid tissue were observed (**Fig. 6H**). The osteoinductive potential of HA1200 was the poorest among the three ceramics, without bone formation at week 6, and with pores filled with fibrous tissue mainly consisting of fibroblasts or fibroblast-like cells. Bone formation was observed after 12 weeks in 10/36 samples (**Fig. 5I and Fig, 7C**) and incidence further increased to 23/36 after 24 weeks (**Fig. 6I**).

Incidence of ectopic bone formation in BCP1100 was statistically higher than in HA1200 at all time points. Further, statistical differences in bone incidence were observed between BCP1200 and HA1200, and also between BCP1100 and BCP1200 after 6 and 12 weeks. Regarding presence of osteoid tissue, HA1200 performed significantly poorer than the other two ceramics at all time points, while no significant differences between BCP1100 and BCP1200 were observed after 12 and 24 weeks of implantation.

In BCP1100, the percentage of bone in the pores increased from 8.8±4.5% at week 6 to 13.7±4.3% at week 12 and 14.7±4.2% after 24 weeks (**Fig. 3E and Table 2**). The percentage of bone in the pores of BCP1200 was lower than that in BCP1100, increasing from 2.4±2.2% at week 6 to 7.0±3.7% and 10.9±4.2% at 12 and 24 weeks, respectively. Finally, the bone tissue area in the pores of HA1200 was 8.5±5.9% and 8.8±6.4% after 12 and 24 weeks, respectively. The area of osteoid tissue inside the pores of samples that did not contain bone (for BCP1100 and BCP1200) was always below 1.5%.

Discussion

In the present study, we explored the dependence of osteinductive potential of three calcium phosphate bioceramics on their physicochemical properties and on the animal model in which osteoinduction was studied. The three ceramics varied in their chemical composition, microstructure or both as these properties have previously been suggested to be of influence on the osteoinductive potential [3,26,32,39,41,45]. To investigate the effect of chemical composition, both phase-pure HA and biphasic calcium phosphate ceramic consisting of HA and TCP were used. In addition, BCP ceramic was sintered at either 1100°C or 1200°C, to vary surface microstructure in terms of grain size, microporosity and surface area. The three ceramics were implanted intramuscularly in rats, rabbits and dogs to obtain information about the species-dependent ectopic bone formation. While a number of studies exists in which synthetic biomaterials were tested for their osteoinductive potential in various animal models, in most of these studies one animal species was used as a model, making the comparison among different studies difficult. In studies where different animal species were used in parallel, either

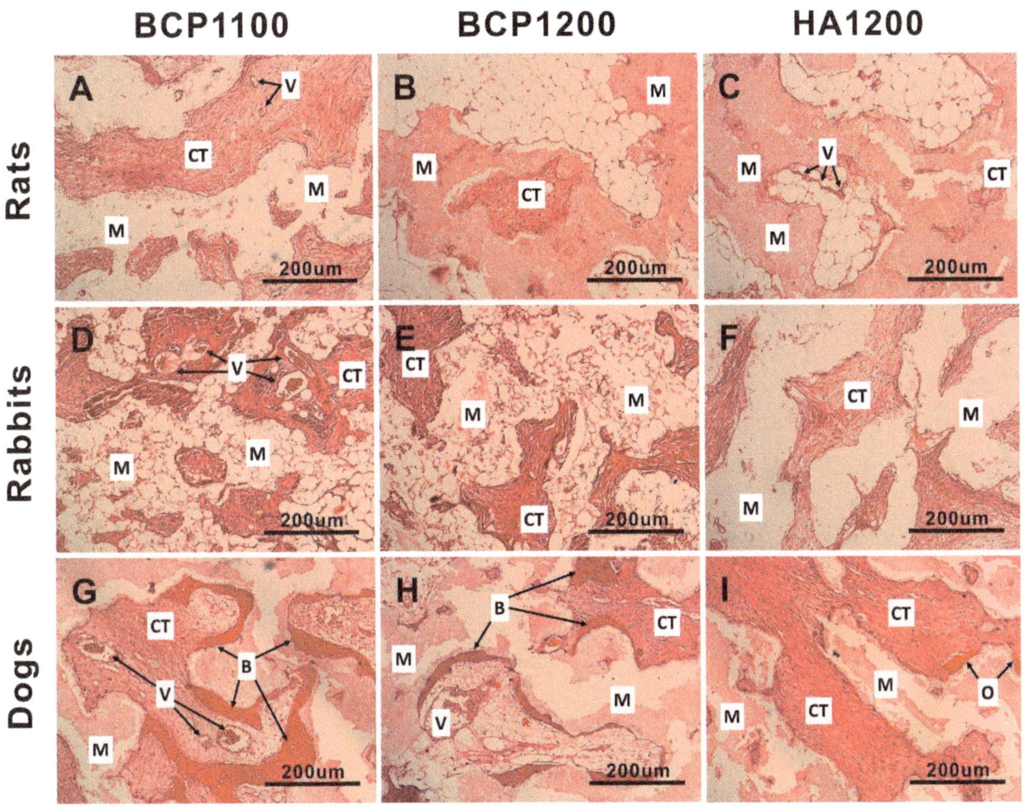

Figure 4. Histological evaluation of tissue formation in the three ceramics upon intramuscular implantation in rats, rabbits and dogs. Representative images showing tissue response to BCP1100 (A, D, G), BCP1200 (B, E, H) and HA1200 (C, F, I) upon implantation in the femoral muscle of rat, dorsal muscle of rabbit and dog for 6 weeks. Decalcified sections, HE Staining, magnification = ×100, scale bar = 200 μm; V→ Blood vessel, CT→Connective tissue, M→Material, O→Osteoid tissue, B→Bone.

a low number of animals per time point was used [46] or bone formation was assessed at a single time point [32]. Besides variation among different species, variations in the amount of bone induced in individuals from the same species have been reported, making the sample size of such studies of great importance [23]. Earlier studies have shown that experiments involving 8–12 large animals (goats, sheep, dogs) allow for a reliable comparison of ectopic tissue formation by different materials [24]. Based on these studies, and using a formula to estimate the minimum sample size for our study design [44], we have selected the sample size for the present study.

Physicochemical properties of osteoinductive materials such as chemical composition, geometry, surface microstructure and implant size have been reported as critical parameters for successful osteoinductivity. As recently reviewed [39], in the past decades, a large number of publications have illustrated osteoinduction by diverse biomaterials, such as synthetic and coral-derived HA ceramics, α-TCP, β-TCP, BCP and calcium phosphate cements (CPC). Besides calcium phosphate-containing biomaterials, osteoinduction was also observed in alumina ceramic, titanium and glass ceramics. However, so far most biomaterials shown to be osteoinductive contained calcium phosphate, which suggests the importance of calcium and phosphate ions in the process of osteoinduction by biomaterials [47–51]. When a biphasic calcium phosphate ceramic, consisting of HA and β-TCP, was compared with a pure HA ceramic with similar macro- and microstructure in a goat model, more bone was found in BCP ceramic. This suggested that the more soluble calcium phosphate phases are more osteoinductive than the less soluble ones [3]. These results are in accordance with our findings, where the biphasic ceramics (BCP1100, BCP1200) showed significantly more osteoid tissue and/or bone than the phase-pure HA (HA1200) in dogs (**Fig. 3–7** and **Table 2**). Furthermore, initiation of ectopic bone formation occurred earlier in biphasic than in HA ceramic in dogs. While in rats and rabbits no bone formation was found in any of the ceramics, the incidence of osteoid tissue formation, believed to precede bone formation was also more prominent in biphasic ceramic than in HA ceramic. The proposed reasons for the observed positive effect of the more soluble calcium phosphate ceramic on ectopic bone formation are multiple, ranging from the sole effect of calcium release of osteogenic differentiation of cells, to the process of dissolution/reprecipitation of the biological apatite layer on the ceramic surface, accompanied by the co-precipitation of endogenous osteoinductive growth factors. In addition, the inflammatory response (especially the chronic inflammation) induced by the implanted ceramics and its solubility products (i.e. cytokines) may also be an initiator of material-induced ectopic bone formation [39,52–55].

Besides chemistry, structural features of the ceramics, at macro- and micro level were suggested to influence osteoinductive potential of biomaterials. For example, Ripamonti showed that geometry is one of the crucial characteristics necessary for rendering a biomaterial osteoinductive [20] and presence of interconnected macropores was suggested to positively influence ectopic bone formation by biomaterials [24,56,57]. In addition to

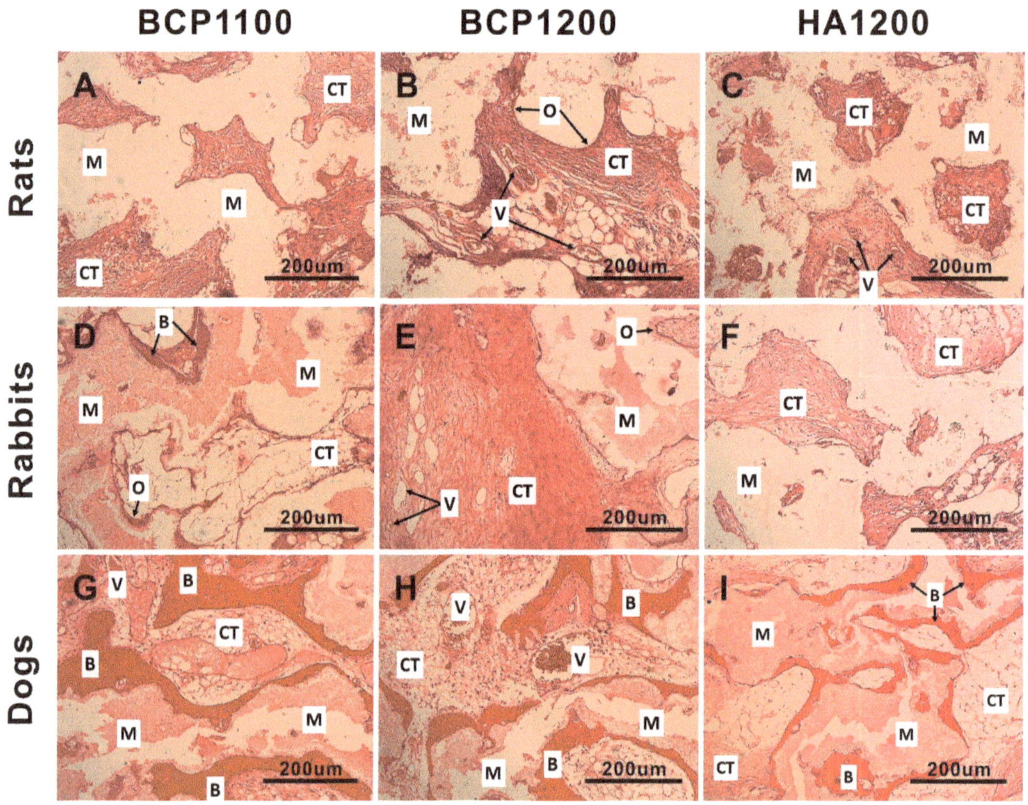

Figure 5. Histological evaluation of tissue formation in the three ceramics upon intramuscular implantation in rats, rabbits and dogs. Representative images showing tissue response to BCP1100 (A, D, G), BCP1200 (B, E, H) and HA1200 (C, F, I) upon implantation in the femoral muscle of rat, dorsal muscle of rabbit and dorsal muscle of dog for 12 weeks. Decalcified sections, HE Staining, magnification = ×100, scale bar = 200 μm; V→Blood vessel, CT→Connective tissue, M→Material, O→Osteoid tissue, B→Bone.

macroporosity, surface features like size of grains and presence of micropores have also been proven important in the bone induction process [58–61]. Micropores enlarge the surface area, which facilitates ion exchange and bone-like apatite formation, and/or accompanied binding of endogenous bone inducing proteins to the surface, all of which may positively affect the adhesion, proliferation, and differentiation of the bone forming cells [23,62]. Furthermore, the microstructure of the materials has been shown to directly influence cell shape and related signaling pathway, and therewith also the final cell fate [63–65]. In the present study, BCP sintered at a relatively low temperature of 1100°C (BCP1100), having a larger specific surface area as well as more micropores than BCP and HA sintered at 1200°C (BCP1200 and HA1200), indeed showed relatively more bone formation than the other two ceramics.

As previously mentioned, osteoinduction by biomaterials has shown large interspecies differences, and numerous studies have demonstrated that material induced bone formation is more likely to occur in large animals [19–36,39,43] such as dogs, goats, sheep, and non-human primates, than in rabbits [31,32], rats and mice [32–34], as shown by Kurashina et al. [31] and Yuan et al. [32] and confirmed by our study. While no bone formation was observed in rats by any of the implanted materials, and bone formation in rabbits was limited to BCP1100 and only a single animal, osteoinduction in dogs occurred as early as after 6 weeks of implantation in all the ceramics tested, though to a different extent as demonstrated by histomorphometry (**Table 2**). Osteoid formation occurred more frequently than bone formation, however, its incidence was also largely dependent on both the ceramic type and the species, following a trend similar to that of bone formation dependence on material and species parameters: increasing from rat to rabbit to dog for all ceramics and from HA1200 to BCP1200 to BCP1100 in all animals. While incidence of osteoid was in general higher than incidence of bone formation, percentage of osteoid tissue inside the pores of the ceramics was rather limited, though it should be mentioned that this analysis was only performed on those samples that did not show any bone formation. Based on these findings, it may be speculated that osteoid formation is a precursor of bone formation during osteoinduction by biomaterials, which in turn suggests surface dynamics of dissolution and reprecipitation to play an important role in this process.

The difference in the species-dependent osteoinductive potential of the materials is not clear as yet, but it could plausibly be related to various factors, such as genetic background and related physiological metabolism of each species, including different substance exchange, degradation and absorption kinetics, levels of BMPs or other osteogenic factors available near the implants and differences in response to osteogenic signals. In a recent study by Barradas et al. [34], in which 11 mouse strains were compared, reproducible ectopic bone formation upon subcutaneous implantation of calcium phosphate ceramics was only found in one strain (FVB mice), indeed suggesting the influence of genetic factors. Besides interspecies variations in osteoinductive potential, also differences in ectopic bone formation among individual animals of the same species have been found in this study, comparable to the

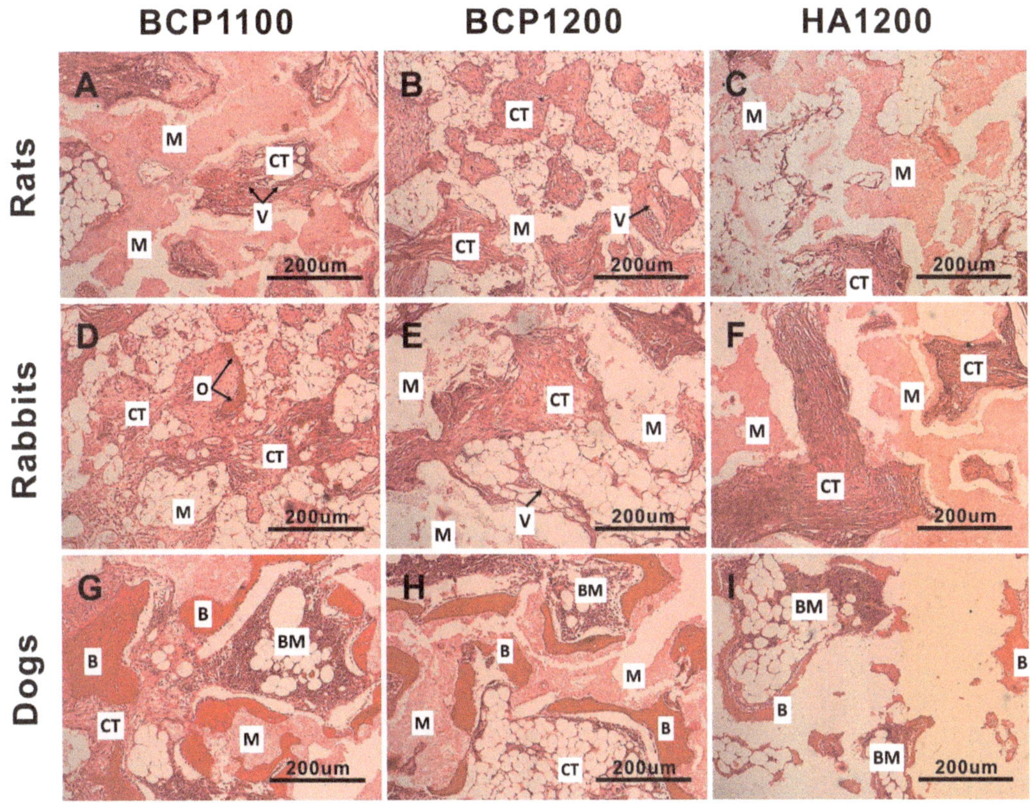

Figure 6. Histological evaluation of tissue formation in the three ceramics upon intramuscular implantation in rats, rabbits and dogs. Representative images, showing tissue response to BCP1100 (A, D, G), BCP1200 (B, E, H) and HA1200 (C, F, I) upon implantation in the femoral muscle of rat, dorsal muscle of rabbit and dorsal muscle of dog for 24 weeks. Decalcified sections, HE Staining, magnification = ×100, scale bar = 200 μm; V→Blood vessel, CT→Connective tissue, M→Material, O→Osteoid tissue, B→Bone, BM→Bone marrow.

differences among individuals reported earlier [23]. Both the differences in the response of the species and the response of the individuals from the same species may have important implications for the clinical use of these synthetic bone graft substitutes as alternatives to natural bone grafts. The growing understanding of the parameters that influence osteoinductive potential of biomaterials can be used to in design new materials, which may be osteoinductive in smaller animals, like rats and mice as well.

With this study, in which we systematically analyzed effects of physicochemical properties of calcium phosphate bioceramics in three animal species and by using large numbers of samples, we have shown that both presence of a more soluble β-TCP phase in the ceramic and a decrease of sintering temperature that affects surface microstructure of the ceramic positively influence its ability to induce ectopic osteoid and bone formation. Furthermore, we have demonstrated that material-induced ectopic bone formation indeed reproducibly occurred in dogs, while in rats and rabbits, only osteoid formation was observed.

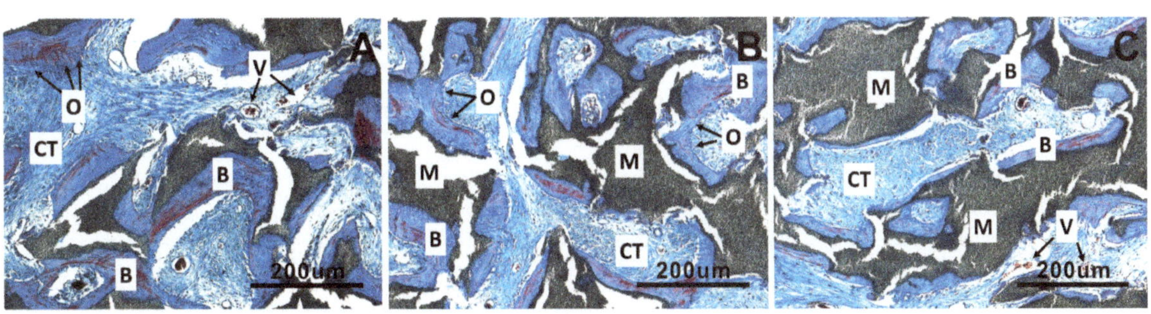

Figure 7. Histological evaluation of bone formation upon implantation of the three ceramics in dorsal muscles of dogs for 12 weeks. The highest amount of osteoid and ectopic bone were observed in BCP1100 (A), followed by BCP1200 (B) an HA1200 (C). In all cases, bone was trabecular in appearance, with laminar bone and osteocytes contained in the lacunae, and infiltrated by blood vessels. Decalcified sections, Masson's trichrome staining, magnification = ×100, scale bar = 200 μm; V→Blood vessel, CT→Connective tissue, M→Material, O→Osteoid tissue, B→Bone.

Conclusion

The results of the present study showed that a BCP ceramic, consisting of HA and TCP, that was sintered at a temperature of 1100°C showed more osteoid and/or bone formation upon ectopic implantation than BCP (or HA) sintered at a higher temperature (1200°C). Concerning the animal species, all tested materials showed more ectopic osteoid and bone formation in dogs than in rabbits or rats. In conclusion, this study demonstrated that both, the type of calcium-phosphate ceramic and the animal species influence the osteoinductive potential of the bioceramics.

Author Contributions

Conceived and designed the experiments: LW CYB JH XDZ PH. Performed the experiments: LW BZ. Analyzed the data: LW BZ. Contributed reagents/materials/analysis tools: CYB JH XDZ. Wrote the paper: LW CYB PH.

References

1. Takahashi Y, Yamamoto M, Tabata Y (2005) Enhanced osteoinduction by controlled release of bone morphogenetic protein-2 from biodegradable sponge composed of gelatin and beta-tricalcium phosphate. Biomaterials 26: 4856–4865.
2. Koyama N, Okubo Y, Nakao K, Osawa K, Bessho K (2011) Experimental study of osteoinduction using a new material as a carrier for bone morphogenetic protein-2. Br J Oral Maxillofac Surg 49: 314–318.
3. Habibovic P, de Groot K (2007) Osteoinductive biomaterials–properties and relevance in bone repair. J Tissue Eng Regen Med 1: 25–32.
4. Langer R, Vacanti JP (1993) Tissue engineering. Science 260: 920–926.
5. Rehman K, Iqbal MJ, Zahra N, Akash MS (2014) Liver stem cells: from preface to advancements. Curr Stem Cell Res Ther 9: 10–21.
6. Trzaska KA, Rameshwar P (2007) Current advances in the treatment of Parkinson's disease with stem cells. Curr Neurovasc Res 4: 99–109.
7. Herrmann RP, Sturm MJ (2014) Adult human mesenchymal stromal cells and the treatment of graft versus host disease. Stem Cells Cloning 7: 45–52.
8. Xiao Y, Mareddy S, Crawford R (2010) Clonal characterization of bone marrow derived stem cells and their application for bone regeneration. Int J Oral Sci 2: 127–135.
9. Robert L, Vacanti JP (1993) Tissue engineering. Science 260: 920–941.
10. Szpalski C, Wetterau M, Barr J, Warren SM (2012) Bone tissue engineering: current strategies and techniques–part I: Scaffolds. Tissue Eng Part B Rev 18: 246–257.
11. Kim TG, Shin H, Lim DW (2012) Biomimetic scaffolds for tissue engineering. Adv Funct Mater 22: 2446–2468.
12. Meijer GJ, de Bruijn JD, Koole R, van Blitterswijk CA (2008) Cell based bone tissue engineering in jaw defects. Biomaterials 29: 3053–3061.
13. Grellier M, Bordenave L, Amedee J (2009) Cell-to-cell communication between osteogenic and endothelial lineages: implications for tissue engineering. Trends Biotechnol 27: 562–571.
14. Holt GE, Halpern JL, Lynch CC, Devin CJ, Schwartz HS (2008) Imaging analysis of the in vivo bioreactor: a preliminary study. Clin Orthop Relat Res 466: 1890–1896.
15. Service RF (2005) Tissue engineering. Technique uses body as 'bioreactor' to grow new bone. Science 309: 683.
16. Stevens MM, Marini RP, Schaefer D, Aronson J, Langer R, et al. (2005) In vivo engineering of organs: the bone bioreactor. Proc Natl Acad Sci U S A 102: 11450–11455.
17. Warnke PH, Wiltfang J, Springer I, Acil Y, Bolte H, et al. (2006) Man as living bioreactor: fate of an exogenously prepared customized tissue-engineered mandible. Biomaterials 27: 3163–3167.
18. Warnke PH, Springer IN, Wiltfang J, Acil Y, Eufinger H, et al. (2004) Growth and transplantation of a custom vascularised bone graft in a man. Lancet 364: 766–770.
19. Gosain AK, Song L, Riordan P, Amarante MT, Nagy PG, et al. (2002) A 1-year study of osteoinduction in hydroxyapatite-derived biomaterials in an adult sheep model: part I. Plast Reconstr Surg 109: 619–630.
20. Ripamonti U (1991) The morphogenesis of bone in replicas of porous hydroxyapatite obtained from conversion of calcium carbonate exoskeletons of coral. J Bone Joint Surg Am 73: 692–703.
21. Ripamonti U (1996) Osteoinduction in porous hydroxyapatite implanted in heterotopic sites of different animal models. Biomaterials 17: 31–35.
22. Ripamonti U, Richter PW, Nilen RW, Renton L (2008) The induction of bone formation by smart biphasic hydroxyapatite tricalcium phosphate biomimetic matrices in the non-human primate Papio ursinus. J Cell Mol Med 12: 2609–2621.
23. Habibovic P, Yuan H, van der Valk CM, Meijer G, van Blitterswijk CA, et al. (2005) 3D microenvironment as essential element for osteoinduction by biomaterials. Biomaterials 26: 3565–3575.
24. Yuan H, Fernandes H, Habibovic P, de Boer J, Barradas AM, et al. (2010) Osteoinductive ceramics as a synthetic alternative to autologous bone grafting. Proc Natl Acad Sci U S A 107: 13614–13619.
25. Barbieri D, Yuan H, de Groot F, Walsh WR, de Bruijn JD (2011) Influence of different polymeric gels on the ectopic bone forming ability of an osteoinductive biphasic calcium phosphate ceramic. Acta Biomater 7: 2007–2014.
26. Le Nihouannen D, Daculsi G, Saffarzadeh A, Gauthier O, Delplace S, et al. (2005) Ectopic bone formation by microporous calcium phosphate ceramic particles in sheep muscles. Bone 36: 1086–1093.
27. Xingdong Z, Pin Z, Jianguo Z, Weiqun C, Wu C (1992) A Study of Hydroxyapatite Ceramics and its Osteogenesis. In: Ravaglioli A, Krajewski A, editors. Bioceramics and the Human Body: Springer Netherlands. 408–416.
28. Klein C, de Groot K, Chen W, Li Y, Zhang X (1994) Osseous substance formation induced in porous calcium phosphate ceramics in soft tissues. Biomaterials 15: 31–34.
29. Kondo N, Ogose A, Tokunaga K, Umezu H, Arai K, et al. (2006) Osteoinduction with highly purified beta-tricalcium phosphate in dog dorsal muscles and the proliferation of osteoclasts before heterotopic bone formation. Biomaterials 27: 4419–4427.
30. Yuan H, Van Blitterswijk C, De Groot K, De Bruijn J (2006) A comparison of bone formation in biphasic calcium phosphate (BCP) and hydroxyapatite (HA) implanted in muscle and bone of dogs at different time periods. J Biomed Mater Res A 78: 139–147.
31. Kurashina K, Kurita H, Wu Q, Ohtsuka A, Kobayashi H (2002) Ectopic osteogenesis with biphasic ceramics of hydroxyapatite and tricalcium phosphate in rabbits. Biomaterials 23: 407–412.
32. Yuan H, van Blitterswijk CA, de Groot K, de Bruijn JD (2006) Cross-species comparison of ectopic bone formation in biphasic calcium phosphate (BCP) and hydroxyapatite (HA) scaffolds. Tissue Eng 12: 1607–1615.
33. Cheng L, Shi Y, Ye F, Bu H (2013) Osteoinduction of calcium phosphate biomaterials in small animals. Mater Sci Eng C Mater Biol Appl 33: 1254–1260.
34. Barradas A, Yuan H, van der Stok J, Le Quang B, Fernandes H, et al. (2012) The influence of genetic factors on the osteoinductive potential of calcium phosphate ceramics in mice. Biomaterials 33: 5696–5705.
35. Yamasaki H, Sakai H (1992) Osteogenic response to porous hydroxyapatite ceramics under the skin of dogs. Biomaterials 13: 308–312.
36. Bodde EW, Cammaert CT, Wolke JG, Spauwen PH, Jansen JA (2007) Investigation as to the osteoinductivity of macroporous calcium phosphate cement in goats. J Biomed Mater Res B Appl Biomater 83: 161–168.
37. Yao J, Li X, Bao C, Fan H, Zhang X, et al. (2009) A novel technique to reconstruct a boxlike bone defect in the mandible and support dental implants with In vivo tissue-engineered bone. J Biomed Mater Res B Appl Biomater 91: 805–812.
38. Fujibayashi S, Neo M, Kim HM, Kokubo T, Nakamura T (2004) Osteoinduction of porous bioactive titanium metal. Biomaterials 25: 443–450.
39. Barradas AM, Yuan H, van Blitterswijk CA, Habibovic P (2011) Osteoinductive biomaterials: current knowledge of properties, experimental models and biological mechanisms. Eur Cell Mater 21: 407–429.
40. Habibovic P, Sees TM, van den Doel MA, van Blitterswijk CA, de Groot K (2006) Osteoinduction by biomaterials–physicochemical and structural influences. J Biomed Mater Res A 77: 747–762.
41. Miron RJ, Zhang YF (2012) Osteoinduction: a review of old concepts with new standards. J Dent Res 91: 736–744.
42. Ozdemir T, Higgins AM, Brown JL (2013) Osteoinductive biomaterial geometries for bone regenerative engineering. Curr Pharm Des 19: 3446–3455.
43. Kondo N, Ogose A, Tokunaga K, Umezu H, Arai K, et al. (2006) Osteoinduction with highly purified β-tricalcium phosphate in dog dorsal muscles and the proliferation of osteoclasts before heterotopic bone formation. Biomaterials 27: 4419–4427.
44. Eng J (2003) Sample Size Estimation: How Many Individuals Should Be Studied? 1. Radiology 227: 309–313.
45. Bianchi M, Urquia Edreira ER, Wolke JG, Birgani ZT, Habibovic P, et al. (2014) Substrate geometry directs the in vitro mineralization of calcium phosphate ceramics. Acta Biomater 10: 661–669.
46. Yang Z, Yuan H, Tong W, Zou P, Chen W, et al. (1996) Osteogenesis in extraskeletally implanted porous calcium phosphate ceramics: variability among different kinds of animals. Biomaterials 17: 2131–2137.
47. Kalita SJ, Bhardwaj A, Bhatt HA (2007) Nanocrystalline calcium phosphate ceramics in biomedical engineering. Materials Science and Engineering: C 27: 441–449.
48. Zayzafoon M (2006) Calcium/calmodulin signaling controls osteoblast growth and differentiation. J Cell Biochem 97: 56–70.
49. Dvorak MM, Riccardi D (2004) Ca2+ as an extracellular signal in bone. Cell calcium 35: 249–255.
50. Klar RM, Duarte R, Dix-Peek T, Dickens C, Ferretti C, et al. (2013) Calcium ions and osteoclastogenesis initiate the induction of bone formation by coral-derived macroporous constructs. J Cell Mol Med 17: 1444–1457.

51. Boonrungsiman S, Gentleman E, Carzaniga R, Evans ND, McComb DW, et al. (2012) The role of intracellular calcium phosphate in osteoblast-mediated bone apatite formation. Proc Natl Acad Sci U S A 109: 14170–14175.
52. Eyckmans J, Roberts SJ, Bolander J, Schrooten J, Chen CS, et al. (2013) Mapping calcium phosphate activated gene networks as a strategy for targeted osteoinduction of human progenitors. Biomaterials 34: 4612–4621.
53. Fellah BH, Delorme B, Sohier J, Magne D, Hardouin P, et al. (2010) Macrophage and osteoblast responses to biphasic calcium phosphate microparticles. J Biomed Mater Res A 93: 1588–1595.
54. Rokn AR, Khodadoostan MA, Reza Rasouli Ghahroudi AA, Motahhary P, Kharrazi Fard MJ, et al. (2011) Bone formation with two types of grafting materials: a histologic and histomorphometric study. Open Dent J 5: 96–104.
55. Weitzmann MN (2013) The Role of Inflammatory Cytokines, the RANKL/OPG Axis, and the Immunoskeletal Interface in Physiological Bone Turnover and Osteoporosis. Scientifica (Cairo) 2013: 125705.
56. Fukuda A, Takemoto M, Saito T, Fujibayashi S, Neo M, et al. (2011) Osteoinduction of porous Ti implants with a channel structure fabricated by selective laser melting. Acta Biomater 7: 2327–2336.
57. Klenke FM, Liu Y, Yuan H, Hunziker EB, Siebenrock KA, et al. (2008) Impact of pore size on the vascularization and osseointegration of ceramic bone substitutes in vivo. J Biomed Mater Res A 85: 777–786.
58. Rosa AL, Beloti MM, Oliveira PT, Van Noort R (2002) Osseointegration and osseoconductivity of hydroxyapatite of different microporosities. J Mater Sci Mater Med 13: 1071–1075.
59. Chan O, Coathup MJ, Nesbitt A, Ho CY, Hing KA, et al. (2012) The effects of microporosity on osteoinduction of calcium phosphate bone graft substitute biomaterials. Acta Biomater 8: 2788–2794.
60. Hing KA, Annaz B, Saeed S, Revell PA, Buckland T (2005) Microporosity enhances bioactivity of synthetic bone graft substitutes. J Mater Sci Mater Med 16: 467–475.
61. Rouahi M, Gallet O, Champion E, Dentzer J, Hardouin P, et al. (2006) Influence of hydroxyapatite microstructure on human bone cell response. J Biomed Mater Res A 78: 222–235.
62. Zhu X, Fan H, Xiao Y, Li D, Zhang H, et al. (2009) Effect of surface structure on protein adsorption to biphasic calcium-phosphate ceramics in vitro and in vivo. Acta biomaterialia 5: 1311–1318.
63. Li X, van Blitterswijk CA, Feng Q, Cui F, Watari F (2008) The effect of calcium phosphate microstructure on bone-related cells in vitro. Biomaterials 29: 3306–3316.
64. Okuda T, Ioku K, Yonezawa I, Minagi H, Kawachi G, et al. (2007) The effect of the microstructure of beta-tricalcium phosphate on the metabolism of subsequently formed bone tissue. Biomaterials 28: 2612–2621.
65. Martinez E, Engel E, Planell JA, Samitier J (2009) Effects of artificial micro- and nano-structured surfaces on cell behaviour. Ann Anat 191: 126–135.

Synthetic Bone Substitute Engineered with Amniotic Epithelial Cells Enhances Bone Regeneration after Maxillary Sinus Augmentation

Barbara Barboni[1,5], Carlo Mangano[3], Luca Valbonetti[1,5], Giuseppe Marruchella[1], Paolo Berardinelli[1], Alessandra Martelli[1], Aurelio Muttini[1,5], Annunziata Mauro[1], Rossella Bedini[2], Maura Turriani[1], Raffaella Pecci[2], Delia Nardinocchi[1], Vincenzo Luca Zizzari[4], Stefano Tetè[4,5]*, Adriano Piattelli[4], Mauro Mattioli[1]

1 Department of Comparative Biomedical Science, University of Teramo, Teramo, Italy, **2** Department of Technologies and Health, Istituto Superiore di Sanità, Rome, Italy, **3** Department of Surgical and Morphological Science, University of Insubria, Varese, Italy, **4** Department of Medical, Oral and Biotechnological Science, University "G. d'Annunzio", Chieti, Italy, **5** Stem TeCh Group, Chieti, Italy

Abstract

Background: Evidence has been provided that a cell-based therapy combined with the use of bioactive materials may significantly improve bone regeneration prior to dental implant, although the identification of an ideal source of progenitor/stem cells remains to be determined.

Aim: In the present research, the bone regenerative property of an emerging source of progenitor cells, the amniotic epithelial cells (AEC), loaded on a calcium-phosphate synthetic bone substitute, made by direct rapid prototyping (rPT) technique, was evaluated in an animal study.

Material And Methods: Two blocks of synthetic bone substitute ($\sim$0.14 cm^3), alone or engineered with 1×10^6 ovine AEC (oAEC), were grafted bilaterally into maxillary sinuses of six adult sheep, an animal model chosen for its high translational value in dentistry. The sheep were then randomly divided into two groups and sacrificed at 45 and 90 days post implantation (p.i.). Tissue regeneration was evaluated in the sinus explants by micro-computer tomography (micro-CT), morphological, morphometric and biochemical analyses.

Results And Conclusions: The obtained data suggest that scaffold integration and bone deposition are positively influenced by allotransplantated oAEC. Sinus explants derived from sheep grafted with oAEC engineered scaffolds displayed a reduced fibrotic reaction, a limited inflammatory response and an accelerated process of angiogenesis. In addition, the presence of oAEC significantly stimulated osteogenesis either by enhancing bone deposition or making more extent the foci of bone nucleation. Besides the modulatory role played by oAEC in the crucial events successfully guiding tissue regeneration (angiogenesis, vascular endothelial growth factor expression and inflammation), data provided herein show that oAEC were also able to directly participate in the process of bone deposition, as suggested by the presence of oAEC entrapped within the newly deposited osteoid matrix and by their ability to switch-on the expression of a specific bone-related protein (osteocalcin, OCN) when transplanted into host tissues.

Editor: Xiaoming He, The Ohio State University, United States of America

Funding: This work was supported by Tercas Foundation and by PRIN 2010–2011 (PRIN 20102ZLNJ5) financed by the Ministry of Education, University and Research (M.I.U.R.), Rome, Italy. The funders had no role in study design, data collection and analysis, decision to publish, or preparation of the manuscript.

Competing Interests: The authors have declared that no competing interests exist.

* E-mail: tete@unich.it

Introduction

Bone regeneration in maxillary sinus is an essential condition for dental implants in atrophic posterior maxilla. Different strategies leading to the replacement of missing bone have been conventionally used for over 30 years [1,2]. Limited availability of autografts, and the risk of disease transmission by allo/xenografts, have increased the demand of synthetic bone substitutes, which have to reproduce the physical/chemical properties of native bone tissues in order to maximize osteointegration, osteoconduction and osteoinduction [2]. Calcium phosphate ceramics, such as hydroxyapatite (HA) and tricalcium-phosphate (TCP), are considered both suitable materials for bone reconstruction since they conjugate a high biocompatibility with an efficient osteoconductivity [3]. The porous architecture and the degree of interconnectivity are additional critical factors to determine the clinical success of biomaterials [4,5]. In fact, the chemical composition and architecture of biomaterials are both crucial to drive and stimulate bone healing and deposition. In order to mimic the structure of native bone and to ensure cell viability and function, the ideal scaffold should exhibit porosity at different length scales: nano-porosity, to allow molecule transport essential for any nutrition, waste removal and signaling; micro-porosity, to ensure cell migration and capillary formation; millimeter-wide porosity to incorporate nerves and blood vessels [6,7]. Scaffold porosity improves mechanical interlocking between the implanted biomaterial and the surrounding host bone [8,9], and positively influences the scaffold degradation rate. During the last few years, innovative technologies, such as three-dimensional (3D) printing and dispense-plotting, allowed to create scaffolds with a controlled 3D architecture [9–13], thus enhancing their biocompatibility [14–17]. However, the latest generation of synthetic bone substitutes still requires a long time to regenerate a large amount of bone tissue thus limiting their surgical use in validated therapeutic protocols such as sinus augmentation [18,19]. Therefore, cell-based therapies are an emerging strategy to improve bone tissue healing and regeneration [20–23]. In this context, increasing attention has been recently addressed to placental components and, in particular, to amnion as a possible reserve of stem/progenitor cells [24–29]. Actually, the therapeutic use of amniotic membrane has been studied for decades. Davis first reported in 1910 the use of fetal membranes as surgical materials in skin transplantation performed on 550 patients [30]. Amniotic membranes showed anti-inflammatory [31–33], antimicrobial [34], antifibroblastic [35] and low immunogenicity properties [36,37]. Several surgical applications for amniotic membranes have been reported, including their use as a biological dressing for the treatment of skin wounds, burn injuries and chronic leg ulcers, as well as in the treatment of tissue adhesion in surgical procedures and ocular burns [26].

More recently, amniotic membranes have been investigated as a possible source of stem/progenitor cells for therapeutic applications. Cells from mesenchymal and epithelial amniotic layers, amniotic mesenchymal stromal and amniotic epithelial cells (AEC), respectively, can be obtained without any ethical concerns, in large amounts and with validated protocols [38,39]. Amnion derived cells have been shown to maintain a remarkable plasticity and to possess a high self-renewing capacity [40–45]. To date, AEC have been more extensively investigated and their potential to differentiate into cell types of all three germ layers has been demonstrated [24,25,46,47]. The mesodermal differentiation attitude of AEC has attracted increasing attention [25,44,48,49], in particular, by considering the usefulness of amniotic derived cells in supporting bone lineage differentiation [25,26,28,29]. Moreover, *in vitro* and *in vivo* studies demonstrated how also isolated AEC maintain no-immunogenic and even immunosuppressive properties [50–52] Even more important, AEC are not carcinogenic after transplantation [53] thus allowing a safe use in allogeneic preclinical settings into immunocompetent animals [44,47,54–58].

Based on previous findings, the present research has been designed to evaluate whether AEC may represent an alternative source of progenitor/stem cells for cell-based therapeutic protocols in dentistry. To this aim, their regenerative properties were studied in an animal model. In detail, ovine AEC (oAEC) were engineered on a custom-made 3D hydroxyapatite (HA) and beta TCP (β-TCP) scaffold and grafted into the maxillary sinus of six adult sheep, before evaluating their influence on bone regeneration at 45 and 90 days post implantation (p.i.).

Materials and Methods

Chemical Agents

All chemical reagents and media were purchased from Sigma Chemical Co (St Louis, MO, USA), unless otherwise specified.

Ovine AEC Isolation And Characterization

Ovine AEC Isolation. As previously described, oAEC were obtained by enzymatic digestion from five sheep slaughtered at approximately 3 months of pregnancy [44]. Freshly isolated oAEC were seeded in flasks containing minimum essential medium eagle alpha (αMEM), supplemented with 20% fetal calfserum (FCS), 1% ultraglutamine, 1% penicillin/streptomycin plus 10 ng/ml epidermal growth factor (EGF) at a concentration of 3×10^3 cells/cm^2. At 70–80% confluence, cells were dissociated by 0.05% trypsin-ethylenediaminetetraacetic acid (EDTA) and plated at the same concentration (3×10^3 cell/cm^2) for 3 consecutive passages. At the end of expansion, oAEC were stored (2.5×10^6 cells/vial) in liquid nitrogen before using For each pool of stored oAEC belonging to each fetus, some vials were randomly thawed to test the maintenance of the following biological properties before using the remaining batch of cells in animal model. More in detail, aliquots of frozen oAEC were assessed for:

- cytokeratin 8 (epithelial marker) and alpha smooth muscle actin (α-SMA: mesodermal marker), expression by immunocytochemistry [46];
- hemopoietic, adhesion, pluripotent and major histocompatibility complex (MHC) markers expression by cytofluorimetry;
- the ability to differentiate into three different mesodermal cell lineages (osteoblast, tenocytes and adipose cells) by molecular biology and immunocytochemistry.

Immunocytochemical Detection Of Cytokeratin 8 And αSMA. Immunocytochemical analyses were carried out on freshly isolated, expanded and thawed oAEC as previously described [46]. Briefly, cells were fixed in 4% paraformaldehyde in phosphate buffered saline solution (PBS) and permeabilized with 0.1% Triton X-100/PBS. After incubation in PBS/1% bovine serum albumine (BSA) for 1 hour, oAEC were incubated with primary and secondary antibodies (Abs, see Table 1 for details). Negative controls were obtained by omitting the primary Ab. Cell nuclei were counterstained with 4′-6-diamidino-2-phenylindole dihydrochloride (DAPI).

Flow Cytometer Analysis. oAEC were screened immediately after isolation, before and after thawing by flow cytometry, according to Barboni *et al.* [46] and using the Abs reported in Table 2. Abs were used for FACSCalibur flow cytometer (BD

Table 1. Details of antibodies used in immunofluorescent analysis.

Primary Abs (Company details)	µg/ml	Secondary Abs (Company details)	µg/ml
Cytokeratin 8	5	**Anti-Mouse Cy3**	2
(Abcam, Cambridge, UK)		(Sigma-Aldrich, Missouri, USA)	
vWF	0.02	**Anti-Rabbit FITC**	5
(Dako, Glostrup, Denmark)		(Sigma-Aldrich, Missouri, USA)	
α-SMA	4	**Anti-Mouse Alexa Fluor488**	5
(Abcam, Cambridge, UK)		(Sigma-Aldrich, Missouri,USA)	
Osteocalcin	20	**Anti Mouse Alexa Fluor488**	5
(Abcam, Cambridge, UK)		(Invitrogen, Oregon, USA)	
VEGF	100	**Anti-Rabbit Alexa Fluor488**	5
(Calbiochem, USA)		(Invitrogen, Oregon, USA)	

Table 2. Details of primary and secondary antibodies used in flow cytometry analysis.

Antigen	Conj. fluorescent probe	Company details	
Hemopoietic markers			
CD14	FITC	LifeSpan Biosci.	Seattle, WA, USA
CD31	FITC	AbD Serotec	Oxford, UK
CD45	FITC	AbD Serotec	Oxford, UK
Adhesion molecules			
CD29		VMRD	Pullman, WA, USA
CD49f		Beckman Coulter	Fullerton, CA, USA;
CD166	FITC	Ancell	MN, USA
MHC antigens			
Class I		Novus Biologicals	Cambridge UK
Class II HLA-DR		Abcam	Cambridge, UK
Stemness markers			
CD117		Abcam	Cambridge, UK
SOX2		Abcam	Cambridge, UK
OCT4	PE	Becton Dickinson	BD, San Jose, CA
TERT		Calbiochem	Gibbstown, NJ
NANOG		Chemicon Int.	Billerica, MA

Bioscience) analyses and results were elaborated on CellQuest™ software (see Table 2 for details). Flow cytometry measurements were carried out by using rainbow calibration particles (6 peaks) and CaliBRITE beads (both from BD Biosciences) as quality control. Debris was excluded from the analysis by gating on morphological parameters (lymphocyte gate); 20.000 non-debris events in the morphological gate were recorded for each sample. All Abs were titrated under assay conditions and optimal photomultiplier (PMT) gains were established for each channel [59]. Data were analysed using FlowJo™ software (TreeStar, Ashland, OR). Mean fluorescence intensity ratio (MFI) was calculated dividing the MFI of positive events by the MFI of negative events [60].

Mesodermal Lineage *In Vitro* Differentiation. oAEC were preliminary differentiated *in vitro* into three different mesodermal cell lineages. Osteogenic differentiation was obtained according to Mattioli *et al.* [44]. Briefly, oAEC (2×10^4 cells) were incubated in osteo-inductive medium (αMEM plus 50 μM ascorbic acid, 10 mM β-glycerol phosphate, 0.2 μM dexamethasone and 10% FCS) for 21 days and osteogenic differentiation was evaluated by Alizarin Red staining [44] and RT-PCR (see Table 3 for details). Tenogenic *in vitro* differentiation was performed co-culturing oAEC with fetal tendon explants isolated from calcaneal tendons, as previously described [46]. Tenogenic differentiation was monitored by analyzing oAEC ability to generate three dimensional tendon-like structures enriched in collagen type I (COL1), as well as to express specific tendon-related-genes (see Table 3 for details). Finally, adipogenic differentiation was investigated by culturing oAEC in αMEM supplemented with 10% FCS, 0.5 mM isobutyl-methylxanthine (IBMX), 200 μM indomethacin, 0.1 μM dexamethasone, 10 ug/ml insulin and 1 μM rosiglitazone/BRL49653 for 4 weeks, according to a validated method [61]. For adipogenesis evaluation, cells were stained with oil-red O and, in parallel, tested with RT-PCR to evaluate the mRNA content for two adipose-related genes (see Table 3 for details).

Total RNA Isolation And RT-PCR. Total RNA was extracted from oAEC, native tissues (bone, tendon and adipose tissue) and from cryosections of explanted sinuses by using TRI Reagent, followed by reverse transcription (RT) reaction, as previously described [46]. RT-PCR was performed in order to define the expression profile of specific genes (Table 3). PCR products were separated on 2% agarose gel stained with ethidium bromide, visualized on a Gel Doc 2000 (Bio-Rad Laboratories) and analyzed with Quantity One 1-D Analysis software (Bio-Rad Laboratories). RT-PCR was normalized by the transcriptional levels of glyceraldehyde 3-phosphate dehydrogenase (GAPDH). Each PCR reaction was carried out in triplicate.

Ovine AEC Labeling. Thawed oAEC displaying a stable proliferation index, phenotype and the ability to undergo mesodermal lineage differentiation were then used for the animal model experiments described below. Before performing sinus augmentation, oAEC were marked with a fluorescence lipophilic dye, the PKH26, as previously described [44,47,62].

Biomaterial Characterization

The ceramic scaffolds used in this study were fabricated by the direct rPT dispense-plotting. A virtual scaffold model was designed with a cylindrical outer geometry by using 3D computer aided design software. The size of the model was adapted to the shrinkage of the ceramic material in the subsequent sintering process. The inner geometry, i.e. the pathway of the material rods, was defined by custom-made software which generates the control commands of the rapid prototyping machine. To build up the green bodies, material rods, consisting of a paste-like aqueous ceramic slurry, were extruded out of a cartridge through a nozzle and deposited using an industrial robot (GLT Pforzheim, Germany). In this study, HA/β-TCP powders (Merck, Germany) were blended to get a powder mixture with a HA/β-TCP weight ratio of 30/70. The characteristic rheological behavior of the aqueous ceramic slurry was achieved by thermal treatment of the raw HA powder at 900°C for 1 h and by adding a compatible

Table 3. Primer sequences used in real time and RT-PCR analyses.

Gene	Accession No.	Primer sequences	size (bp)	Cycles
COL1^{b,t}	AF129287.1	F: CGTGATCTGCGACGAACTTAA	212	40
	Ovine	R: GTCCAGGAAGTCCAGGTTGT		
SCXB^{t}	XM_866422.2	F: AACAGCGTGAACACGGCTTTC	299	45
	Bos Taurus	R: TTTCTCTGGTTGCTGAGGCAG		
OCN^{b,t}	DQ418490.1	F: AGACACCATGAGAACCCCCAT	234	40
	Ovine	R: TTGAGCTCACACACCTCCCT		
LPL^{a}	NM_001009394.1	F: GTCACGGGCCCAGCAGCATT	313	40
	Ovine	R: GCCAGGTGACCCCCTGGTGA		
PPAR^{a}	AY137204.1	F: CGCATGCCACAGGCCGAGAA	265	40
	Ovine	R: CCTGCAGGGGGCTGATGTGC		
TNMD^{t}	NM_001099948.1	F: GTCACGGGCCCAGCAGCATT	352	40
	Bos Taurus	R: GCCAGGTGACCCCCTGGTGA		
*GAPDH**	AF030943.1	F: CCTGCACCACCAACTGCTTG	224	40
	Ovine	R: TTGAGCTCAGGGATGACCTTG		
*VEGF**	AF250375.1	F: GTGCCCACTGAGGAGTTCAA	208	40
	Ovine	R: GTCTGGTTCCCGAAACCCTG		

^{b,t,a}superscripts indicate specific tissue-related genes expressed in bone, tendon and adipose tissues, respectively.
*primers used in real time PCR to quantify the VEGF expression. The other primers were used instead under RT-PCR.

binder/dispersant system of organic additives of 10.5 wt% relative to the mass of ceramic powder. The rod deposition was controlled in x, y and z direction to assemble 3D scaffolds layer by layer on a building platform. By rotating the direction of the rod deposition by 60° from layer to layer a three-dimensional network with an interconnecting pore structure was generated. The assemblies made of ceramic slurry were dried at room temperature and subsequently sintered at 1250°C for 1 h. Finally the sintered scaffolds were reduced to smaller blocks with a volume of about 0.14 cm^3 in order to remove the solid rim that resulted from the turning points at the edge of the printed pathways. Images of the sintered dispense-plotted scaffolds taken with a light microscope (MZM1, Askania, Mikroskop Technik Rathenow, Germany) were analysed using an image analysis system (analySIS, Soft Imaging System, Germany) to determine rod diameters and pore sizes. Density of the sintered scaffolds was measured using weight/volume method in which volume was determined geometrically and by using a helium pycnometer (AccuPyc 1330, micromeritics, Germany). Semi-quantitative X-ray diffraction measurements (XRD, 3000P, Seifert, Germany), and scanning electron microscopy (SEM; Quanta 200, FEI, The Netherlands) were used to characterize the surface of material rods and fracture faces of sintered samples.

Ovine AEC Scaffold Loading

In order to achieve 3D scaffold loading, 1×10^6 PKH26-labeled oAEC were seeded on single blocks of synthetic bone substitute (~0.14 cm^3) and incubated in 35 mm Petri dish over a roller apparatus (Wheaton, Millville, NJ). Each culture was taken under agitation for 3 days at a speed of 6 rpm. In order to verify the efficiency of loading process, the incubation medium was collected and centrifuged to quantify the concentration of unloaded cells. In addition, samples of HA/β-TCP biomaterials were, preliminarily, incubated with 1×10^6 oAEC to evaluate the incidence and the distribution of vital cells on the biomaterial surface. To this aim, the vital cytoplasmic fluorescent dye Calcein-AM was used [44]. The scaffolds used in control group (CTR) were incubated without cells for 3 days under analogous cultural conditions.

Animal Model

Animals. The present study was approved by the Ethics Committee of the Universities of Teramo and Chieti-Pescara (prot. 05/2012/CEISA/PROG/32). Six adult sheep, two years old and 40–50 Kg of weight, were bred according to the European community guidelines (E.D. 2010/63/UE) before performing bilateral sinus augmentation. Animals were quarantined for 2 weeks to check the general healthy status. Surgical procedures were then carried out in an authorized veterinary hospital. The animals were randomly divided into two groups and euthanized to explant grafted sinuses at 45 and 90 days p.i..

Sinus Augmentation Surgical Procedure. Sheep were operated under general anesthesia to carry out a bilateral sinus augmentation as previously described in Berardinelli *et al* [63]. The lateral wall of the sinus was approached through an oval ostectomy (1 cm up and 1 cm caudal to tuber facial tuberosity) carried out using a piezoelectric unit (Biosafin Easy Surgery, Italy). The *Schneiderian* membrane was elevated and the cavity between the mucosa and the inferior osseous septum of the sinuses was augmented with two blocks of biomaterial alone (CTR) or previously loaded with PHK26 labeled oAEC. The animals were treated i.v. with 20 mg/kg of ampicillin (Vetamplius®, Fatro) every 12 h for 3 days. Surgical wounds were inspected daily. Groups of three animals were euthanized at 45 and 90 days p.i. by applying an overdose of thiopental (Pentothal Sodium, Intervet) and embutramide (Tanax®, Intervet).

Morphological Analysis Of Sinus Explants

Sinus explants were fixed overnight in 4% paraformaldehyde solution (PBS, pH 7.4) and then analyzed by micro-CT technique in order to evaluate the biomaterial integration into the host tissue. The fixed explants were then processed for histological, immunohistological and biochemical analyses.

Micro-CT analysis. The bi-dimensional (2D) micro-CT analysis was performed using the SkyScan1072 (SkyScan, Belgium) operating at 98 μA currents and 100 kV voltages. The samples were analyzed at 20 mm×20 mm image size reaching a sample thickness of 19.5 μm. Since the large-size explanted samples were acquired at low resolution, in order to compare micro-CT images with the histological sections, no quantitative measures have been obtained.

The data sets were acquired over a rotation range of 180° (with 0.45° rotation step) and reconstructed with a software (NRecon v1.6.3; SkyScan, Belgium) based on the cone beam algorithm. 3D reconstruction was carried out by using ANT 3D Creator (v2.4; SkyScan, Belgium).

Histological Analysis. The explants were decalcified for at least 30 days in 14% EDTA [64–67] and placed overnight in 30% sucrose-PBS before freezing. Each explant was completely sectioned at 10 μm of thickness. For morphological and morphometric analysis, serial sections were collected every ~400 μm of distance. Two sections were stained with hematoxylin–eosin (HE) to evaluate the architecture of newly formed tissues and the presence of inflammation. Histological grading of inflammation was carried out by two independent investigators blind to the experimental conditions, using a semi-quantitative scoring system adapted from Cargnoni *et al.* [68]. For each group, at least 6 sections were analyzed. More in detail, 20 fields at ×200 final magnification were assessed for each tissue section under study, and the severity of inflammation was scored from 0 to 3 to reflect absent, mild, moderate or extensive inflammatory cell infiltration. Scores were summed and reported as mean value. For polarized light microscopy, HE-stained sections were observed using two orthogonally polarized light waves (Leica, Switzerland).

Morphometric and Biochemical Quantitative Analyses

Morphometric and biochemical analyses were performed on at least 10 different serial sections/animals (3 sections for HE, 3 for vascular area, VA, and 4 for vascular endothelial growth factor, VEGF, expression analysis) [69]. Morphometric data were calculated on at least 8 different fields/sections, randomly distributed in the central (C) or peripheral (P) portion of the implanted area, were blind evaluated.

A light microscopy (Axioscop 2 plus microscope) provided with a digital camera (Axiovision Cam, Zeiss, Germany) was used for image capture, while software ImageJ® 1.44 (National Institutes of Health, USA) was used to quantify the extension of newly deposited bone, scaffold and soft tissues within the implanted area. In detail, HE analysis were carried out at ×50 magnification and images obtained were processed to digitally convert scaffold, soft and newly deposited bone tissues in different colors as described in Berardinelli *et al.* [63]. The relative tissues/scaffold extension was calculated on the digitally converted images. The quantification of newly deposited bone was performed exclusively by considering the vital bone as the amount of mineralized and vascularized tissues displaying osteocytes within their lacunae [64,70–71].

In addition, serial sections were immunolabeled by using an anti-von Willebrand factor (vWF) Ab (Table 2), in order to detect blood vessel remodeling and to calculate the total vascular area (VA: see morphometric analyses), according to Martelli *et al.* [69]. In detail, primary Ab (Table 2) was applied at room temperature overnight after hotbox oven at 95°C before exposing the cells to secondary Ab (Table 2). Ovine aorta tissue was used as positive control. Negative controls were obtained by omitting the primary Ab. Tissue sections were counterstained with DAPI to visualize cell nuclei. Blood vessel analyses were carried out by using an Axioscop 2 plus microscope (Zeiss, Germany) and a digital camera (Axiovision Cam, Zeiss, Germany). Data were processed with a KS300 computed image analysis system (Zeiss, Germany), as previously described [69,70]. The total VA was quantified at ×200 magnification and the digitized fluorescent vessel signals (vWF positive area) were obtained using a semi-automated algorithm [70]. The VA was calculated as the extension expressed in μm^2 of vWF-positive area/field (15000 μm^2).

Real time PCR was performed to evaluate the VEGF expression on grafted sinuses at 45 and 90 days p.i.. More in detail, laser capture micro dissection (LCM) was performed using a MMI Cellcut apparatus to isolate the implanted area from native tissues. Frozen sections (n = 12/group) were briefly air dried on the glass slides and then kept on dry ice until they were subjected to

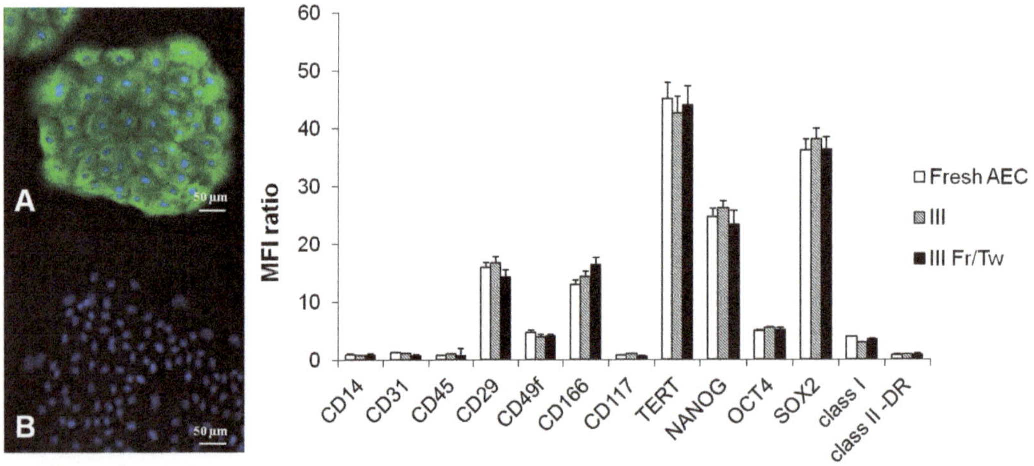

Figure 1. Molecular characterization of cultured and cryopreserved oAEC. Immunocytochemestry investigations show that freshly isolated oAEC displayed cytokeratin-8 positivity (A) while αSMA expression is not detectable (B). oAEC nuclei are counterstained with DAPI (blue fluorescence). Histograms show the average of surface, MHC and pluripotency markers expressed as MFI ratio levels obtained with flow cytometry analysis (3 replicates ± SD). The analyses were performed on oAEC expanded *in vitro* for III passages, before and after freezing and thawing (Fr/Tw).

LCM. Just before the procedure, the sections were exposed to 70% ethanol for 10 seconds and stained with HE. The settings of the laser were: spot diameter 20 μm, pulse duration 50 ms and power 50 mW. The area to micro dissect was identified under a light microscope at ×40 of magnification. The micro dissected area including the implantation site was dropped onto a separate cap before going on to RNA extraction and RT reaction as previously described. Real-time quantitative PCR was performed in the Stratagene MX3000P instrument using SYBR Green I dye detection. For VEGF and GAPDH genes the primers used were reported in Table 3. The following reaction components were prepared to the indicated end-concentrations: 2.5 μl forward (0.25 μM) primer, 2.5 μl reverse (0.25 μM) primer (Table 3), 3.5 μl water and 12.5 μl Brilliant SYBR Green QPCR Master Mix 2X (Stratagene). 3 μl of cDNA were added to 22 μl of the master mix. The real-time protocol employed was: initial denaturation for 10 min at 95°C, 40 cycles at 95°C for 45 sec, 60°C for 45 sec, 72°C for 45 sec. The specificity of amplicons was confirmed both by dissociation curve and by gel electrophoresis. Each sample was run in triplicate and for quantitation of the VEGF gene target the Comparative *Ct* Method was applied for all samples normalized to the control housekeeping GAPDH gene using the formula: $2^{-\Delta\Delta C(t)} = 2^{-(\Delta Ct \text{ control gene} - \Delta Ct \text{ target gene})}$. For statistical analyses the mean of three independent experiments was considered.

Ovine AEC Detection In Sinus Explants

The PKH26-labelled oAEC were identified on cryosections, by using an Axioskop 2 plus (Zeiss, Germany) epifluorescence microscope (excitation: 551 nm, emission: 567 nm). In order to count the endogenous and transplanted cells, the cryosections were processed with the KS300 computed image analysis system (Zeiss, Germany) by performing a guided program (macros for KS300, Zeiss, Germany) created to count the cells inside a standard field of 15000 μm^2, at ×200 magnification. In detail, for the present study two classes of cells were considered:

a) PKH26 positive oAEC are expressed by the number of cells displaying the DAPI counterstained nucleus surrounded by the red fluorescence PKH26 dye;
b) endogenous cells were the remaining cells identified exclusively by the DAPI stained nuclei.

Finally, to evaluate the contribution of oAEC to the process of tissue regeneration the co-labeling between OCN, VEGF or vWF and PKH26 were analyzed on cryosections collected at 45 and 90 days p.i.. Moreover, *in vitro* cultured PKH26 labeled cells were fixed in 4% paraformaldehyde/PBS and permeabilized with 0.05% Tween 20/PBS as described in the previous materials and methods paragraph. After incubation in PBS/1% BSA, OCN or VEGF primary and secondary Ab were applied (Table 2). According to the data sheet instructions, ovine calvaria tissue was used as positive control and the omission of primary Ab as negative control.

Statistical Analysis

The data were checked for normal distribution by D'Agostino and Pearson test and were compared, after arctan (x) transformation, by two way-ANOVA test to compare the effect exerted by different treatments. Finally, the post-hoc Tukey test (GraphPad Prism 5, GraphPad Software, USA) was carried out in order to evaluate the "individual" effect on each examined variable. The data are reported as mean ± Standard Deviation (± S.D).

Results

Ovine AEC Isolation And Molecular Characterization

The oAEC isolated from amniotic membranes were characterized by a uniform cell population characterized by a high cytokeratin 8 expression and by the absence of/low αSMA positivity (Fig. 1AB). The oAEC expanded in *vitro* for 3 passages did not substantially modify their molecular and antigenic profile (Fig. 1) as confirmed by flow cytometry analysis. In detail, oAEC did not display any hemopoietic markers (CD14, CD31 and CD45) and MHC class II antigens (Fig. 1). On the contrary, oAEC expressed several surface adhesion molecules (CD29, CD49f and CD166) and, to a lesser degree, MHC class I antigen. Freshly, expanded and cryopreserved oAEC showed similar levels of pluripotency markers (TERT, SOX2, OCT4 and NANOG; see Fig. 1). Moreover, RT-PCR demonstrated that oAEC had

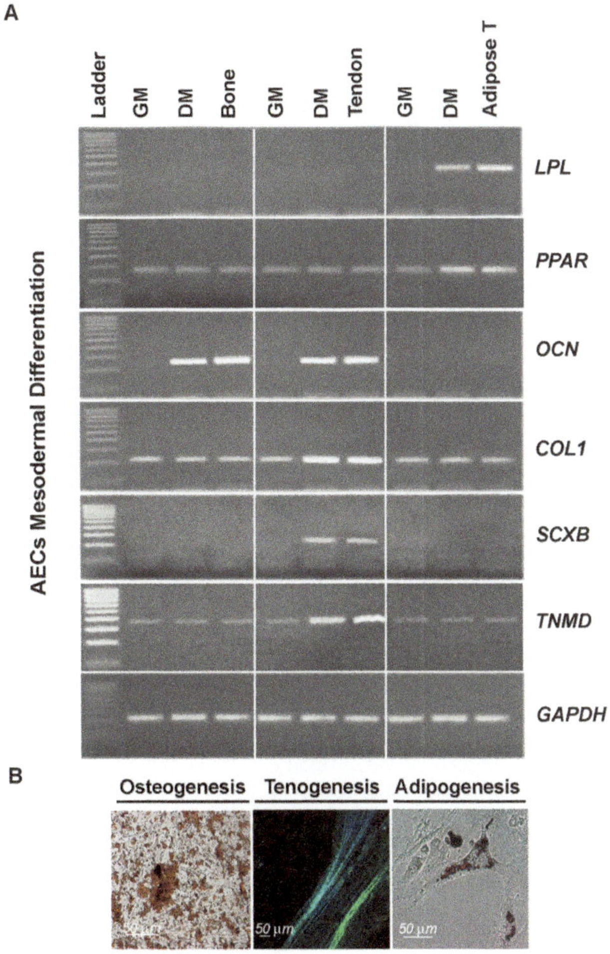

Figure 2. Expression profile of tissue specific-related genes after *in vitro* oAEC mesodermal differentiation. A) The mRNA content in *LPL, OCN, COL1, SCXB, PPAR, TNMD* and, *GAPDH* was analysed by RT-PCR in oAEC cultured for 28 days in growth medium (GM) or in adipogenic, osteogenic and tenogenic differentiation media (DM). B) Image examples showing the *in vitro* mesodermal differentiation properties of oAEC: left, the mineralization of extracellular matrix evidenced by Alizarin Red staining, middle, COL1 protein (green fluorescence) distribution recorded by immunohistochemistry in cell aggregates obtained during oAEC differentiation toward tenogenic lineage (blue nuclei counterstained with DAPI), right, intracellular lipid droplets evidenced by oil-red O staining.

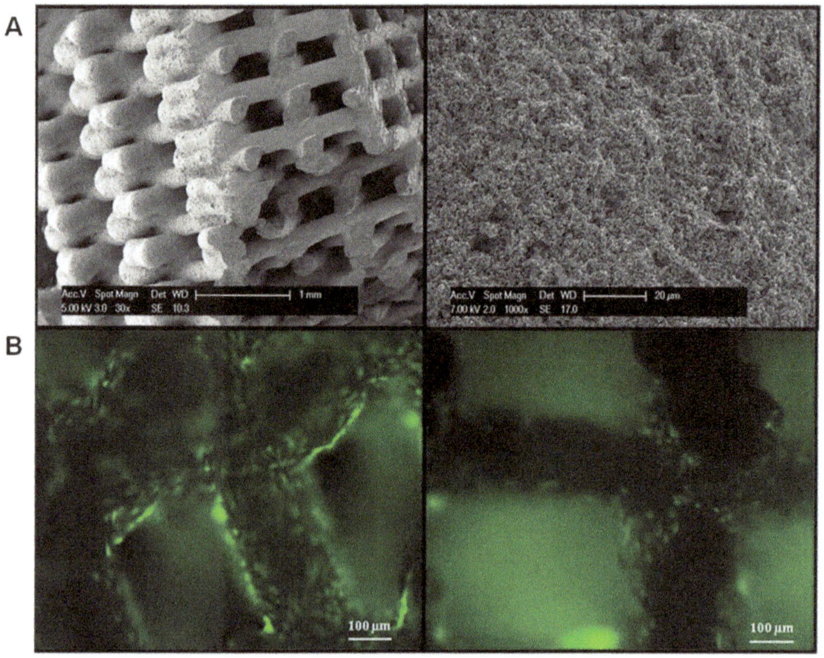

Figure 3. Ovine AEC engineered bone synthetic scaffold. A) Scanning electron microscopy images showing the surface and fracture face of direct rPT HA/β-TCP scaffold in different magnifications. Left image: fracture through ceramic scaffold. Right image: surface (white star) and fracture face (white arrow) of material rod. B) An example of HA/β-TCP scaffold's loaded with 1×10^6 oAEC under rotating culture for 3 days. The high concentration of cells loaded on biomaterial surface were visualized by the vital fluorescence permeable dye Calcein-AM (green emission).

undetectable mRNA level of *SCXB*, *OCN* and *LPL* and, a low expression of *COL1*, *TNMD* and *PPAR* (Fig. 2A). *In vitro* differentiation tests confirmed the mesodermal attitude of oAEC. In detail, osteogeneic differentiation was suggested either by the prompt ability of oAEC to mineralize the extracellular matrix demonstrated by Alizarin Red staining (Fig. 2B) or by the increased levels of *COL1* mRNA and the appearance of *OCN* gene expression (Fig. 2A). Similarly, oil-red O staining (Fig. 2B) and *LPL* and *PPAR* gene expressions (Fig. 2A) demonstrated the ability of oAEC to undergo *versus* adipose tissue lineage. Finally, the oAEC co-cultured with fetal tendon explants underwent tenogenic differentiation developing 3D elongated structures constituted of fusiform aligned cells displaying a high intracellular content of COL1 (Fig. 2B). In parallel, oAEC modified their gene profile displaying an increased expression of tendon related genes such as *COL1*, *TMND* and *SCX* (Fig. 2A).

Scaffold Characterization and Loading

The sintered dispense-plotted assemblies had a characteristic mesh-like structure with rod diameters of 300±30 µm, and pore sizes between the rods of about 370±25 µm.. By measuring the geometrical density of the sintered scaffolds, a total porosity of about 60% was calculated. Relative density of the sintered samples determined by helium pycnometry was about 99% th.d., which indicates only a small amount of closed porosity inside the material rods. Two main material phases of the sintered ceramic were detected by semi-quantitative XRD measurements: 30% HA, 60% β-TCP. Also a slight peak of α-TCP (10%) was identified (Fig. S1). SEM micrographs of the specimens revealed the anisotropic inner structure of the ceramic scaffolds as a result of the layer by layer process of dispense-plotting (Fig. 3A). The scaffolds were easily loaded with oAEC as indicated by the low concentration of cells recorded in the cultural media at the end of the incubation

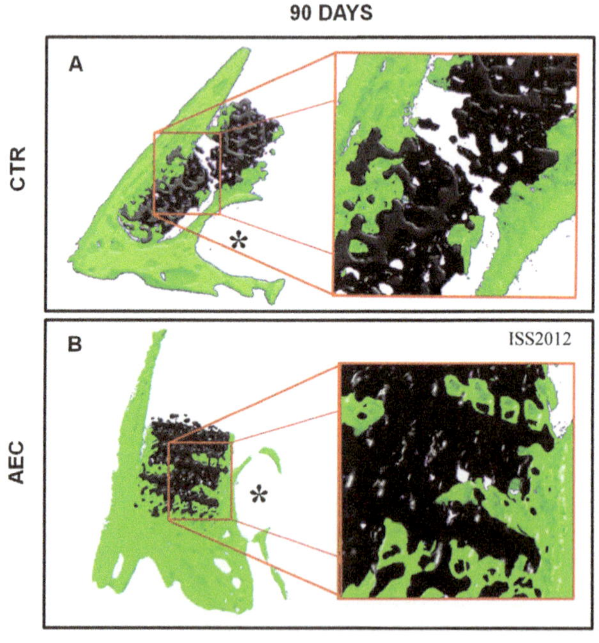

Figure 4. Micro-CT analysis: 3D reconstruction. Examples of 3D micro-CT analysis of sinus explants isolated after 90 days p.i. performed with the scaffold alone (CTR) or after engineering with oAEC (AEC). The different tissue densities were described by an arbitrary color scale showing in green and black colors the bone tissue and the biomaterial, respectively. *Asterisks indicate the infraorbital canal.

periods, which never exceeded 0.01% of added cells (1×10^6), as well as by the high concentration of oAEC recorded over the surface of the block bone substitutes (Fig. 3B).

Sinus Augmentation Clinical Procedure

Animals did not show any post-operative complications nor clinical symptoms of maxillary sinusitis. Extra-oral surgical windows healed in all groups after 45 and 90 days p.i. Sinus explants appeared as uniform blocks of tissue with the scaffolds firmly inserted.

Micro-CT Analysis

The 3D micro-CT analysis distinguished between bone tissue (green color) and bone substitute (black color) thus allowing to describe the process of bone generation and scaffold integration after sinus implants. The micro-CT revealed several newly deposited bone bridges, distributed between the native bone and the periphery of the scaffold in both CTR and oAEC. Foci of bone formation were also detected in the core of oAEC-treated sinuses differently from the CTR (Fig. 4).

Histological Analysis

Scaffolds of CTR explants resulted constantly surrounded by fibrotic reaction. At 45 days p.i., a mild inflammation (score 1–2) was seen, mainly characterized by the presence of scattered "foreign-body type" multinucleate giant cells (Fig. 5A, black arrow). Then, the inflammation was from mild to moderate (score 1–2) and characterized by clusters of macrophages, usually containing scaffold particles, and to a lesser degree, by small aggregates of lymphocytes. The infiltration of lymphocytes, sometimes organized in follicle-like structures, was more prominent beneath the *Schneiderian* membrane, where the "mucosa-associated lymphoid tissue" is found constitutively (Fig. 6A, black arrow).

In oAEC-transplanted animals, fibrous tissue combined with newly deposited bone foci were found either at 45 or 90 days p.i.. Inflammation was absent or very mild (score 0–1) in both interval points and, almost exclusively supported by macrophages (Fig. 5 and 6).

As shown in Fig. 5 and Fig. 6, a good integration of scaffolds was always obtained. In oAEC-treated animals, foci of newly deposited bone were observed in the center and at the periphery of the scaffold (Fig. 5B and 6B). In CTR, the newly deposited bone became evident at the periphery of the grafted area only at 90 days p.i. (Fig. 6A). Osteogenesis was generated, in both groups, by the endosteal apposition of bone tissue deposited next to the native bone or the scaffold surface (Fig. 5 and 6). More in detail, *trabeculae* of bone tissue lined by a continuous layer of osteoblasts were clearly observed at 45 days p.i. in oAEC-treated animals (Fig. 5B). Bone foci were then diffusely observed in the core and at the periphery of implanted areas (Fig. 6B). Polarized light microscopy showed that the woven deposited bone (Fig. 6) recorded 90 days

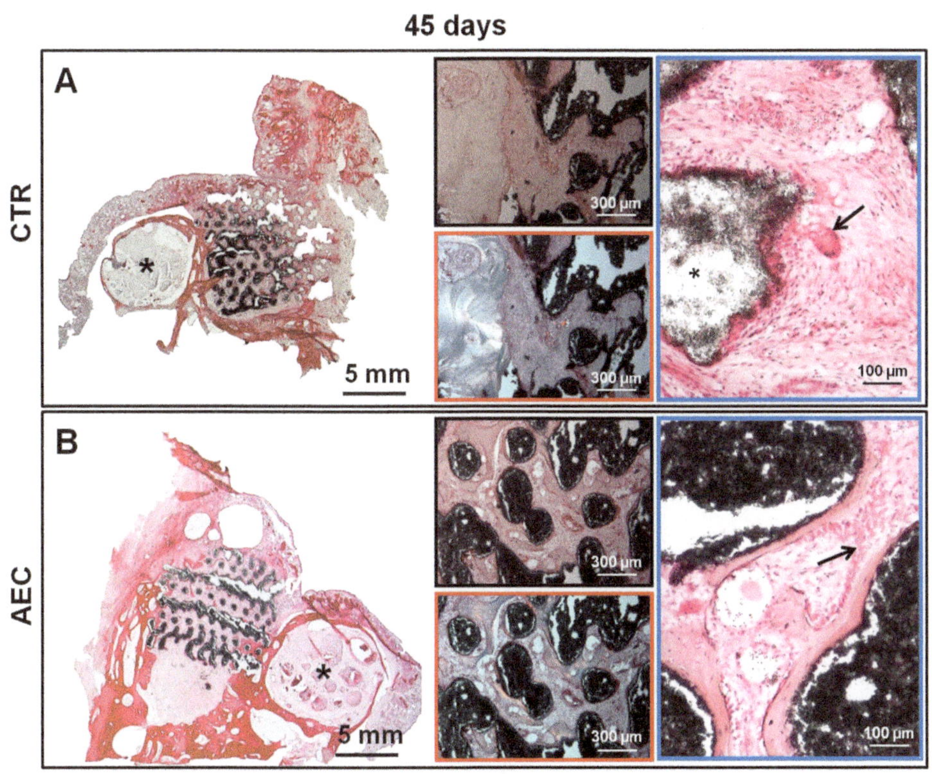

Figure 5. Morphological analysis of sinus explants 45 days p.i.. Examples of histological sinus explants cryosections stained with HE collected at 45 days p.i. performed with A) HA/β-TCP scaffold alone (CTR), and B) HA/β-TCP scaffold engineered with oAEC (AEC). Left big images: the lower magnification images show the whole implantation area demonstrating the integration of the scaffolds (dark structure) within the newly generated tissues. *Asterisks indicate the infraorbital canal. Middle images: in the top, representative examples of light images displaying the newly deposited matrix localized in the peripheral and central zones of CTR and oAEC-treated sinuses, respectively: in the bottom, polarized microscope images showing in blue and orange the immature and mature osteoid matrix, respectively. Right images: the higher magnifications show in CTR a multinucleate giant cell (top image: arrow) within the abundant fibrous connective tissue and in oAEC-treated explants, *trabeculae* of newly deposited bone tissue lined by a continuous layer of osteoblasts (bottom image: arrows).

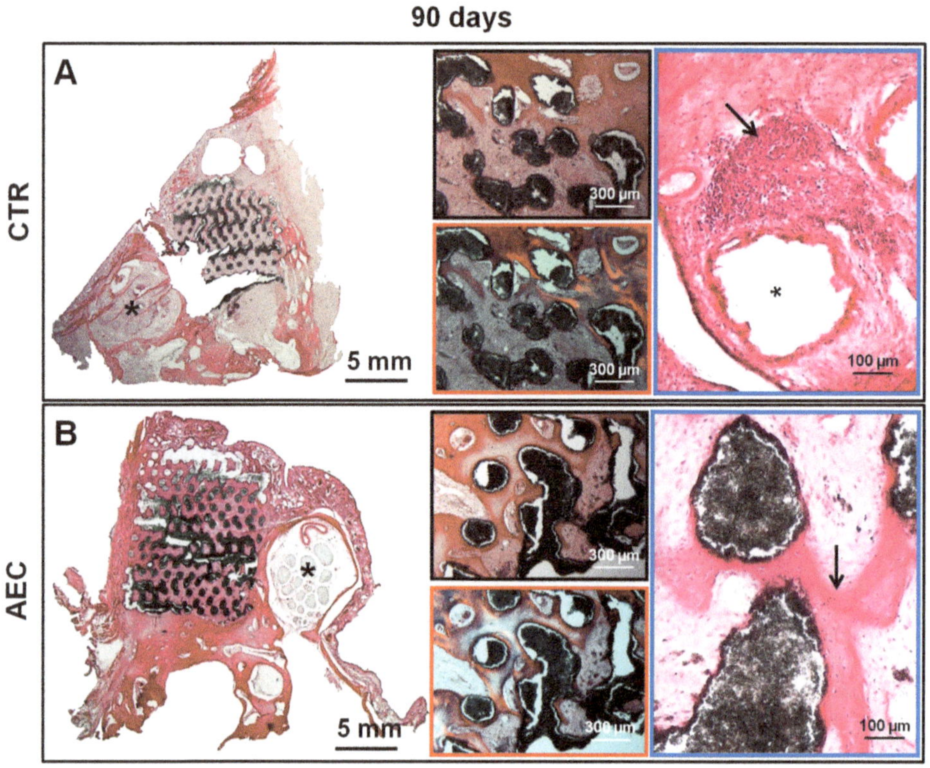

Figure 6. Morphological analysis of sinus explants 90 days after sinus lift. Examples of HE stained cryosections of sinus explants isolated at 90 days p.i. using A) HA/β-TCP scaffold alone (CTR), and B) HA/β-TCP scaffold engineered with oAEC (AEC). Left big images: the lower magnification shows the whole implantation area demonstrating the difference in bone matrix deposition. *Asterisks indicate the infraorbital canal. Middle images: in the top, light images displaying the foci of trabecular bone matrix localized in the peripheral zones of a CTR grafted sinus and the woven bone tissue distributed in the central area of an oAEC-treated sinus; at the bottom, polarized microscope images (bottom images) showing the higher level of maturation of newly deposited osteoid matrix recorded in oAEC-treated sinuses demonstrated by the prevalence of orange zones. Right images: the higher magnifications reveal in a CTR sinus (top image), beneath the *Schneiderian* membrane, a lymphoid follicle (top image: arrow) develops close to the scaffold cavity (top image: asterisk) surrounded by abundant fibrous tissue and newly deposited bone matrix. In oAEC-treated sinus, the presence of newly-formed bone tissue (bottom image: arrow) without evident inflammatory cell infiltrates was observed.

after oAEC transplantation is characterized by an abundant osteoid matrix (Fig. 6). By contrast, the foci of bone tissue, visualized mainly at the periphery of CTR sinuses at 90 days p.i. displayed an osteoid matrix with a different degree of maturation (Fig. 6 blue and orange areas). Noteworthy, the epithelial surface of the *Schneiderian* membranes was undamaged and formed by a continuous layer of ciliated cells, both in CTR and oAEC treated sinuses.

Ovine AEC Detection In Sinus Explants

Grafted oAEC were steadily recovered up to the 90th day (p>0.05: Fig. 7A). In detail, oAEC represented, in both the experimental intervals, ~10% of total cells (total cells = . endogenous cells plus PKH26 positive cells). On the contrary, the total cells recorded in CTR tissues at day 45 were significantly greater than those visualized at day 90 p.i. (p<0.05; Fig. 7A). Interestingly, oAEC transplantation was able to affect tissue cellularity: in fact, total cells detected in CTR groups were significantly greater than those recorded in oAEC-treated sinuses in both intervals (p<0.05 CTR *vs* oAEC at day 45 and p<0.05 CTR *vs* oAEC at day 90: Fig. 7A). Scattered isolated oAEC or, more frequently, oAEC aggregates were observed in proximity of blood vessels although none of them co-expressed vWF (Fig. 7B). By contrast, some PKH26 labeled cells displayed a cytoplasmic immune reactivity for OCN (Fig. 7C and C_1), a specific bone related protein unexpressed in oAEC before transplantation (Fig. 7C, small box). OCN-immune reactivity was clearly detected in oAEC interspersed within the fibrous tissue (Fig. 7C_1), but it was undetectable within the bone matrix because of the intense autofluorescence. PKH26-stained oAEC were also observed within the newly deposited bone matrix (Fig. 7D and D_1) close to endogenous osteocytes. In order to evaluate the angiogenic role exerted by transplanted oAEC, VEGF distribution was also analyzed. Immunohistochemistry revealed that more than 50% oAEC displayed the ability to synthetize the angiogenic factor (Fig. 8C small box). VEGF/PKH26 co-localization was then, visualized in ~35% and in ~15 % of transplanted oAEC recorded at 45 (data not shown) and at 90 days p.i. (Fig. 8C arrows), respectively. Moreover, VEGF positivity displayed, first, a widespread distribution in oAEC-treated sinuses (data not shown) to become mainly surrounding the blood vessels at 90 days p.i. (Fig. 8C). Differently, VEGF positivity was very faint initially in CTR tissues (45 days, data not shown) while it increased at day 90 p.i. when it assumed a diffused distribution (Fig. 8C).

Morphometric Analysis

The process of bone deposition progressively increased after sinus lift in both groups (Fig. 9A) even if the whole bone area always resulted significantly greater in oAEC-treated sinuses. The largest area of deposited bony tissue was recorded in oAEC grafted

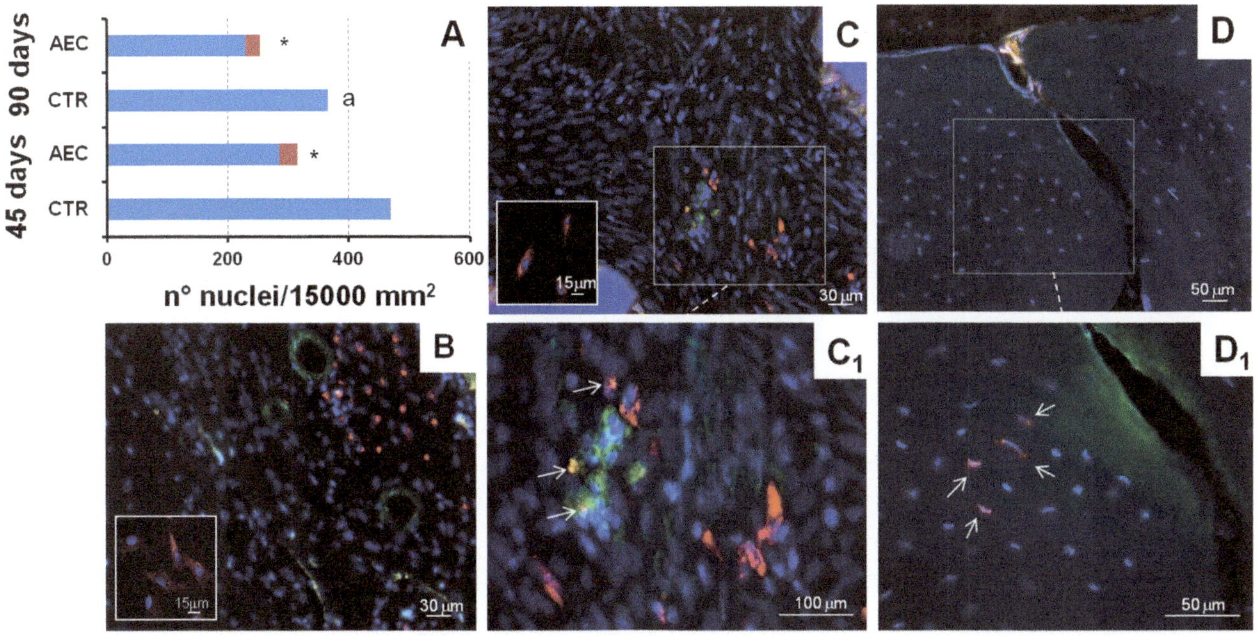

Figure 7. Detection of oAEC in transplanted sinuses. A) Histogram indicates the number of endogenous (blue bars) and transplanted cells (red bars) recorded in allotransplanted sinuses at 45 and 90 days p.i.. Endogenous cells or oAEC were identified by the DAPI counterstained nucleus visualized alone or surrounded by PKH26 dye, respectively. The values are expressed as mean. The SD never exceeded the 4%. * Values of total cells statistically different in oAEC *vs* CTR groups within the same p.i. interval, $p<0.05$ [a] Values of total cells recorded in oAEC or CTR groups significantly different when compared at 45 *vs* 90 days p.i., $p<0.05$; B) PKH26 labelled oAEC (red fluorescence) recorded in grafted sinuses at 45 days p.i.. The cells were recorded amongst fibrous matrix containing several endogenous cells identified by the bleu nuclei (DAPI fluorescence), newly deposited bone identified by the autofluorescence, blood vessels visualized by the endothelial marker, vWF (green fluorescence). Note how neither PKH26 labelled oAEC before (small box) and after transplantation express vWF. C-C_1) PKH26-labelled oAEC (red fluorescence) and endogenous cells recorded in grafted sinuses at 90 days p.i.. The cells were recorded within the fibrous tissue at lower (C) and higher magnification (C_1). Some of them are able to express OCN (green fluorescence), a typical bone-related protein. The OCN positivity was observed both in endogenous cells identified by the counterstained blue nuclei (DAPI) and in PKH26 positive cells (white arrows) that assume a typical orange fluorescent merge (green *plus* red). The PKH26 cells before transplantation did not express OCN (small box in C). D-D_1) Osteocytes embedded within the newly deposited bone tissue observed at lower (D) and higher magnification (D_1). The bone tissue, identified by the autofluorescence (green autofluorescence) of its matrix, showed several osteocytes. The blue nuclei (DAPI) identified the endogenous cells and the PKH26 dye displaying a typical spot fluorescence localized around the blue nuclei (arrows) identified oAEC transplanted cells.

sheep at 90 days p.i., when a, parallel, reduction in fibrous tissue was observed (Fig. 9A). Newly formed bone was mainly localized in the periphery of the scaffold (Fig. 9B) both in oAEC and CTR sinuses. However, osteogenesis, resulted more diffuse in oAEC-treated sinuses where a greater extension of bony foci was observed in the central zones in both the interval times (Fig. 9B).

Grafted oAEC significantly affected the extension of VA that resulted significantly greater at 45 days. (Fig. 10A) and lower at 90 days p.i.. By contrast, VA progressively increased in CTR group, thus gaining significantly higher values than those recorded in oAEC-treated animals at 90 days p.i. ($p<0.01$ Fig. 10A). Consistent with the pattern of bone deposition, topographic analysis of VA showed that angiogenesis was more uniform in oAEC–treated sinuses (Fig. 10B) than in CTR, where it was mainly localized in the periphery of the scaffold, in particular at 90 days p.i. (Fig. 10B).

The angiogenic response in the host tissue was studied, in addition, by analyzing the mRNA VEGF expression extracted from the implanted area selectively isolated with the aid of LMC techniques as summarized in Fig. 8A. The VEGF mRNA levels (see Fig 8) were significantly higher in oAEC than in CTR transplanted sinuses at day 45 p.i. (oAEC 45 *vs* CTR 45 days $p<0.01$) and became lower later (oAEC 90 *vs* CTR 90 days $p>0.05$). Consistent with VA results, in fact, the VEGF expression progressively increased in CTR grafted sinuses (CTR 45 *vs* 90 days, $p<0.01$) while it progressively decreased in the presence of oAEC (oAEC 45 *vs* 90 days $p<0.01$).

Discussion

The clinical outcome of dental implants strongly depends upon the regeneration of bone tissue, which should be quali-quantitatively adequate and rapidly produced. To achieve those results, two technologies should be strictly integrated: biomaterial manufacturing and regenerative medicine. Previous and extensive studies demonstrated that chemical composition and physical structure of biomaterials are both crucial for cell ingrowth and bone deposition [72–74]. Our data confirm the high biocompatibility of HA/TCP ceramic scaffold made using rPT technique [15,16]. In addition, the scaffold demonstrated to be an ideal substrate to support oAEC loading, thus facilitating the delivery of cells into the grafted sinuses. The scaffold used in our investigations reached a high degree of integration at 90 days p.i., but it was unable to stimulate a widespread process of bone deposition as demonstrated by the micro-CT and morphometric analyses.

Stated the poor osteo-regenerative properties induced by scaffolds transplantation, the *in vivo* osteogenic attitude of oAEC was easier to evaluate. On the basis of the present results, oAEC could be considered as a useful source of progenitor cells able to conjugate a good mesenchymal plasticity [25,44,46,48,49] with

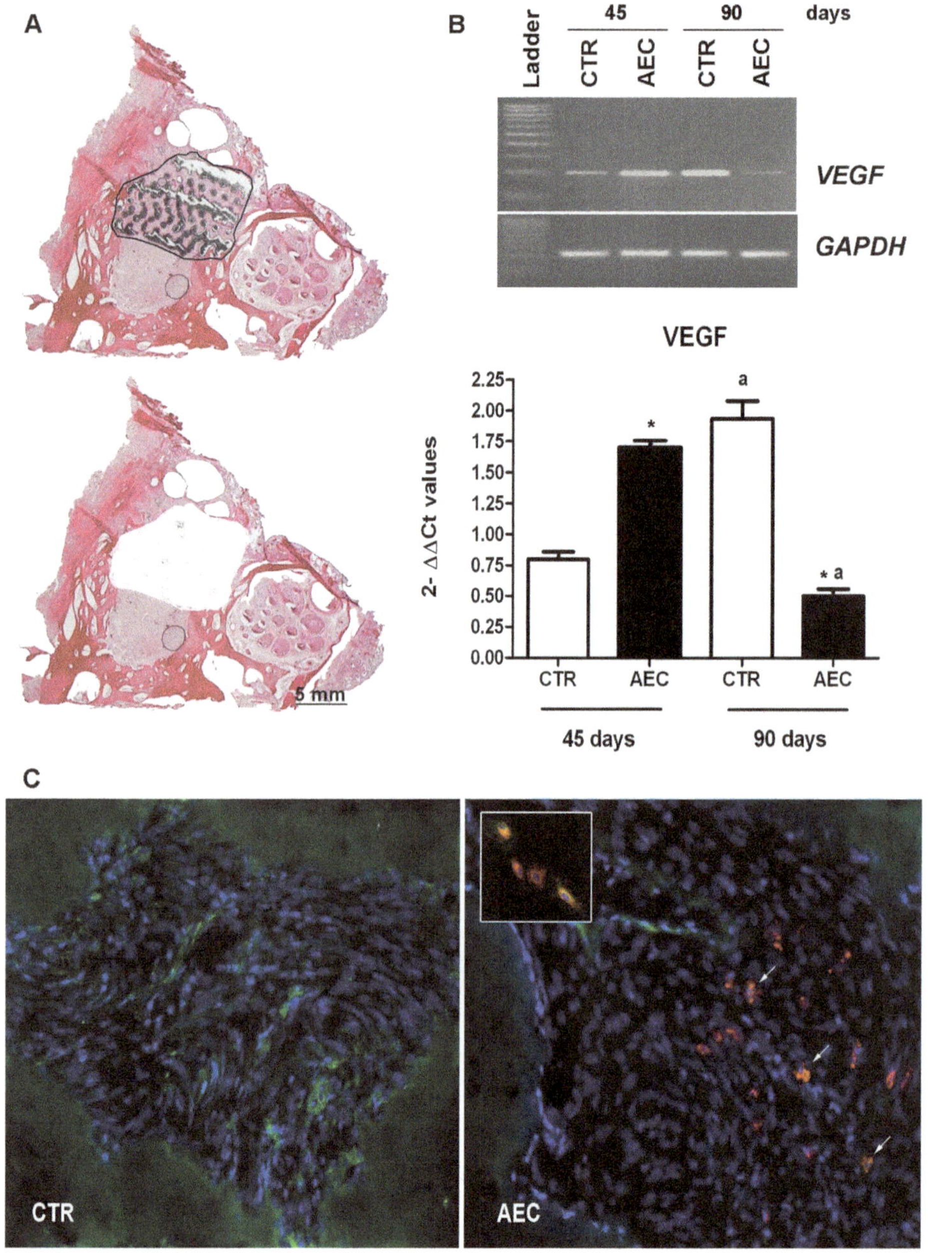

Figure 8. VEGF expression in grafted sinuses. A) Laser capture micro dissection (LCM) isolation of the implanted area from sinus explant. The top image shows an example of cryosection stained with HE where a black line delimits the implanted area that must be isolated by using LCM technique. The bottom image shows the same section after the implanted site was removed by LCM. Scale bar = 5 mm. B) VEGF mRNA expression. Top: image is a representative example of VEGF amplification. Bottom: quantitative VEGF analysis expression performed in real time PCR (values are mean of 3 replicates ±SD). *indicates values significantly different between CTR and oAEC groups within the same time interval ($p<0.01$), [a] indicates values significantly different in CTR or oAEC sinuses between the two time intervals ($p<0.01$). C) VEGF protein distribution in grafted sinuses 90 days after p.I.. Two representative images displaying the VEGF protein (green fluorescent signal) within the implanted area of a CTR (left image) and oAEC-treated sinuses (right image) explanted 90 days p.i. The nuclei are counterstained with DAPI in bleu and PKH26 labeled cells are visualized by the red fluorescence. Arrows indicate oAEC transplanted cells expressing VEGF (merge = orange fluorescence). Scale bars = 40 μm. The small box in the corner shows a representative example of VEGF expressing oAEC before grafting. The VEGF content of PKH26 labeled oAEC were analyzed on plated cultured cells. The merged orange fluorescence indicates the co-presence of PKH26 dye (red fluorescence) with the Alexa Fluor488 conjugated secondary antibody (green fluorescence). Scale bar = 60 μm.

other useful biological properties. In fact, AEC are uncontroversial and largely available, are stable and can be easily expanded *in vitro*, thus overcoming the practical limitation still linked to the use of the mesenchymal stromal cells (MSC) isolated from adult tissues such as dental pulp, periodontal ligament, dental follicle and bone marrow aspirates [20,22,23,75–82]. Furthermore, oAEC confirmed pro-angiogenetic activity and no tumurogenic attitude [29,44,47,54,83].

oAEC easily colonized the bone substitute when cultured under dynamic conditions and significantly improved bone tissue

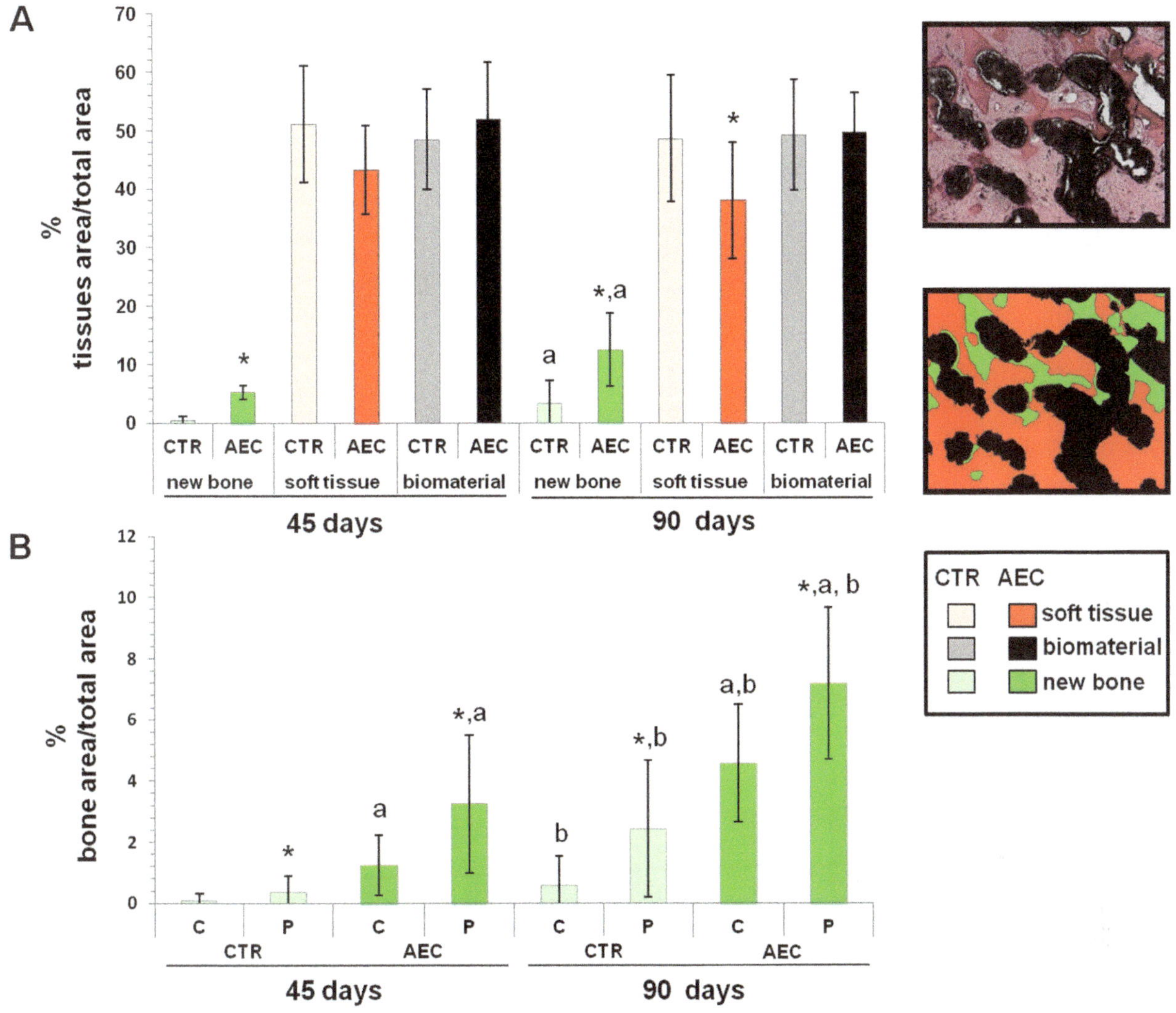

Figure 9. Morphometric analysis of grafted sinuses. A) Morphometric analysis of bone, biomaterials and fibrous tissues (soft tissue) area recorded within the grafted sinuses. The two images placed in the right corner show the digital conversion of the HE stained image used to quantify tissues/biomaterial area (green, new bone; red, fibrous tissue, and black, biomaterial) reported in the histograms. The data are expressed as mean ± SD. *Asterisks indicate values that resulted significantly different ($p<0.05$) when the two experimental groups were compared within the same interval point (i.e. CTR 45 days p.i. *vs* AEC 45 days p.i.). [a] letter indicates data significantly different ($p<0.05$) when area values at the different interval times within each experimental group were compared (i.e. AEC 45 days p.i. *vs* AEC 90 days p.i.). B) Histogram showed the morphometric data of newly deposited bone tissue recorded in the central (C) and peripheral zones (P) of grafted areas at 45 and 90 days p.i.. The data are expressed as mean ± SD. *Asterisks indicate C and P bone area values of CTR and AEC explants significantly different ($p<0.05$) when compared at each interval point (i.e. C AEC 45 day p.i. *vs* P AEC 45 days p.i.). [a] letter indicates C and P bone area values significantly different ($p<0.05$) when compared between the two experimental group at each interval point (i.e. P AEC 45 days p.i. *vs* P CTR 45 days p.i.). [b] letter indicate C or P bone area values recorded in CTR or oAEC-treated sinuses significantly different ($p<0.05$) when comparing the two interval points (i.e. C CTR 45 days p.i. *vs* C CTR 90 days p.i.).

regeneration after *in vivo* allotransplantation. Both micro-CT and histological analysis demonstrated that the process of bone deposition was accelerated and enhanced by oAEC starting to identify the mechanisms involved.

oAEC were never shown to express endothelial markers vWF and, as a consequence, they apparently do not directly contribute to neo-angiogenesis. Notwithstanding this, oAEC showed a modulatory effect on neoangiogenesis, which could be mediated by their ability to secrete VEGF [26,46] as well as by their modulatory effect on the VEGF expression of host maxillary tissues.

By contrast, molecular and morphological evidence suggests a direct role of oAEC in the process of osteogenesis. In fact, oAEC transplanted into the maxillary sinus acquired the ability to express OCN, a typical bone related protein, which is completely absent in cells before transplantation. A direct contribution of oAEC to the process of bone deposition may be, moreover, confirmed by the evidence that some PKH26 positive cells were detected within the newly deposited bone matrix. In fact, the typical red spots of PKH26 fluorescence surrounding flattened nuclei were recorded within bone matrix close to other endogenous osteocytes. However, beside a direct oAEC commitment into osteoblast lineage, the present results also suggest a paracrine role of oAEC involved in supporting tissue repair/regeneration. In fact, oAEC modulated the process of tissue regeneration by influencing either the cellularity or the inflammatory response in the transplanted sinuses. This last finding confirmed the immunomodulatory and a long term antifibrotic activity already demonstrated for the

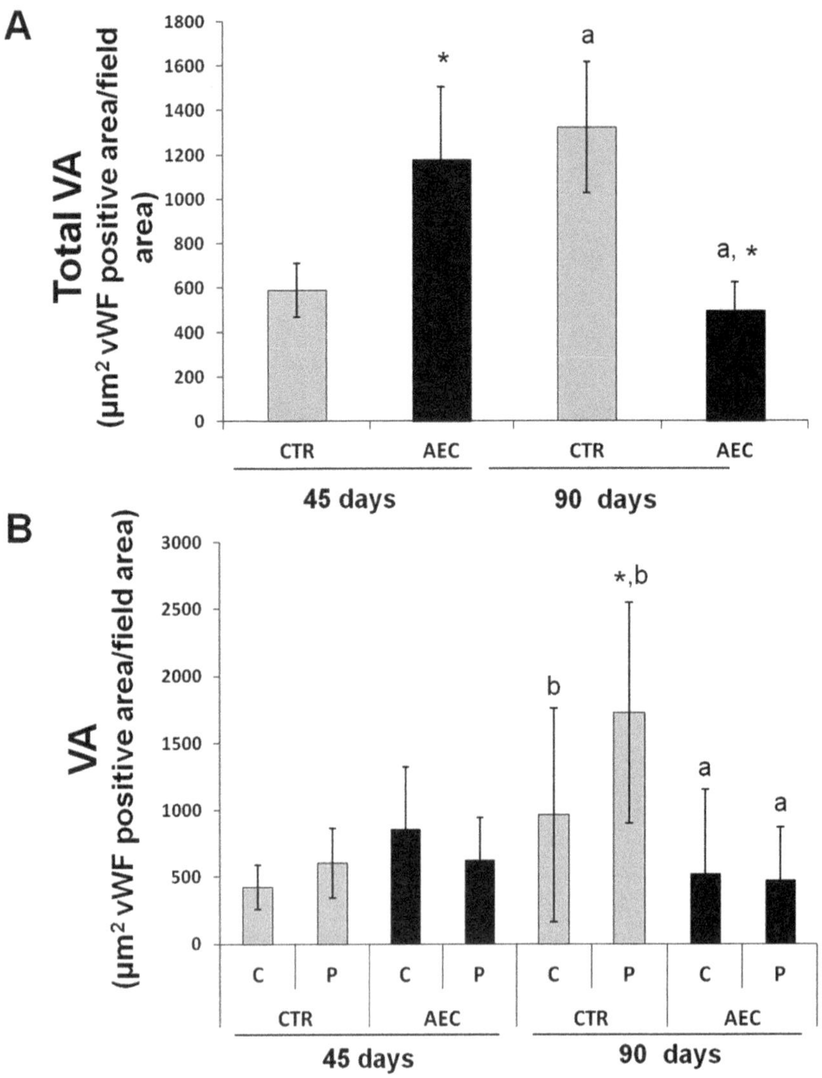

Figure 10. Morphometric analysis of vascular area (VA) in grafted sinuses. A) Morphometric analysis of total VA recorded in CTR and oAEC-grafted sinuses at 45 and 90 days p.i.. The histograms show the total VA quantified after the identification of blood vessel walls performed by using the vWF endothelial marker. The values of total VA were expressed as mean ± SD. *Asterisks indicate CTR *vs* AEC VA, significantly different ($p<0.05$) when recorded at each interval point (i.e. CTR 45 days p.i. *vs* AEC 45 days p.i.). [a] letter indicates values of VA recorded within each experimental group that resulted significantly different ($p<0.05$) when comparing the two time intervals (i.e. CTR 45 days p.i. *vs* CTR 90 days p.i.). B) Histogram showing VA recorded in the central (C) and peripheral (P) zones of grafted area analyzed at 45 and 90 days p.i.. The VA are expressed as mean ± SD. *Asterisks indicate C and P VA recorded in CTR or oAEC-treated sinuses significantly different ($p<0.05$) when compared at each interval point (i.e. C AEC 45 days p.i. *vs* P AEC 45 days p.i.). [a] letter indicates C or P VA recorded in CTR *vs* AEC significantly different ($p<0.05$) when compared at each interval point (i.e. P AEC 45 days p.i. *vs* P CTR 45 days p.i.). [b] letter indicates C or P VA recorded in CTR or oAEC-treated sinuses significantly different ($p<0.05$) when comparing the two interval points (i.e. C CTR 45 days p.i. *vs* C CTR 90 days p.i.).

amniotic derived cells in another tissue model [84]. In particular, fibrous tissue in grafted oAEC sinuses was positively substituted after 90 days p.i by the newly deposited bone. Sinus explants revealed that the scaffolds themselves induced mild to moderate inflammatory reactions, characterized by the infiltration of clusters of macrophages and small lymphocytic aggregates always associated with a severe surrounding fibrotic reaction. In oAEC grafted sinuses, the fibrous tissue remained localized along the newly-formed *trabeculae* of bone tissue and the inflammation response was absent or very mild. The anti-inflammatory and anti-fibroblastic role exerted by the transplanted oAEC may have exerted a positive influence on the regenerative processes. Our scaffolds were fabricated to obtain a low degradation rate and robust mechanical property. However, as a consequence of its high porosity, the scaffold presents a large surface that, interacting with the host tissue, may potentially accelerate degradation due to macrophages via oxidation and/or hydrolysis. However, differences in scaffold degradation did not become evident in 90 days p.i. even if a higher infiltration of inflammatory cells/macrophages was observed in CTR tissues. The timing of scaffold degradation remains still to be defined since probably it requires longer periods than 90 days according to previously described tests [85,86]. Ideally, scaffold degradation rate must be appropriately combined with the growth rate of the new-tissue, in such a way that the scaffold will be totally degraded by the time the implantation site is totally regenerated.

The oAEC influence on neoangiogenesis is more evident. By a simple histological examination, the scaffolds appeared completely colonized by a densely vascularized tissue in both experimental groups starting from 45 days p.i.. The morphometric analysis, however, revealed that the process of blood vessel rearrangement was significantly enhanced by oAEC at the beginning, to subsequently decline on values lower than in CTR. oAEC, in addition, stimulated a diffused vascular network that resulted equally distributed between the core and the periphery of the implanted area. This blood vessel organization may support a stable and homogenous trophic supply, thus increasing the more uniform bone deposition within the grafted sinus. By contrast, the VA progressively increased in the CTR explants, thus becoming significantly greater after 90 days p.i., mainly at the periphery of the grafted area. This different angiogenic behaviour may be interpreted as a consequence of the higher inflammatory reaction and VEGF expression recorded in CTR tissues at 90 days.

These results clearly indicate that the surgical outcome of a common approach such as the sinus augmentation procedure could be markedly improved by combining the innovative techniques in producing bone graft substitutes with the use of an adequate source of progenitor cells. Since bone regeneration proceeds always centripetal after the transplantation of a good synthetic bone substitute, the use of an adequate source of progenitor cells such as oAEC is determinant to enhance and make more extent the foci of bone nucleation, thus increasing, strengthening and accelerating the alveolar bone reconstruction.

Although the role of oAEC remains unclear and questionable, their transplantation clearly supported an overall increase of bone deposition either directly contributing to osteogenesis or indirectly modulating the major mechanisms (inflammation, cellularity and angiogenesis) involved in tissue regeneration. Both these mechanisms have also been previously described in bone marrow derived MSC as well as in other MSC sources [20–22,75–77,80–81] where, however, the reduced availability of progenitor cells still remains overcome problem. Actually, in the absence of experimental studies addressed to compare the regenerative properties of MSC *vs* AEC any speculation becomes difficult even if the results obtained in the present research together with the high biological properties of amniotic derived cells [28,29] seems to suggest their potential relevance in dentistry.

In particular, the ability of oAEC in modulating the angiogenic remodeling and in reducing inflammatory reaction may be essential to drive crucial mechanisms involved in improving scaffold persistence and integration within the host tissue, as well as to guarantee an adequate trophic support to the different cells actively engaged in the process of tissue regeneration. Altogether, the present results provide the first evidence in favor of an efficient and safe therapeutic role of AEC. These results support the hypothesis that AEC hold much promise for the development of cell-based therapies in craniofacial surgical applications, leading to the idea of their safe use under allotransplantation settings.

Supporting Information

Figure S1 Biomaterial Characterization. Semi-quantitative x-rays diffraction measurement of HA/β-TCP custom made scaffold showed that the bone substitute is composed of three co-existing mineral phases from the mineral phases point of view: β-TCP (60%), HA (30%) and αTCP (10%).

Acknowledgments

The authors gratefully acknowledge the help of Sabine Hamisch, Uwe Hamhaber, BioCer Entwicklungs-GmbH, Ludwig-Thoma-Str. 36c, 95447 Bayreuth, Germany, in providing the materials, Oriana Di Giacinto and Renato Peli for their precious technical support.

Author Contributions

Conceived and designed the experiments: BB MM. Performed the experiments: ST A. Muttini LV VLZ. Analyzed the data: GM PB A. Martelli A. Mauro LV AP DN MT BB. Contributed reagents/materials/analysis tools: BB MM. Wrote the paper: BB. Micro CT analysis: RB RP LV. Manufactured and supported the biomaterials: CM.

References

1. Stevens B, Yang Y, Mohandas A, Stucker B, Nguyen K (2008) A review of materials, fabrication methods, and strategies used to enhance bone regeneration in engineered bone tissues. J Biomed Mater Res B 85: 573–582.
2. Hing K (2005) Bioceramic bone graft substitutes: influence of porosity and chemistry. Int J of Appl Ceram Technol 2: 184–199.
3. Dorozhkin S, Epple M (2002) Biological and medical significance of calcium phosphates. Angew Chem Int Ed Engl 41: 3130–3146.
4. Eggli PS, Müller W, Schenk RK (1988) Porous hydroxyapatite and tricalcium phosphate cylinders with two different pore size ranges implanted in the cancellous bone of rabbits. A comparative histomorphometric and histologic study of bony ingrowth and implant substitution. Clin Orthop Relat Res 232: 127–138.
5. Kühne JH, Bartl R, Frisch B, Hammer C, Jansson V, et al. (1994) Boneformation in corallinehydroxyapatite. Effects of pore size studied in rabbits. Acta Orthop Scand 65: 246–252.
6. Weiss P, Obadia L, Magne D, Bourges X, Rau C, et al. (2003) Synchrotron X-ray microtomography (on a micron scale) provides three-dimensional imaging representation of bone ingrowth in calcium phosphate biomaterials. Biomaterials 24: 4591–4601.
7. Fierz FC, Beckmann F, Huser M, Irsen SH, Leukers B, et al. (2008) The morphology of anisotropic 3D-printed hydroxyapatite scaffolds. Biomaterials 29: 3799–3806.
8. Kuboki Y, Takita H, Kobayashi D, Tsuruga E, Inoue M, et al. (1998) BMP-induced osteogenesis on the surface of hydroxyapatite with geometrically feasible and nonfeasible structures: topology of osteogenesis. J Biomed Mater Res 39: 190–199.
9. Hollister S (2005) Porous scaffold design for tissue engineering. Nat Mater 4: 518–524.
10. Hutmacher DW (2000) Scaffolds in tissue engineering bone and cartilage. Biomaterials 21: 2529–2543.
11. Chu T, Orton D, Hollister S, Feinberg S, Halloran J (2002) Mechanical and *in vivo* performance of hydroxyapatite implants with controlled architectures. Biomaterials 23: 1283–1293.
12. Lee M, Dunn J, Wu B (2005) Scaffold fabrication by indirect three dimensional printing. Biomaterials 26: 4281–4289.
13. Adachi T, Osako Y, Tanaka M, Hojo M, Hollister SJ (2006) Framework for optimal design of porous scaffold microstructure by computational simulation of bone regeneration. Biomaterials 27: 3964–3972.
14. Yeong WY, Chua CK, Leong KF, Chandrasekaran M (2004) Rapid prototyping in tissue engineering: challenges and potential. Trends Biotechnol 22: 643–652.
15. Deisinger U, Hamisch S, Schumacher M, Uhl F, Detsch R, et al. (2008) Fabrication of tailored hydroxyapatite scaffolds: comparison between a direct and an indirect rapid prototyping technique. Key Engineering Materials 361–363: 915–918.
16. Schumacher M, Deisinger U, Detsch R, Ziegler G (2010) Indirect rapid prototyping of biphasic calcium phosphate scaffolds as bone substitutes: influence of phase composition, macroporosity and pore geometry on mechanical properties. J Mater Sci Mater Med 21: 3119–3127.
17. Schumacher M, Uhl F, Detsch R, Deisinger U, Ziegler G (2010) Static and dynamic cultivation of bone marrow stromal cells on biphasic calcium phosphate scaffolds derived from an indirect rapid prototyping technique. J Mater Sci Mater Med 21: 3039–48.
18. Sorní M, Guarinós J, García O, Peñarrocha M (2005) Implant rehabilitation of the atrophic upper jaw: a review of the literature since 1999. Med Oral Patol Oral Cir Bucal 10: E45–E56.
19. Aghaloo TL, Moy PK (2007) Which hard tissue augmentation techniques are the most successful in furnishing bony support for implant placement? Int J Oral Maxillofac Implants 22: 49–70.
20. Beyth S, Schroeder J, Liebergall M (2011) Stem cells in bone diseases: current clinical practice. Br Med Bull 99: 199–210.

21. Mendonça JJ, Juiz-Lopez P (2010) Regenerative facial reconstruction of terminal stage osteoradionecrosis and other advanced craniofacial diseases with adult cultured stem and progenitor cells. Plast Reconstr Surg 126: 1699–1709.
22. Runyan CM, Taylor JA (2010) Clinical applications of stem cells in craniofacial surgery. Facial Plast Surg 26: 385–395.
23. Ward BB, Brown SE, Krebsbach PH (2010) Bioengineering strategies for regeneration of craniofacial bone: a review of emerging technologies. Oral Dis 16: 709–716.
24. Tamagawa T, Ishiwata I, Saito S (2004) Establishment and characterization of a pluripotent stem cell line derived from human amniotic membranes and initiation of germ layers *in vitro*. Hum Cell 17: 125–130.
25. Miki T, Lehmann T, Cai H, Stolz DB, Strom SC (2005) Stem cell characteristics of amniotic epithelial cells. Stem Cells 23: 1549–59.
26. Parolini O, Soncini M, Evangelista M, Schmidt D (2009) Amniotic membrane and amniotic fluid-derived cells: potential tools for regenerative medicine? Regen Med 4: 275–291.
27. Ilancheran S, Moodley Y, Manuelpillai U (2009) Human fetal membranes: a source of stem cells for tissue regeneration and repair? Placenta 30: 2–10.
28. Evangelista M, Soncini M, Parolini O (2008) Placenta-derived stem cells: new hope for cell therapy? Cytotechnology 58: 33–42.
29. Miki T (2011) Amnion-derived stem cells: in quest of clinical applications. Stem Cell Res Ther 2: 25.
30. Davis JW (1910) Skin transplantation of 550 cases at Johns Hopkins Hospital. Johns Hopkins Med J 15: 307–396.
31. Kim JS, Kim JC, Na BK, Jeong JM, Song CY (2000) Amniotic membrane patching promotes healing and inhibits proteinase activity on wound healing following acute corneal alkali burn. Exp Eye Res 70: 329–337.
32. Hao Y, Ma DH, Hwang DG, Kim WS, Zhang F (2000) Identification of antiangiogenic and antiinflammatory proteins in human amniotic membrane. Cornea 19: 348–352.
33. Solomon A, Rosenblatt M, Monroy D, Ji Z, Pflugfelder SC, et al. (2001) Suppression of interleukin 1alpha and interleukin 1beta in human limbal epithelial cells cultured on the amniotic membrane stromal matrix. Br J Ophthalmol 85: 444–449.
34. Talmi YP, Sigler L, Inge E, Finkelstein Y, Zohar Y (1991) Antibacterial properties of human amniotic membranes. Placenta 12: 285–288.
35. Tseng SC, Li DQ, Ma X (1999) Suppression of transforming growth factor-beta isoforms, TGF-beta receptor type II, and myofibroblast differentiation in cultured human corneal and limbal fibroblasts by amniotic membrane matrix. J Cell Physiol 179: 325–335.
36. Hori J, Wang M, Kamiya K, Takahashi H, Sakuragawa N (2006) Immunological characteristics of amniotic epithelium. Cornea 25: S53–S58.
37. Toda A, Okabe M, Yoshida T, Nikaido T (2007) The potential of amniotic membrane/amnion-derived cells for regeneration of various tissues. J Pharmacol Sci 105: 215–228.
38. Miki T, Marongiu F, Dorko K, Ellis EC, Strom SC (2010) Isolation of amniotic epithelial stem cells. Curr Protoc Stem Cell Biol Chapter 1:Unit 1E.3.
39. Marongiu F, Gramignoli R, Sun Q, Tahan V, Miki T, et al. (2010) Isolation of amniotic mesenchymal stem cells. Curr Protoc Stem Cell Biol Chapter 1: Unit 1E.5.
40. Parolini O, Alviano F, Bagnara GP, Bilic G, Bühring HJ, et al. (2008) Concise review: isolation and characterization of cells from human term placenta: outcome of the first international Workshop on Placenta Derived Stem Cell. Stem Cells 26: 300–311.
41. Miki T, Strom SC (2006) Amnion-derived pluripotent/multipotent stem cells. Stem Cell Rev 2: 133–142.
42. Marcus AJ, Coyne TM, Rauch J, Woodbury D, Black IB (2008) Isolation, characterization, and differentiation of stem cells derived from the rat amniotic membrane. Differentiation 76: 130–144.
43. Elwan MA, Thangavel R, Ono F, Sakuragawa N (1998) Synthesis and release of catecholamines by cultured monkey amniotic epithelial cells. J Neurosci Res 53: 107–113.
44. Mattioli M, Gloria A, Turriani M, Mauro A, Curini V, et al. (2011) Stemness characteristics and osteogenic potential of sheep amniotic epithelial cells. Cell Biol Int 36: 7–19.
45. Lange-Consiglio A, Corradetti B, Bizzaro D, Magatti M, Ressel L, et al. (2011) Characterization and potential applications of progenitor-like cells isolated from horse amniotic membrane. J Tissue Eng Regen Med DOI: 10.1002/term.465.
46. Barboni B, Curini V, Russo V, Mauro A, Di Giacinto O, et al. (2012) Indirect co-culture with tendons or tenocytes can program amniotic epithelial cells towards stepwise tenogenic differentiation. PLoS One 7: e30974.
47. Barboni B, Russo V, Curini V, Mauro A, Martelli A, et al. (2012) Achilles tendon regeneration can be improved by amniotic epithelial cells allotransplantation. Cell Transplant DOI:10.3727/096368912X638892
48. Stadler G, Hennerbichler S, Lindenmair A, Peterbauer A, Hofer K, et al. (2008) Phenotypic shift of human amniotic epithelial cells in culture is associated with reduced osteogenic differentiation *in vitro*. Cytotherapy 10: 743–752.
49. Bilic G, Zeisberger SM, Mallik AS, Zimmermann R, Zisch AH (2008) Comparative characterization of cultured human term amnion epithelial and mesenchymal stromal cells for application in cell therapy. Cell Transplant. 17: 955–968.
50. Bailo M, Soncini M, Vertua E, Signoroni PB, Sanzone S, et al. (2004) Engraftment potential of human amnion and chorion cells derived from term placenta. Transplantation 27: 1439–1448.
51. Wolbank S, Peterbauer A, Fahrner M, Hennerbichler S, van Griensven M, et al. (2007) Dose-dependent immunomodulatory effect of human stem cells from amniotic membrane: a comparison with human mesenchymal stem cells from adipose tissue. Tissue Eng 13: 1173–1183.
52. Banas RA, Trumpower C, Bentlejewski C, Marshall V, Sing G, et al. (2008) Immunogenicity and immunomodulatory effects of amnion-derived multipotent progenitor cells. Hum Immunol 69: 321–328.
53. Takahashi S, Ohsugi K, Yamamoto T, Shiomi M, Sakuragawa N (2001) A novel approach to ex vivo gene therapy for familial hypercholesterolemia using human amniotic epithelial cells as a transgene carrier. Tohoku J Exp Med 193: 279–292.
54. Parolini O, Caruso M (2011) Review: Preclinical studies on placenta-derived cells and amniotic membrane: an update. Placenta 32: S186–S195.
55. Sant'Anna LB, Cargnoni A, Ressel L, Vanosi G, Parolini O (2011) Amniotic membrane application reduces liver fibrosis in a bile duct ligation rat model. Cell Transplant 20: 441–453.
56. Kubo M, Sonoda Y, Muramatsu R, Usui M (2001) Immunogenicity of human amniotic membrane in experimental xenotransplantation. Invest Ophthalmol Vis Sci 42: 1539–1546.
57. Sankar V, Muthusamy R (2003) Role of human amniotic epithelial cell transplantation in spinal cord injury repair research. Neuroscience 118: 11–7.
58. Colosimo A, Curini V, Russo V, Mauro A, Bernabò N, et al. (2012) Characterization, GFP gene nucleofection and allotransplantation in injured tendons of ovine amniotic fluid-derived stem cells. Cell Transplant DOI:10.3727/096368912X638883.
59. Perfetto SP, Chattopadhyay PK, Lamoreaux L, Nguyen R, Ambrozak D, et al. (2006) Amine reactive dyes: an effective tool to discriminate live and dead cells in polychromatic flow cytometry. J Immunol Methods 313: 199–208.
60. Miscia S, Ciccocioppo F, Lanuti P, Velluto L, Bascelli A, et al. (2009) Abeta (1–42) stimulated T cells express P-PKC-delta and P-PKC-zeta in Alzheimer disease. Neurobiol Aging 30: 394–406.
61. Fink T, Zachar V (2011) Adipogenic differentiation of human mesenchymal stem cells. Methods Mol Biol 698: 243–251.
62. Muttini A, Mattioli M, Petrizzi L, Varasano V, Sciarrini C, et al. (2010) Experimental study on allografts of amniotic epithelial cells in calcaneal tendon lesions of sheep. Vet Res Commun 34: S117–S120.
63. Berardinelli P, Valbonetti L, Muttini A, Martelli A, Peli R, et al. (2012) Role of amniotic fluid mesenchymal cells engineered on MgHA/collagen-based scaffold allotransplanted on an experimental animal study of sinus augmentation. Clin Oral Investig DOI 10.1007/s00784-012-0857-3.
64. Froum SJ, Tarnow DP, Wallace SS, Rohrer MD, Cho SC (1998) Sinus floor elevation using anorganic bovine bone matrix (OsteoGraf/N) with and without autogenous bone: a clinical, histologic, radiographic, and histomorphometric analysis-Part 2 of an ongoing prospective study. Int J Periodontics Restorative 18: 528–543.
65. Liu LS, Thompson AY, Heidaran MA, Poser JW, Spiro RC (1999) An osteoconductive collagen/hyaluronate matrix for bone regeneration. Biomaterials 20: 1097–1108.
66. Jiang X, Kalajzic Z, Maye P, Braut A, Bellizzi J, et al. (2005) Histological analysis of GFP expression in murine bone. J Histochem Cytochem 53: 593–602.
67. Artzi Z, Weinreb M, Carmeli G, Lev-Dor R, Dard M, et al. (2008) Histomorphometric assessment of bone formation in sinus augmentation utilizing a combination of autogenous and hydroxyapatite/biphasic tricalcium phosphate graft materials: at 6 and 9 months in humans. Clin Oral Implants Res 19: 686–692.
68. Cargnoni A, Di Marcello M, Campagnol M, Nassuato C, Albertini A, et al. (2009) Amniotic membrane patching promotes ischemic rat heart repair. Cell Transplant 18: 1147–1159.
69. Martelli A, Palmerini MG, Russo V, Rinaldi C, Bernabò N, et al. (2009) Blood vessel remodeling in pig ovarian follicles during the periovulatory period: an immunohistochemistry and SEM-corrosion casting study. Reprod Biol Endocrinol 7: 72.
70. Barboni B, Martelli A, Berardinelli P, Russo V, Turriani M, et al. (2004) Ovarian follicle vascularization in fasted pig. Theriogenology 62: 943–957.
71. Crespi R, Mariani E, Benasciutti E, Capparè P, Cenci S, et al. (2009) Magnesium-enriched hydroxyapatite versus autologous bone in maxillary sinus grafting: combining histomorphometry with osteoblast gene expression profiles ex vivo. J Periodontol 80: 586–593.
72. Roach P, Eglin D, Rohde K, Perry CC (2007) Modern biomaterials: a review - bulk properties and implications of surface modifications. J Mater Sci Mater Med 18: 1263–1277.
73. Hollister SJ, Maddox RD, Taboas JM (2002) Optimal design and fabrication of scaffolds to mimic tissue properties and satisfy biological constraints. Biomaterials 23: 409–4103.
74. Graziano A, d'Aquino R, Cusella-De Angelis MG, De Francesco F, Giordano A, et al. (2008) Scaffold's surface geometry significantly affects human stem cell bone tissue engineering. J Cell Physiol Jan 214: 166–172.
75. Huang GT, Gronthos S, Shi S (2009) Mesenchymal stem cells derived from dental tissues vs. those from other sources: their biology and role in regenerative medicine. J Dent Res 88: 792–806.
76. Razzouk S, Schoor R (2011) Mesenchymal stem cells and their challenges for bone regeneration and osseointegration. J Periodontol. 83: 547–550.

77. Bruder SP, Kraus KH, Goldberg VM, Kadiyala S (1998) The effect of implants loaded with autologous mesenchymal stem cells on the healing of canine segmental bone defects. J Bone Joint Surg Am 80: 985–996.
78. Tai K, Pelled G, Sheyn D, Bershteyn A, Han L, et al. (2008) Nanobiomechanics of repair bone regenerated by genetically modified mesenchymal stem cells. Tissue Eng Part A 14: 1709–1720.
79. Zhao L, Weir MD, Xu HH (2010) An injectable calcium phosphate-alginate hydrogel-umbilical cord mesenchymal stem cell paste for bone tissue engineering. Biomaterials 31: 6502–6510.
80. Laino G, d'Aquino R, Graziano A, Lanza V, Carinci F, et al. (2005) A new population of human adult dental pulp stem cells: a useful source of living autologous fibrous bone tissue (LAB). J Bone Miner Res 20:1394–1402.
81. d'Aquino R, De Rosa A, Lanza V, Tirino V, Laino L, et al. (2009) Human mandible bone defect repair by the grafting of dental pulp stem/progenitor cells and collagen sponge biocomplexes. Eur Cell Mater 18: 75–83.
82. Tetè S, Mastrangelo F, Carone L, Nargi E, Costanzo G, et al. (2007) Morphostructural analysis of human follicular stem cells on highly porous bone hydroxyapatite scaffold. Int J Immunopathol Pharmacol 20: 819–826.
83. Parolini O, Alviano F, Bergwerf I, Boraschi D, De Bari C, et al. (2010) Toward cell therapy using placenta-derived cells: disease mechanisms, cell biology, preclinical studies, and regulatory aspects at the round table. Stem Cells Dev 19: 143–154.
84. Manuelpillai U, Moodley Y, Borlongan CV, Parolini O (2011) Amniotic membrane and amniotic cells: potential therapeutic tools to combat tissue inflammation and fibrosis? Placenta 32: S320–S325.
85. Piattelli A, Scarano A, Mangano C (1996) Clinical and histologic aspects of biphasic calcium phosphate ceramic (BCP) used in connection with implant placement. Biomaterials 17: 1767–1770.
86. Ripamonti PW, Richter RWN, Nilen L, Renton J (2008) The induction of bone formation by smart biphasic hydroxyapatite tricalcium phosphate biomimetic matrices in the non-human primate Papio ursinus. J Cell Mol Med 3: 1–15.

Repair of Abdominal Wall Defects with Biodegradable Laminar Prostheses: Polymeric or Biological?

Gemma Pascual[2], Sandra Sotomayor[2], Marta Rodríguez[1], Bárbara Pérez-Köhler[2], Juan M. Bellón[1]*

1 Department of Surgery, Networking Research Center on Bioengineering, Biomaterials and Nanomedicine (CIBER-BBN), Faculty of Medicine, Alcalá University, Alcalá de Henares, Madrid, Spain, **2** Department of Medical Specialties, Networking Research Center on Bioengineering, Biomaterials and Nanomedicine (CIBER-BBN), Faculty of Medicine, Alcalá University, Alcalá de Henares, Madrid, Spain

Abstract

Introduction: Biological and synthetic laminar absorbable prostheses are available for the repair of hernia defects in the abdominal wall. They share the important feature of being gradually degraded in the host, resulting in place the formation of a neotissue. This study was designed to assess the host tissue's incorporation of collagen bioprostheses and a synthetic absorbable prosthesis.

Methods: Partial defects were created in the abdominal walls of 72 New Zealand rabbits and repaired using collagen bioprostheses Tutomesh® and Strattice® or a synthetic prosthesis Bio-A®. Specimens were collected for light microscopy, collagens gene and protein expression, macrophage response and biomechanical resistance at 14, 30, 90 and 180 days post-implantation.

Results: Tutomesh® and Bio-A® were gradually infiltrated by the host tissue and almost completely degraded by 180 days post-implantation. In contrast, Strattice® exhibited material encapsulation, no prosthetic degradation and low cell infiltration at earlier timepoints, whereas at later study time, collagen deposition could be observed within the mesh. In the short term, Bio-A® exhibited higher level of collagen 1 and 3 mRNA expression compared with the two other biological prostheses, which exhibited two peaks of higher expression at 14 and 90 days. The expression of collagen III was homogeneous throughout the study and collagen I deposition was more evident in Strattice®. Macrophage response decreased over time in biomeshes. However, in the synthetic mesh remained high and homogeneous until 90 days. The biomechanical analysis demonstrated the progressively increasing tensile strength of all biomaterials.

Conclusions: The tissue infiltration of laminar absorbable prostheses is affected by the structure and composition of the mesh. The synthetic prosthesis exhibited a distinct pattern of tissue incorporation and a greater macrophage response than did the biological prostheses. Of all of the laminar, absorbable biomaterials that were tested in this study, Strattice® demonstrated the optimal levels of integration and degradation.

Editor: Samuel J. Lin, Harvard Medical School, United States of America

Funding: This study was supported by a grant from the Fundación Mutua Madrileña 2008 (FMM08), Madrid, Spain, and by the Spanish Ministry of Science and Technology through the research project DPI2011-27939-C02-02. The funders had no role in study design, data collection and analysis, decision to publish, or preparation of the manuscript.

Competing Interests: The authors have declared that no competing interests exist.

* E-mail: juanm.bellon@uah.es

Introduction

The study and development of prosthetic materials for the repair of abdominal wall defects has evolved and progressed during the past several years with the ultimate goal of discovering the "ideal prosthesis."

The classic polymeric materials (such as polyester, polypropylene and expanded polytetrafluoroethylene), despite providing satisfactory results, have been replaced by materials of natural origin, the latter mainly from animal sources. These implants, called "biomeshes," which are primarily composed of collagen, can not only repair but can also regenerate new tissue that is similar to that of the human recipient [1]. During this process, the biomeshes undergo a progressive degradation in the host. Two types of materials exist: those with cross-links that stabilize the collagen molecule, thus preventing its rapid degradation, and those noncrosslinked, which undergo a progressive and variable degradation over time [2]. Clearly, the process for which these prostheses are designed is not feasible for the majority of the synthetic polymeric prostheses, which remain for life in the recipient organism; in certain instances, these synthetic prostheses elicit inflammatory and foreign body reactions with the potential for more diverse post-implant complications [3].

The advantage of using biomeshes is that the repair mechanisms approach optimal conditions. However, there may also be inconveniences, including adverse effects that have been described after implantation [4,5]. One of the areas for improvement and research is the control of the prosthetic degradation times, particularly of noncross-linked prostheses.

Alternatively, polymeric, midterm biodegradable materials have emerged, which are indicated more for reinforcement than for tissue replacement. These materials constitute a commitment to the future within the scope of new prosthetic developments. *Bio-A*® is composed of polyglycolic acid and trimethylene carbonate. These polymers are widely known for their biocompatibility and have been predominantly used in sutures. Prior experience with this prosthesis is very limited; one *in vitro* study compares its elicited immune reaction in humans with those of biological prostheses [6], and another study relates to hernia repair [7].

Considering the hypothesis that biodegradable synthetic laminar prostheses provide advantages over certain collagen biomeshes, our objective was to study the behavior of this new prosthetic material. Noting its biodegradation characteristics (3–6 months), we compared the synthetic laminar prosthesis with collagen noncross-linked bioprostheses; our goal was to evaluate the repair and/or regenerative capacity at the receptor-tissue level in a partial abdominal wall defect model.

Materials and Methods

Experimental Animals

The experimental animals included 72 male New Zealand White rabbits (weighing approximately 2500 g). This study was carried out in strict accordance with the recommendations in the Guide for the Care and Use of Laboratory Animals of the National and European Institutes of Health (Spanish law 32/2007, Spanish Royal Decree 1201/2005, European Directive 2010/63/UE and European Convention of the Council of Europe ETS123). All the procedures were performed at the Animal Research Center of Alcalá University. The protocol was approved by the Committee on the Ethics of Animal Experiments of the University of Alcalá (registered code: ES280050001165).

Prosthetic Materials

The following biomaterials were used. Characterization, before the implant was performed by scanning electron microscopy and Sirius red staining (Fig. 1):

- *Bio-A*® *(Bio-A)* (W.L. Gore, Flagstaff, AZ, USA): a synthetic laminar bioabsorbable material (1.3 mm thick), composed of polyglycolic acid:trimethylene carbonate (PGA:TMC) fibers.
- *Tutomesh*® *(Tuto)* (Tutogen Medical GmbH, Nümberg, Germany): a xenogenic collagen I membrane (0.5 mm) from bovine pericardium (noncross-links).
- *Strattice*® *(St)* (LifeCell Corporation, Branchburg, NJ, USA): a porcine dermal, biological noncross-linked tissue matrix (1.2 mm thick).

Surgical Technique

To minimize pain, all of the animals were administered 0.05 mg/kg buprenorphine (Buprecare®, Divasa Farmavic, Barcelona, Spain) 1 hour before and 3 days after the surgical procedure. Anesthesia was induced with a mixture of ketamine hydrochloride (Ketolar, Parke-Davis, Spain) (70 mg/kg), diazepam (Valium, Roche, Spain) (1.5 mg/kg) and chlorpromazine (Largactil, Rhone-Poulenc, Spain) (1.5 mg/kg) administered intramuscularly.

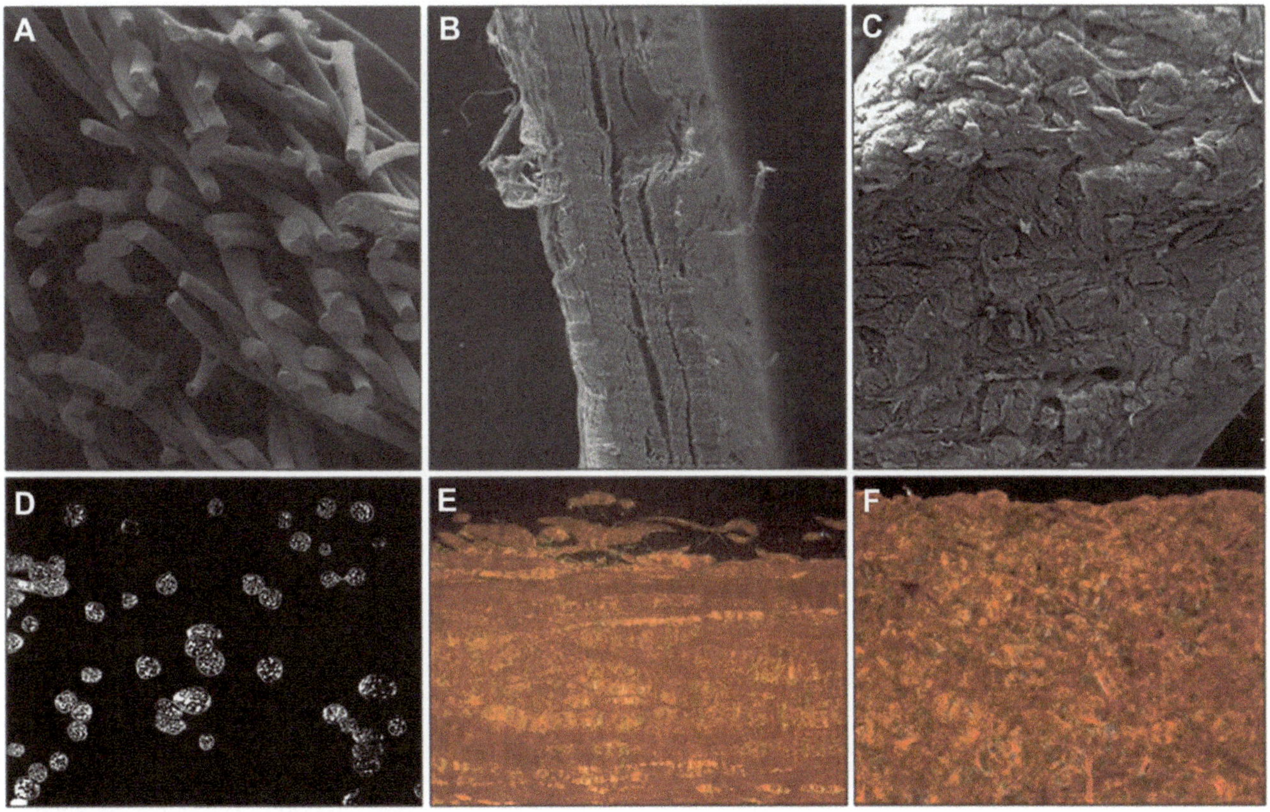

Figure 1. Used biomaterials. Scanning electron microscopy images (100×) showing the aspect and the thickness of *Bio-A* (A), *Tuto* (B) and *St* (C). Polarized light images, with collagen fibers displayed in red after Sirius Red staining (200×). *Bio-A* (D), *Tuto* (E) and *St* (F).

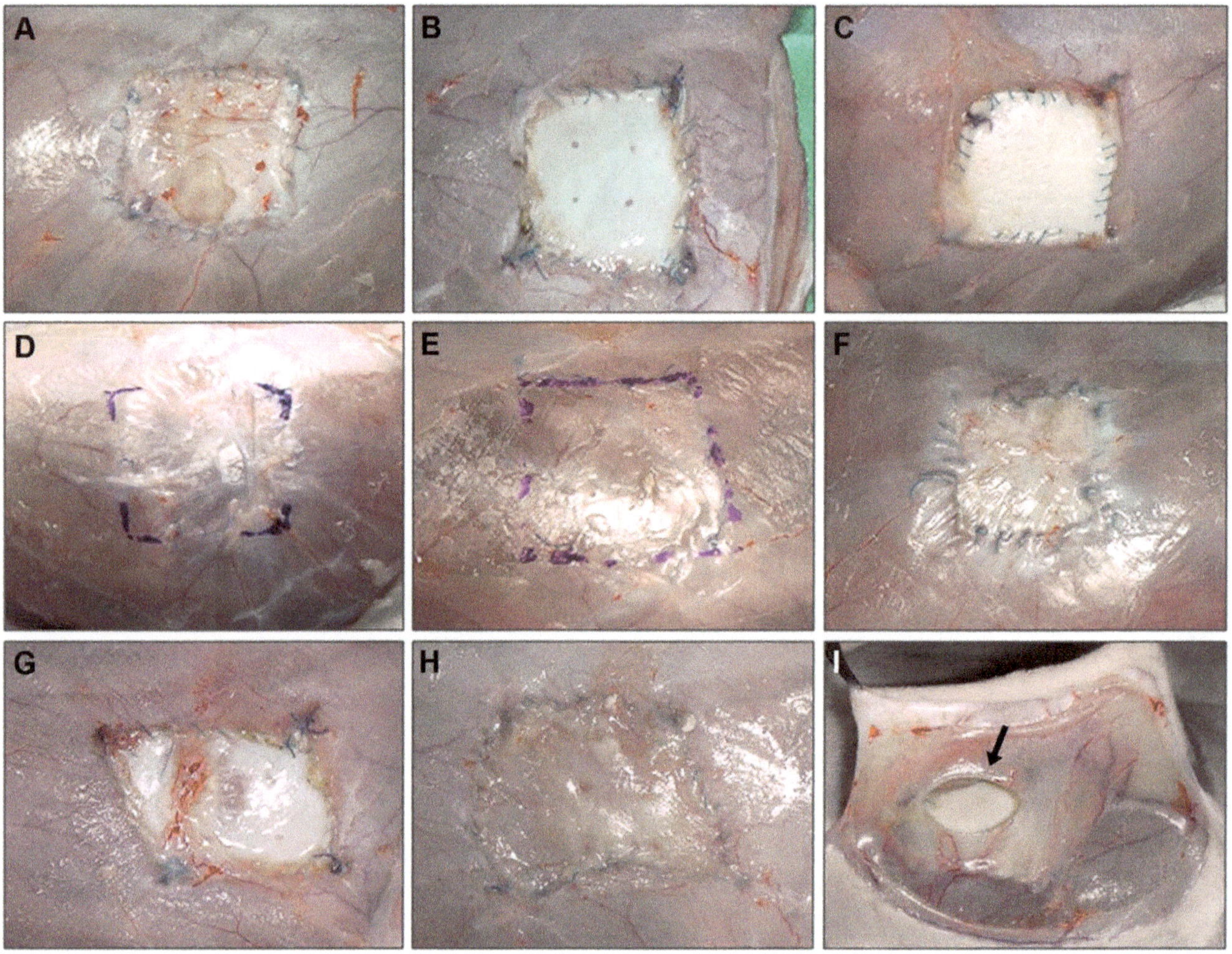

Figure 2. Macroscopic biomaterial images. *Bio-A* implants after 14 days (A) and 30 days postimplantion (D). *Tuto,* 14 days (B) and 30 days (E). *St,* 14 days (C) and 30 days (F). Partial degradation of *Tuto* at 14 (G) and 30 days (H). *St* encapsulation (arrow) at 30 days post-implantation (I).

Using a sterile surgical technique and after making an approximately 6-cm-long incision in the skin, 3×3-cm defects were created in the lateral wall (right side) of the abdomen, transecting the planes of the external and internal oblique muscles and sparing the transversalis muscle and parietal peritoneum. The defects were then repaired by fixing a mesh of the same size to the edges of the defect, using a 4/0 polypropylene running suture interrupted at the four corners. The skin was closed using a 3/0 polypropylene running suture.

During all the study, the animals were visually inspected for signs of dehiscence of the skin wound, seroma formation, wound infection and/or mesh incompatibility.

Experimental Design

A total of 72 implants, divided into three study groups, were made:

- *Bio-A®* (n = 24): synthetic laminar bioabsorbable material, composed of PGA:TMC fibers.
- *Tutomesh®* (n = 24): xenogenic collagen I membrane from bovine pericardium.
- *Strattice®* (n = 24): porcine dermal, biological tissue matrix.

The animals were sacrificed following the protocols for experimental animal euthanasia, in a CO_2 chamber at 14, 30, 90 and 180 days after implant. Study times were established with the objective of carrying out a proper follow-up of the evolution of the prosthesis, once implanted in the animal, from short to long term. Specimens of each prosthesis with some surrounding host tissue, were obtained for the different analysis. From each sample, one 1.5 cm wide and 5 cm long strip (taking the 3 cm of the sample and an additional 1 cm on each side of receptor tissue), were used for biomechanical studies. The rest of the sample was used for histology and molecular biology.

Morphological Analyses

For light microscopy, the specimens were stained with Massons trichrome (Goldner-Gabe) staining and examined under a Zeiss light microscope (Carl Zeiss, Oberkochen, Germany).

RNA Isolation and Real-time PCR (RT-PCR)

Tissue fragments that were 1 cm^2 in size were obtained from the implant area and stored at $-80°C$ until use. The RNA was extracted by guanidine-phenol-chloroform isothiocyanate procedures using TRIzol (Invitrogen, Carlsbad, CA, USA). The RNA was recovered from the aqueous phase by precipitation. The

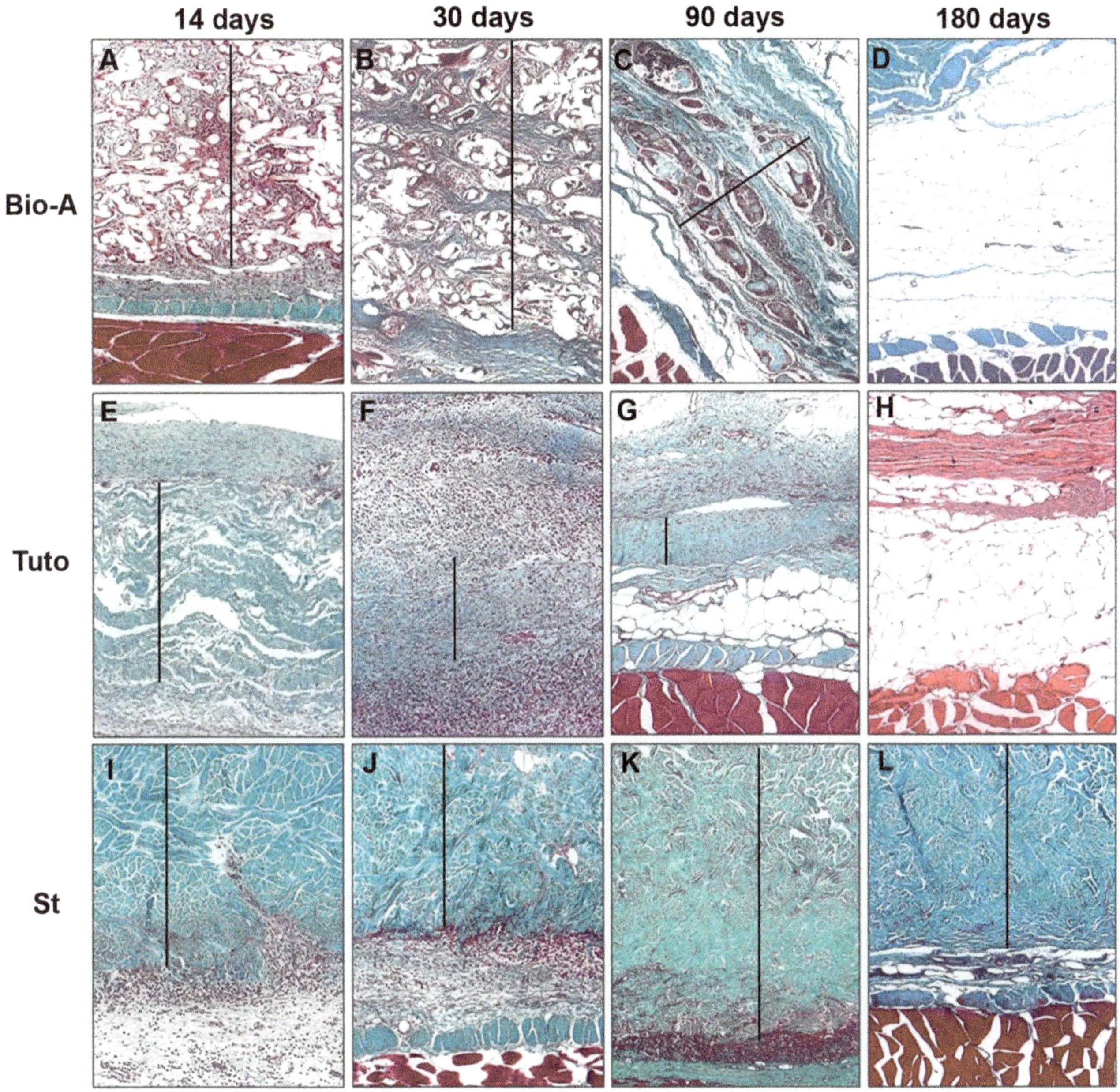

Figure 3. Light microscopy images. Tissue integration and prostheses degradation in the different timepoints (100×). *Bio-A* (A–D) and *Tuto* (E–H) showed a gradual infiltration of host tissue and was completely degraded by 180 days post-implantation. *St* exhibited material encapsulation, without signs of degradation and low cell infiltration at 180 days (I–L). (– Prosthesis).

amount and purity were measured using the optical density at 260/280 nm and 260/230 nm in a NanoDrop ND-1000 spectrophotometer (Thermo Fisher Scientific Inc., DE, USA).

Complementary DNA was synthesized from 200 ng of the total RNA by reverse transcription (RT) with oligo dT primers (Amersham, Fairfield, USA) and the M-MLV reverse transcriptase enzyme (Invitrogen). The cDNAs were amplified using the following primers: collagen 1 (col 1) (sense 5′-GAT GCG TTC CAG TTC GAG TA-3′ and antisense 5′-GGT CTT CCG GTG GTC TTG TA-3′), collagen 3 (col 3) (sense 5′-TTA TAA ACC AAC CTC TTC CT-3′ and antisense 5′-TAT TAT AGC ACC ATT GAG AC-3′) and GAPDH (sense 5′-TCA CCA TCT TCC AGG AGC GA-3′ and antisense 5′-CAC AAT GCC GAA GTG GTC GT-3′).

The RT-PCR mixture contained 5 μl of the inverse transcription product (cDNA) diluted 1:20, 10 μl of iQ SYBR Green Supermix (Bio-Rad Laboratories, Hercules, CA, USA) and 1 μl (6 μM) of each primer in a final reaction volume of 20 μl. The RT-PCR was performed in a StepOnePlus Real-Time PCR System (Applied Biosystems, Foster City, California, USA). The samples were subjected to an initial stage of 10 minutes at 95°C. The conditions for cDNA amplification were as follows: 40 cycles of 95°C for 15 s, 60°C (col 1 and 3) or 55°C (GAPDH) for 30 s and 72°C for 1 minute. The products were subjected to 2% agarose gel electrophoresis, stained with a SYBR Green II RNA gel stain (Invitrogen) and visualized by UV light. The gene expression was normalized against the expression recorded for the

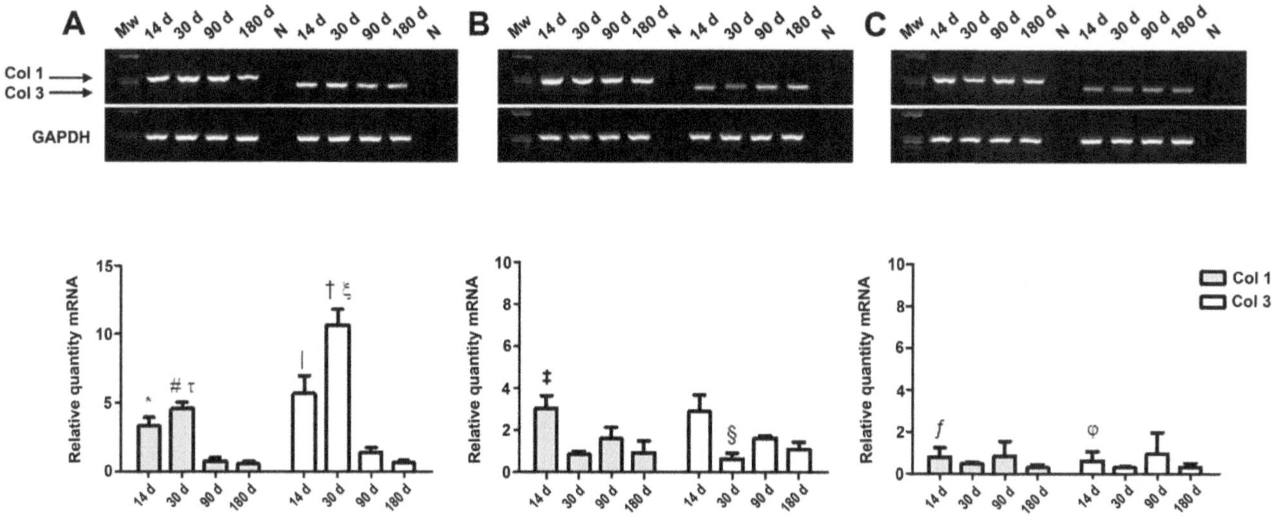

Figure 4. Collagen 1 and 3 mRNA expression determined by RT-PCR. Agarose gel product and relative mRNA quantity of *Bio-A* (A), *Tuto* (B) and *St* (C) after 14, 30, 90 and 180 days post-implantation. The results are expressed as the mean ± SEM of three experiments. Gene expression was normalized with the GAPDH gene. A) *Bio-A*: Collagen (Col) 1: *, vs. 90 days ($P<0.05$) and 180 days ($P<0.01$); #, vs. 90 days ($P<0.05$) and 180 days ($P<0.01$); τ, vs. *Tuto* ($P<0.05$) and *St* ($P<0.01$) at 30 days. Col 3: |, vs. 30 days and 90 days ($P<0.05$) and 180 days ($P<0.001$); †, vs. 90 days ($P<0.05$) and 180 days ($P<0.001$); ξ, vs. *Tuto* ($P<0.05$) and *St* ($P<0.01$) at 30 days. B) *Tuto*: Col 1: ‡, vs. 30 days ($P<0.05$). Col 3: §, vs. 14 and 90 days ($P<0.05$). C) *St*: Col 1: *f*, vs. *Bio-A* and *Tuto* at 14 days ($P<0.01$). (N = negative control; Mw = molecular weight markers).

constitutive gene glyceraldehyde 3-phosphate-dehydrogenase (GAPDH).

Immunofluorescence Microscopy/collagen Expression

The collagen content was detected by immunofluorescence. The monoclonal antibodies anti-collagen I (Sigma, St. Louis, MO, USA) and anti-collagen III (Medicorp, Montreal, Canada) were used as primary antibodies. A secondary antibody conjugated with rhodamine was used in the study. The negative controls were subjected to 3% BSA instead of the primary antibody. The cell nuclei were counterstained with DAPI. The samples were examined under a Leica SP5 confocal microscope (Leica Microsystems, Wetzlar, ***Germany***) to detect fluorescence.

Immunohistochemistry/macrophage Response

For immunohistochemistry, a specific monoclonal antibody to rabbit macrophages, RAM-11 (DAKO M-633, USA), was used as primary antibody. The antigen-antibody reaction was detected by alkaline phosphatase avidin-biotin procedures. The chromogenic substrate contained alpha-napthol and fast-red. The nuclei were counterstained with acid hematoxylin. RAM-11-labeled macrophages were quantified by performing counts in 20 microscopic fields (×20) for each biomaterial.

Biomechanical Strength

To determine the biomechanical strength and modulus of elasticity of the meshes after implant, strips of the different biomaterials (1.5 cm wide and 5 cm long) including the mesh and the infiltrated host tissue, were analyzed using an INSTRON 3340 testing system (static load 500N) (Instron Corp., UK). The crosshead speed was 5 cm/minute, and the recording speed was 2 cm/minute.

All of the tests were conducted immediately after the animals were sacrificed at the different study times.

Statistical Analysis

The statistical analysis was performed using the Graph Pad Prism 5 package (GraphPad Soft-ware, Inc., La Jolla, CA, USA). The percentages of mRNA expression and RAM-11 positive cells, as well as the measurements of biomechanical strength, were compared among the three study groups using the Mann-Whitney U test. The results were expressed as the mean ± SEM. The level of statistical significance was set at $P<0.05$.

Results

Macroscopic Analysis

At 14 days, seroma formation was observed in three of the six *Bio-A* implants. At 30 days, the biomaterial presented areas of degradation. By 90 days, the prosthesis had almost completely disappeared. After 6 months, the area of the implant was covered by newly formed tissue and appeared to be partially distended.

No seromas were observed in *Tuto*, but signs of degradation of this biomaterial were observed in 2 of the samples at 14 days. 30 days after implant at least half of the samples showed signs of degradation and/or disinsertion of the biomaterial. By 180 days, the prosthesis had completely disappeared, and the implant area appeared to be distended.

In the implants using *St*, seromas were observed in two of the six implants at 2 weeks and in one at 30 days. No signs of prosthetic degradation were observed at any of the timepoints (Fig. 2).

Morphology

At 14 days and 30 days (Fig. 3A and B), the filaments of the *Bio-A* prostheses were surrounded by an intense inflammatory reaction and loose connective tissue. At 30 days, the polymer showed initial signs of degradation, and the neoformed tissue was growing denser. At 3 months (Fig. 3C), an evident degradation of the filaments was observed; these areas were occupied by connective tissue composed of large bundles of collagen fibers. The filaments appeared to be surrounded by inflammatory cells. By 6 months

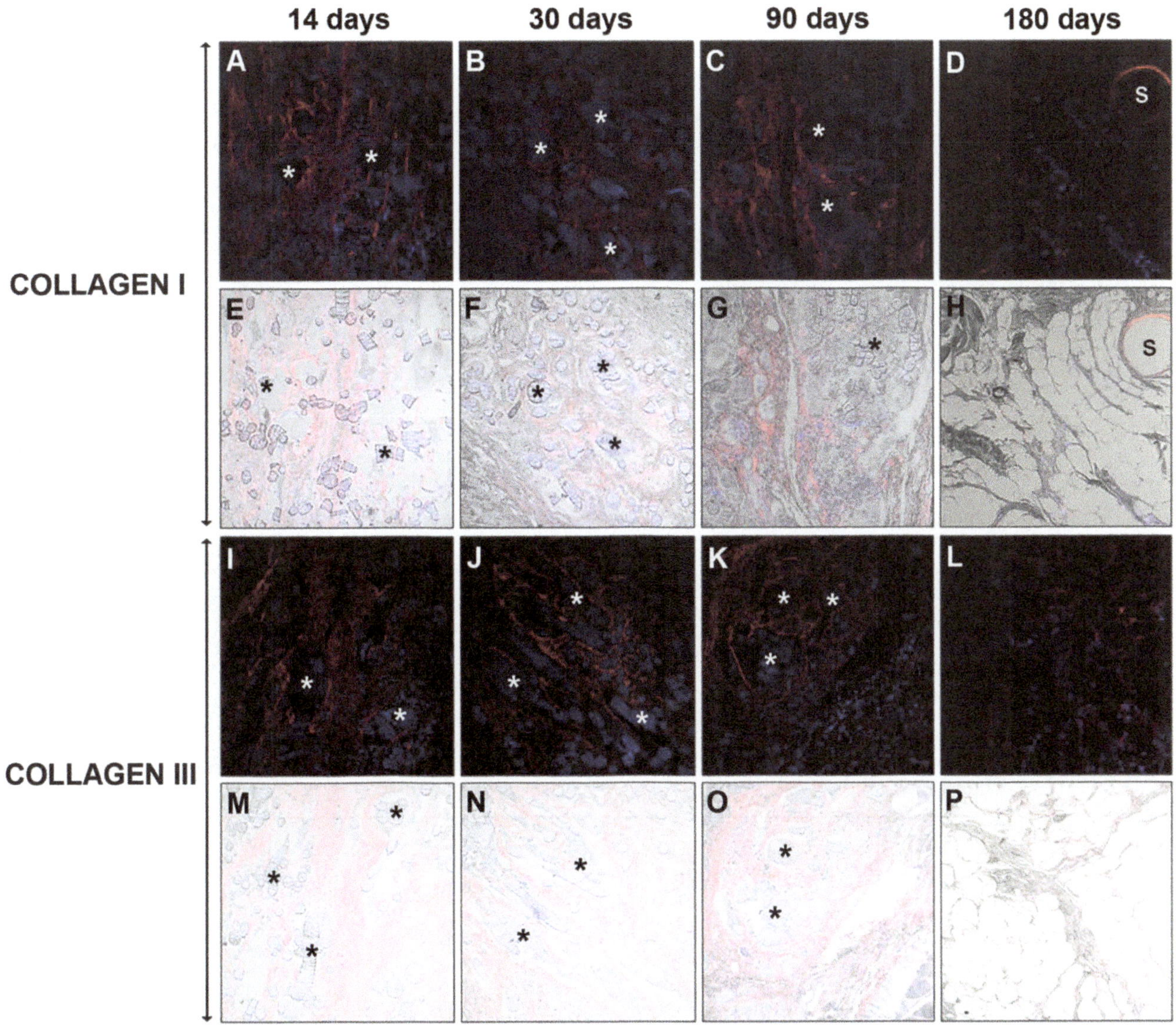

Figure 5. Immunodetection of neoformed collagen I and III in *Bio-A*®. Mature Collagen I (A–H) and immature Collagen III (I–P) immunofluorescence at 14, 30, 90 and 180 days post-implantation. The neoformed collagen appears in red and the cell nuclei (stained with DAPI) appear in blue. The DIC images that identify the biomaterial appear translucent (E–H and M–P). Confocal light microscopy (200×). (* Synthetic mesh).

post-implantation (Fig. 3D), the prosthetic components–as well as the inflammatory cells–had disappeared, and the connective tissue had been almost completely substituted by adipose tissue.

At 2 weeks (Fig. 3E), the *Tuto* prosthesis appeared to be infiltrated by a small population of receptor-tissue cells throughout its thickness. A moderate inflammatory response occurred, which significantly increased at 30 days (Fig. 3F). This response decreased significantly at 90 days post-implantation (Fig. 3G), along with a significant degradation of the prosthesis and a decrease in its thickness. By 180 days (Fig. 3H), as in the case of *Bio-A*, the biomaterial was completely reabsorbed and replaced by adipose tissue and areas of dense connective tissue.

The *St* implant was initially observed to be surrounded by a capsule of highly vascularized connective tissue and to present an intense inflammatory reaction (Fig. 3I,J). 90 days after implant, the neoformed connective tissue and the inflammatory cells colonized the inferior and superior thirds of the bioprosthesis (Fig. 3K). No evident signs of reabsorption of the collagen lamina were observed at the different timepoints. Subsequently, 180 days after implant, a significant decrease in the inflammatory reaction was observed (Fig. 3L), compared with the previous timepoints (14, 30 and 90 days) where the inflammatory reaction was more intense.

Real-time RT-PCR

We studied the mRNA expression of collagens 1 and 3 in the neoformed tissue (Fig. 4). The synthetic mesh exhibited a higher expression of both collagens versus the biological prostheses. The behavior of the immature (collagen 3) and mature (collagen 1) was similar and independent of the type of prosthesis used.

The pattern of mRNA expression for both types of collagen was similar in the two bioprostheses but completely different in the *Bio-A* implant. *Bio-A* exhibited a significant increase in collagen 1 and 3 mRNA expression at 14 and 30 days, that significantly decreased at later timepoints. However, the biological meshes yielded two peaks demonstrating the higher expression of both types of collagen at 14 and 90 days.

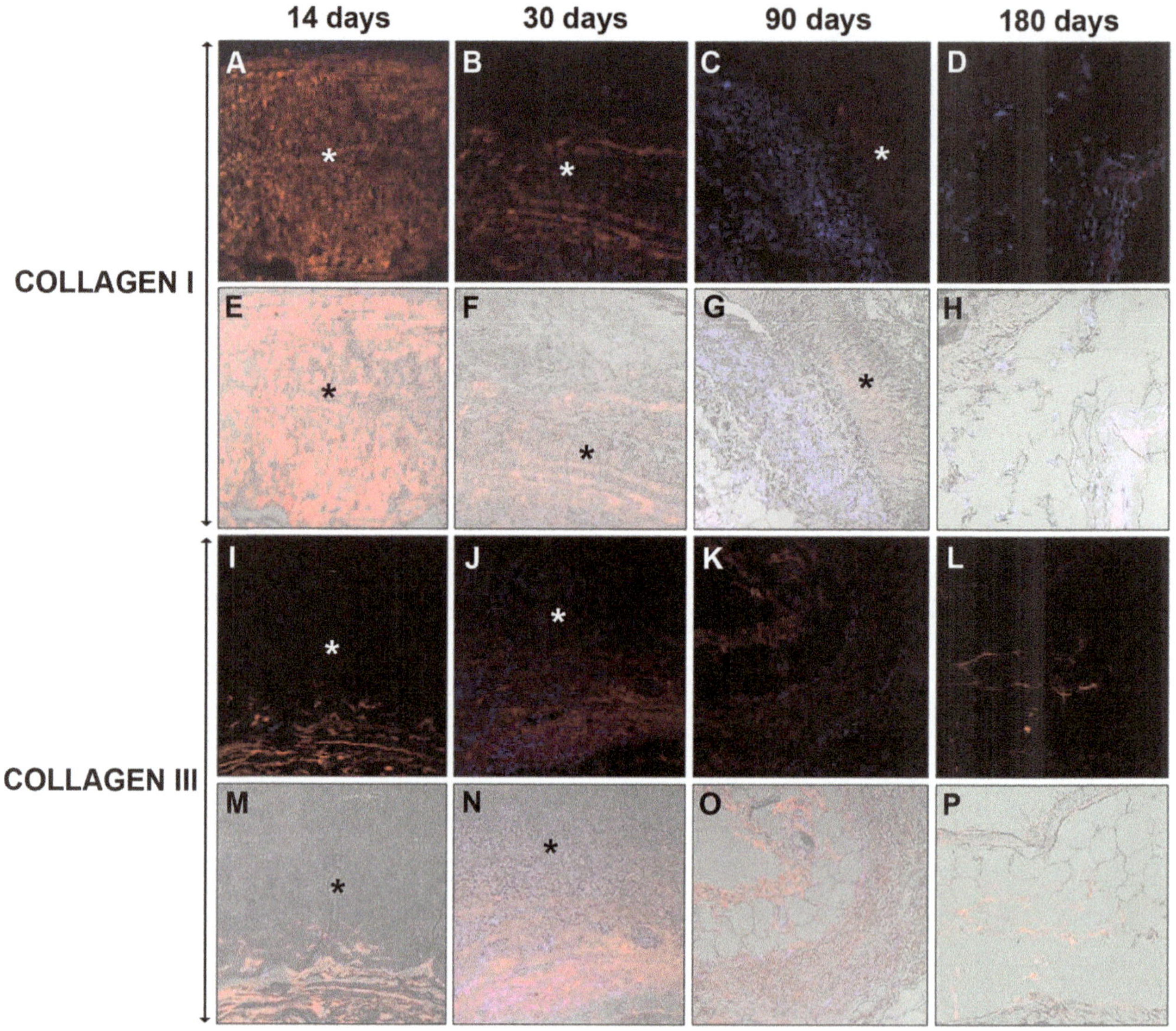

Figure 6. Immunodetection of neoformed collagen I and III in *Tutomesh*®. Mature Collagen I (A–H) and immature Collagen III (I–P) immunofluorescence at 14, 30, 90 and 180 days post-implantation. The neoformed collagen appears in red and the cell nuclei (stained with DAPI) appear in blue. The DIC images that identify the biomaterial appear translucent (E–H and M–P). Confocal light microscopy (200×). (* Biomesh).

For *Tuto*, the expression of collagen 1 and 3 at 14 days was significantly higher than at 30 days, and the collagen 3 mRNA demonstrated differences between 30 and 90 days. For *St*, we did not observe any significant difference over time.

When the different types of meshes were compared at 14 and 30 days, *Bio-A* exhibited a greater collagen 1 and 3 mRNA expression compared with the two other biological prostheses. Significant differences were observed at 14 days in both types of collagens when *Bio-A* and *St* were compared and when *St* was compared with *Tuto*. At 30 days, significant differences were observed in both types of collagens when the synthetic mesh was compared with the two bioprostheses. Any significant differences were observed at 90/180 days.

Immunofluorescence Microscopy/collagen Expression

To identify the neoformation of native collagen in the bioprostheses, we generated differential interference contrast (DIC) images, superimposing the immunofluorescence confocal images. Consequently, the new collagen was labeled by red fluorescence, and the collagen or synthetic prostheses appeared translucent.

In the *Bio-A*, the expression of mature collagen was first observed at 14 days post-implantation, around the filaments of the prosthesis. The labeling was maintained between 30 and 90 days, but a decrease occurred at 180 days, when the adipose tissue occupied the greatest area of the implant. The intensity of the collagen III labeling was similar at all of the timepoints and was localized to the tissue invading the prosthesis. At 180 days, the collagen III was mainly found around the adipose tissue (Fig. 5).

In *Tuto*, the intensity of the collagen I labeling was greater at the earlier timepoints and subsequently decreased. It was confirmed that the unimplanted prosthesis exhibited an immunoreactivity for collagen I similar to that observed 14 days post-implantation, which prevented the differentiation of the prosthetic collagen I from the newly formed native collagen.

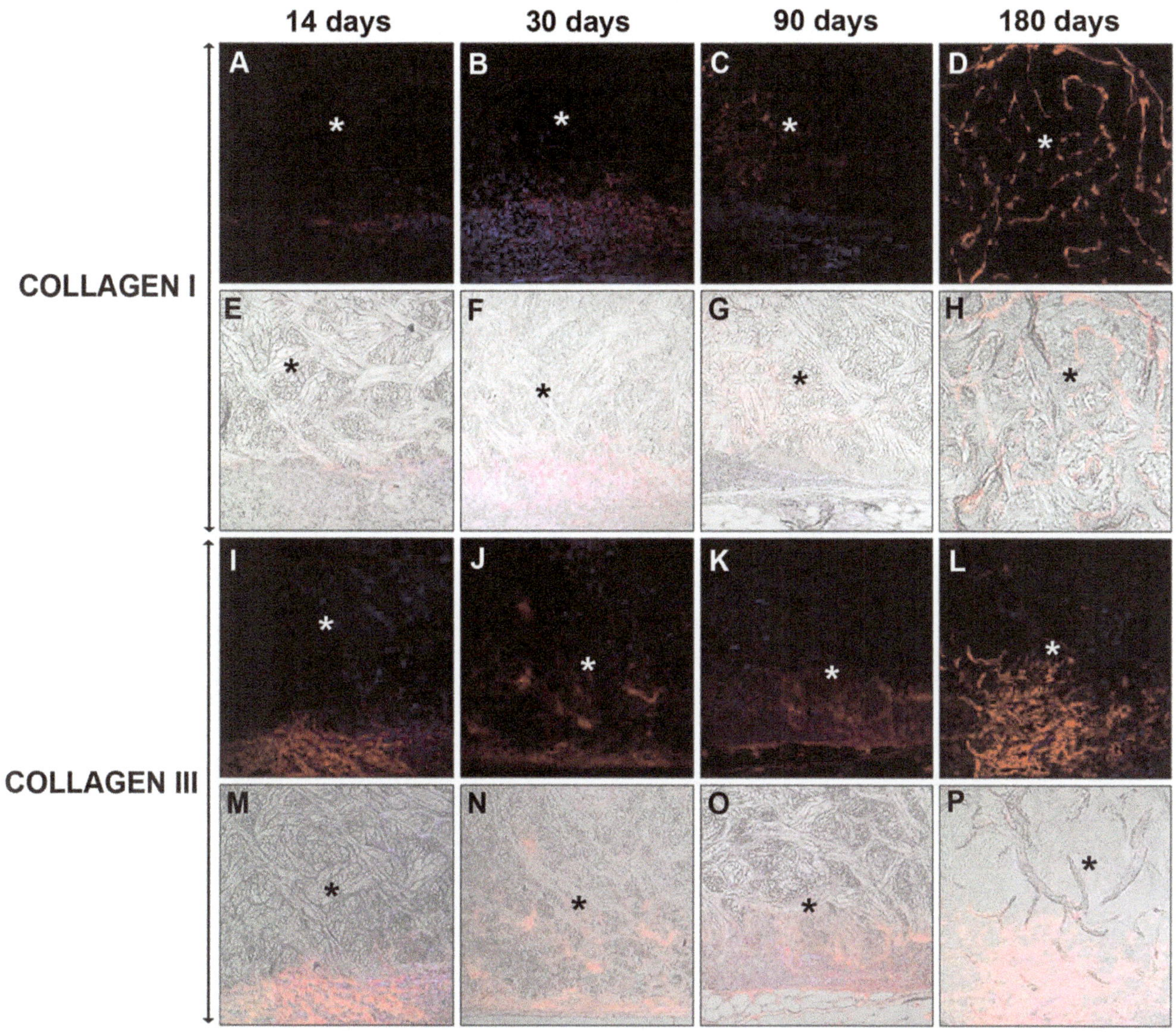

Figure 7. Immunodetection of the neoformed collagen I and III in *Strattice*®. Mature Collagen I (A–H) and immature Collagen III (I–P) immunofluorescence at 14, 30, 90 and 180 days post-implantation. The neoformed collagen appears in red and the cell nuclei (stained with DAPI) appear in blue. The DIC images that identify the biomaterial appear translucent (E–H and M–P). Confocal light microscopy (200×). (* Biomesh).

Compared with collagen I, the expression of collagen III was predominant, as determined by the more intense labeling of the latter at all of the timepoints (Fig. 6).

In *St*, collagen I expression increased in the long term. At 14 days, the fluorescence was faint and limited to the capsule of tissue that formed between the prosthesis and the muscle. No labeling was observed in the interior of the prosthesis. At day 30, the labeling was observed in certain areas of the interior of the prosthesis that had been infiltrated by the neoformed tissue; moreover, there was increased labeling of the capsular tissue. By 180 days, the fluorescence intensity had increased inside the prosthesis. The expression of collagen III was evident at all of the timepoints and followed a pattern of expression that was similar to that of collagen type I (Fig. 7).

Immunohistochemistry/macrophage Response

Both collagen meshes elicited a similar expression of macrophages, which decreased over time. However, the synthetic mesh exhibited a different pattern of expression, which remained at high and homogeneous levels until 90 days post-implantation, then decreased significantly thereafter. The *Bio-A* synthetic prosthesis presented a significantly greater percentage of positive cells compared with the biological prostheses (except for the *Tuto* at 14 days). The active macrophages primarily formed foreign-body giant cells around the absorbable filaments.

Both biological meshes exhibited a short-term localization of positive cells to the inflammatory tissue that formed between the prosthesis and the recipient muscle; at later timepoints, these cells formed visible colonies within the prosthesis. Subsequently, this response significantly decreased, almost disappearing in the case of *St* at 180 days (Fig. 8).

Biomechanics

At the earliest timepoints (14 and 30 days), the resistance to breakage of the different prostheses was similar. A slight increase was observed at 90 days, with no significant differences prior to

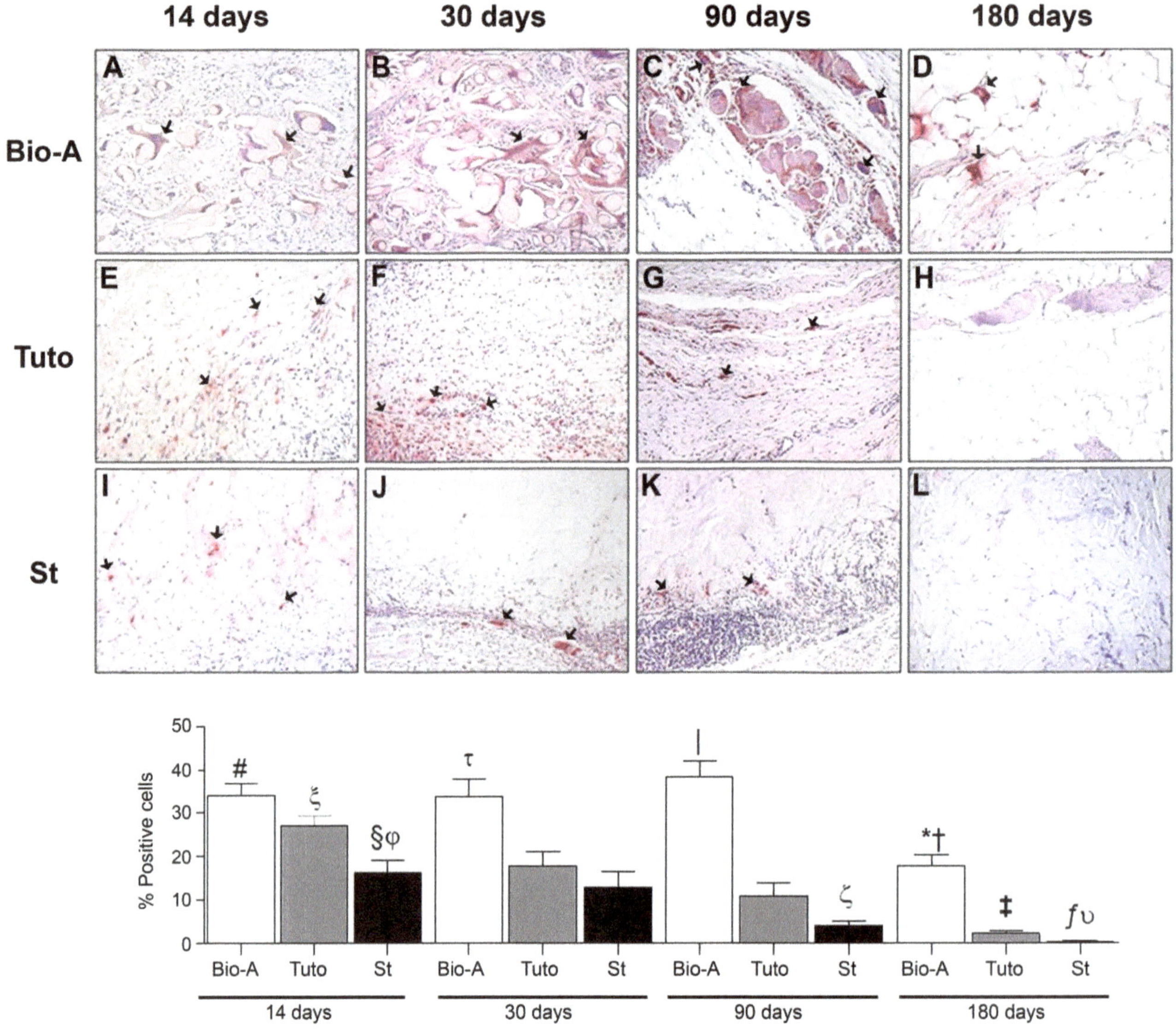

Figure 8. Foreign-body reaction of the different meshes. Immunohistochemical labeling of rabbit macrophages (red color, arrows) using the RAM-11 monoclonal antibody (200×) (top panel): *Bio-A* (A–D), *Tuto* (E-H) and *St* (I–L). Percentage of positive cells for the RAM-11 antibody in the different prostheses after 14, 30, 90 and 180 days post-implantation (bottom panel). The results were expressed as the mean ± SEM. A) *Bio-A*: *, vs. 14 days and 90 days ($P<0.001$) and 30 days ($P<0.01$); #, vs. *St* ($P<0.001$) at 14 days; τ, vs. *Tuto* and *St* ($P<0.01$) at 30 days; |, vs. *Tuto* and *St* ($P<0.001$) at 90 days; †, *Tuto* and *St* ($P<0.001$) at 180 days post-implantion. B) *Tuto*: ξ, vs. 30 days ($P<0.05$) and 90 days ($P<0.001$); ‡, vs. 14 days and 30 days ($P<0.001$) and 90 days ($P<0.01$). C) *St*: §, vs. 90 days ($P<0.001$); *f*, vs. 14, 30 and 90 days ($P<0.001$); ϕ, vs. *Tuto* ($P<0.01$) at 14 days; ξ, vs. *Tuto* ($P<0.05$) at 90 days; υ, vs. *Tuto* ($P<0.001$) at 180 days post-implantation.

this timepoint. The greatest gain in biomechanical resistance was reached at 180 days, when resistance to breakage was significantly increased in *Bio-A* ($p<0.01$) and *St* ($p<0.05$) implants at 14 and 30 days (Fig. 9).

Discussion

The use of collagen-derived prostheses for elective hernia repair are feasible alternatives in contaminated abdominal defects for which synthetic prostheses are contraindicated [8]. However, the most important limitations to the clinical use of collagen-derived prostheses are their costs and the cultural groups of patients in whom they may be implanted [9].

The goal of using these biomaterials in tissue-engineering is to achieve not only repair but also tissue regeneration [1]. To this end, once implanted, the biomaterials promote angiogenesis and the formation of new tissue, which are processes involving the stimulation of growth factors and the synthesis of extracellular matrix elements.

The ideal characteristic of biomeshes is that they do not rapidly degrade, remaining stable until they are gradually and fully incorporated into the recipient tissue. For this function, it is necessary for the triple-helix links that constitute the collagen molecule to be efficient; otherwise, the mechanical firmness will be compromised. The success of the repair would thus depend on the balance between the tissue-regeneration processes and the degradation of the prosthesis.

Resembling these collagen-derived biological prostheses, the synthetic prosthesis (*Bio-A*) used in this study is formed by a PGA

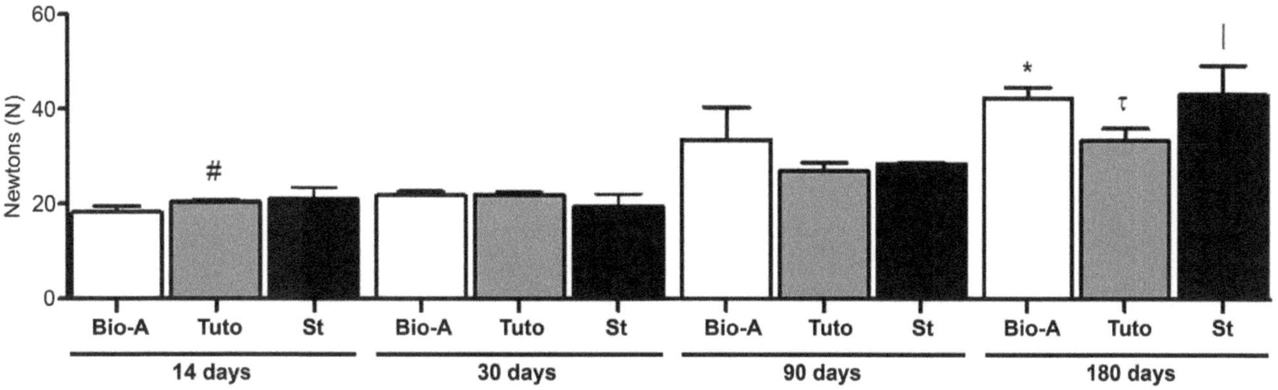

Figure 9. Biomechanical strength of the different meshes. The results (Newtons) were expressed as the mean ± SEM at 14, 30, 90 and 180 days post-implantation. *Bio-A*: *, vs. 14 days and 30 days ($P<0.01$). *Tuto*: #, vs. 90 days ($P<0.01$); τ, vs. 14 days and 90 days ($P<0.05$) and 30 days ($P<0.01$). *St*: |, vs. 14 days and 30 days ($P<0.05$).

and TMC copolymer. This material can undergo a progressive biodegradation through hydrolytic and enzymatic processes that do not leave any permanent residues in the body.

Various investigators have attempted to demonstrate the possible advantages of biomeshes over synthetic prostheses. A recently published *in vitro* study [10] reveals that human mononuclear cells are activated by porcine crosslinked bioprostheses, which induce a greater expression of cytokines compared with those noncross-linked. This finding has been attributed to the processing methods and/or the degree of collagen cross-linking.

St belongs to a new generation of porcine bioprostheses that are derived from the dermis and are processed without chemical cross-links; in it the galactose-alpha 1,3 antigen–which is the greatest elicitor of the immune responses associated with acellular xenografts–has been enzymatically eliminated [11,12]. When implanted into the ventral hernial defects of non-human primates, *St* does not provoke a xenogeneic immune response [12]. This fact was confirmed in the present study, in which *St* elicited the lowest macrophage response, despite being derived from a different animal species.

A recent study [13] has demonstrated that cross-links materials may be more durable during the remodeling process, as suggested by the progressive and significant thinning and weakening of *St*. By contrast, we found [14] that the behavior of *St* was similar to that of the other prostheses with similar thickness and cross-links (*Collamend®/Permacol®*) however, over time, we observed a greater cellular infiltration and deposits of neoformed collagen for St. Consistent with our results, the Butler group [15] has tested *Collamend®* and *St* and affirms that these prostheses exhibit rapid tissue and vascular infiltration, indicating the clinical advantages of an abdominal reconstruction that does not compromise the resistance to the implant's traction area. However, compared with bioprostheses of human origin, *Collamend®* and *St* exhibit greater cellular and vascular infiltration [16] and provide stronger support [17].

Tuto, is a non-crosslinked prosthesis created from bovine pericardium that has been used in the repair of multiple structures, such as the pericardium, various types of hernias and the pelvic floor.

An investigation using bovine pericardium [2] in a rat subcutaneous implant has revealed how the degree of crosslinkage determines the index of prosthetic degradation and thus significantly affects the pattern of tissue regeneration. Another experimental study [18] of total abdominal defects (also in rats) has revealed that the bovine pericardium and ePTFE elicits a minimal foreign-body reaction and provides adequate mechanical resistance at 2 weeks post-implantation. These results are consistent with our study, in which both collagen meshes elicited a similar macrophage response that decreased over time. Clinical studies of contaminated tissues [19,20] have demonstrated that the acellular bovine pericardium may be a useful tool for these situations.

Despite the good behavior showed by biological prostheses, several clinical complications like evisceration, disintegration, poor mesh integration, infection or seroma, associated with xenograft biologic mesh implantation in abdominal wall reconstruction, have been described [5]. A systematic review recently published, has stated that wound infection and seroma formation are the most common postoperative complications related to biological prostheses implantation [21].

Bio-A, is a biodegradable material that has been used for applications other than the abdominal wall, such as anal fistulas [22,23]. Regarding hernia repair, a clinical pilot study of 10 patients [7] demonstrated the efficiency of *Bio-A* in this type of surgery. Other authors [24] have reported the suitability of this material in a single clinical case that involved the repair of a large inguinal hernia.

In our experimental study, *Tuto* and *Bio-A* macroscopically exhibited a rapid degradation, being complete between 90 and 180 days post-implantation. This did not occur with the *St* implants, which were significantly more stable over time, remaining intact at 6 months post-implantation. Areas of tissue "relaxation" at the level of the defects could be observed in *Tuto* and *Bio-A*, without herniation of the intra-abdominal contents, that have been reported in clinical practice [25].

In our experimental study, the noncross-linked collagen implants exhibited the early overexpression of the collagen 1 and 3 genes, consistent with a previous study [14]. Two expression peaks occurred at 14 and 90 days. However, *Bio-A* demonstrated a completely different pattern of expression, with a significant increase in collagen 1 and 3 mRNA at 30 days, whereas this expression significantly decreased at later timepoints.

Regarding the inflammatory reaction, *Bio-A* elicited a significantly greater macrophage response until 90 days post-implantation, compared with the collagen implants. This finding was inconsistent with another study [6] that has demonstrated an attenuated inflammatory response in *Bio-A* compared with a biomesh of human origin.

From the biomechanical point of view, the existing areas of "relaxation" at the implant sites did not correlate with the final mechanical resistance that was obtained. This result indicated a progressive increase in resistance to breakage in each of the prosthetic materials throughout the timepoints in the study. Most likely, in the materials with the greatest biodegradation, the tissue support of the implant region (partial defect) restricts the exact analysis of the resistance of the various materials to breakage, which should obviously be considered when evaluating these parameters.

In summary, we can affirm that prostheses of the absorbable laminar type are among the newly available tools within a surgeon's therapeutic arsenal. They provide a scaffold upon which the host's fibroblasts can form well organized and adequately vascularized new connective tissue. However, these prostheses are still inadequate with respect to their precise timepoints of biodegradation, which fundamentally affects the repair process.

In the present study, *St* yielded optimal results during the repair process, with a progressive collagenization and minimal foreign-body reactions. We considered that the slower degradation of *Bio-A*, which is a completely polymeric prosthesis, could offer excellent reparative results compared with the prostheses of biological origin, particularly regarding the cost-benefit of the biomaterial.

Ultimately, we conclude that in the present study, the *St* prostheses exhibited the optimal tissue behavior. However, additional and long-term studies of this biomaterial are necessary to corroborate its total degradation after generating new tissue within the implant area.

Author Contributions

Conceived and designed the experiments: JMB. Performed the experiments: JMB GP SS MR BP. Analyzed the data: GP SS. Contributed reagents/materials/analysis tools: SS BP. Wrote the paper: JMB GP SS.

References

1. Bellows CF, Albo D, Berger DH, Awad SS (2007) Abdominal wall repair using human acellular dermis. Am J Surg 194: 192–198.
2. Liang HCh, Chang Y, Hsu ChK, Lee MH, Sung HW (2004) Effects of crosslinking degree of an acellular biological tissue on its tissue regeneration pattern. Biomaterials 25: 3541–3552.
3. Losanoff JE, Richman BW, Jones JW (2002) Entero-colocutaneous fistula: a late consequence of polypropylene mesh abdominal wall repair: case report and review of the literature. Hernia 6: 144–147.
4. Petter-Puchner AH, Fortelny RH, Walder N, Mittermayr R, Öhlinger W, et al. (2008) Adverse effects associated with the use of porcine cross-linked collagen implants in an experimental model of incisional hernia repair. J Surg Res 145: 105–110.
5. Harth KC, Rosen MJ (2009) Major complications associated with xenograft biologic mesh implantation in abdominal wall reconstruction. Surg Innnov 16: 324–329.
6. Orenstein SB, Qiao Y, Kaur M, Klueh U, Kreutzer DL, et al. (2010) Human monocyte activation by biologic and biodegradable meshes in vitro. Surg Endosc 24: 805–811.
7. Efthimiou M, Symeonidis D, Koukoulis G, Tepetes K, Zacharoulis D, et al. (2011) Open inguinal hernia repair with the use of a polyglycolic acid-trimethylene carbonate absorbable mesh: a pilot study. Hernia 15: 181–184.
8. Hiles M, Record Ritchie RD, Altizer AM (2009) Are biologic grafts effective for hernia repair?: a systematic review of the literature. Surg Innov 16: 26–37.
9. Jenkins ED, Yip M, Melman L, Frisella MM, Matthews BD (2010) Informed consent: cultural and religious issues associated with the use of allogenic and xenogenic mesh products. J Am Coll Surg 210: 402–410.
10. Orenstein SB, Qiao Y, Klueh U, Kreutzer DL, Novitsky YW (2010) Activation of human mononuclear cells by porcine biologic meshes in vitro. Hernia 14: 401–407.
11. Connor J, McQuillan D, Sandor M, Wan H, Lombardi J, et al. (2009) Retention of structural and biochemical integrity in a biological mesh supports tissue remodeling in a primate abdominal wall model. Regen Med 4: 185–195.
12. Sandor M, Xu H, Connor J, Lombardi J, Harper JR, et al. (2008) Host response to implanted porcine-derived biologic materials in a primate model of abdominal wall repair. Tissue Eng Part A 14: 2021–2031.
13. Mulier KE, Nguyen AH, Delaney JP, Marquez S (2011) Comparison of Permacol™ and Strattice™ for the repair of abdominal wall defects. Hernia 15: 315–319.
14. Pascual G, Rodríguez M, Sotomayor S, Moraleda E, Bellón JM (2012) Effects of collagen prosthesis cross-linking on long-term tissue regeneration following the repair of an abdominal wall defect. Wound Repair Regen 20: 402–413.
15. Butler CE, Burns NK, Campbell KT, Mathur AB, Jaffari MV, et al. (2010). Comparison of cross-linked and non-cross-linked porcine acellular dermal matrices for ventral hernia repair. J Am Coll Surg 211: 368–376.
16. Campbell KT, Burns NK, Rios CN, Mathur AB, Butler CE (2011) Human versus non-cross-linked porcine acellular dermal matrix used for ventral hernia repair: comparison of in vivo fibrovascular remodeling and mechanical repair strength. Plast Reconstr Surg 127: 2321–2332.
17. Ngo MD, Aberman HM, Hawes ML, Choi B, Gertzman AA (2011) Evaluation of human acellular dermis versus porcine acellular dermis in an in vivo model for incisional hernia repair. Cell Tissue Bank 12: 135–145.
18. Kapan S, Kapan M, Goksoy E, Karabicak I, Oktar H (2003) Comparison of PTFE, pericardium bovine and fascia lata for repair of incisional hernia in rat model, experimental study. Hernia 7: 39–43.
19. Cavallaro A, Lo Menzo E, Di Vita M, Zanghì A, Cavallaro V, et al. (2010) Use of biological meshes for abdominal wall reconstruction in highly contaminated fields. World J Gastroenterol 21: 1928–1933.
20. Limpert JN, Desai AR, Kumpf AL, Fallucco MA, Aridge DL (2009) Repair of abdominal wall defects with bovine pericardium. Am J Surg 198: 60–65.
21. Bellows CF, Smith A, Malsbury J, Helton WS (2012) Repair of incisional hernias with biological prosthesis: a systematic review of current evidence. Am J Surg 3: [Epub ahead of print].
22. Ratto C, Litta F, Parello A, Donisi L, Zaccone G, et al. (2012) Gore Bio-A(®) Fistula Plug, a New Sphincter-Sparing Procedure For Complex Anal Fistula. Colorectal Dis 14: 264–269.
23. Buchberg B, Masoomi H, Choi J, Bergman H, Mills S, et al. (2010) A tale of two (anal fistula) plugs: is there a difference in short-term outcomes?. Am Surg 76: 1150–1153.
24. Burgess PL, Brockmeyer JR, Johnson EK (2011) Amyand hernia repaired with Bio-A: a case report and review. J Surg Educ 68: 62–66.
25. Gupta A, Zahriya K, Mullens PL, Salmassi S, Keshishian A (2006) Ventral herniorrhaphy: experience with two different biosynthetic mesh materials, Surgisis and AlloDerm. Hernia 10: 419–425.

15

Tissue Engineering in Animal Models for Urinary Diversion

Marije Sloff[1]*, Rob de Vries[2], Paul Geutjes[1], Joanna IntHout[3], Merel Ritskes-Hoitinga[2], Egbert Oosterwijk[1¶], Wout Feitz[1¶]

1 Department of Urology, Nijmegen Centre for Molecular Life Sciences, Radboud University Medical Center, Nijmegen, The Netherlands, **2** SYRCLE (SYstematic Review Centre for Laboratory animal experimentation), Central Animal Laboratory, Radboud University Medical Center, Nijmegen, The Netherlands, **3** Department for Health Evidence, Radboud University Medical Center, Nijmegen, The Netherlands

Abstract

Tissue engineering and regenerative medicine (TERM) approaches may provide alternatives for gastrointestinal tissue in urinary diversion. To continue to clinically translatable studies, TERM alternatives need to be evaluated in (large) controlled and standardized animal studies. Here, we investigated all evidence for the efficacy of tissue engineered constructs in animal models for urinary diversion. Studies investigating this subject were identified through a systematic search of three different databases (PubMed, Embase and Web of Science). From each study, animal characteristics, study characteristics and experimental outcomes for meta-analyses were tabulated. Furthermore, the reporting of items vital for study replication was assessed. The retrieved studies (8 in total) showed extreme heterogeneity in study design, including animal models, biomaterials and type of urinary diversion. All studies were feasibility studies, indicating the novelty of this field. None of the studies included appropriate control groups, i.e. a comparison with the classical treatment using GI tissue. The meta-analysis showed a trend towards successful experimentation in larger animals although no specific animal species could be identified as the most suitable model. Larger animals appear to allow a better translation to the human situation, with respect to anatomy and surgical approaches. It was unclear whether the use of cells benefits the formation of a neo urinary conduit. The reporting of the methodology and data according to standardized guidelines was insufficient and should be improved to increase the value of such publications. In conclusion, animal models in the field of TERM for urinary diversion have probably been chosen for reasons other than their predictive value. Controlled and comparative long term animal studies, with adequate methodological reporting are needed to proceed to clinical translatable studies. This will aid in good quality research with the reduction in the use of animals and an increase in empirical evidence of biomedical research.

Editor: Wei-Chun Chin, University of California, Merced, United States of America

Funding: This work is funded by Fonds NutsOhra: Kunststoma 1102-56(http://www.stichtingnutsohra.nl/). The funders had no role in study design, data collection and analysis, decision to publish or preparation of the manuscript.

Competing Interests: The authors have declared that no competing interests exist.

* Email: marije.sloff@radboudumc.nl

¶ These authors contributed equally to this work.

Introduction

Urinary diversion with gastrointestinal (GI) tissue remains the gold standard treatment for patients suffering from end-stage bladder disease caused by bladder cancer or congenital malformations, e.g. bladder exstrophy or spina bifida [1,2]. There are three approaches to create a urinary diversion in these patients. The first and most commonly used type among surgeons is the incontinent ileocutaneostomy; a urinary conduit with a skin-outlet. Alternatively, continent diversions can be formed non-orthotopically with a skin-outlet or orthotopically as a neobladder [3,4]. Although the use of GI tissue provides a satisfactory outcome in most cases, it can be associated with severe complications. These can be either related to the bowel surgery (obstruction, infections, fistulas, etc.) or to the urostomy implantation (metabolic disorders, stone formations, infections, etc.) [5,6].

A tissue engineering and regenerative medicine (TERM) approach may provide new possibilities by creating a man-made construct to replace GI tissue for urinary diversions. The implementation of such constructs could prevent invasive bowel surgery and the potentially life-threatening complications, therefore reducing health care costs. Several investigators have focused on the development of new materials for this purpose, including naturally derived materials, synthetic polymers and decellularized scaffolds [7–9]. These biomaterials can be applied with and without autologous cells, using the regenerative capacity of the body [10,11].

In the field of urogenital reconstruction, bladder domes for cystoplasty and uretheral reconstruction with man-made constructs have already been used in patients [12,13]. However, despite the progress in *in vitro* research and animal experimentation, clinical translation of TERM approaches for urinary diversion has been negligible. Translation from bench to bedside for these tissue engineered constructs starts with the analysis of biodegradability, biocompatibility and foreign body response, which is usually performed in small rodents. To engineer and regenerate specific tissues, evaluation should preferably be performed at relevant anatomical sites with appropriately sized constructs to permit easy clinical translation. Large animal models closely mimicking the human body are therefore desirable, but their use might be ethically debatable [14]. In general, the choice of an animal model is dependent on financial considerations, the investigators experience, ethical sensitivity and practical limitations. Even though other and better translatable models might be available [15]. To our knowledge a superior animal for tissue engineering and urinary diversion has not been identified yet.

To decide on the most suitable type of animal model an evidenced-based systematic review is essential, since it potentially increases the chance of successful clinical translation [16,17]. We therefore systematically searched the current literature for all types of studies on the efficacy of tissue engineered constructs in animal models for urinary diversion. The results were analyzed with respect to survival, side effects, functionality and urothelium formation, to investigate whether there was sufficient evidence to decide if any animal model was superior for evaluating tissue engineered constructs for urinary diversion applications.

Materials and Methods

1. Search strategy

We identified relevant studies, including peer reviewed articles and (congress) abstracts, through a systematic search of PubMed, EMBASE (OvidSP) and Web of Science up until the 23rd of January 2013, following the approach as described by de Vries et al., and Leenaars et al. [16,17]. In all three databases synonyms for tissue engineering (e.g. tissue engineering, tissue engineered, regenerative medicine or biomaterials) were combined with synonyms for urinary diversion (e.g. orthotopic diversion, neobladders, continent or incontinent stomas). MeSH terms (PubMed) and EMTREE terms (EMBASE) were used when available and were combined with additional free-text words from titles or abstracts ([tiab] or/ti,ab.). For the complete strategy, see Table S1. In PubMed and Embase (OvidSP), the results were filtered for animal studies, using previously designed 'animal filters' [18,19]. The included primary studies and relevant reviews on the subject were screened for additional relevant references.

2. Study selection

Only primary studies that evaluated tissue engineered construct for urinary diversion in animal models were included. From the retrieved set of papers, duplicates and triplicates were manually deleted from EndNote, considering the preference PubMed> EMBASE>Web of Science. Based on title/abstract, primary screening was performed by a single review author (MS), deleting articles that clearly did not involve tissue engineering or urinary diversion. In case of any doubt, articles were included for further screening. Secondary screening of title/abstract was independently performed in Early Review Organizing Software (EROS, IECS, Buenos Aires, Argentina, www.eros-systematic-review.org) by two review authors (MS and RdV). The following inclusion criteria were used: 1) urinary diversion, 2) tissue engineering, 3) (living) animals of any species, and 4) primary articles. In this step, a procedure was considered to be a urinary diversion if it involved a total/radical cystectomy or implantation of a stoma or pouch connected to at least one ureter. Articles that described a partial cystectomy, hemi-cystectomy or bladder augmentation were excluded, because they do not relate to urostomy. Tissue engineered constructs were defined as biomaterials or polymers that aided the (re)construction of tissues. Articles were either categorized as 'included', 'excluded' or 'more information necessary' if important details were not included in the abstract. Any discrepancies were discussed and re-evaluated until consensus was reached. Full-text articles were retrieved and evaluated for definite inclusion/exclusion, based on the same criteria used for the secondary screening. The reference lists of the included studies and manually identified reviews on the subject were screened for any missed references. Unfortunately, one of the included studies was published in Korean and we did not have the resources to have it translated [20]. The article was therefore excluded from this review.

3. Data extraction

From every included study, basic information (author, year of publication, etc.), animal characteristics (species, sex, etc) and study characteristics (biomaterial, follow-up, etc.) were extracted and tabulated by MS and RdV after reaching consensus (Table 1). The outcome of the studies for the meta-analysis was assessed using extracted data on mortality, adverse effects, occlusion (blockade of urinary flow) and the formation of urothelium on the implanted construct (Table 2).

4. Methodological quality assessment

All included studies were feasibility studies only, i.e. no comparison was made between the new (tissue engineering) and classical treatment (GI tissue) or any other relevant control group. Therefore, performing a risk of bias-assessment was not possible, and we consequently focused on the quality of the reporting of data and outcomes of the studies (Table 3).

5. Meta-analyses

Meta-analyses were performed for the outcome measures functionality (absence of occlusion) and formation of urothelium in seven studies. One study did not describe the animal species and was therefore excluded [26]. Since appropriate control groups were not included in any of the studies, it was not possible to perform a standard meta-analysis using, for example, odds ratios. We therefore performed a meta-analysis of proportions, more specifically of the number of animals in which a functional construct or urothelium was formed as a proportion of the total number of treated animals. First, exact binomial confidence intervals were calculated for the individual studies. To circumvent continuity corrections (some studies had 0 events), an arcsine transformation of the proportions was carried out for the meta-analyses [21]. Because high heterogeneity was expected, the individual proportions were pooled using a random effects model. Given the low number of studies, a Hartung-Knapp adjustment for random effects models was applied [22]. I^2 was used as a measure of heterogeneity. The analyses were conducted in R (version 3.0.1; R Core Team 2012), using the metafor package [23,24]. To explore the potential influence of animal size on the effect, the studies were ranked according to the subgroups "large" (rabbits and larger) and "small" (rats and mice) models in the forest plots. The small group sizes prevented calculation of an overall effect per subgroup and therefore only visually derived tendencies are presented.

Results and Discussion

1. Study inclusion

Database searches yielded 573 references for PubMed, 855 references for Embase and 315 references for Web of Science (Figure 1). After removal of duplicates and triplicates 1157 references remained. During the primary screening in EndNote, 883 references that did not meet our inclusion criteria were

Table 1. Study characteristics.

	Reference	Publication	Species (strain)	Sex	weight/ age	Group size	Type of intervention	Biomaterial	Size scaffold	Cell type/ amount	Culture period	Study Design	Evaluation time point	outcome measures
1	Basu 2012	Research paper	Minipigs (Göttingen swine)	M/F	12–16 kg	gr 1: 8 gr 2: 8 gr 3: 8 gr 4: 8	Urinary conduit 2 ureters	PLGA-coated PGA	*	1 cm2 bladder biopsy, 2 cm2 adipose biopsy 50 mL peripheral blood All: 30–40×10^6 SMC	6 days	gr 1: bladder SMC gr 2: adipose SMC gr 3: blood SMC gr 4: unseeded	84+/−5 days	Occlusion? Histology Immunohistochemistry
2	Bertram 2009	Poster	Canines	*	*	gr 1: 8 gr 2: 8 gr 3: 8 gr 4: 8	Neo-bladder	Tengion Autologous Neo-bladder (PLGA)	*	SMC	*	gr 1: 4×10^6 SMC gr 2: 12×10^6 SMC gr 3: 25×10^6 SMC gr 4: reimplanted bladder	9 months	Histology Contractile response Electrical field stimulation
3	De Filippo 2009	Abstract	*	*	*	*	Neo-bladder	PLGA-based scaffold	*	a	*	gr 1: cystecomized animals gr 2: weight/age matched human patients	6 months	Occlusion? Histology Immunohistochemstry Voiding intervals Bladder capacity
4	Dorflinger 1985	Research paper	Dogs (Mongrel)	F	gr 1: 17–21 kg gr 2: 22–27 kg[b]	gr 1: 5 gr 2: 5	Neo-bladder	Silicone prosthesis with PGA mesh	65 cc	NA	NA	gr 1: thin prosthesis 2 months preimplant gr 2: thick prosthesis 1 month preimplant	6 months	Occlusion Histology Urogram Macroscopic evaluation Colony-forming assay
5	Drewa 2007	Research paper	Rats (Wistar)	M	300 g, 6 months	gr 1: 3 gr 2: 3	Urinary conduit 1 ureter	Small intestinal submucosa (SIS), porcine	3 cm 3-layered [c]	Fibroblast 3T3 2×10^8 cells in alginate gel	*	gr 1: 2×10^8 3T3 cells gr 2: unseeded	2 and 4 weeks	Occlusion Histology Pyelogram Macroscopic evaluation
6	Geutjes 2012	Research article	Pigs (Landrace)	F	50 kg	gr 1: 6 gr 2: 4	Urinary conduit 1 ureter	Collagen and Vypro polymer	l = 12 cm, d = 15 mm	UC from 4 cm2 bladder biopsy: 10×10^6 cells	6 days	gr 1: bladder UC gr 2: unseeded	1 month	Occlusion Histology Immunohistochemistry Loopogram Macroscopic evaluation
7	Kloskowski 2012	Abstract	Rats (Wistar)	*	*	gr 1: 12 gr 2: 2	Urinary conduit 1 ureter	Decellularized aortic arch or PCL scaffold	*	NA	NA	gr 1: aortic arch gr 2: PCL	3 weeks	Occlusion Histology
8	Liao 2013	Research paper	Rabbits (New Zealand White)	M	2,0–2,5 kg	gr 1: 24 gr 2: 6	Urinary conduit 2 ureters	Rabbit bladder acellular matrix (BAM)	l = 4 cm d = 0,8 cm	UC from 4 cm2 bladder biopsy: 80×10^7 cells	7 days	gr 1: bladder UC gr 2: unseeded	1, 2, 4 and 8 weeks [d]	Occlusion? Histology Immunohistochemistry

SMC = smooth muscle cells, PLGA = poly(lactic-co-glycolic acid), PCL = polycaprolactone, PGA = polyglycolic acid, UC = urothelial cells, * = not mentioned, N.A. = not applicable, ? = is implied in the text, but not specifically mentioned, a = cells are used, type and amount are not mentioned, b = group 3 included only partial cystetomies, as well as 1 dog in group 1 (not included), c = diameter of a 12-Fr catheter, d = evaluation time point of control group is unclear.

Table 2. Scoring of the included studies.

	Reference	Follow-up	Mortality	Adverse effects	Formation of UD	Urothelium formation
1	Basu 2012	84+/−5 days	0%[a]	*	32/32[a]	32/32[b]
2	Bertram 2009	9 months	0%?	none	24/24?[c]	24/24?[c]
3	De Filippo 2009	6 months	0%?	*	*	*
4	Dorflinger 1985	6 months	60%	hydronephrosis (5#), hydroureter (4#), pyonephrosis (2#), inflammation (3#), leakage (2#), infection (1#), ulcers (1#), reflux (2#)	2/10	2/10
5	Drewa 2007	2 or 4 weeks	0%	adhesion (3#), inflammation (4#), leakage (1#), pseudocyst (1#), hydronephrosis (4#), hydroureter (3#)	3/6	1/6
6	Geutjes 2012	1 month	11%	stenosis (3#), leakage (2#), hydroureteronephrosis (all), hydroureter (all)	8/9	6/9
7	Kloskowski 2012	3 weeks	28%[d]	Inflammation (all),	0/14	1/14[e]
8	Liao 2013	1, 2, 4 or 8 weeks	13%[f]	scarring (4#), atresia (4#), hydronephrosis (4#), fistulas (2#), inflammation(2#)[g]	26/30[h]	26/30[h]

* = not mentioned, ? = it is implied that all animals survived and formed a functional conduit with urothelial layers, a = all animals were euthanized at indicated time points, animals remained healthy, no explicit mentioning of occlusions, b = no mentioning of place of sampling. Unclear whether it covers the entire conduit, c = group 4 does not include TE, leaving 24 animals for UD, d = deaths were only in aortic arch group, e = mentioning of formation of cell layers, not specific on type of cell layer, f = all animals died in the experimental group, g = complications were observed in the control group only, h = unclear if this accounts for all animals.

removed. Secondary screening of the remaining 274 references in EROS led to the removal of 206 references. Full text analyses of the remaining 68 references resulted in the inclusion of only 8 studies: two abstracts, one poster and five full-text papers (Figure S1). Screening of the reference lists of these papers and manually identified relevant reviews on the subject did not yield any new references. Thus, the final set included: Geutjes et al. (2012), Basu et al. (2012), Kloskowski et al. (2012), Liao et al. (2013), De Filippo et al. (2009), Bertram et al. (2009), Drewa et al. (2007), and Dørflinger et al. (1985) [11,25–31]. Although the number of published studies on tissue engineering for urinary diversion in animal models appeared to be substantial, only these selected studies applied tissue engineered constructs for urinary diversion at relevant clinical sites in animal models. The recent increase in published papers on this subject is remarkable. Although all relevant databases were explored, we cannot exclude that some data reside within company protected domains. Moreover, studies tend to be published only when results are positive and statistically significant [32]. These two factors may have resulted in an incomplete data set and they may have introduced a publication bias in this systematic review. Due to the limited number of included studies, we were not able to estimate the risk and effect of this publication bias.

2. Study characteristics

Analysis of study characteristics revealed extraordinary diversity (Table 1). Various animal species, including pigs, minipigs, dogs, rats and rabbits, were used. They were implanted with constructs of either biodegradable polymers (poly(lactic-*co*-glycolic acid) (PLGA), polyglycolic acid (PGA) or collagen) or decellularized material (small intestinal submucosa (SIS) and bladder acellular matrix (BAM)). Kloskowski et al., used a biodegradable polymer polycaprolactone (PCL) and decellularized aortic arches. Although three different approaches for urinary diversion are known (urinary conduit, abdominal pouch and neobladder), the included studies only created urinary conduits (5 studies) or neobladders (3 studies). An animal model for an abdominal pouch was not described.

There are three different study designs within the pool of included studies: studies that use cellular or acellular scaffolds and studies that compare these two constructs (Table 1 and 4). The most frequently used study design was the comparison between cellular and acellular constructs (4 studies), investigating whether (pre-)seeding of the biomaterial resulted in a superior outcome. These studies generally isolated their cells from a bladder biopsy, although alternative sources like adipose tissue, peripheral blood and cell lines were investigated (Basu et al., Drewa et al., Geutjes et al. and Liao et al.). The other studies compared different materials (Kloskowski et al.), cell concentrations (Bertram et al.) or different experimental designs (Dørflinger et al.). Filippo et al. compared the behavior of a cell-seeded PLGA-based scaffold in cystectomized animals with its behavior in children enrolled in a Phase II trial for bladder augmentation. This diversity in animal species, biomaterial or study design complicates the interpretation of the results of this systematic review.

Since tissue engineering for urinary diversion is a relatively new area of research, the main focus of the studies was to determine the feasibility of implantation and subsequent behavior of the designed constructs. Follow-up time was less than a year in all cases, whereas constructs need to be functional throughout the remainder of a patients' life time. Stoma complications can occur after several years in patients, and short follow-up in animals will not provide evidence on late complications. This indicates the necessity of appropriate control groups with the classical techniques and longer follow-up for at least 1 year.

3. Quality assessment

The lack of control groups precluded a risk of bias-analysis. We therefore focused on the reporting quality of the studies (Table 3). Our specific interests were the animal characteristics, the composition, dimensions and preparations of the construct, the

Table 3. Quality assessment.

		Abstracts/Poster			Research Articles				
		Bertram 2009	**De Filippo 2009**	**Kloskowski 2012**	**Basu 2012**	**Dorflinger 1985**	**Drewa 2007**	**Geutjes 2012**	**Liao 2013**
Study Design	**Q1:** Is the animal species described?	yes	no	yes	yes	yes	yes	yes	yes
	Q2: Is the specific strain described?	no	no	yes	yes	?	yes	yes	yes
	Q3: Is the sex of the animal specified?	no	no	no	yes	yes	yes	yes	yes
	Q4: Is the age or weight of the animals specified?	no	no	no	yes	yes	yes	yes	yes
	Q5: Is the number of animals specified?	yes	no	yes	yes	yes	yes	yes	yes
	Q6: Is the creation of the urinary diversion clearly described?	yes	?	yes	yes	yes	yes	yes	yes
	Q7: Is the composition of the tissue-engineered construct clearly described?	no	no	yes	yes	yes	yes	yes	yes
	Q8: Are the dimensions of the implanted construct clearly described?	no	no	no	no	yes	yes	yes	yes
	Q9: Is the preparation or culture period clearly described?	no	no	no	yes	N.A.	?	yes	yes
Outcomes	**Q10:** Are the used outcome measures clearly described?	yes	yes	yes	yes	yes	yes	yes	yes
	Q11: Is the follow-up time after implantation clearly described?	yes	yes	yes	yes	yes	yes	yes	?[a]
	Q12: Are there any comments on the representativeness of the results?	no	no	no	no	yes	yes	?	no
	Q13: Is the location of sampling for histology clearly described?	no	no	no	no	N.A.	yes	yes	no
	Q14: Is it explicitly indicated that there were any drop-outs?	no	yes	yes	?	yes	yes	yes	yes
	Q15: Is the number of drop-outs described?	N.A.	N.A.	yes	N.A.	yes	N.A.	yes	yes
	Q16: Are the reasons for dropping out specified?	N.A.	N.A.	no	N.A.	yes	N.A.	yes	yes
	Q17: Is it stated that there were any adverse effects?	yes	?	?	no	yes	yes	yes	yes

? = not clearly described in the text, N.A. = not applicable, a = unknown for the control group.

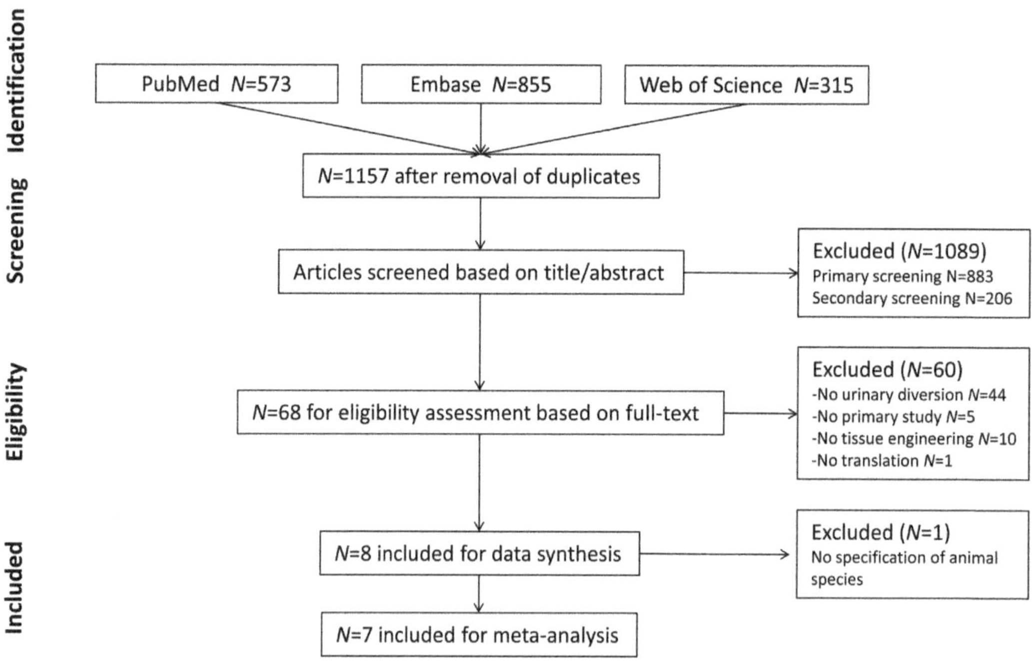

Figure 1. Flow-chart of search and screening process. Primary screening exclusion was performed in End-Note. Criteria included: no urinary diversion, no tissue engineering or reconstruction of ureter or urethra. Secondary inclusion was performed in EROS. Criteria included: no urinary diversion, no tissue engineering, no animal study or no primary study.

representativeness of the results and the adequate reporting of drop-outs.

The reporting was relatively poor in the abstracts compared to the full-text papers. Many abstracts omitted important details that are essential to compare different studies, including animal species, strain, number and sex, type of tissue engineered scaffolds, description of the composition, dimensions and preparations of the construct (Q1-Q9). This might have been partly the consequence of the word and space limitation for abstracts.

All included studies appropriately described the predefined outcome measures and stated at which time evaluation took place. Only two studies commented on the representativeness of the figures for the overall study outcomes (Dørflinger et al. and Drewa et al.) (Q10-Q13). In total, three papers were complete in their reporting of the study design (Drewa et al., Geutjes et al., and Liao et al.).

Although guidelines for standardized reporting of animal experimentation have been described, these have not yet been generally accepted [33,34]. We observed that in some studies, especially in the abstracts, these guidelines were not implemented. It is crucial to further improve methodological reporting to aid future research, even in abstracts.

4. Data synthesis

To determine the efficacy of the tissue engineered constructs, we focused on the outcome measures: formation of a functional conduit or reservoir and the formation of urothelium in the regenerated tissue, evidenced by histology (Table 2). Some studies performed additional analyses, including immunohistochemistry, urograms or pyelograms (Table 1), but these were not considered here. Because constructs can only be functional in the absence of major complications (including mortality), we first looked at the survival of the animals and adverse effects. Survival of the animals was regarded as the first indication for the safety of a construct and represented a condition for the efficacy of a construct. Secondly, since urine needs to exit the body adequately, the formation of a reservoir or conduit with a urothelial lining was deemed essential.

4.1 Adverse effects and mortality. Experience shows that even in some situations animals might die of unrelated causes, with more likelihood in larger experimental groups. A clear and detailed description of the drop-outs will increase the credibility of

Table 4. Cellular vs. Acellular.

	Acellular	Cellular
Basu et al 2012	8/8	24/24
Drewa et al 2007	2/3	1/3
Geutjes et al 2012*	4/4	5/6
Liao et al 2013	4/6	24/24

Amount of functional conduits formed in comparative studies with cellular and acellular groups. * was tabulated after correspondence with the first author.

the study. Surprisingly, two studies did not explicitly report on the mortality rate, but the studies suggest that all animals survived the procedure in good health (Basu et al, and Bertram et al.) They imply that a functional urinary conduit with urothelial linings is formed in all animals without any adverse effects (Table 2 and Table 3, Q14-Q17). Such a successful score was not described in any of the other studies, which raises the possibility that the success rate was overestimated. The reported mortality of the other studies ranged between 60% in the pioneering study in 1985 (Dørflinger et al.) to 13% in the most recently published paper (Liao et al.), suggesting the application of improved constructs or improved surgical techniques over the years. Geutjes et al. is the only study that reports on both related and unrelated deaths.

All studies report on inflammation of the construct and the surrounding tissues, although this does not necessarily constitute a negative effect. Some degree of inflammation may guide tissue regeneration, remodeling and the formation of blood vessels. The formation of a vasculature structure in large constructs still remains a major problem in tissue engineering [35,36]. Other common adverse effects found in the majority of the studies were hydronephrosis and hydroureters, prevention of which remains the biggest challenge for tissue engineers. The designed constructs were not able to control urinary pressure, leading to reflux to the kidney and the aforementioned conditions.

The formation of stones in urinary diversions is one of the complications in using GI tissue. Remarkably, only one study, performed in pigs reported the formation of calcifications (Geutjes et al.). The absence of stones, particularly in the included rabbit study, is unexpected (Liao et al.). Implantation of a tissue engineered patch in the bladder wall of rabbits resulted in a high incidence of stone formation and encrustation, indicating that this animal model in particular is prone to develop stones [37,38]. Although a different biomaterial was used in a different setting, no stones were formed up until two months in the study performed by Liao et al. In humans, it takes months to years to develop urinary stones and perhaps the follow-up time in these studies is too short (<1 year) to detect stones [39]. Obviously, due to the difference in diet composition, flow speed, composition and pH of urine in different species, urinary stones might not develop in some animals. Long-term follow-up is necessary to exclude encrustation or stone formation.

4.2 Efficacy. We conducted meta-analyses for the outcome measures functionality and urothelium formation (Figures 2A and B). The aim of these meta-analyses was not to obtain a precise point estimate, but rather to get an impression of the quantitative relations between the results of the individual studies and to detect trends [40]. The heterogeneity between the studies was very high ($I^2 = 96\%$ for functionality and $I^2 = 95\%$ for urothelium formation). Although this did not justify pooling of the results, nevertheless, overall proportions were calculated for functionality and urothelium formation (69% [CI: 18–100%] and 65% [CI: 17–98%], respectively). Due to the large heterogeneity and small number of studies, these overall proportions should be interpreted with caution.To prevent misinterpretations, overall proportions are only shown as dotted vertical lines in the forest plots.

Successful formation of functional conduits with full urothelial linings, the primary goal of all studies, varied between the studies. Only small numbers of functional conduits and urothelial coverage were observed by Dørflinger et al., Drewa et al. and Kloskowski et al. In contrast, Liao et al. showed that in 4/5 cases a functional conduit was formed with urothelial linings. The large heterogeneity and small number of studies complicates identification of the underlying cause of the different results. We can only speculate whether these were the consequence of the biomaterial, the animal model or confounding factors.

4.3 Animal models. The small number and heterogeneity of the studies did not permit subgroup analysis for different animal species. However, the forest plots (Figures 2A and B) from the meta-analyses suggested a tendency towards better results in large animal models compared to small animals. Although we cannot exclude the effect of confounding factors, this supports the idea that future research should focus on larger animals. Larger animals have a more similar anatomy to the human body than, for example, rodents, and allow for evaluation at clinical relevant sites with constructs of a comparable size. Moreover, surgical techniques and materials are more comparable to the human setting, therefore mimicking the human situation as closely as possible. In 1985, the first report on the use of a construct to tissue engineer a urinary diversion as a replacement for GI tissue in large animals was published (Dørflinger et al.). Even though the mortality rate was high, a functional reservoir could be formed showing the feasibility of this approach. In later studies small animals were used (Kloskowski et al. and Drewa et al.) but the success rate in these studies was low. This might be explained by the surgical challenge with a higher risk of complications in smaller animals.

The extraordinary heterogeneity in animal studies has previously been reported by Roosen et al. [41], who reviewed different animal models for the classical types of urinary diversion using GI tissue. A uniform conclusion regarding the most suitable animal model could not be drawn and the author advised to view the results with caution, when translating these for clinical implementation.

4.4 Cellular and acellular scaffolds. The studies that compared acellular and cellular scaffolds investigated the effect of (pre-) seeding on tissue regeneration (Table 1 and 4). Basu et al., Liao et al. and Geutjes et al. expanded autologous cells from a biopsy. Basu et al. stated that (pre-)seeding of the scaffolds provided an additional advantage, although this was not substantiated. The same conclusion was drawn by Liao et al., but here the authors clearly report that the drop-outs and adverse effects in the acellular group were higher. Geutjes et al. observed no difference between cellular and acellular constructs.

In the study by Drewa et al., scaffolds were seeded with 3T3 fibroblast from mouse origin, followed by implantation in rats. Not surprisingly, this resulted in an excessive inflammatory response and consequently a better outcome for the unseeded group.

The presence of a cellular component can trigger M1 macrophages, resulting in fibroblast deposition. In contrast, acellular scaffolds activate the M2 macrophages, which leads to reconstruction and regeneration of tissues [35]. Some researchers have reported that scaffold (pre-)seeding provides an additional advantage for tissue regeneration, since regeneration is less dependent on cellular in-growth [42,43]. Since the effect preseeding is outside the scope of this systematic review, a meta-analyses was not performed. Therefore, based on the available information, it is unclear whether (pre-)seeding of the scaffolds provides an advantage for tissue reconstruction in urinary diversions.

4.5 Concluding remarks. Based on this systematic review the most adequate animal model for urinary diversion is still undefined. Only a limited amount of studies could be identified despite the comprehensive search strategy, all showing large heterogeneity. It appears that the predictive value of a particular animal model was not a decisive factor in the studies performed. Nevertheless, the forest-plots suggested a trend towards successful experimentation in larger animal models, supporting the idea that

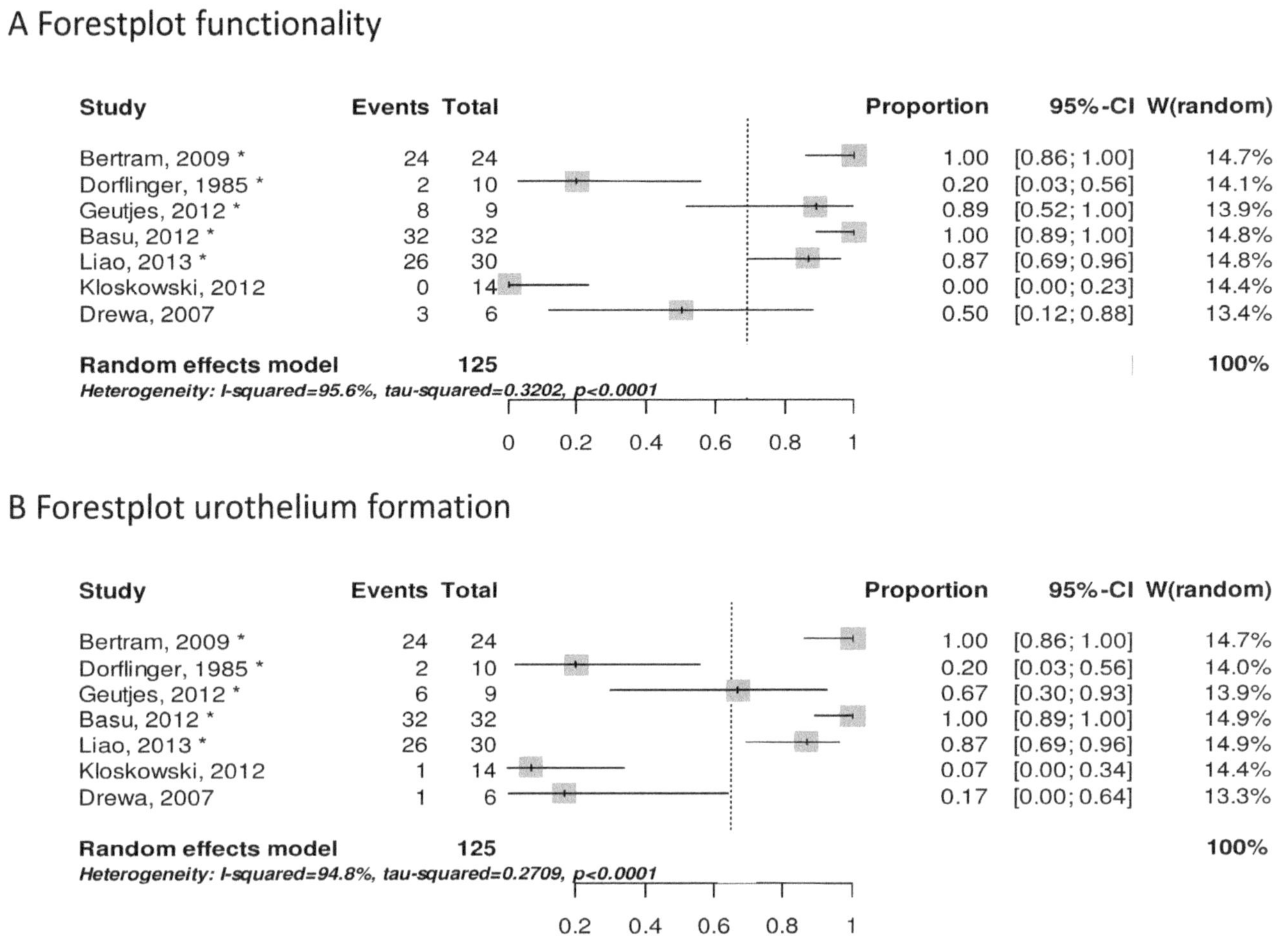

Figure 2. Forestplots for functionality (A) and urothelium formation (B). Forest plot of 7 studies. Filippo et al. was excluded for both meta-analyses since the study does not report on the type of animal, functionality and urothelium formation. * indicates studies with large animal species.

future research should focus on the evaluation of constructs in larger animal models which more closely mimic the human body than small (rodent) animal models. Bladder augmentations and urethral replacements have been successful in dogs [44–47]. Although the results of the earliest study were disappointing (Dorflinger) this might be explained by the tissue engineering strategy. Use of a more sophisticated tissue engineering approach did lead to satisfactory results (Bertram). In view of the former, evaluation of urinary diversion constructs in dogs might be a valid alternative.

Despite the limited amount of data available on this subject, a phase I clinical trial was initiated for implantation of a (pre-)seeded tissue-engineered urinary conduit in cystectomized patients [48]. Whether it is acceptable to expose patients to a new device which has not been investigated in a long term and multicenter animal study remains a matter of debate. Critics of animal experimentation might say that it is difficult to find a representative animal model to improve clinical translation. In order to improve translational practice, a re-evaluation of preclinical practice is warranted, in which systematic reviews and meta-analyses of animal studies can provide a valuable tool [49].

This systematic review focused on the identification of an appropriate animal model to investigate tissue engineering for urinary diversion. Remarkably, only feasibility studies were identified, which are necessary to evaluate potentially valuable techniques. To obtain a more thorough estimation of the feasibility and applicability, a large variety of approaches should be investigated and should include different biomaterials, growth factors and large animal models. Evaluation of the functionality and advantages of newly developed constructs should include appropriate controls and evaluate long-term effects and outcomes to investigate safety and efficacy of these constructs. Moreover, standardized assessment methods are desirable. The ultimate control would be the use of GI tissue, the golden standard technique currently used in the clinic. Assessment of these preclinical experiments should focus on both functionality and tissue regeneration, using urodynamic measurements and histological evaluations. To continue to clinically translatable studies, standardization of (large) controlled and comparative studies with adequate methodological reporting is required as laid down in existing guidelines [33,34].

Animal experimentation remains subject to ethical debate and -among others- for animal welfare reasons it is therefore essential that research is properly conducted and reported. Nevertheless, the reporting of the methodology and representativeness of results was rather poor in the reviewed studies. Improvements in this area are urgently needed for ethical, scientific and economic reasons, and will allow researchers to repeat studies reliably and make an unbiased decision on the most correct animal model for their future experiments. The scientific community should adopt the existing guidelines for standardized reporting of animal experimentation and implement them in journal author guidelines, which should be enforced by editors, to improve standardized reporting in full-text articles. This will aid in good quality research with more relevant output for the clinic and reduction of the use of animals in biomedical research.

Supporting Information

Figure S1 Inclusion and exclusion of papers during secondary and full-text screening in EROS. After primary screening 274 papers were analyzed in EROS, resulting in the inclusion of 8 papers. Papers were excluded which were without urinary diversion, tissue engineering or animals or when they were reviews. For one study, we did not have the resources for translation.

Table S1 Complete search strategy for PubMed, EMBASE and Web of Science. MeSH and EMTREE terms were used when possible and combined with free-text words from title or abstract ([tiab] or.ti,ab.).

Author Contributions

Conceived and designed the experiments: MS RdV PG EO MRH WF. Performed the experiments: MS RdV. Analyzed the data: MS RdV JH. Contributed reagents/materials/analysis tools: JH. Wrote the paper: MS RdV PG JH MRH EO WF.

References

1. National Cancer Institute (2010) What you need to know about bladder cancer. NIH publiciation No. 10–1559.
2. Pannek J, Senge T (1998) History of urinary diversion. Urol Int 60:1–10
3. Hautmann RE (2003) Urinary diversion: ileal conduit to neobladder. J Urol 169: 834–842.
4. Kaefer M, Hendren WH, Bauer SB, Goldenblatt P, Peters CA, et al. (1998) Reservoir calculi: a comparison of reservoirs constructed from stomach and other enteric segments. J Urol 160: 2187–2190.
5. Hautmann RE, Hautmann SH, Hautmann O (2011) Complications associated with urinary diversion. Nat Rev Urol 8: 667–677.
6. Nieuwenhuizen JA, de Vries RR, Bex A, van der Poel HG, Meinhardt W, et al. (2008) Urinary diversions after cystectomy: the association of clinical factors, complications and functional results of four different diversions. Eur Urol 53: 834–842.
7. Drewa T, Adamowicz J, Sharma A (2012) Tissue Engineering for the oncologic urinary bladder. Nat Rev Urol 9: 561–572.
8. Kim BS, Baez CE, Atala A (2000) Biomaterials for tissue engineering. World J Urol 18: 2–9.
9. Beiko DT, Knudsen BE, Watterson JD, Cadieux PA, Reid G, et al. (2004) Urinary tract biomaterials. J Urol 171: 2438–2444.
10. Drewa T, Sir J, Czajkowski R, Wozniak A (2006) Scaffold seeded with cells is essential in urothelium regeneration and tissue remodeling in vivo after bladder augmentation using in vitro engineered graft. Transplant Proc 38: 133–135.
11. Geutjes P, Roelofs L, Hoogenkamp H, Walraven M, Kortmann B, et al. (2012) Tissue engineered tubular construct for urinary diversion in a preclinical porcine model. J Urol 188: 653–660.
12. Raya-Rivera A, Esquiliano DR, Yoo JJ, Lopez-Bayghen E, Soker S, et al. (2011) Tissue-engineered autologous urethras for patients who need reconstruction: an observational study. Lancet 377: 1175–1182.
13. Atala A, Bauer SB, Soker S, Yoo JJ, Retik AB (2006) Tissue-engineered autologous bladders for patients needing cystoplasty. Lancet 367:1241–1246.
14. Wood MW, Hart LA (2007) Selectin appropriate animal models and strains: making the best use of research, information and outreach. AATEX 14: 303–306.
15. Nordgren A (2004) Moral imagination in tissue engineering research on animal models. Biomaterials 25: 1723–1734.
16. De Vries RB, Buma P, Leenaars M, Ritskes-Hoitenga M, Gordijn B (2012) Reducing the number of laboratory animals used in tissue engineering research by restricting the variety of animal models. Articular cartilage tissue engineering as a case study. Tissue Eng Part B Rev 18: 427–435.
17. Leenaars M, Hooijmans CR, van Veggel N, ter Riet G, Leeflang M, et al. (2012) A step-by-step guide to systematically identify all relevant animal studies. Lab Anim 46: 24–31.
18. Hooijmans CR, Tillema A, Leenaars M, Ritskes-Hoitenga M (2010) Enhancing search efficiency by means of a search filter for finding all studies on animal experimentation in PubMed. Lab Anim 44: 170–175.
19. De Vries RB, Hooijmans CR, Tillema A, Leenaars M, Ritskes-Hoitinga M (2011) A search filter for increasing the retrieval of animal studies in Embase. Lab Anim 45:168–270.
20. Lee YS, Cho SY, Kim HW, Kang SH, Kim HY, et al. (2009) Preliminary study of tissue-engineered ileal conduit using poly (e-caprolactone) (PCL) nano-sheet seeded with muscle-derived stem cells. Korean J Urol. 50: 282–287.
21. Rücker G, Schwarzer G, Carpenter J, Olkin I (2009) Why add anything to nothing? The arcsine difference as a measure of treatment effect in meta-analysis with zero cells. Stat Med 28: 721–728.
22. Knapp G, Hartung J (2003) Improved tests for a random effects meta-regression with a single covariate. Stat Med 22: 2693–2710.
23. R Core Team (2012) R: a language and environment for statistical computing. R foundation for statistical computing. ISBN 3-900051-07-0.
24. Viechtbauer W (2010) Conducting meta-analyses in R with the metafor package. J Stat Soft 36: 1–48.
25. Liao W, Yang S, Song C, Li Y, Meng L, et al. (2013) Tissue-engineered tubular graft for urinary diversion after radical cystectomy in rabbits. J Surg Res 182: 185–191.
26. De Filippo R, Bertram TA, Jayo MJ, Seltzer E (2009) Adaptive regulation of regenerated bladder size after implantation with tengion neo-bladder augment^{TM} early clinical outcomes and preclinical evidence. J Urol 181: 267.
27. Drewa T (2007) The artificial conduit for urinary diversion in rats: a preliminary study. Transplant Proc 39: 1647–1651.
28. Dørflinger T, Frimodt-Møller PC, England DM, Madsen PO, Bruskewitz R (1985) Prosthetic urinary bladder implantation to facilitate bladder regeneration. Neurourol Urodyn 4: 47–59.
29. Basu J, Jayo MJ, Ilagan RM, Guthrie KI, Sangha N, et al. (2010) Regeneration of native-like neo-urinary tissue from non-bladder cell sources. Tissue Eng Part A 18: 1025–1034.
30. Kloskowski T, Jundzill A, Gurtowska N, Olkowska J, Kowalczuk T, et al. (2012) Urinary conduit construction using tissue engineering. J Regen Med Tissue Eng 6: 23.
31. Bertram TA, Christ GJ, Andersson KE, Aboushwareb T, Fuellhase C, et al. (2009) Pharmacologic response of regenerated bladders in a preclinical model. FASEB J 23: 291.
32. Knight J (2003) Negative results: Null and void. Nature 422: 554–555.
33. Kilkenny C, Browne WJ, Cuthill IC, Emerson M, Altman DG (2010) Improving bioscience research reporting: the ARRIVE guidelines for reporting on animal research. PLoS Biol 8: e1000412.
34. Hooijmans C, Leenaars M, Ritskes-Hoitenga M (2010) A gold standard publication checklist to improve the quality of animal studies, to fully integrate the Three Rs, and to make systematic reviews more feasible. Altern Lab Anim 38: 167–182.
35. Brown BN, Valentin JE, Stewart-Akers AM, McCabe GP, Badylak SF (2009) Macrophage phenotype and remodeling outcomes in response to biologic scaffolds with and without cellular component. Biomaterials 30: 1482–1491.
36. Badylak SF, Valentin JE, Ravindra AK, McCabe GP, Stewars-Akers AM (2008) Macrophage phenotype as a determinant of biologic scaffold remodeling. Tissue Eng Part A 14: 1835–1842.
37. Nuininga JE, van Moerkerk H, Hanssen A, Hulsbergen CA, Oosterwijk-Wakka J, et al. (2004) A rabbit model to tissue engineer the bladder. Biomaterials 25: 1657–1661.
38. Grover PK, Ryall RL (1994) Urate and calcium oxalate stones: from repute to rhetoric to reality. Miner Electrolyte Metab 20: 361–370.
39. Evans K, Costabile RA (2005) Time to development of symptomatic urinary calculi in a high risk environment. J.Urol 173: 858–861.
40. Vesterinen HM, Sena ES, Egan KJ, Hirst TC, Churolov L, et al. (2014) Meta-analyis of data from animal studies: a practical guide. J Neurosci Methods 221: 92–102.
41. Roosen A, Woodhouse CR, Wood DN, Stief CG, McDougal WS, et al. (2011) Animal models in urinary diversion. BJU Int 109: 6–23.
42. De Filippo RE, Yoo JJ, Atala A (2002) Urethral replacement using cell seeded tabularized collagen matrices. J Urol 168: 1789–1792.
43. Zhang Y, Kropp BP, Lin HK, Cowan R, Cheng EY (2004) Bladder regeneration with cell-seeded small intestinal submucosa. Tissue Eng 10:181–187.
44. Sievert KD, Fandel T, Wefer J, Gleason CA, Nunes L, et al. (2006) Collagen I:III ratio in canine heterologous bladder acellular matrix grafts. World J Urol 24:101–109.
45. Probst M, Piechota HJ, Dahiya R, Tanagho EA (2000) Homologous bladder augmentation in dog with the bladder acellular matrix graft. BJU Int 85: 362–371.

46. Orabi H, Aboushwareb T, Zhang Y, Yoo JJ, Atala A (2013) Cell-seeded tabularized scaffolds for reconstruction of long urethral defects: a preclinical study. Eur Urol 63: 531–538.
47. Roth CC, Mondalek FG, Kibar Y, Ashley RA, Bell CH, et al. (2011) Bladder regeneration in a canine model using hyaluronic acid-poly(lactic-co-glycolic-acid) nanoparticle modified porcine small intestinal submucosa. BJU Int 108: 148–155.
48. ClinicalTrials.gov (2014) Incontinent urinary diversion using an autologous neo-urinary conduit. Identifier NCT 01087697.
49. Hooijmans CR, Ritskes-Hoitinga M (2013) Progress in using systematic reviews of animal studies to improve translational research. PLoS Med 10: e1001482.

16

Effect of Hydrophobic Acrylic versus Hydrophilic Acrylic Intraocular Lens on Posterior Capsule Opacification: Meta-Analysis

Ying Li[1], Jiaxing Wang[2]**, Zhuo Chen[1], Xin Tang[1]***

1 Clinical College of Ophthalmology, Tianjin Medical University, Tianjin Eye Hospital, Tianjin, China, **2** Department of Ophthalmology, Tianjin Medical University General Hospital, Tianjin, China

Abstract

Purpose: This meta-analysis aims to evaluate the differences in performance of posterior capsular opacification (PCO) between hydrophobic acrylic intraocular lens (IOLs) and hydrophilic acrylic IOLs.

Setting: Tianjin Medical University, Tianjin Eye Hospital, Tianjin Key Laboratory of Ophthalmology and Visual Science, Tianjin, China.

Design: Systematic review of randomized controlled trials (RCTs) or meta-analysis.

Methods: An electronic literature search was performed using the PubMed, EMBASE and Cochrane Library database before May in 2013 to identify prospective RCTs comparing hydrophobic acrylic IOLs and hydrophilic acrylic IOLs in patients after phacoemulsification with IOL implantation with a follow-up time of at least 1 year. Pertinent studies were selected by meeting predefined criteria and reviewed systematically by meta-analysis. The PCO scores and YAG capsulotomy rate, as indicator of PCO, were measured and discussed in a meta-analysis. Standardized mean differences (SMD), relative risk ratio (RR), and the pooled estimates were computed according to a random effect model or fixed effect model.

Results: Nine prospective RCTs involving 861 eyes were included in the current meta-analysis. The hydrophobic acrylic IOLs were favored and the pooled SMD of PCO severity was1.72 (95% confidence interval (CI), 0.20 to 1.23, $P = 0.0002$) and 1.79 (95% CI, 0.95 to 2.64, $P < 0.0001$) with 1-year follow-up and 2-year follow-up respectively. The pooled RR of Nd:YAG laser capsulotomy rates at postoperative 2-year follow-up was 6.96 (95% CI, 3.69 to 13.11, $P < 0.00001$) comparing hydrophilic acrylic IOLs with hydrophobic acrylic IOLs.

Conclusions: Compared with hydrophilic acrylic IOLs, the hydrophobic acrylic IOLs showed superior reduction in rates of PCO and laser capsulotomy in 2-year follow-up. More RCTs with standard methods for longer follow-up are needed to validate the association.

Editor: Helge Bruns, University Hospital Heidelberg, Germany

Funding: The authors have no support or funding to report.

Competing Interests: The authors have declared that no competing interests exist.

* E-mail: xtang2013@gmail.com

These authors contributed equally to this work.

Introduction

With the development of surgical techniques and biomaterial science, cataract surgery with intraocular lens (IOL) implantation has brought great benefits for patients. However, posterior capsule opacification (PCO), remains the most frequent long-term complication [1], decreasing the visual performance in 1 or 2 years after cataract surgery. Although treatment with Nd:YAG laser capsulotomy is effective, the complications, such as retina detachment, macular edema, intraocular pressure elevation [2], cannot be ignored.

Intraocular lens, with various designs and materials, have been observed in playing a vital role in the developmentof PCO. Two areas of concern are the biomaterials and the edge design of IOLs. Studies have shown that the rate of PCO with sharp edge designed IOLs was lower due to the inhibition of lens epithelial cells (LECs) migration [3,4]. Acrylic IOLs with hydrophilic or hydrophobic surfaces, as two types of biocompatibility materials, safe for intraocular implantation, have a long history of clinical practice and have shown significantly lower rates of PCO and less Nd:YAG laser capsulotomy [5–8]. Studies found that acrylic material has a relatively low propensity to induce cell proliferation in the capsular bag [9]. Yet whether the hydrophilic or hydrophobic IOLs are better for PCO prevention remains controversial.

Numerous studies have compared on PCO to different designs and materials combinations of IOLs. Few comparative studies, however have evaluated the differences between hydrophobic acrylic IOLs and hydrophilic acrylic IOLs with the same edge design specifically. The aim of this meta-analysis is to investigate the differences between hydrophobic and hydrophilic IOLs with the same edge design in the development of PCO and the rate of Nd:YAG laser capsulotomy in a 2 year period.

Materials and Methods

Literature Search

This review was conducted following the QUOROM guideline standards [10]. Reports of randomized controlled trials (RCTs) comparing hydrophobic acrylic and hydrophilic acrylic IOLs in patients after phacoemulsification with IOL implantation were identified through a computerized literature search. The systematic search was conducted in the PubMed, EMBASE, and Cochrane Controlled Trials Register database up to the end of May 2013 by using the search terms "*posterior capsular opacification*" "*intraocular lens*" "*hydrophilic*" "*hydrophobic*" and limiting the search to reports of randomized controlled trials. The abstract of all potentially relevant articles were screened to determine their relevance followed by evaluation of candidate full articles. In addition, literature reference proceedings were scanned manually to obtain extra eligible trials until no more relevant trials were found in databases. For data collected from duplicate patient groups, only the most recent studies were included in each part of meta-analysis. The process of trials selection is shown in Figure 1. Two independent investigators performed the literature search (YL, JXW).

Selection Criteria

Selected trials fulfilling the following inclusion criteria were used in this analysis: (1) prospective design, randomized controlled trial; (2) population, patients with senile cataract undergoing cataract surgery; (3) intervention, phacoemulsification and IOL implantation; (4) comparison, hydrophobic acrylic and hydrophilic acrylic IOLs; all involved IOLs are designed with sharp edge;(5) outcome variables, at least one of the following primary outcome variables: PCO score, Nd:YAG capsulotomy rate.

Exclusion criteria were as follows: (1) patients with a history of ocular diseases other than senile cataract; (2) patients with a history of intraocular surgery other than IOL; (3) Cases with intraoperative complications, such as incomplete continuous curvilinear capsulorhexis (CCC), posterior capsule rupture or whose with postoperative complications (eg, iris synechia); (4) follow-up time of less than one year.

Data Extraction and Quality Assessment

Two reviewers (YL, JXW) completed searches independently according to the above the criteria, assessed the methodological quality of trials and extracted data from each eligible randomized clinical trial results. Differences were resolved by discussion to reach consensus between the investigators and results were checked by the third author (ZC) when required. The following items were collected from each trial: author's name, year of publication, design of study, patients' mean age, gender, group size, IOL biomaterials and special designs, evaluation of PCO, and Nd:YAG capsulotomy (number or rate), follow-up period, respectively.

Outcome Measures

The primary outcome measure was the difference in PCO intensity or PCO score from software between the 2 types of IOLs at 1-year and 2-year follow-up post-operation. The secondary outcome measures were Nd:YAG capsulotomy rate at 2-year follow-up post-operation. For those patients who underwent Nd:YAG capsulotomy, the PCO value just before Nd:YAG capsulotomy was used for further statistical analysis. If there was more than 1 published report on the same population or group of patients, the most recent results with complementary data from previous articles were used for statistical analysis.

Statistical Analysis and Assessment of Heterogeneity

According to various measurement scales, PCO severity outcomes, as continuous data, were pooled using standardized mean differences (SMD) with 95% confidence intervals (CIs)[11]. The data of Nd:YAG capsulotomy rate, as dichotomous data,

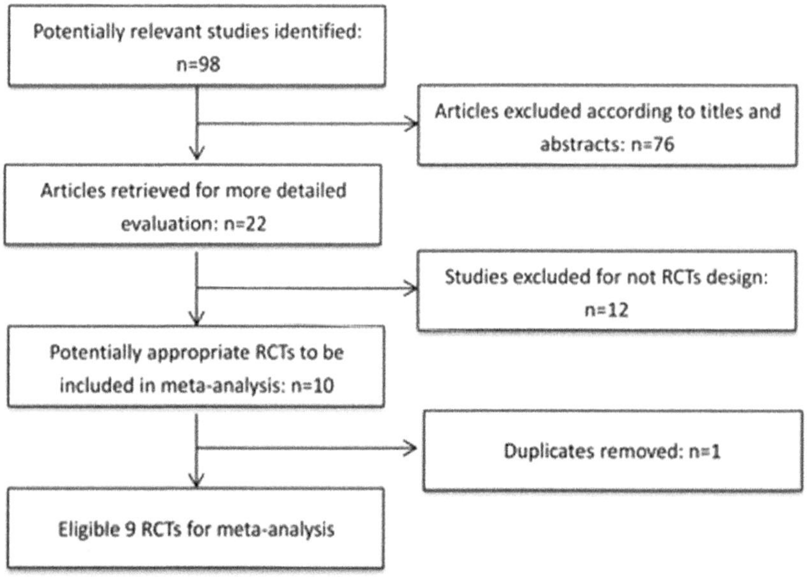

Figure 1. Flow chart of literature search and study selection.

were tabulated into 2×2 tables and the relative risk ratio (RR) and 95% CIs of the results were compared. Heterogeneity was also assessed through chisquare test, and an I^2 value greater than 50%, $P<0.1$ was considered significant. We evaluated the pooled summary effect by using random-effect model. Otherwise(I^2 value≤50%, $P\geq0.1$), data were combined using fixed-effect model to reduce the heterogeneity between studies. Finally, publication bias was assessed visually with funnel plots. The statistical analyses were carried out with RevMan software (version 5.0, The Cochrane Collaboration).

Results

Study included

The results of our research strategy are shown in Fig. 1. A total of 132 potentially relevant publications were identified through the literature search from multiple databases before May in 2013, 20 of which were based on their titles and abstracts. Only 9 random controlled trials [2,8,12–18] were eligible for this meta-analysis.

Quality assessment of eligible studies and characteristics of included IOL

The methodological quality of trials that were identified and the characteristics of 9 potential RCTs included in the current meta-analysis are presented in Table 1. In these 9 RCTs included in this meta-analysis, the investigators described the random component in a sequence generation process such as: using an envelope [2,12,17], or referring to a computer random number generator[14,15]. Only 1 of 9 studies used double-blinding[12], while the others used single-blinding[2,8,13–18]. Recruited RCTs were carried out in many countries including United Kingdom, Japan, India, Sweden and Korea. The length of the studies was between 1 and 2 years. All studies included described the dropout patients' number and reasons respectively. Characteristics of IOLs included in the recruited studies are presented in Table 2.

Efficacy analysis

Effects of hydrophobic acrylic versus hydrophilic acrylic IOLs on development of posterior capsule opacification in 1-year follow-up. Based on 7 studies (620 total eyes) that evaluated PCO after a 1-year follow-up period[2,8,12,14–16,18], hydrophobic acrylic IOLs were associated with significantly lower PCO scores than hydrophilic acrylic IOLs; the SMD was 1.72 (95% CI, 0.82 to 2.63, $P=0.0002$). The data showed that they had heterogeneity of effect size ($P<0.00001$, $I^2=96\%$), so the random effect model was used for meta-analysis. The results are shown in Figure 2.

Effects of hydrophobic acrylic versus hydrophilic acrylic IOLs on development of posterior capsule opacification in 2-year follow-up. Seven studies involving 525 eyes used different scales to report the outcomes for PCO after 2-year follow-up [8,12–14,16–18]. They also had heterogeneity of effect size ($P<0.00001$, $I^2=94\%$), so the random effect model was used for meta-analysis. A significant difference was found between the hydrophobic acrylic and hydrophilic acrylic IOLs; the SMD was 1.79 (95% CI, 0.95 to 2.64, $P<0.0001$), indicating that hydrophobic acrylic IOLs were associated with lower PCO score in 2-year follow-up. The results are shown in Figure 3.

Effects of hydrophobic acrylic versus hydrophilic acrylic IOLs on rate of Nd:YAG capsulotomy in 2-year follow-up. Seven studies[8,12–14,16–18] involving 546 eyes compared the Nd:YAG capsulotomy rate of hydrophobic acrylic IOLs with hydrophilic acrylic IOLs in a 2-year follow-up period. No statistical heterogeneity was detected between studies ($P=0.85$, $I^2=0\%$). Therefore, the fixed effect model was analyzed for this research. The results from analysis suggest that hydrophobic acrylic IOLs had a lower Nd:YAG capsulotomy rate; the RR was 6.96(95% CI, 3.69 to 13.13, $P<0.00001$). The results are shown in Figure 4.

Table 1. Evaluation of the quality of RCTs included in the meta-analysis.

Study	Random	Blind	Withdraw	Jadad score(0–5)
Gangwani 2011	Appropriate	Double-blind	Described	5
Iwase 2011	Yes	Yes	Described	3
Vasavada 2011	Appropriate	Yes	Described	4
Kang 2009	Yes	Yes	Described	3
Cleary 2009	Appropriate	Yes	Described	4
Kugelberg 2008	Appropriate	Yes	Described	4
Hancox 2007	Yes	Yes	Described	3
Kugelberg 2006	Appropriate	Yes	Described	4
Heatley 2005	Yes	Yes	Described	3

Publication Bias

The publication bias was assessed graphically for each outcome independently using funnel plots. The funnel shaped with the apex near the symmetry, which suggesting publication bias is less of a concern.

Discussion

At present, PCO remains the most common complication of modern cataract surgery. There is considerable interest in the impact of IOL on the development of PCO since characteristics and designs of IOLs play a crucial role in preventing PCO. Furthermore, differences in PCO performance between IOLs were likely to reflect their distinction in biomaterials and designs. Both hydrophilic and hydrophobic have been commonly used in cataract surgeries. Hydrophobic acrylic IOLs have a long track record of good PCO rate [19]. However, due to the mechanical properties, hydrophilic acrylic IOLs may be more suitable for implantation through smaller IOL injection systems [12]. Additionally they may have superior biocompatibility and less macrophage adhesion especially when a blood-aqueous barrier breakdown has occurred, such as those with glaucoma, uveitis, and diabetes mellitus [20]. Although higher PCO rates have been reported for hydrophilic materials [2,17,21], controversy remains over whether this finding is a result of differences in material properties or variation in the optic edge design [22]. The potential mechanism that IOLs with sharp optic edge prevent PCO, including cell migration prevention [23], pressure atrophy [15] and contact inhibition [24], have gained acceptance widely. The PCO rate in hydrophilic acrylic IOLs implantation, which have an improved 360-degree sharp edge, is reported lower than with older hydrophilic models [25,26]. It should not be ignored that many studies have a significant limitation, comparing hydrophilic IOLs with round optic edges and hydrophobic IOLs with sharp optic edges [8,27,28].

Since complete surgical removal lens epithelial cells (LECs) is not possible with modern surgical technology, the migration of remaining equatorial LECs which form PCO may be difficult to

Table 2. Characteristics of IOLs included in the meta-analysis.

IOLs	Optic Material	Haptic Material	Lens Type	Optic Diameter	Sharp Edge Design	Optic Shape	Haptic Angulation	Distinctive Feature	PCO evaluation system
Acrysof SA60AT	Hydrophobic Acrylic	Acrylic	1-piece	6 mm	Posterior	Anterior Asymmetric biconvex	0	N/A	AQUA Scheimpflug POCOman
Acrysof SN60WF	Hydrophobic Acrylic	Acrylic	1-piece	6 mm	Posterior	Posterior biconvex	0	N/A	EPCO
Sensar AR40e	Hydrophobic Acrylic	PMMA	3-piece	6 mm	Posterior	Equal biconvex	5	N/A	POCOman
Acrysof MA60AC	Hydrophobic Acrylic	PMMA	3-piece	6 mm	Posterior	Anterior Asymmetric Biconvex	10	N/A	POCOman
Acrysof MA30AC	Hydrophobic Acrylic	PMMA	3-piece	5.5 mm	Double	Anterior Asymmetric Biconvex	5	N/A	POCOman
Idea 613XC	Hydrophilic Acrylic	Acrylic	1-piece	6 mm	Double	Biconvex	9	Broad optic-haptic junction	AQUA
Meridian HP60M	Hydrophilic Acrylic	PMMA	1-piece	6 mm	Double	Anterior biconvex	6	N/A	Scheimpflug
C-flex 570C	Hydrophilic Acrylic	Acrylic	1-piece	5.75 mm	Double	N/A	0	N/A	EPCO
Akreos Adapt	Hydrophilic Acrylic	Acrylic	1-piece	6 mm	Double	Equal biconvex	0	N/A	EPCO
Bio Vue3	Hydrophilic Acrylic	PVDF+	3-piece	6 mm	Double	Equal biconvex	5	Heparin surface modification	POCOman
MC611M	Hydrophilic Acrylic	Acrylic	1-piece	6 mm	Double	N/A	0	Broad optic-haptic junction	POCOman
BL27	Hydrophilic Acrylic	Acrylic	1-piece	6 mm	Posterior	N/A	0	N/A	POCOman
1CU	Hydrophilic Acrylic	Acrylic	1-piece	5.5 mm	Double	Equal biconvex	0	4 haptics Accommodation	POCOman
Centerflex 570H	Hydrophilic Acrylic	Acrylic	1-piece	5.75 mm	Posterior	Equal biconvex	0	N/A	POCOman

Study or Subgroup	hydrophilic acrylic IOLs Mean	SD	Total	hydrophobic acrylic IOLs Mean	SD	Total	Weight	Std. Mean Difference IV, Random, 95% CI	Year
Heatley 2005	50.3	15.365	45	4.9	1.56	45	12.7%	4.12 [3.38, 4.86]	2005
Kugelberg 2006	0.18	0.055	57	0.055	0.2075	59	13.5%	0.81 [0.43, 1.19]	2006
Hancox 2007	87.3	11.75	20	3.1	0.9	28	7.2%	10.92 [8.57, 13.27]	2007
Kang 2009	6.12	5.21	28	5.91	4.91	30	13.3%	0.04 [-0.47, 0.56]	2009
Cleary 2009	17.03	24.15	27	5.104	9.286	27	13.2%	0.64 [0.09, 1.19]	2009
Gangwani 2011	19.2	15	63	15	11.5	63	13.6%	0.31 [-0.04, 0.66]	2011
Vasavada 2011	0.23	0.47	33	0.02	0.11	33	13.3%	0.61 [0.11, 1.10]	2011
Vasavada 2011	0.31	0.49	31	0.01	0.32	31	13.3%	0.72 [0.20, 1.23]	2011
Total (95% CI)			**304**			**316**	**100.0%**	**1.72 [0.82, 2.63]**	

Heterogeneity: $Tau^2 = 1.55$; $Chi^2 = 164.97$, df = 7 ($P < 0.00001$); $I^2 = 96\%$
Test for overall effect: $Z = 3.73$ ($P = 0.0002$)

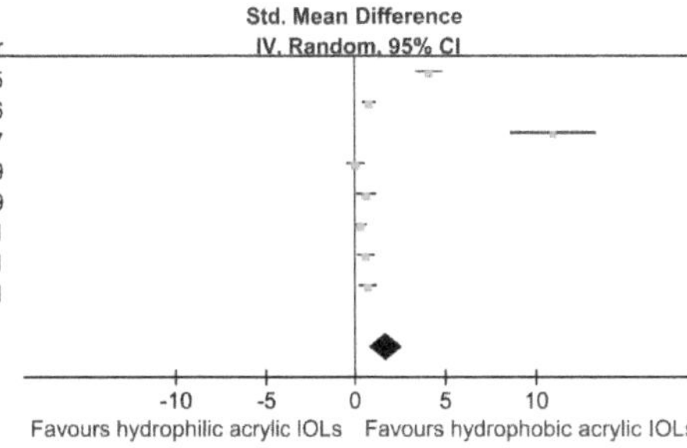

Figure 2. Forest plots describing effects of hydrophobic acrylic versus hydrophilic acrylic IOLs on development of posterior capsule opacification in 1-year follow-up. (Chi^2 = chi square statistic, CI = confidence interval, df = degrees of freedom, I^2 = I-square heterogeneity statistic, IV = inverse variance, SMD = standard mean difference, Z = Z-statistic).

avoid. As a proliferative pathological process, there is a close relationship between the severity of PCO and post-operative follow-up time. The longer follow-up period, the better clinicians understand the effect of IOLs on the development of PCO. Some studies have shown no differences in PCO rates between the two different materials IOLs after cataract surgery 1 year [13,15]. Others have shown significantly different results [14,16]. Indeed, some researches have addressed the influence of various IOLs on the incidence of PCO in different lengths of follow-up [1,29]. Therefore long term randomized controlled trials, especially with a multi center large sample size, are needed to evaluate further effects of various IOL biomaterials with similar optic edge designs in decreasing PCO and Nd:YAG capsulotomy rates.

This meta analysis evaluated the 1-year and 2-year postoperative PCO results and rate of Nd:YAG laser capsulotomy in 2-year follow-up between hydrophilic acrylic and hydrophobic acrylic IOLs implantation respectively. All the IOLs involved in this analysis were designed to prevent PCO by incorporating a sharp edge. Therefore, the differences shown in this analysis may be interpreted primarily based on material effects. The result of this meta-analysis support the theory that compared to hydrophilic acrylic IOLs, hydrophobic acrylic IOLs led to significantly less PCO in 1-year and 2-year follow-up periods. Meanwhile, the rates of Nd:YAG laser capsulotomy were also reduced following hydrophobic acrylic IOLs implantation 2-year post-operation. A reasonable interpretation for the difference in this meta-analysis may be that hydrophobic acrylic IOLs can adhere to collagen membrane [30], leading to tight apposition of IOLs in posterior capsular bag, and advanced adhesiveness through fibronectin [8]. This may result in less space between IOLs and posterior capsule where the LECs could migrate. On the other hand, the hydrophilic surface properties were found to promote proliferation and migration of LECs from the equatorial area to the visual region [31]. Moreover, studies compared electron microscope images and found the edge of hydrophilic IOLs to be less sharp than hydrophobic IOLs at several optical powers [13]. The potential reason could be hydrophilic acrylic IOLs are machined in the dehydrated state and then rehydrated, which can lead to loss of edge sharpness [32]. These differences in manufacturing techniques may explain why hydrophobic IOLs appear to have relatively better PCO performance. The recent study [33] suggest a new aspect to consider in lens material, hybrid technique (hydrophilic center and hydrophobic surface coated IOLs), indicating that hybrid IOLs are less susceptible than hydrophilic IOLs to cell adhesion and less susceptible than hydrophobic IOLs to glistening formation. The copolymer hybrid IOLs may present certain important advantages and should therefore be further evaluated with PCO performance in clinical studies.

The studies in this meta-analysis used a variety of different evaluation systems for PCO analysis, such as Scheimpflug photography system [13], POCOman system [2,8,15–18], AQUA (Automated Quantification of After-Cataract system) 12, and EPCO (Evaluation of Posterior Capsule Opacification software) [14]. Each of these software systems has particular features that made them vulnerable. POCOman and AQUA system are objective, however, they do not incorporate whether PCO is peripheral or central location and show limited points [29]. EPCO evaluates the construct validity [34], but it is a subjective style. On account of the different analysis systems for PCO across the studies, SMDs were used in this meta-analysis as in previous studies [29].

Study or Subgroup	Hydrophilic acrylic IOLs Mean	SD	Total	Hydrophobic acrylic IOLs Mean	SD	Total	Weight	Std. Mean Difference IV, Random, 95% CI	Year
Heatley 2005	58.9	16.41	13	1.7	0.125	13	9.9%	4.77 [3.17, 6.38]	2005
Hancox 2007	96.4	10.6	10	2.2	0.8	26	3.2%	16.76 [12.59, 20.92]	2007
Kugelberg 2008	46	25	57	4.5	17.75	58	14.7%	1.90 [1.46, 2.35]	2008
Cleary 2009	25.45	34.51	24	7.82	13.35	24	14.2%	0.66 [0.08, 1.25]	2009
Gangwani 2011	2.6	2	23	1.9	1.9	23	14.2%	0.35 [-0.23, 0.94]	2011
Vasavada 2011	0.46	1.21	33	0.02	0.14	33	14.5%	0.50 [0.01, 1.00]	2011
Vasavada 2011	1.3	1.35	31	0	0.13	31	14.3%	1.34 [0.78, 1.89]	2011
Iwase 2011	24	14.5	63	16	12.3	63	14.9%	0.59 [0.23, 0.95]	2011
Total (95% CI)			**254**			**271**	**100.0%**	**1.79 [0.95, 2.64]**	

Heterogeneity: $Tau^2 = 1.22$; $Chi^2 = 108.60$, df = 7 ($P < 0.00001$); $I^2 = 94\%$
Test for overall effect: $Z = 4.15$ ($P < 0.0001$)

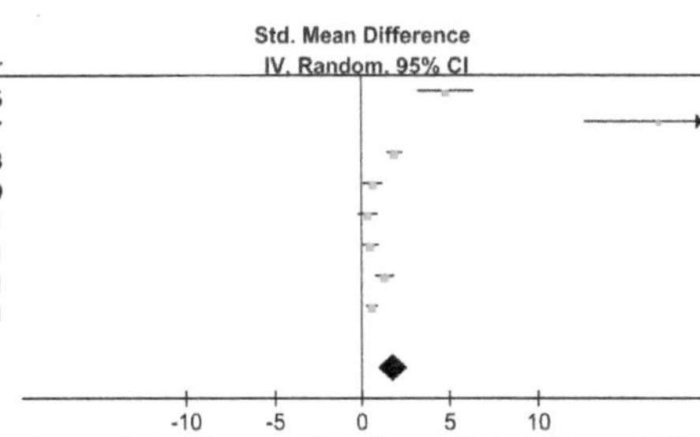

Figure 3. Forest plots describing effects of hydrophobic acrylic versus hydrophilic acrylic IOLs on development of posterior capsule opacification in 2-year follow-up. (Chi^2 = chi square statistic, CI = confidence interval, df = degrees of freedom, I^2 = I-square heterogeneity statistic, IV = inverse variance, SMD = standard mean difference, Z = Z-statistic).

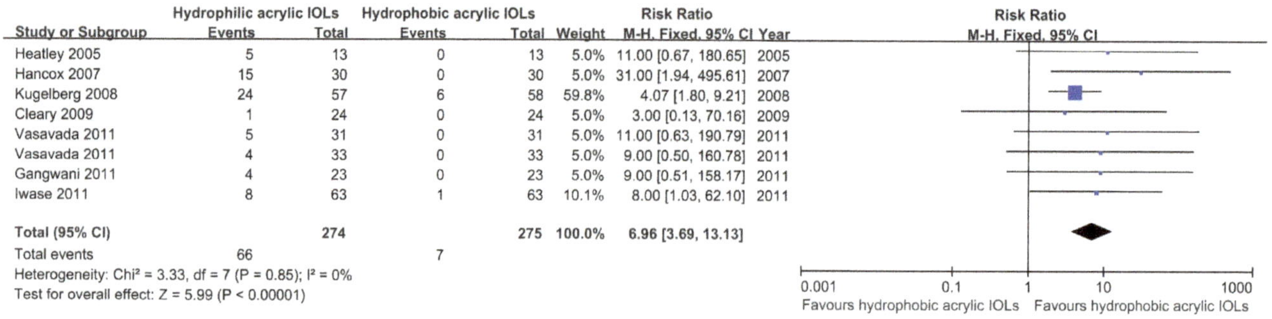

Figure 4. Forest plots describing Effects of hydrophobic acrylic versus hydrophilic acrylic IOLs on rate of Nd:YA Gcapsulotomy in 2-year follow-up. (Chi^2 = chi square statistic, CI = confidence interval, df = degrees of freedom, I^2 = I-square heterogeneity statistic, IV = inverse variance, RR = risk ratio, Z = Z-statistic).

Limitation of this meta-analysis may be found. First, even though all the IOLs in this meta-analysis were designed with sharp edge, as a key factor in retarding PCO, there were various extra factor could make an impact. For example, the IOLs with broad optic-haptic junctions [16], the heparin-surface-modified [15], the asphericity on the posterior surface [35], the degree of haptic angulation [36], and optic size [37] appear to exert influence upon PCO formation. The pathophysiology of PCO is multifactorial. Since dissociation of each factor in PCO development is almost impossible, it is very difficult to observe individual elements in clinical practice. Second, although we conducted a thorough electronic search and a manual search of the references of relevant results to minimize selection and publication bias, there were not sufficient studies included to verify if asymmetry exists in a funnel plot. Consequently, long-term postoperative follow-up of multi-center large-sample size randomized controlled trials for PCO development after cataract surgery are necessary.

Author Contributions

Conceived and designed the experiments: YL ZC XT. Performed the experiments: YL JXW. Analyzed the data: YL JXW ZC. Contributed reagents/materials/analysis tools: YL JXW ZC. Wrote the paper: YL JXW.

References

1. Cheng JW, Wei RL, Cai JP, Xi GL, Zhu H, et al. (2007) Efficacy of different intraocular lens materials and optic edge designs in preventing posterior capsular opacification: a meta-analysis. Am J Ophthalmol. 143:428–436.
2. Kugelberg M, Wejde G, Jayaram H, Zetterström C (2006) Posterior capsule opacification after implantation of a hydrophilic or a hydrophobic acrylic intraocular lens: one-year follow-up. Journal of cataract and refractive surgery. 32:1627–1631.
3. Buehl W, Findl O (2008) Effect of intraocular lens design on posterior capsule opacification. Journal of cataract and refractive surgery. 34:1976–1985.
4. Buehl W, Findl O, Menapace R, Sacu S, Kriechbaum K, et al. (2005) Long-term effect of optic edge design in an acrylic intraocular lens on posterior capsule opacification Journal of cataract and refractive surgery. 31:954–961.
5. Findl O, Menapace R, Sacu S, Buehl W, Rainer G (2005) Effect of optic material on posterior capsule opacification in intraocular lenses with sharp-edge optics: randomized clinical trial. Ophthalmology. 112:67–72.
6. Beltrame G, Salvetat ML, Chizzolini M, Driussi GB, Busatto P, et al. (2002) Posterior capsule opacification and Nd:YAG capsulotomy rates after implantation of silicone, hydrogel and soft acrylic intraocular lenses: a two-year follow-up study. European journal of ophthalmology. 12:388–394.
7. Hollick EJ, Spalton DJ, Ursell PG, Pande MV, Barman SA, et al. (1999) The effect of polymethylmethacrylate, silicone, and polyacrylic intraocular lenses on posterior capsular opacification 3 years after cataract surgery. Ophthalmology. 106:49–54; discussion 54–45
8. Heatley CJ, Spalton DJ, Kumar A, Jose R, Boyce J, et al. (2005) Comparison of posterior capsule opacification rates between hydrophilic and hydrophobic single-piece acrylic intraocular lenses. Journal of cataract and refractive surgery. 31:718–724.
9. Apple DJ, Peng Q, Visessook N, Werner L, Pandey SK, et al. (2001) Eradication of posterior capsule opacification: documentation of a marked decrease in Nd:YAG laser posterior capsulotomy rates noted in an analysis of 5416 pseudophakic human eyes obtained postmortem. Ophthalmology. 108:505–518.
10. Ramos Macias A, de Miguel Martinez I, Martin Sanchez AM, Gómez González JL, Martín Galán A (1992) The incorporation of acyclovir into the treatment of peripheral paralysis. A study of 45 cases. Acta otorrinolaringologica espanola. 43:117–120.
11. Zhu XF, Zou HD, Yu YF, Sun Q, Zhao NQ (2012) Comparison of blue light-filtering IOLs and UV light-filtering IOLs for cataract surgery: a meta-analysis. PloS one. 7:e33013.
12. Gangwani V, Hirnschall N, Koshy J, Crnej A, Nishi Y, et al. (2011) Posterior capsule opacification and capsular bag performance of a microincision intraocular lens. Journal of cataract and refractive surgery. 37:1988–1992.
13. Iwase T, Nishi Y, Oveson BC, Jo YJ (2011) Hydrophobic versus double-square-edged hydrophilic foldable acrylic intraocular lens: effect on posterior capsule opacification. Journal of cataract and refractive surgery. 37:1060–1068.
14. Vasavada AR, Raj SM, Shah A, Shah G, Vasavada V, et al. (2011) Comparison of posterior capsule opacification with hydrophobic acrylic and hydrophilic acrylic intraocular lenses. Journal of cataract and refractive surgery. 37:1050–1059.
15. Kang S, Choi JA, Joo CK (2009) Comparison of posterior capsular opacification in heparin-surface-modified hydrophilic acrylic and hydrophobic acrylic intraocular lenses. Japanese journal of ophthalmology. 53:204–208.
16. Cleary G, Spalton DJ, Hancox J, Boyce J, Marshall J (2009) Randomized intraindividual comparison of posterior capsule opacifcation between a microincision intraocular lens and a conventional intraocular lens. Journal of cataract and refractive surgery. 35:265–272.
17. Kugelberg M, Wejde G, Jayaram H, Zetterström C (2008) Two-year follow-up of posterior capsule opacification after implantation of a hydrophilic or hydrophobic acrylic intraocular lens. Acta ophthalmologica. 86:533–536.
18. Hancox J, Spalton D, Heatley C, Jayaram H, Yip J, et al. (2007) Fellow-eye comparison of posterior capsule opacification rates after implantation of 1CU accommodating and AcrySof MA30 monofocal intraocular lenses. Journal of cataract and refractive surgery. 33:413–417.
19. Leydolt C, Davidovic S, Sacu S, Menapace R, Neumayer T, et al. (2007) Long-term effect of 1-piece and 3-piece hydrophobic acrylic intraocular lens on posterior capsule opacification: a randomized trial. Ophthalmology. 114:1663–1669.
20. Richter-Mueksch S, Kahraman G, Amon M, Schild-Burggasser G, Schauersberger J, et al. (2007) Uveal and capsular biocompatibility after implantation of sharp-edged hydrophilic acrylic, hydrophobic acrylic, and silicone intraocular lenses in eyes with pseudoexfoliation syndrome. Journal of cataract and refractive surgery. 33:1414–1418.
21. Findl O, Buehl W, Bauer P, Sycha T (2010) Interventions for preventing posterior capsule opacification. Cochrane database of systematic reviews. 2:CD003738.
22. Werner L, Tetz M, Feldmann I, Bücker M (2009) Evaluating and defining the sharpness of intraocular lenses: microedge structure of commercially available

square-edged hydrophilic intraocular lenses. Journal of cataract and refractive surgery. 35:556–566.
23. Nagamoto T, Fujiwara T (2003) Inhibition of lens epithelial cell migration at the intraocular lens optic edge: role of capsule bending and contact pressure. Journal of cataract and refractive surgery. 29:1605–1612.
24. Nishi O, Yamamoto N, Nishi K, Nishi Y (2007) Contact inhibition of migrating lens epithelial cells at the capsular bend created by a sharp-edged intraocular lens after cataract surgery. Journal of cataract and refractive surgery. 33:1065–1070.
25. Nishi Y, Rabsilber TM, Limberger IJ, Reuland AJ, Auffarth GU (2007) Influence of 360-degree enhanced optic edge design of a hydrophilic acrylic intraocular lens on posterior capsule opacification. Journal of cataract and refractive surgery. 33:227–231.
26. Khandwala MA, Marjanovic B, Kotagiri AK, Teimory M (2007) Rate of posterior capsule opacification in eyes with the Akreos intraocular lens. Journal of cataract and refractive surgery. 33:1409–1413.
27. Abela-Formanek C, Amon M, Schauersberger J, Kruger A, Nepp J, et al. (2002) Results of hydrophilic acrylic, hydrophobic acrylic, and silicone intraocular lenses in uveitic eyes with cataract: comparison to a control group. Journal of cataract and refractive surgery. 28:1141–1152.
28. Hayashi K, Hayashi H (2004) Posterior capsule opacification after implantation of a hydrogel intraocular lens. The British journal of ophthalmology. 88:182–185.
29. Li N, Chen X, Zhang J, Zhou Y, Yao X, et al. (2008) Effect of AcrySof versus silicone or polymethyl methacrylate intraocular lens on posterior capsule opacification. Ophthalmology. 115:830–838.
30. Nagata T, Minakata A, Watanabe I (1998) Adhesiveness of AcrySof to a collagen film. Journal of cataract and refractive surgery. 24:367–370.
31. Dorey MW, Brownstein S, Hill VE, Mathew B, Botton G, et al. (2003) Proposed pathogenesis for the delayed postoperative opacification of the hydroview hydrogel intraocular lens. Am J Ophthalmol.. 135:591–598.
32. Hancox J, Spalton D, Cleary G, Boyce J, Nanavaty MA, et al. (2008) Fellow eye comparison of posterior capsule opacification with AcrySof SN60AT and AF-1 YA-60BB blue-blocking intraocular lenses. J Cataract Refract Surg. 34:1489–1494
33. Fujita S, Tanaka T, Miyata A, Hirose M, Usui M (2012) Cell adhesion and glistening formation in hybrid copolymer intraocular lenses. Ophthalmic Res. 48:102–108
34. Aslam TM, Aspinall P, Dhillon B (2003) Posterior capsule morphology determinants of visual function. Graefe's archive for clinical and experimental ophthalmology = Albrecht von Graefes Archiv fur klinische und experimentelle Ophthalmologie. 241:208–212.
35. Biber JM, Sandoval HP, Trivedi RH, de Castro LE, French JW, et al. (2009) Comparison of the incidence and visual significance of posterior capsule opacification between multifocal spherical, monofocal spherical, and monofocal aspheric intraocular lenses. Journal of cataract and refractive surgery. 35:1234–1238.
36. Sacu S, Menapace R, Wirtitsch M, Buehl W, Rainer G, et al. (2004) Effect of anterior capsule polishing on fibrotic capsule opacification: three-year results. Journal of cataract and refractive surgery. 30:2322–2327.
37. Wren SM, Spalton DJ, Jose R, Boyce J, Heatley CJ (2005) Factors that influence the development of posterior capsule opacification with a polyacrylic intraocular lens. Am J Ophthalmol. 139:691–695.

QHREDGS Enhances Tube Formation, Metabolism and Survival of Endothelial Cells in Collagen-Chitosan Hydrogels

Jason W. Miklas[1], Susan M. Dallabrida[2], Lewis A. Reis[1], Nesreen Ismail[2], Maria Rupnick[3], Milica Radisic[1,4]*

1 The Institute of Biomaterials and Biomedical Engineering, University of Toronto, Toronto, Ontario, Canada, **2** Center for Vascular Biology Research, Beth Israel Deaconess Medical Center, Boston, Massachusetts, United States of America, **3** Brigham and Women's Hospital, Cardiovascular Division, Boston, Massachusetts, United States of America (Affiliates of Harvard Medical School, Boston, Massachusetts, United States of America), **4** Department of Chemical Engineering and Applied Chemistry, University of Toronto, Toronto, Ontario, Canada

Abstract

Cell survival in complex, vascularized tissues, has been implicated as a major bottleneck in advancement of therapies based on cardiac tissue engineering. This limitation motivates the search for small, inexpensive molecules that would simultaneously be cardio-protective and vasculogenic. Here, we present peptide sequence QHREDGS, based upon the fibrinogen-like domain of angiopoietin-1, as a prime candidate molecule. We demonstrated previously that QHREDGS improved cardiomyocyte metabolism and mitigated serum starved apoptosis. In this paper we further demonstrate the potency of QHREDGS in its ability to enhance endothelial cell survival, metabolism and tube formation. When endothelial cells were exposed to the soluble form of QHREDGS, improvements in endothelial cell barrier functionality, nitric oxide production and cell metabolism (ATP levels) in serum starved conditions were found. The functionality of the peptide was then examined when conjugated to collagen-chitosan hydrogel, a potential carrier for *in vivo* application. The presence of the peptide in the hydrogel mitigated paclitaxel induced apoptosis of endothelial cells in a dose dependent manner. Furthermore, the peptide modified hydrogels stimulated tube-like structure formation of encapsulated endothelial cells. When integrin $\alpha v\beta 3$ or $\alpha 5\beta 1$were antibody blocked during cell encapsulation in peptide modified hydrogels, tube formation was abolished. Therefore, the dual protective nature of the novel peptide QHREDGS may position this peptide as an appealing augmentation for collagen-chitosan hydrogels that could be used for biomaterial delivered cell therapies in the settings of myocardial infarction.

Editor: Maria A. Deli, Biological Research Centre of the Hungarian Academy of Sciences, Hungary

Funding: The authors wish to acknowledge the following funding sources: NSERC Strategic Grant (STPGP 381002-09), NSERC-CIHR Collaborative Health Research Grant (CHRPJ 385981-10), NSERC Discovery Grant (RGPIN 326982-10) and Discovery Accelerator Supplement (RGPAS 396125-10), ORF GL2 grant and McLean Award (to MR). The funders had no role in study design, data collection and analysis, decision to publish, or preparation of the manuscript.

Competing Interests: The authors have declared that no competing interests exist.

* E-mail: m.radisic@utoronto.ca

Introduction

In order to stimulate repair of damaged organs, many research groups have looked towards various cell therapies. However, in the vast majority of cases, injecting cells into a damaged organ/tissue results in a large amount of the cells not remaining at the injury site, while the cells that do remain undergo apoptosis due to lack of nutrients and oxygen [1]. Consequently, novel methods are required to keep cells localized to the injury site and ensure their survival until a vascular network is created [2].

Hydrogels are used as the base biomaterial in many cell therapy strategies as they provide a non-invasive option for cell injection [3]. Since hydrogels can be mixed with cells in liquid state and allowed to gel upon injection, these hydrogel cell suspensions can be administered through a minimally invasive catheter injection procedure. However, the problem of creating a stable vasculature to ensure nutrient and oxygen supply to the injected cells still remains.

To address this, efforts have focused on implanting endothelial cells (EC) within or around biomaterials to help induce vascular formation as well as the use of angiogenic growth factors to help promote vessel formation [4,5,6,7]. The main caveat with using growth factors in a clinical or commercial setting is their high cost and susceptibility to denaturation [8]. Furthermore, it is difficult to quality-control biomaterials modified with growth factors due to the variable shelf-life of growth factors. Covalent or physical immobilization may result in protein inactivation and blocking of the active site. To solve this problem, it is possible to use short peptide sequences that are derived from a specific growth factor to stimulate a similar response in cells. These peptide sequences would be more stable than their growth factor counterparts while, also being less susceptible to conformational changes during binding or encapsulation in biomaterials as they are short linear chains of amino acids. Most importantly, synthetic peptides are significantly more cost effective to produce than recombinant human proteins.

One such peptide sequence of interest for use in biomaterials is the novel angiopoietin-1 (ang1) based peptide QHREDGS [9]. This peptide sequence is based on the fibrinogen-like domain of ang1. Previously, we reported that this peptide sequence was able to promote the survival of neonatal rat cardiomyocytes during paclitaxel (taxol) treatment on glass slides treated with the QHREDGS peptide conjugated to photocrosslinkable chitosan [10]. Our laboratory further characterized this peptide's function in a collagen-chitosan hydrogel with encapsulated cardiomyocytes finding that the peptide enhanced neonatal rat cardiomyocyte morphology, viability and metabolic activity in a dose dependent manner [11]. The cardio-protective effects of the peptide in concert with the ease with which it can be incorporated into a hydrogel for minimally invasive delivery make this platform appealing as a potential therapy for post myocardial infarction rehabilitation.

Upon myocardial infarction, the ventricular wall loses functional vasculature as well as cardiomyocytes. With time, a non-contractile scar tissue composed of dense extracellular matrix and fibroblasts will be formed. As a result, even if cells were implanted with the aid of a biomaterial for cell localization, the lack of vasculature would result in cell death due to the lack of oxygen and nutrients motivating the need for the development of new biomolecules that will be cardioprotective and enhance blood vessel formation simultaneously. We hypothesised that the QHREDGS peptide would exhibit beneficial effects on endothelial cell survival and tube formation.

Here, we evaluated the effect of soluble QHREDGS peptide on metabolism, permeability and nitric oxide (NO) release of endothelial cell monolayers. Next, the peptide was covalently immobilized to chitosan to create a collagen-chitosan hydrogel, previously shown suitable for myocardial cell injection [11]. Monolayer studies of endothelial cells cultured on the surface of the peptide modified collagen-chitosan hydrogel were performed. These studies allowed us to determine the combined effect of the hydrogel and peptide on endothelial cell phenotype, such as CD31 expression, vascular endothelial cadherin (VE-CAD) expression, cell size and survival. Finally, studies of endothelial cells encapsulated within the peptide modified hydrogel were performed to determine which integrins mediate EC-peptide interactions and to model implantation studies *in vitro*. Collectively our findings suggest that QHREDGS stimulates endothelial cell-cell interactions, enhances cell metabolism and promotes survival.

Materials and Methods

Endothelial Cell Maintenance

Dermal human microvascular endothelial cells (HMVEC-D) were obtained (Lonza) and cultured as recommended in the manufacturer's instructions in full endothelial growth media. Cells were cultured on plates coated with 50 μg/ml collagen I (10 min) (Advanced Biomatrix). Microvascular endothelial cells (MS1 EC) were obtained (ATCC) and cultured as previously described [9]. The human umbilical vein endothelial cells (HUVECs) were received from Lonza (product number C2519A). The cells were then expanded in endothelial cell growth medium-2 (EGM2), which was obtained from Lonza, and frozen down at passage 3. The final concentration of fetal bovine serum in the media was 2%. There was no bovine brain extract and a number of additives were included in the culture media before use as per manufacturer's instructions. They included: human epidermal growth factor (hEGF), hydrocortisone, GA-1000, vascular endothelial growth factor (VEGF), human basic fibroblast growth factor (hFGF-B), insulin-like growth factor-1 (R^3-IGF-1), ascorbic acid, and heparin. The media was stored at −20°C and heated to 37°C before use.

Peptide Modified Chitosan (UP-G113-QHREDGS)

QHREDGS peptide was conjugated to chitosan using 1-ethyl-3-(3-dimethylaminopropyl) carbodiimide HCl (EDC) chemistry in a similar fashion as previously described [10,11]. To determine the dose response of the peptide, two different concentrations were implemented, a 100 μM and 650 μM concentration of the peptide in proportion to the amount of chitosan present. Chitosan (UP-G113, Novamatrix) was dissolved at 20 mg/ml in 0.9% normal saline and peptide at 10 mg/ml in PBS (Lonza). This solution was then mixed with EDC and *N*-hydroxysulfosuccinimede (S-NHS) dissolved in PBS to a final solution concentration of 5 mg/ml chitosan and the two peptide concentrations: 100 μM for the low peptide (0.5 mg/ml) concentration and 650 μM for the high peptide (3.0 mg/ml) concentration. It should be noted that the ratio of [EDC]/[peptide] and [S-NHS]/[EDC] was kept constant in the two reaction mixtures at 0.8 and 2.75.

Chitosan-collagen Hydrogel Formulation

The chitosan-collagen hydrogel was created in a similar fashion as previously described [11]. Briefly, the hydrogel had a final concentration of 2.5 mg/ml each of chitosan and collagen; consisted of 203.1 μl pure or peptide-modified chitosan mixed with 100 μl 10X PBS, 681.2 μl of 3.67 mg/ml collagen (BD Biosciences), and 15.7 μl 1N NaOH on ice to make 1 ml of gel solution. The hydrogel was micropipetted onto a tissue culture plate and placed within a humidified incubator at 5% CO_2 and 37°C to allow the hydrogel to gel.

Effects of Soluble QHREDGS on Endothelial Cell Monolayers

Transwell assay. HMVEC-D were plated on collagen I-coated 12 well Transwell plates (0.4 μm pore size polyester membrane, 6.5 mm diameter inserts) in full EGM2 and grown to confluence N = 3/group with experiments done in triplicate. Some plates were stained using Giemsa to confirm confluence via inverted phase contrast tissue culture microscopy. In wells containing non-stained confluent cells, full media was removed and changed to serum starvation media. Serum starvation media was composed of EGM2 diluted 1 to 5 with endothelial basal media 2 (EBM2) (Lonza). This resulted in a final serum concentration of 1%. Cells were maintained in serum starvation media for 1 hr (37°C, 5% CO_2). Next, FITC-albumin (Sigma) (1 mg/ml) was added to the top wells (1 hr, 37°C, 5% CO_2). Linear peptide QHREDGS, scrambled peptide DQSHGER, or matched vehicle control (PBS) was added to a final concentration of 400 μM of peptides (1 hr), human plasma-derived thrombin was added at 2 U/ml (Calbiochem), and cells were incubated (3 hr, 37°C, 5% CO_2). Fluorescence was quantified using a SpectraMax plate reader (excitation 485 nm/emission 538 nm). The coefficient of albumin permeability (P_a) was calculated as described [12] and P_a/Hr determined. The assay was performed at least 4 times with N = 2/group for each repetition.

Transendothelial electrical resistance. Collagen I-coated (10 min) ECIS electrode array eight chamber slides were rinsed 5X in Hank's balanced salt solution (HBSS). HMVEC-D were trypsinized and resuspended in phenol red free EBM (Lonza) plus 1X concentrations of EGM2 (Lonza). Then, ECs were plated and cultured on collagen I-coated chamber slides (37°C, 5% CO_2) in full growth media in the ECIS 1600R chamber (Applied Biophysics). When ECs were confluent, half of the full media was removed and replaced with phenol red free EBM three times, resulting in a serum starvation media containing 1/6 of the growth factor concentration of full growth media. Linear peptide

QHREDGS, scrambled peptide DQSHGER, or matched vehicle control (PBS) to a final concentration of 400 μM peptides was added (1 hr, 37°C, 5% CO_2). Next, thrombin was added at 2 U/ml and cells incubated up to 3 days (37°C, 5% CO_2). Transendothelial electrical resistance was monitored throughout the study (ECIS 1600) using N = 2/group with studies conducted in triplicate. The assay was initiated without thrombin for 1 hr to examine the permeability among conditions at baseline. The permeability was assessed for up to 70 hr to demonstrate the peptide's effect at longer times.

Fluorometric nitric oxide assay. The MS1 ECs were cultured in 96-well plates and rinsed twice in phenol red-free DMEM (Invitrogen). Linear QHREDGS peptide, scrambled QHREDGS peptide (DQSHGER), GRGDSP peptide (SynBioSci), angiopoietin-2 (ang2), or ang1 (R&D Systems) or matched vehicles (PBS) were added (N = 6/group with studies done in duplicate) in phenol red-free DMEM/0.5% BSA and incubated (0.5 hr, 37°C, 10% CO_2). Then, 100 μM L-arginine (Sigma) was added. Where indicated, 100 μM of S-Nitroso-*N*-acetylpenicillamine (SNAP) (Calbiochem) was also added and cells were incubated (1.5 hr, 37°C, 10% CO_2). A quantitative fluorometric extracellular NO assay (Calbiochem) was performed as per manufacturer's instructions. Since the final products of NO are nitrates and nitrites, this assay uses nitrate reductase to convert nitrates to nitrites, then uses 2,3-diaminonapthalene and NaOH to convert nitrites to a fluorescent compound. Plates were read using a Wallac 1420 Multilabel Victor^3 microplate reader (Perkin Elmer), with excitation at 355 nm and emission at 430 nm.

Survival and energetics assays. The MS1 ECs were cultured in full media on 96 well plates. Cells were rinsed twice in basal DMEM media (Invitrogen) and linear QHREDGS peptide, scrambled QHREDGS peptide (DQSHGER), or vehicle (PBS) were added in each cell's basal media plus 0.5% BSA. Cells were incubated for 1 day at 37°C, 10% CO_2. MTT assays were conducted as described [9]. ATP levels were quantified using the CellTiter-Glo Luminescent Assay (Promega) as per manufacturer's instructions. NADH/NADPH levels were measured using the CellTiter-96 Aqueous One Assay (Promega). Viability/NAD(P)H levels were used following manufacturer's instructions. For both assays, N = 6/group, with studies done in duplicate. Luminescence was measured using a Wallac 1420 microplate reader for ATP studies and the optical density (OD) at 490 nm was measured to assess NADH/NADPH levels.

Effects of QHREDGS Immobilized In Collagen-chitosan Hydrogel on Endothelial Cells

Live/dead staining. Live/dead staining was performed on N = 3 gels/group with a HUVEC seeding density of 187,500 cells per hydrogel coated well of a 12 well plate and a working volume of 1 ml during the 4 days of cultivation. The culture media was changed fully every other day. Prior to application of the staining solution, the gels were washed once with 1×PBS (Gibco). The live/dead staining solution (500 μl) consisted of 89% of the total volume being culture media, 10.8% of the final volume being 1.5 μM propidium iodide (Invitrogen), and 0.2% of the final volume being 10 μM 5-(and-6-)-Carboxyfluorescein Diacetate, Succinimid – Mixed Isomers (CFDA-SE) (Invitrogen). After 30 min at 37°C, the staining solution was removed, the gels were washed once with PBS, 200 μl of PBS was added back to the solution and the samples were then observed under a fluorescence microscope (IX81, Olympus). Live and dead cells were counted using the cell counter plug in on ImageJ to determine cell viability.

CD31 staining and VE-CAD staining. Immunostaining was performed on N = 3 gels/group with a HUVEC seeding density of 187,500 cells per hydrogel coated well of a 12 well plate and a working volume of 1 ml during the 4 days of cultivation. The cells were fixed in 4% paraformaldehyde for 20 min. The samples were blocked in 10% normal goat serum (NGS) for 40 min at room temperature followed by the application of the primary antibody CD31 (Abcam, 1:200) or VE-CAD (Abcam, 1:100) for 18 hr at 4°C. For CD31 staining, the secondary antibody solution consisted of 10% normal goat serum in PBS with 1:100 anti-mouse antibody Alexa Fluor 488 (AbCam) and 1:100 DAPI added. For VE-CAD staining the secondary antibody solution consisted of 1.5% normal goat serum in PBS with 1:100 anti-rabbit antibody Rhodamine (Jackson ImmunoResearch) and 1:100 DAPI added. The hydrogels were then covered for one hr and incubated at room temperature. The hydrogels were then imaged under a fluorescence microscope (IX81, Olympus). CD31 images were used to determine the average cell number and cell area per condition. Cells were counted using the ImageJ cell counter feature for N = 3 gels/group. For area measurements, cells were traced using the ImageJ free hand selection tool and the diameter was determined using the ImageJ measure feature for N = 3 gels/group.

Caspase-3/7 assay. The amount of caspase-3/7 activity after 3 days of culture on the three different hydrogel-peptide concentrations was determined using the Apo-ONE caspase assay (Promega) as per the manufacturer's protocol. Briefly, 30,000 HUVECs were seeded onto the three different collagen-chitosan hydrogel conditions in a 96 well plate. After 3 days of culture, the cells had one of the three treatments applied in the culture media for 17 hr: DMSO (1:1000), taxol dissolved in DMSO (2.5 μM) and culture media control (no treatment). Following treatment the caspase-3/7 assay was performed.

XTT assay. To normalize the caspase-3/7 activity to cell number, an XTT assay was used to determine the number of cells after three days of culture and 17 hr of treatment in the various hydrogel-peptide groups and treatment conditions. The XTT assay (Biotium) was performed according to the manufacturer's instructions and as we described previously [11]. A standard curve of known cell numbers was created to correlate the colorimetric data to cell number.

Encapsulation of Endothelial Cells in Collagen-chitosan QHREDGS Hydrogels

Encapsulation time course studies: kinetics and tube-formation. HUVEC encapsulation was performed with the three hydrogel-peptide groups generated as described above. Once the hydrogels had been created, before they were pipetted onto a 96 well plate, they were mixed with a specific number of cells so that 35 μl of gel had 170,000 HUVECs ($5 \cdot 10^6$ cells/ml of gel). Once the hydrogel and cells had been mixed on ice, 35 μl of the hydrogel/cell mixture was pipetted into a 96 well plate. The plate was then placed into the incubator for 30 min (37°C and 5% CO_2) to allow the hydrogel to gel. After 30 min, 200 μl of culture media was added to each sample and the plate was placed back into the incubator with media changed every 2 days.

To image the HUVECs within the hydrogel, samples were stained using CFDA-SE and imaged at days 2, 4 and 6 as described in the above live/dead staining methods or with CellTracker dye (Invitrogen) and imaged at 6, 18 and 24 hours after seeding. Using CellTracker, encapsulated cells were stained with a 2 μM solution of CellTracker dye in serum free media for 30 min (37°C and 5% CO_2). The staining solution was then removed and the encapsulated cells were cultured in EGM2 for 30 min (37°C and 5% CO_2). The media was then removed and the encapsulated cells were washed twice with PBS for 5 min. The encapsulated cells stained with CellTracker dye were then fixed in 4% paraformaldehyde for 15 min and washed in PBS and imaged.

To determine the average longest construct path length, ImageJ was used to draw the longest continuous path along a HUVEC construct without back tracing on the current path. The longest path was determined for all constructs within a single image and then averaged. The average total path length was determined through the addition of all the average longest construct path lengths in an image.

At the end of cultivation of encapsulated cells, the samples were fixed and stained for CD31 as described above. Confocal images were taken using the Olympus FluoView 1000 Laser Scanning Confocal Microscope with FV10-ASW 3.0 Viewer software.

Encapsulation integrin blocking assay. HUVEC encapsulation was performed with the three hydrogel-peptide groups generated as described above. The hydrogel solution and the cells were combined so that 35 μl of gel had 170,000 HUVECs ($5 \cdot 10^6$ cells/ml of gel). Once the hydrogel and cells had been mixed on ice, 35 μl of the hydrogel/cell mixture was pipetted into a 96 well plate. The plate was then placed into the incubator for 30 min (37°C and 5% CO_2) to allow the hydrogel to gel. After 30 min, control Igg antibody cardiac troponin-T (Thermo Scientific) at 6.75 μg/ml, anti-αvβ3 integrin (BD Biosciences) or anti-α5 integrin and anti-β1 integrin (BD Biosciences) were added at 50 μg/ml, concentrations proven effective in previous studies [13], into the culture media and cultured for two days (37C and 5% CO_2).

1.7 Statistical Analysis

Statistical analysis was performed using SigmaPlot 12.0. Differences between experimental groups were analyzed using one-way and two-way ANOVA depending on the experiment. An $\alpha = 0.05$ was used for all tests and a normality test (Shapiro-Wilk) and pairwise multiple comparison procedures (Holm-Sidak method) were used for all ANOVA tests. If the equal variance test failed, Kruskal-Wallis One Way Analysis of Variance on Ranks was performed with all pairwise multiple comparisons performed using the Tukey Test. $P<0.05$ was considered significant for all statistical tests.

Results

QHREDGS Decreased Endothelial Cell Permeability upon Thrombin Treatment

The effect that the peptide had on the permeability of HMVEC-D was determined using a transwell assay. **Figure 1A** shows that the peptide treated HMVEC-D significantly decreased ($P<0.004$) the permeability of the confluent cell layer in comparison to the control and the scrambled peptide group. Next, the effect of the peptide on endothelial barrier function was evaluated at baseline and in response to thrombin using a transendothelial electrical resistance (TEER) assay, **Figure 1B**. At time 0, all three groups had similar resistance. However, in the group treated with the peptide QHREDGS, resistance rapidly increased. At 1 hr, thrombin was added to induce endothelial cell permeability and disrupt barrier function. Thrombin, which disrupts endothelial cell barrier function, was added to each of the three conditions of HMVEC-D culture and, resulted in a decrease in resistance of the endothelial layer of all three conditions. However, the resistance of the QHREDGS peptide treated group recovered to a greater extent and remained at a higher resistance level compared to the scrambled and control treated HMVEC-D groups throughout the 70 hr study. Peptide QHREDGS increased HMVEC-D barrier function at baseline and with thrombin. Thus, similar to ang1 [14,15], peptide QHREDGS attenuated EC permeability and improved intercellular barrier integrity.

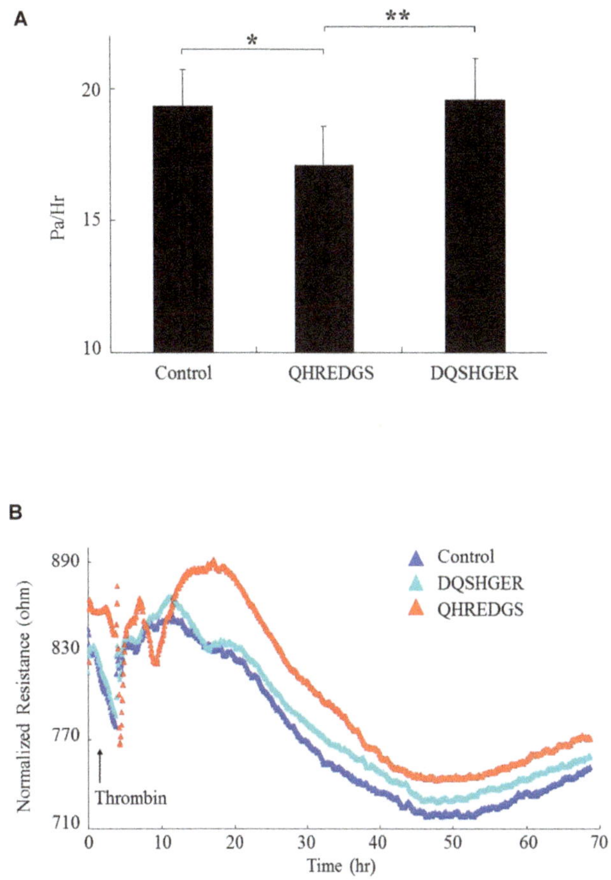

Figure 1. Linear QHREDGS Peptide Decreases Endothelial Cell Permeability. (A) HMVEC-D were cultured on collagen I and permeability was assessed using a Transwell assay (N = 3/group). Endothelial cells were serum starved (1 hr) and FITC labeled albumin was added to the top well and cells were incubated. QHREDGS or scrambled (DQSHER) peptides at 400 μM or control (PBS) were added (1 hr). Thrombin was added to induce permeability and the amount of FITC-linked albumin in the bottom well was quantified (Pa/hr) (3 hr). Peptide QHREDGS reduced HMVEC-D permeability (*P = 0.004) compared to control and scrambled peptide (**P = 0.001) groups. (B) Transendothelial electrical resistance was measured in HMVEC-D cultured on collagen I. HMVEC-D were serum starved and QHREDGS or scrambled (DQSHER) peptides at 400 μM or control (PBS) were added (1 hr). Thrombin was added to reduce resistance and resistance was measured as a function of time (N = 2/group). Peptide QHREDGS increased QHREDGS resistance compared to control and scrambled peptides. Red coloured triangles indicate normalized resistance values for the QHREDGS treated group.

QHREDGS Enhanced Endothelial Cell NO Production and Metabolism

The addition of the peptide to MS1 endothelial cell cultures increased the release of nitric oxide in a dose dependent manner (**Figure 2A**). Examining the effect ang1 had on nitric oxide release, a statistically significant increase in comparison to the PBS control was found at an ang1 concentration of 200 nM. The control, RGD-based peptide sequence GRGDSP as well as ang2 did not show a statistically significant increase in nitric oxide release in comparison to the PBS controls. Next, the peptide QHREDGS, which is derived from ang1, was tested at two different concentrations, 100 μM and 500 μM. Both concentrations showed a statistically significant increase in nitric oxide production in comparison to their respective PBS controls and there was a dose response found with the 500 μM QHREDGS peptide inducing a greater amount of nitric oxide produced compared to the 100 μM condition. **Figure 2B** shows that when MS1 endothelial cells were pre-treated with NO donor S-Nitroso-N-acetylpenicillamine, the cells treated with either 200 nM of ang1 or 500 μM of QHREDGS peptide released a significantly greater amount of nitric oxide compared to their respective PBS controls. No statistically significant difference was found between the scrambled peptide group or the 100 μM QHREDGS and their respective PBS controls.

The effect that the peptide had on MS1 endothelial cell survival and metabolic activity in serum-starved conditions is shown in **Figure 3** through the measurements of MTT, ATP and NAD(P)H levels. **Figure 3A** shows that in serum-starved conditions, the peptide group had a significantly higher ($P = 0.001$) amount of MTT being reduced, meaning more cells were viable and active, in comparison to PBS controls. The scrambled peptide group, DQSHGER, had similar MTT levels as the PBS control. Additionally, in serum-starved conditions, the QHREDGS peptide group had a significantly higher ($P \leq 0.03$) ATP level in comparison to the PBS control (**Figure 3B**). The scrambled peptide group had a similar ATP level to the PBS control. The QHREDGS peptide also significantly increased the NAD(P)H levels in MS1 ECs in serum starved conditions compared to the PBS control and scrambled peptide ($P = 0.001$; **Figure 3C**).

Cultivation of HUVECs on QHREDGS Modified Collagen-Chitosan Hydrogels Led to Improved Cell Viability

HUVECs cultured on the QHREDGS modified collagen-chitosan hydrogels for 4 days had 84% viability for the control hydrogel and 97% and 95% for the 100 μM and 650 μM peptide modified hydrogel conditions (**Figure 4A, C**). Using One-Way ANOVA, a statistically significant difference between the control hydrogel and the 100 μM and 650 μM peptide modified hydrogel conditions were found ($P = 0.036$ and $P = 0.046$).

CD31 is an integral membrane protein that mediates cell-to-cell adhesion. CD31 is expressed constitutively on the surface of adult and embryonic endothelial cells and mediates endothelial cell-cell interactions and angiogenesis. CD31 staining for HUVECs cultivated as monolayers on top of the collagen-chitosan hydrogels, shown in **Figure 5A**, was positive for all three of the hydrogel groups. There was a closer association of HUVEC membranes in the 650 μM peptide hydrogel group, while the 100 μM peptide hydrogel and control hydrogel groups showed HUVECs intercellular connections were not as tight. Furthermore, the cells cultivated on the 100 μM and 650 μM hydrogel had more pronounced staining for VE-CAD compared to those cultivated on the control hydrogel (**Figure 5B**). Cell area for HUVECs on each of the three hydrogel groups, shown in **Figure 5C**, were found to be comparable at 1360, 1270 and 1260 μm^2 for the control, 100 μM and 650 μM peptide hydrogel groups. The average cell density (**Figure 5D**) in each of the three hydrogel groups was also found to be similar with values of 60,000, 70,000 and 65,000 cells/cm^2 for the control, 100 μM and 650 μM peptide hydrogel groups. These data are consistent with known actions of ang1. Ang1 does not promote EC proliferation [16] and as such, QHREDGS was not expected to increase EC density or cell area. However, similar to ang1 [17], peptide QHREDGS did increase fortification of intercellular connections, as evidenced by the change in CD31 and VE-CAD staining showing less space between cells and more cell to cell connections.

A

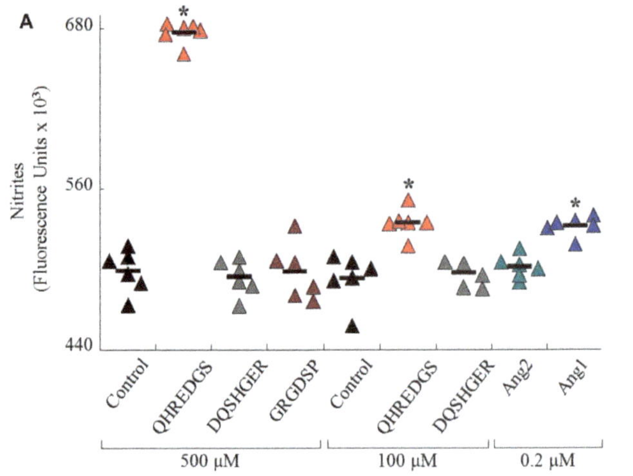

B

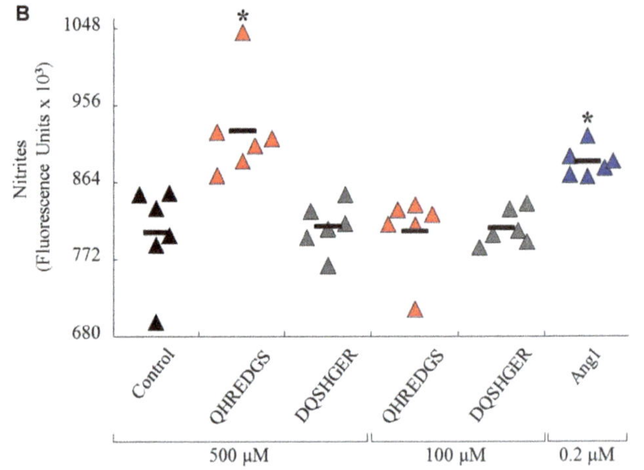

Figure 2. Linear QHREDGS Peptide Increases Nitric Oxide Release from Endothelial Cells. (A) Mouse MS1 endothelial cells were incubated with linear QHREDGS peptide, scrambled QHREDGS peptide (DQSHGER), GRGDSP peptide, control (PBS), or 200 nM angiopoietin-2 (ang2) or angiopoietin-1 (ang1) (0.5 hr). Then, 100 μM L-arginine was added and cells incubated (1.5 hr). Extracellular nitric oxide (NO) levels were measured ($N = 6$/group). In graphs, the mean value is indicated with a bar. Linear QHREDGS peptide increased NO levels at 500 and 100 μM (*$P \leq 0.00001$) and ang1 also increased NO levels (**$P = 0.0002$). Ang2, GRGDSP, and scrambled QHREDGS peptides had no effect on NO levels compared to controls. (B) MS1 endothelial cells were incubated with peptides and proteins as above, except SNAP (NO donor) was added to the 1.5 hr incubation with L-arginine. In graphs, the mean value is indicated with a bar. Linear QHREDGS peptide at 500 μM and ang1 increased NO levels (*$P < 0.002$). Scrambled QHREDGS peptide and 100 μM QHREDGS did not affect NO levels. Triangle color code: control (black), scrambled peptide (gray), QHREDGS peptide (red), GRGDSP peptide (burgundy), ang2 (green) and ang1 (blue).

QHREDGS Promotes Endothelial Cell Survival

Upon treatment with paclitaxel (taxol), capsase-3/7 activation was significantly increased in cells cultivated on control and peptide modified collagen-chitosan hydrogels with 650 μM QHREDGS peptide in comparison to their respective PBS control

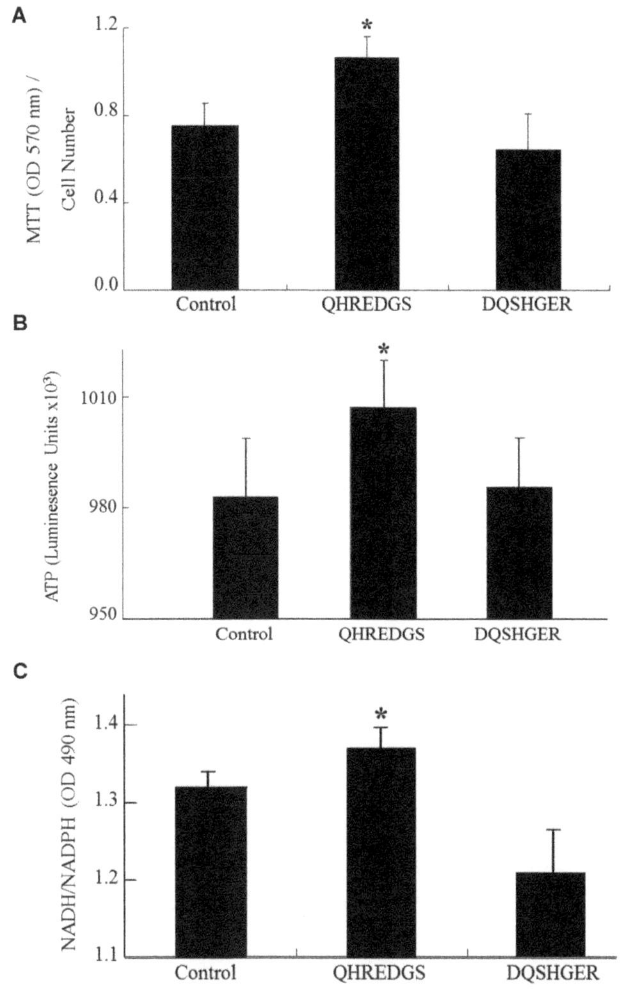

Figure 3. Linear QHREDGS Peptide Increases Endothelial Cell Survival and Energetics. (A-B) MS1 endothelial cells were incubated with 500 μM QHREDGS or scrambled QHREDGS peptide (DQSHER) or control (PBS) overnight. (A) Cells were incubated for 4 hr with MTT, absorbances measured at 570 nm, and values normalized to cell number (N = 5/group). Peptide QHREDGS increased MS1 endothelial cell viability (*P = 0.001) as indicated by the increased processing of MTT by mitochondria. (B) QHREDGS peptide increased endothelial cell ATP levels, whereas scrambled QHREDGS had no such effect (*P = 0.03). (C) MS1 endothelial cells were incubated with 500 μM QHREDGS or scrambled QHREDGS peptide or control (PBS) for 3 hr. QHREDGS peptide increased endothelial cell NADH/NADPH levels (*P = 0.05) suggesting increased cellular energetics. Optical density (OD).

as well as their DMSO carrier alone (**Figure 6**). Interestingly, when cells were cultivated on the peptide modified hydrogels with 100 μM QHREDGS, there was no statistically significant increase in caspase-3/7 activation between the taxol, DMSO and PBS conditions.

Lower Concentrations of QHREDGS Led to more Robust Tube-Like Structure Formation by Encapsulated Endothelial Cells

Upon encapsulation of HUVECs in QHREDGS modified collagen-chitosan hydrogels, changes in tube-formation kinetics, morphology, and organization were observed. We have performed early stage vascular morphogenesis studies to better understand the peptide's effect on promoting EC tube-like structures (**Figure 7**). The 18 and 24 hr time points for the peptide groups exhibited more robust tube-like formations with the 100 μM peptide group having the densest networks of tube-like structures that were wider compared to those in the control or 650 μM conditions (**Figure 7**). Over the course of 6 days, the HUVECs found within the control hydrogel continued to show minor changes in cell morphology and organization (**Figure 8**). The majority of the cells remained rounded throughout the time course. In the 100 μM peptide hydrogel group, the largest degree of change in cell structure and morphology was observed. By day 2, many of the HUVECs had clustered together to form elongated tube-like structures within the hydrogel with the peak of this phenomenon occurring on day 4. By day 6, these tube-like structures began to regress in both number and size. The 650 μM peptide hydrogel displayed a faster and less pronounced effect than the 100 μM peptide hydrogel group, with peak construct size and elongation found on day 2 and regression beginning to occur on day 4. Some HUVEC tubular structures remained at day 6 but were smaller in comparison to those observed on day 2.

Quantification of the tube-like constructs was then performed on day 4 hydrogels. **Figure 8B** shows the average longest single path that can be drawn, without overlapping paths, on a HUVEC tube-like structure. This result showed that the peptide hydrogel groups had on average longer tube constructs in comparison to the control hydrogel group with, the 100 μM peptide hydrogel group having the largest average constructs of the three hydrogel groups. A statistically significant difference in the longest path length was found between the 100 μM peptide modified hydrogel and the control hydrogel ($P<0.05$). Quantification of the size and frequency of HUVEC constructs formed was performed by the addition of all the total path lengths for a single image which was then averaged across multiple images to give an average total path length per image. The results, **Figure 8C**, showed that the peptide modified hydrogels had a higher average total path length, or in other words a higher abundance of HUVEC constructs, in comparison to the control hydrogel group with the 100 μM peptide hydrogel group having the highest average total path length of the three groups. A statistically significant difference in the total path length was found between the 100 μM peptide hydrogel and the control hydrogel ($P<0.05$) as well as the 650 μM peptide hydrogel and control hydrogel ($P<0.05$).

Confocal images of CD31 stained cells helped to better elucidate the structure of the HUVEC tube constructs seen in the encapsulation experiments (**Figure 9**). As was previously stated, the peptide hydrogels had larger constructs in comparison to the control hydrogel group. The control hydrogel constructs were comprised of few cells with minimal cell elongation and organization in both the day 4 and 6 images. In contrast, the 100 μM peptide hydrogel group had extensive tubes on day 4, consisting of many elongated and connected cells staining strongly for CD31. On day 6, the tube constructs began to regress even in the 100 μM group. The 650 μM peptide group mirrored the 100 μM peptide hydrogel group, however, in general the construct size and cell elongation at days 4 and 6 were smaller in the 650 μM peptide group compared to the control.

Integrins αvβ3 and α5β1 Mediate Tube Formation by HUVECs in Peptide Modified Hydrogels

When the integrin receptor αvβ3 was blocked, the ability of HUVECs to form robust tube-like structures in the peptide modified hydrogels was hindered (**Figure 10**). When integrin receptor α5β1 was blocked in the cells encapsulated in the peptide modified hydrogels, HUVECs did not form tube-like structures,

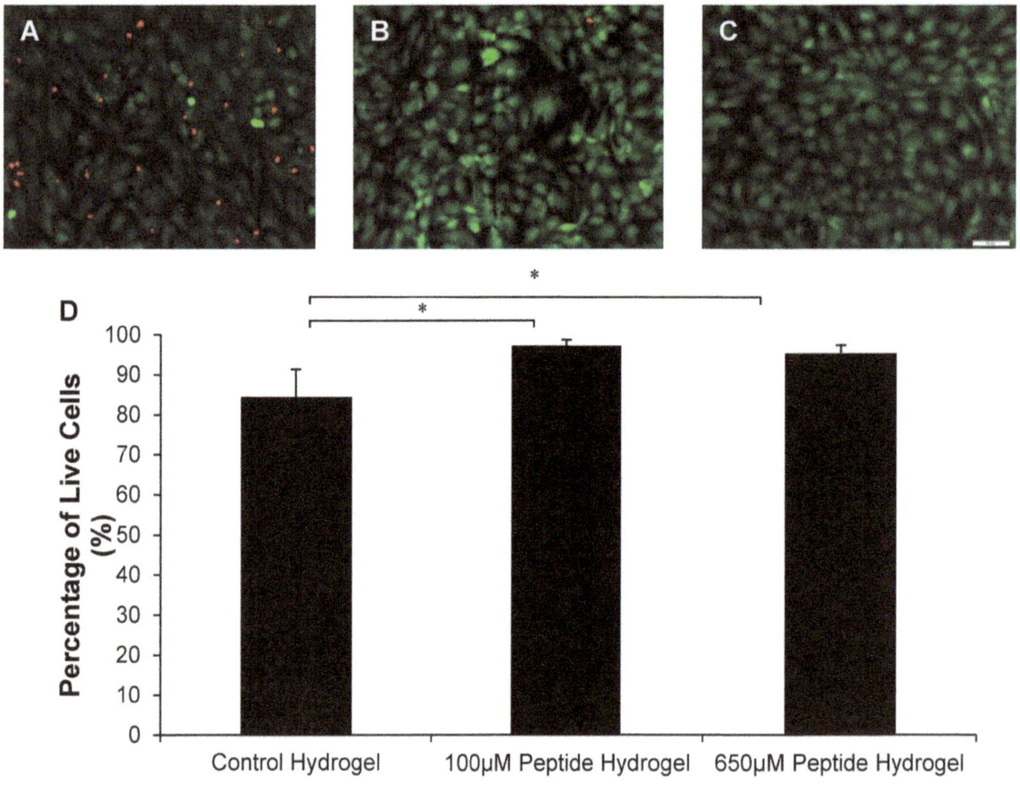

Figure 4. Cultivation of HUVECs on QHREDGS Modified Collagen-Chitosan Hydrogel Leads to Improved Viability. Characterization of HUVECs on collagen-chitosan hydrogels with different peptide concentrations after 4 days of cultivation. 187,500 cells were seeded on the freshly made hydrogels. (A) Control hydrogel, (B) 100 µM peptide hydrogel and (C) 650 µM peptide hydrogel with HUVECs stained for live/dead imaging with CFDA-SE for live cells (green) and PI for dead cells (red). (D) Percentage of live cells. All images are 20X magnification with scale bar of 50 µm. *$P<0.05$.

elongate or cluster and instead remained as round isolated cells. The cells in the control hydrogel with blocked $\alpha v\beta 3$ integrin, exhibited a similar appearance as those cultivated with the control Igg antibody, while $\alpha 5\beta 1$ integrin blocking completely abolished tube formation in the control hydrogel. Since integrin receptor $\alpha v\beta 3$ is specific to fibrin we did not expect a significant effect of its blocking on the HUVECs ability to interact with the control hydrogel which, is comprised of only collagen and chitosan. $\alpha 5\beta 1$ is the shared integrin receptor for both collagen and fibrin, thus the use of $\alpha 5\beta 1$ integrin antibody prevented HUVEC's interaction with the major components of the hydrogel (collagen and QHREDGS peptide), while blocking of $\alpha v\beta 3$ integrin most likely prevented HUVECs from interacting with the peptide only.

Discussion

Modified biomaterials are required to provide protection for cells introduced into the hostile environment of a wounded tissue. Consequently, many groups are focusing on discovering novel peptides, derived from growth factors, to enhance the properties of biomaterials and protect cell activity during the initial stage of implantation. Presented in this paper is the use of a novel peptide, QHREDGS, based on the fibrinogen-like domain of angiopoietin-1, to enhance endothelial cell survival and preserve endothelial cell functionality. We examined the effects of the soluble peptide on endothelial cells cultured on tissue culture plastic as well as the peptide covalently immobilized in a hydrogel suitable for myocardial cell injection [11]. We demonstrated previously that QHREDGS promoted survival and functional properties of cardiomyocytes cultured in an *in vitro* collagen-chitosan hydrogel [11]. The current study revealed that the peptide QHREDGS had a dual protective effect for both cardiomyocytes and endothelial cells. QHREDGS is thought to signal both cardiomyocytes and endothelial cells through integrin binding domains since the peptide is based on the fibrinogen-like domain of ang1. The cells are thought to bind to QHREDGS via the $\alpha_V\beta_3$ and/or $\alpha_5\beta_1$ integrins [9,18], due to the homologous nature of the fibrinogen-like domains of angptl3, ang1 and ang2 [9,19]. The dual protective nature of QHREDGS positions the peptide as a beneficial modifier of biomaterials for myocardial cell therapy.

Peptide sequences are being used in favour of growth factors, for the modification of biomaterials, due to the plethora of issues surrounding the use of growth factors. These include: a) undesirable immune responses as the growth factors are often isolated from non-human organisms, b) the proteins are subject to proteolytic degradation, with further degradation occurring during an inflammatory response, and c) their binding to biomaterials could block the domain of interest making the growth factor ineffective [20]. Specifically for ang1, the production of this recombinant protein has been hindered by its insolubility and the variability of the protein's activity [21]. One peptide sequence of much interest has been the cell binding motif Arg-Gly-Asp (RGD) [22]. RGD has been used in many biomaterials to improve endothelial cell adhesion, proliferation and migration [20,23,24,25]. While the peptide that we present is thought to also interact with endothelial cells through integrin binding

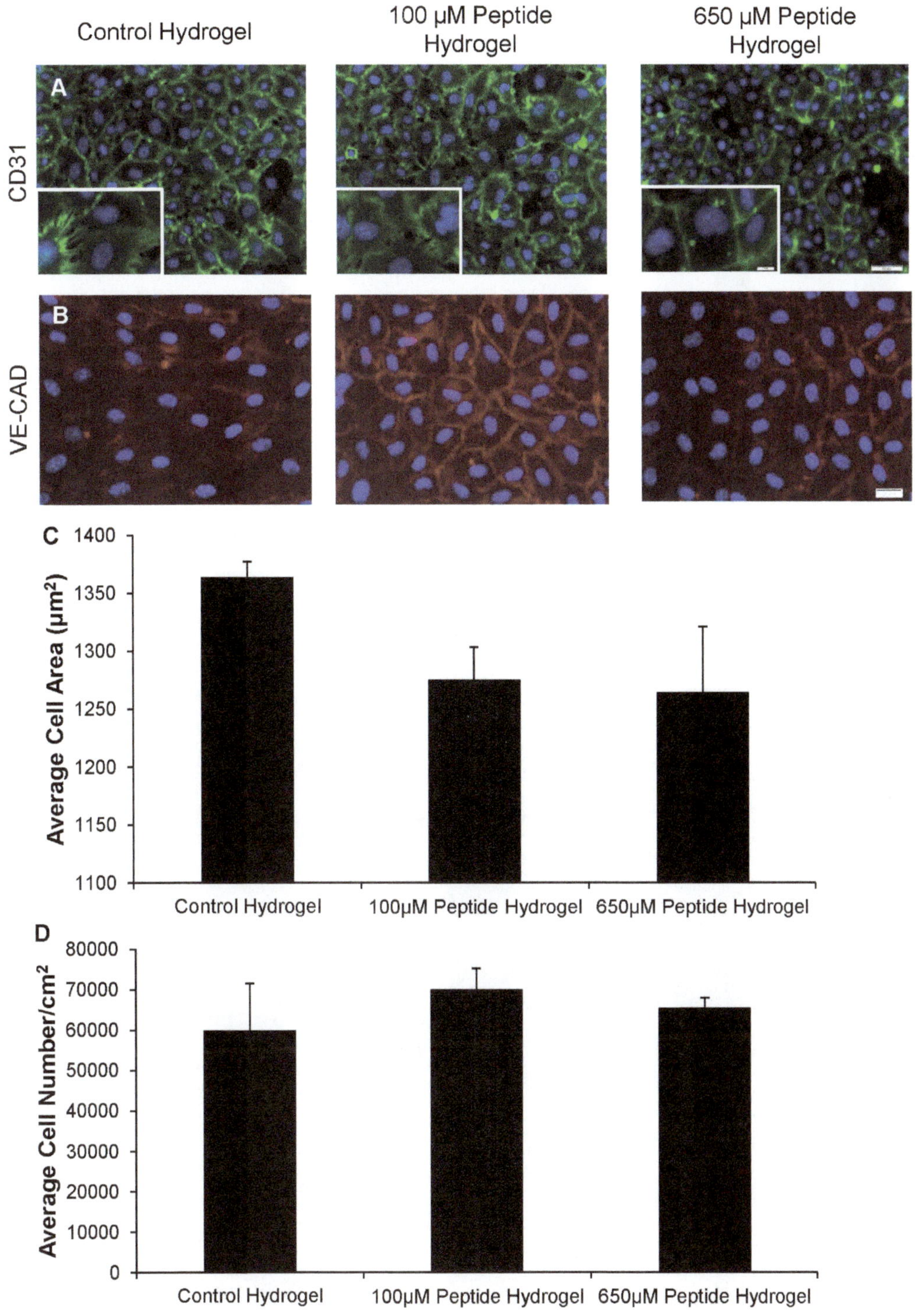

Figure 5. QHREDGS Modified Collagen-Chitosan Hydrogels Support Expression of CD31 and VE-CAD in Endothelial Cells Characterization of HUVECs on collagen-chitosan hydrogels with different peptide concentrations after 4 days of cultivation. 187,500 cells were seeded on the freshly made hydrogels. (A) HUVECs stained for CD31 (green) and cell nucleus (blue) in control, 100 μM peptide and 650 μM peptide hydrogels. 20X magnification with scale bar of 50 μm and higher magnification inset scale bar of 10 μm. (B) HUVECs stained for VE-CAD (red) and cell nucleus (blue) stained in control, 100 μM peptide, and 650 μM peptide hydrogels. 40X magnification with scale bar of 20 μm. (D) Average cell area. (E) Average cell number per cm^2.

domains, QHREDGS also has the added functionality in its ability to mitigate cell apoptosis.

Biomaterials implanted into patients need to ensure survival of the encapsulated cells and promote the formation of a stable vasculature with the host in order for cells suspended within the biomaterial to survive implantation in the long term. We have shown previously that the peptide QHREDGS can be effectively conjugated to chitosan to create a peptide modified collagen-

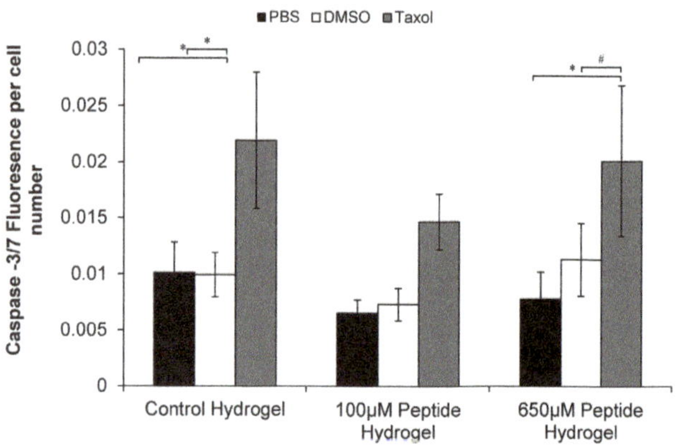

Figure 6. QHREDGS Provides a Protective Effect on Endothelial Cells Undergoing Apoptosis on Collagen-Chitosan Hydrogels. Caspase-3/7 activity, normalized to cell number, is shown for each of the three hydrogel-peptide concentrations. After three days of HUVEC cultivation, each hydrogel group was treated three different ways for 15 hr. The treatments were: PBS control, DMSO carrier and Taxol 2.5 μM. * $P<0.05$, # $P=0.051$.

chitosan hydrogel [11]. The prior study by Rask *et al.* used a photocrosslinkable Az-chitosan with the peptide QHREDGS bound to chitosan. The material was spun coated and UV-crosslinked onto glass cover slips [10]. The reason behind moving away from the photocrosslinkable Az-chitosan to a collagen-chitosan hydrogel was that we wanted to create a material that could be easily injected into the heart and gel *in situ* rather than requiring UV crosslinking, to avoid limitations of UV penetration depth. The new *in situ* gelable formulation was demonstrated to be suitable for culture of encapsulated heart cells and injection into infarcted hearts [11].

Two other available alternatives to recombinant ang1 are COMP-Ang1, which is a modified ang1 recombinant protein containing a minimal coiled-coil domain [21], and Vasculotide, which is an ang-based peptide-mimetic compound [26]. COMP-Ang1 has been shown to improve endothelial cell barrier function, mitigate radiation induced endothelial cell apoptosis and induce the postcapillary venule and venous ends of capillaries to enlarge due to endothelial cell proliferation leading to enhanced blood flow [21,27,28]. However, COMP-Ang1 is still a recombinant protein that could lead to a foreign body immune response. Vasculotide has been shown to induce cell migration, MMP2 release, and protection from serum withdrawal-induced apoptosis in endothelial cells [26]. Furthermore, Van Slyke *et al.* found that

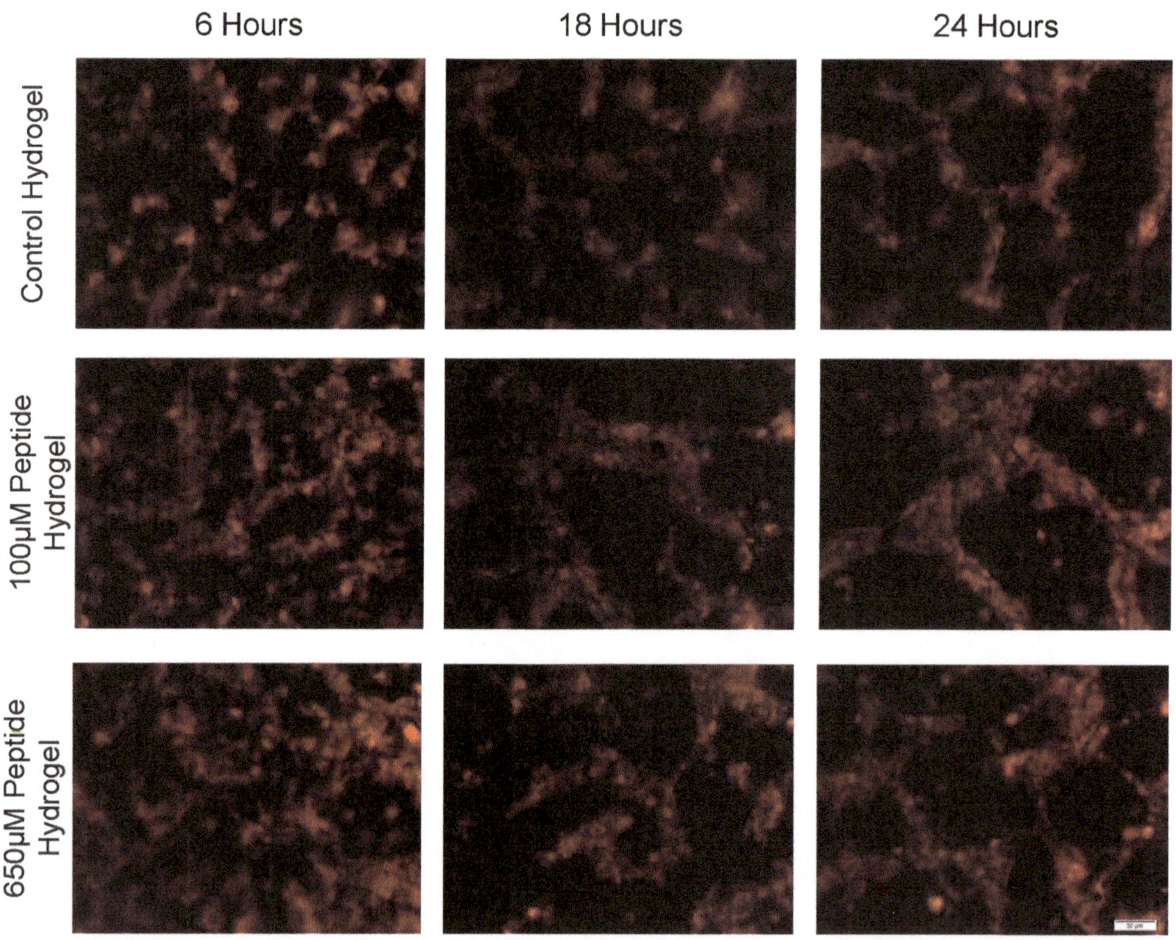

Figure 7. QHREDGS Modified Collagen-Chitosan Hydrogels Expedite Kinetics of Endothelial Cell Tube Formation. Encapsulation of HUVECs within the control, 100 μM peptide and 650 μM peptide hydrogel groups with HUVEC membranes stained with CellTracker dye (red). Images taken at 6, 18, and 24 hr to show the progression of HUVEC constructs forming within the hydrogel. All images were taken at 20X magnification. Scale bar is 50 μm.

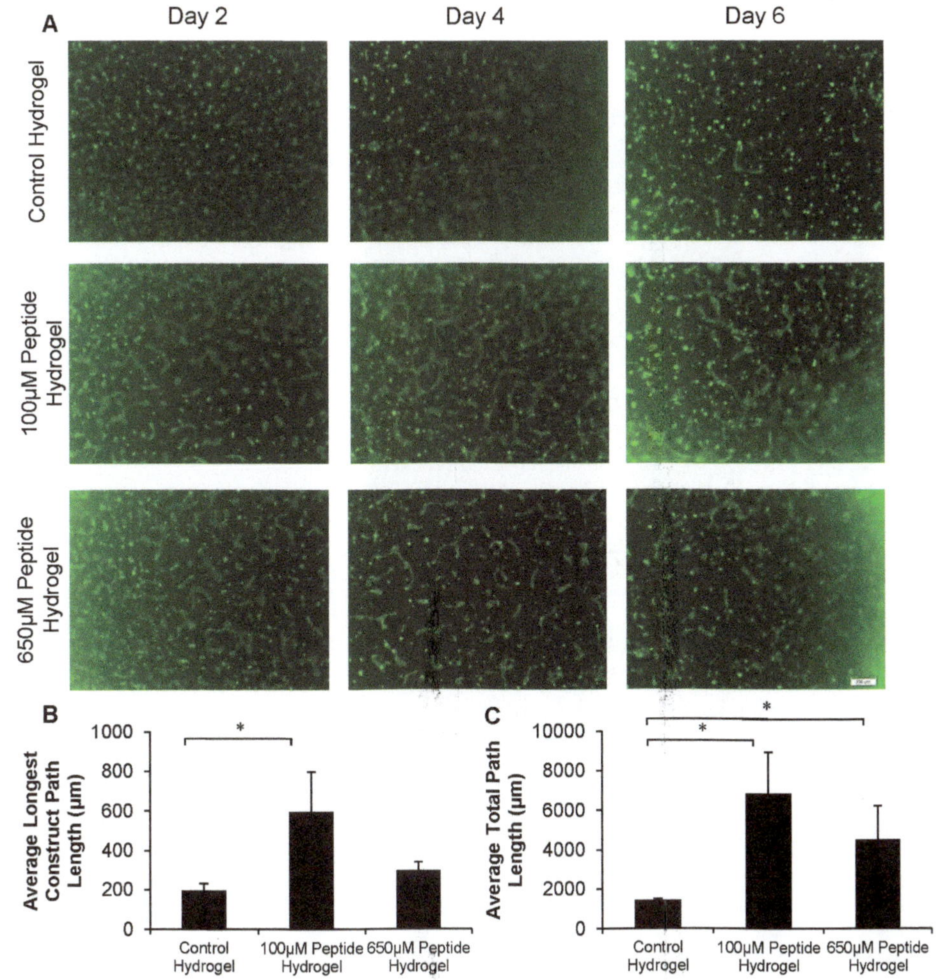

Figure 8. QHREDGS Modified Collagen-Chitosan Hydrogels Enhance Tube-Like Structure Formation of Encapsulated HUVECs. Encapsulation of HUVECs within the control hydrogel and two peptide modified hydrogel groups. (A) HUVECs stained with CFDA-SE for live cells (green). Images taken every other day to show the progression of HUVEC constructs forming within the hydrogel. (B) Average longest construct path length determined on day 4. (C) Average total path length of constructs determined on day 4. * $P<0.05$. All images are taken at 4X magnification. Scale bar is 200 μm.

Vasculotide induced neovascularization *in vivo*, but its effect on cardiomyocytes has not been explored. QHREDGS peptide could be a more appealing modifier for cardiac cell therapy biomaterials due to its dual effect of improving both cardiomyocyte [11] and endothelial cell functionality.

Ang1 has been shown to have a variety of influences on EC migration, permeability, inflammation and vascular maturation [29] activating a number of signalling cascades through Tie2 receptor. Splitting ECs into two broad categories: 1) quiescent ECs that are in contact with one another or 2) motile non-contacting ECs; differences in the Tie2 receptor location on the cell were discovered and lead to a greater understanding of ang1 situation specific signalling [30,31]. In confluent ECs, ang1 induces Tie2 translocation to cell-cell contacts which results in the formation of homotypic Tie2-Tie2 trans-associated complexes [31,32]. These complexes then preferentially activate the AKT pathway resulting in increased survival signals [33], increased endothelial NO synthases (eNOS) synthesis [29] and reduced permeability through the sequestration of Src by mDia which reduces the internalization of VE-CAD [34]. When motile and non-contacting ECs are stimulated by ang1 the Tie2 receptor is instead located preferentially at cell-substratum contacts which facilitates ECM bound ang1 to trigger Dok-R, ERK1/2 and MAPK signalling [30,31]. When ECs are stimulated by ang1 in this context, cell migration is stimulated. Cascone *et al.* also demonstrated that there was an association between Tie2 and the integrin $\alpha5\beta1$ on ECs and that this relationship was regulated through the engagement of $\alpha5\beta1$ to fibronectin [18]. The activation of $\alpha5\beta1$ resulted in Tie2 phosphorylating at low ang1 concentrations yet, attenuated activation at high ang1 concentrations. This combined $\alpha5\beta1$-Tie2 activation at a low concentration of ang1 resulted in promoting EC survival and inducing persistent EC motility [18].

Preservation of endothelial cell barrier functional integrity, protection against vascular leakiness investigated in microvascular endothelial cells, and overall, improved endothelial cell health, survival, and energetics are all characteristics associated with peptide QHREDGS (**Figure 1–3**). Thus, QHREDGS has overlapping functional effects on cardiomyocytes as shown in our previous studies [10,11] and endothelial cells, as shown here.

NO is a short-lived molecular messenger formed from L-arginine by NO synthases (NOS). NO made by endothelial NOS in particular is vital to EC function and protects against

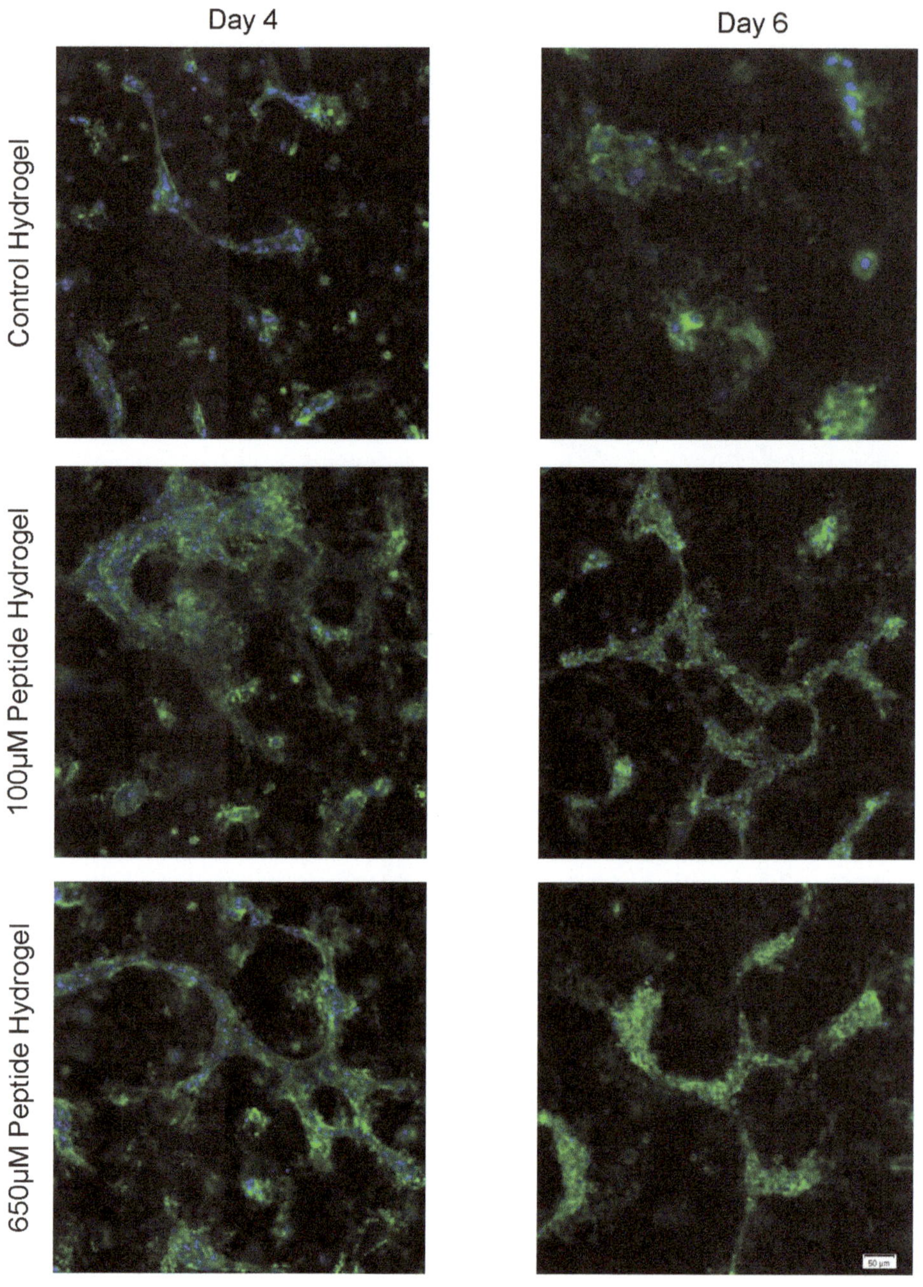

Figure 9. CD31 Staining of Tube-Like Structures is Enhanced in QHREDGS Peptide Modified Hydrogels. Confocal microscopy of encapsulated HUVECs within the control hydrogel and two peptide modified hydrogels after 4 and 6 days of cultivation. CD31 (green) and cell nucleus (blue). All images were taken at 20X magnification. Scale bar is 50 μm.

hypertension and cardiovascular injury [35,36]. It is a potent vasodilator and inhibits smooth muscle cell (SMC) proliferation and migration, and matrix turnover [37]. NO is therefore instrumental in both acute and chronic regulation of vascular tone. Ang1 enhances NO production through Tie2 phosphorylation which leads to the phosphorylation of the p85 subunit of phosphoinositide 3-kinase (PI3K), resulting in the activation of AKT [33,38]. AKT has been shown to promote endothelial nitric oxide synthase production [29].

By closely observing the y-axis scale in **Figures 2A and B**, it is possible to see that the levels of NO were lower in all groups at baseline condition (**Figure 2A**) compared to the levels in the presence of SNAP, the NO donor (**Figure 2B**). QHREDGS peptide and ang1, exhibited enhanced NO production upon addition of L-arginine, compared to the control groups. In the presence of a competing source for NO generation, namely SNAP, the effects of QHREDGS peptide were dampened, as expected. However, without SNAP-induced NO contributions, the effect of QHREDGS peptide and ang1 alone were elaborated. The RGD peptide had no effect on nitric oxide release at 500 μM (**Figure 2A**). It also had no effect on cardiomyocyte survival at any of the concentrations we tested in previous work [10,39], thus we concluded it was not necessary to systematically use this control in all of our experiments. It is possible that a greater effect of the

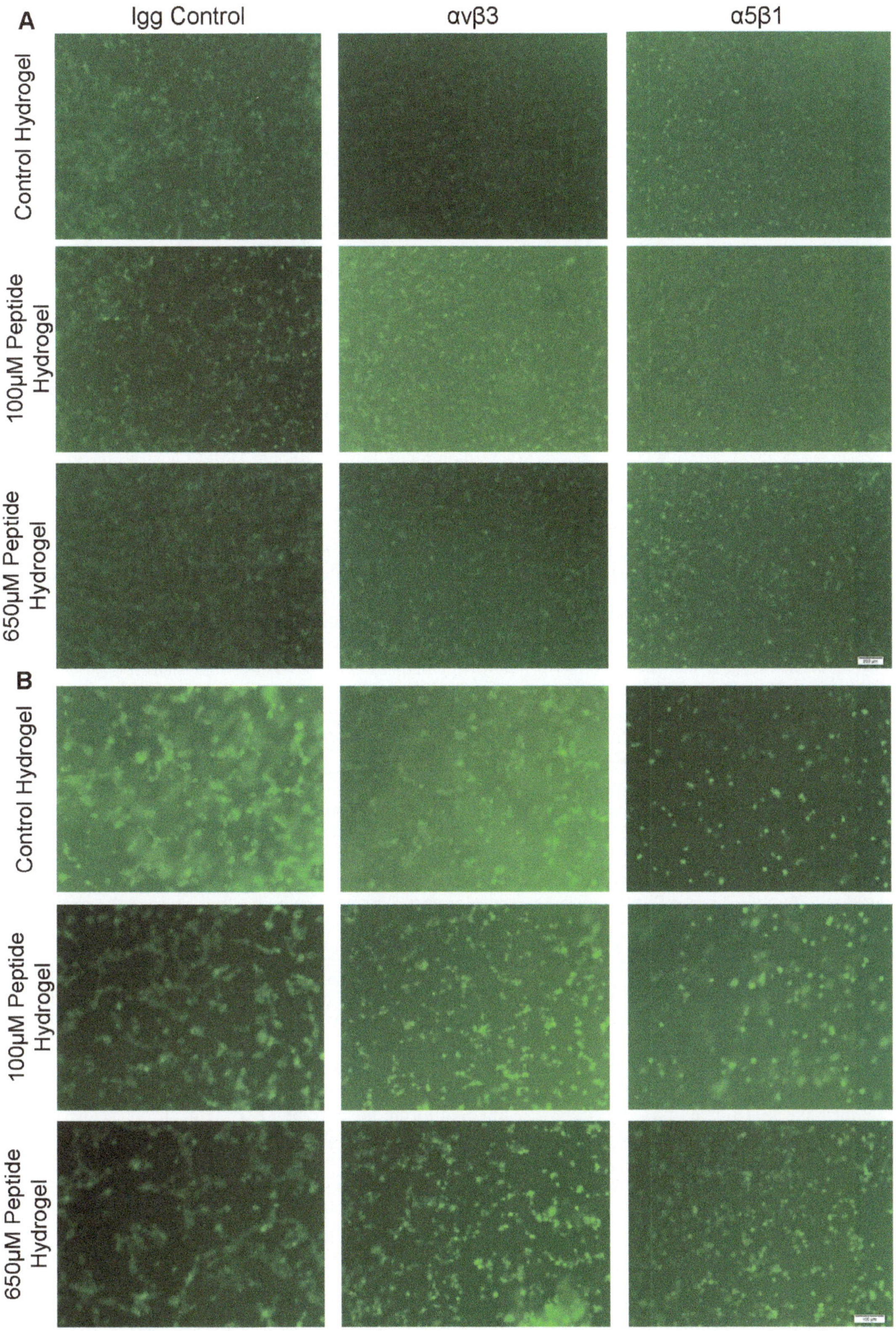

Figure 10. Blocking αvβ3 or α5β1 Integrin Receptor Inhibits Formation of QHREDGS Tube-Like Structures by Encapsulated HUVECs. Encapsulation of HUVECs within the control hydrogel and two peptide modified hydrogel groups for two days in the presence of either an Igg control, αvβ3 or α5β1 antibody. HUVECs stained with CFDA-SE for live cells (green). Images taken at the end of the two day cultivation to examine tube-like structure formation. (A) Images taken at 4X magnification. Scale bar is 200 μm. (B) Images taken at 10X magnification. Scale bar is 100 μm.

QHREDGS peptide on NO release could be gained by using an even higher concentration of the QHREDGS peptide, however, these higher concentrations begin to move out of physiologically relevant doses. We believe that the concentration ranges presented provide the necessary information about the peptide's ability to instigate the release of nitric oxide and these also fall into the ranges we investigated previously for their effects on cardiomyocytes [11].

Ang1 has been shown to improve skin cell metabolism and viability [40]. Consistent with this functional role, we showed that ang1-derived peptide QHREDGS also promoted cell viability and metabolism in ECs and in cardiomyocytes [9,10]. Furthermore, ang1 did not promote EC proliferation, but it did increase EC tube formation and survival, promoted tight junction formation and reduced EC permeability [11]. Studies of both immobilized and soluble peptide QHREDGS showed that this peptide reduced EC permeability, increased electrical resistance, and fortified endothelial cell-cell connections. Thus, the peptide QHREDGS has the potential to prevent vascular leakage in pathologic settings wherein increases in vascular permeability are problematic (e.g. sepsis, tumor angiogenesis, proliferative diabetic retinopathy, age-related macular degeneration, blood brain barrier breakdown).

The rationale for using 100 μM and 650 μM concentrations of the peptide was based on previously reported data in our work with cardiomyocytes [11]. These concentrations were found beneficial for culture of cardiomyocytes compared to the peptide-free control hydrogels. Since our ultimate goal is to create a cell-delivery vehicle that would promote cardiomyocyte survival and vascular tube formation, we investigated the same peptide-hydrogel compositions with endothelial cells.

Here we showed that the peptide increased the viability of HUVECs at both the 100 μM and 650 μM concentrations in comparison to the control hydrogel. These studies were performed using monolayers of endothelial cells on peptide modified collagen-chitosan hydrogels (**Figure 4–6**) to ensure that we observed the effect of the peptide irrespective of the effects related to cell encapsulation such as diffusional limitations in oxygen and nutrient supply [3]. As such, these studies enabled us to assess if the peptide covalently immobilized into the hydrogel retained its functionality in promoting cell survival. To ensure that this statistically significant difference in viability (**Figure 4**) between the peptide modified and non-peptide modified hydrogels was due to the protective nature of the peptide and not due to the peptide increasing proliferation or differential adhesion, cell number and area were examined. It was found that the peptide likely did not enhance endothelial cell proliferation as the average number of cells per area was consistent between the three hydrogel groups (**Figure 5**). As previously mentioned, cell area and cell density assays were performed as monolayer studies of ECs on top of the control and peptide modified collagen-chitosan hydrogels. Since ECs were seeded at a high density and quickly established a confluent monolayer, the effects of the peptide, we believe, would be to promote quiescence in the cells and promote survival similarly to what ang1 has been shown to do in this context specific situation of a confluent monolayer of ECs [31,32].

HUVECs expressed CD31 in all three hydrogel conditions (**Figure 5**). CD31 staining was performed to ensure that we had an appropriate endothelial cell phenotype on our biomaterial at the various QHREDGS concentrations. If CD31 staining was absent, this would have indicated the hydrogel formulation was not supportive for culture of endothelial cells and expression of their phenotypic markers. CD31 has been shown to be involved in a plethora of endothelial cell functions including: maintenance of adherens junction integrity and permeability, organization of the intermediate filament cytoskeleton, regulation of catenin localization and transcriptional activities, participation in STAT isoform signalling and control of apoptotic events [42,43].

However, in the 650 μM peptide group we saw that the junction between HUVECs was stronger and less fenestrated in comparison to the 100μM and control hydrogel. This dose dependent relationship the peptide had on creating a less permeable layer of endothelial cells was also seen during 2D tissue culture plastic culture (**Figure 1**). It had been previously shown that ang1 requires the IQ domain GTPase-activating protein 1 as an activator of Rac 1 to regulate barrier function and that this signalling cascade is initiated through the interaction of ang1 and the Tie2 receptor [15,44].

Since the viability test combined with the cell number analysis pointed towards the peptide's role in cell survival as opposed to cell proliferation, the protective effect that the peptide had on endothelial cells stressed with paclitaxel was examined. We have shown that there was no significant difference between the taxol treated 100 μM peptide group and the basal levels of caspase-3/7 activation found in the PBS and DMSO controls at this peptide concentration (**Figure 6**). Furthermore, we demonstrated a statistically significant difference in the control and 650 μM peptide groups, between the taxol treated cells and their respective PBS controls (**Figure 6**). Since the most pronounced effect was observed at 100 μM, there was no classical dose-dependency but rather a maximum effect was observed at an intermediate concentration. This kind of behaviour could parallel some of the effects observed in previous studies with ang1. Cascone *et al.* showed enhanced Tie2 activation and signalling at low ang1 concentrations through fibronectin sensitization [18]. They demonstrated that ECs had a 4.4 fold increase in Tie2 phosphorylation with 20 ng/ml of ang1 as compared to only a 2.2 fold increase in Tie2 phosphorylation with 100 ng/ml of ang1 in the presence of a fibronectin based ECM [18]. They also demonstrated that ang1 increased cell adhesion, independent of ECM concentration (collagen, fibronectin or fibrinogen), with an intermediate ang1 concentration being optimal [18]. These findings correlate well with our findings of the 100 μM peptide concentration exhibiting optimal effects in both the caspase-3/7 assay and the encapsulation experiments examining tube formation and cell kinetics.

Ang1 has been found in previous studies to help improve cell survival by preventing apoptosis of endothelial cells when they are under chemical stress or serum deprived conditions [6]. In cardiomyocytes, the protective effect of ang1 was mediated by Akt phosphorylation [9]. The anti-apoptotic effect of ang1 was also found to be dose dependent in endothelial cells and was further enhanced in the presence of VEGF [45]. In our study, we used EGM2 media which contained 2% serum along with VEGF and other endothelial growth factors. Our results, therefore, show a combined protective effect that the peptide had at the lower concentration along with VEGF and serum present in the media. For confluent ECs, which is how our caspase assay was set-up, ang1 was reported in previous studies to induce Tie2 translocation to cell-cell contacts which resulted in the formation of homotypic Tie2-Tie2 trans-associated complexes [31,32]. These complexes were reported to then preferentially activate PI3K/Akt [46], which resulted in the up-regulation of a broad spectrum apoptosis inhibitor survivin [33].

To determine the behaviour of HUVECs cultured in a 3D environment with the peptide, HUVECs were encapsulated within the three hydrogel groups (**Figure 7–10**). We found that the peptide modified hydrogels had an increased number of tube-like structures and form these structures at a faster rate in comparison to the control hydrogel with the optimal concentration and time point being 100 μM and 4 days. Ang1 has been shown to promote HUVEC migration in the absence of any other ECM proteins, macromolecules or serum [47]. Furthermore, EC migration has been shown to be Tie2 independent but rather, based on one of the β_1 integrins [47]. While collagen was present within our hydrogel and could be used to promote migration, the control hydrogel showed a diminished ability to promote cells to organize into tubes.

Tube formation typically involves EC proliferation and sprouting in which tip cells acquire motile behaviour. The extended cellular processes connect by lumen propagation through intercellular and intracellular vacuole fusion [48]. Thus there are many processes involved in tube formation and neovascularization besides cell proliferation [49]. The role ang1 plays in signalling various functionalities in endothelial cells is context dependent based on whether or not ECs are isolated or surrounded by ECs, as previously discussed [31,32].

In all three encapsulation conditions, regression of the tube-like structures was observed at longer time points. Many primary cells that are cultured *in vitro* cannot be sustained indefinitely. This is due to the lack of a microenvironment that can provide all the necessary mechanical and chemical cues. In this case, the lack of many growth factors [50], absence of supporting cells such as smooth-muscle cells and pericytes [51] and lack of proper matrix stiffness or shear stress [52] for the ECs led to the eventual regression of the tube-like structures.

To further evaluate which integrin this peptide may engage, we performed a two day blocking experiment using integrin antibodies. We chose to block either α5β1 or αvβ3 integrins of encapsulated ECs in the three hydrogel conditions. By blocking integrins α5β1 or αvβ3 for two days immediately after encapsulation, we examined the role that these integrins may play in enabling the cells to attach to the matrix and form tube like structures. Since the QHREDGS peptide is from the fibrinogen-like domain of ang1 [9], we were expecting to see ECs interact with the peptide using the same integrins they use to interact with fibrin. Furthermore, since the matrix we used was collagen, we expected to see the most appreciable changes in tube formation when both collagen and fibrin binding were blocked. We chose to block the integrin receptor αvβ3 which is specific to EC interaction with fibrin to tease out a specific role of the peptide, while not hindering the EC's ability to interact with collagen [13]. The other integrin receptor blocked was α5β1 since the β1 unit may interact with both collagen and fibrin [13,53]. Consequently, we expected to see a graded response wherein blocking the integrin receptor α5β1 would more strongly disrupt EC tube formation due to the dual disruption of EC binding to the peptide and collagen matrix while, blocking αvβ3 would only disrupt ECs ability to interact with the peptide while still allowing for ECs to interact freely with the collagen matrix. These integrin receptor blocking experiments have 1) shown that the EC peptide interaction can be inhibited through blocking integrin receptor αvβ3 and that 2) blocking integrin receptor α5β1 resulted in isolated and round ECs after two days of culture possibly due to blocking ECs from interacting with both the peptide and collagen.

We believe that the differences in optimal concentrations for monolayer studies (**Figures 1–4**) and the encapsulation studies (**Figures 7–10**) are due to the way in which ECs were cultivated and different ways in which the peptide was presented to the cells. Specifically, **Figures 1–3**, deal with ECs that were cultured in a monolayer and the peptide was applied in the soluble form. When peptide was applied in soluble from, mass transport limitations could be present which would effectively lower the peptide concentration in the vicinity of the cell. When cells were cultivated encapsulated in a hydrogel and the peptide was covalently immobilized to the hydrogel material, the cells were surrounded by the matrix and peptide which allowed for direct and sustained interactions. Too high of a peptide concentration (e.g. 650 μM) could induce cells to adhere too strongly, limiting their ability to migrate and form tubes (as observed in **Figure 8**), thus intermediate peptide concentrations may be most beneficial in terms of the readouts we assessed in this paper. Similar effects were reported previously for other peptides [54,55]. Our previous work [11], focused on cardiomyocytes, found the optimal concentrations to be 650 μM thus, the differences in the optimal peptide concentrations may be cell type specific (cardiomyocytes vs. endothelial cells).

This study showed that in a dose dependent manner, the peptide QHREDGS can increase cell survival, decrease cell permeability, and induce the formation of tube like structures. We used several different endothelial cell lines (HMVEC-D, MS1 EC, HUVEC) in order to confirm that the peptide acted on different endothelial cell types, not just one type, in order to make sure the peptide had a wide applicability. Furthermore, in conjunction with previous studies of QHREDGS and its ability to improve cardiomyocyte survival and function, this peptide has now been shown to have a dual protective nature on endothelial cells and cardiomyocytes. When conjugated to hydrogels, the peptide's function would be to help ensure initial survival of cells through mitigation of cell apoptosis while stimulating the formation of a primitive vasculature to rapidly form during the first few days to provide oxygen and nutrients to the cells in the hydrogel. Further studies are required to examine the functionality of a combined endothelial cell and cardiomyocyte therapy for myocardial infarction using a QHREDGS modified collagen-chitosan hydrogel *in vivo*.

Conclusion

We concluded from these experiments that the angiopoietin-1 based peptide QHREDGS was able to promote endothelial cell connectivity, metabolism and survival of endothelial cells cultured in monolayers with peptide applied in soluble form or when covalently immobilized in collagen-chitosan hydrogels. When endothelial cells were encapsulated within a peptide modified collagen-chitosan hydrogel, the peptide was able to promote a more rapid and robust response to form tube-like structures in comparison to the control hydrogel. We propose that the peptide QHREDGS would be an attractive candidate to modify solutions and biomaterials for endothelial cell applications, in order to promote cell survival and migration of cells leading to the formation of new vasculature.

Acknowledgments

The authors would like to thank Yun Xiao for her kind assistance with confocal microscopy imaging.

Author Contributions

Conceived and designed the experiments: JWM SMD M. Rupnick M. Radisic. Performed the experiments: JWM SMD NI. Analyzed the data: JWM SMD NI M. Radisic. Contributed reagents/materials/analysis tools: LAR. Wrote the paper: JWM SMD M. Radisic.

References

1. Rustad KC, Sorkin M, Levi B, Longaker MT, Gurtner GC (2010) Strategies for organ level tissue engineering. Organogenesis 6: 151–157.
2. Khan OF, Sefton MV (2011) Endothelialized biomaterials for tissue engineering applications in vivo. Trends in biotechnology 29: 379–387.
3. Mikos AG, Herring SW, Ochareon P, Elisseeff J, Lu HH, et al. (2006) Engineering complex tissues. Tissue engineering 12: 3307–3339.
4. Lee K, Silva EA, Mooney DJ (2011) Growth factor delivery-based tissue engineering: general approaches and a review of recent developments. Journal of the Royal Society, Interface/the Royal Society 8: 153–170.
5. Chiu LL, Radisic M (2010) Scaffolds with covalently immobilized VEGF and Angiopoietin-1 for vascularization of engineered tissues. Biomaterials 31: 226–241.
6. Chiu LL, Radisic M, Vunjak-Novakovic G (2010) Bioactive scaffolds for engineering vascularized cardiac tissues. Macromolecular bioscience 10: 1286–1301.
7. Phelps EA, Landazuri N, Thule PM, Taylor WR, Garcia AJ (2010) Bioartificial matrices for therapeutic vascularization. Proceedings of the National Academy of Sciences of the United States of America 107: 3323–3328.
8. Place ES, Evans ND, Stevens MM (2009) Complexity in biomaterials for tissue engineering. Nature materials 8: 457–470.
9. Dallabrida SM, Ismail N, Oberle JR, Himes BE, Rupnick MA (2005) Angiopoietin-1 promotes cardiac and skeletal myocyte survival through integrins. Circulation research 96: e8–24.
10. Rask F, Dallabrida SM, Ismail NS, Amoozgar Z, Yeo Y, et al. (2010) Photocrosslinkable chitosan modified with angiopoietin-1 peptide, QHREDGS, promotes survival of neonatal rat heart cells. Journal of biomedical materials research Part A 95: 105–117.
11. Reis LA, Chiu LL, Liang Y, Hyunh K, Momen A, et al. (2012) A peptide-modified chitosan-collagen hydrogel for cardiac cell culture and delivery. Acta biomaterialia 8: 1022–1036.
12. Tinsley JH, De Lanerolle P, Wilson E, Ma W, Yuan SY (2000) Myosin light chain kinase transference induces myosin light chain activation and endothelial hyperpermeability. American journal of physiology Cell physiology 279: C1285–1289.
13. Hanjaya-Putra D, Bose V, Shen YI, Yee J, Khetan S, et al. (2011) Controlled activation of morphogenesis to generate a functional human microvasculature in a synthetic matrix. Blood 118: 804–815.
14. Mochizuki N (2009) Vascular integrity mediated by vascular endothelial cadherin and regulated by sphingosine 1-phosphate and angiopoietin-1. Circulation journal : official journal of the Japanese Circulation Society 73: 2183–2191.
15. van der Heijden M, van Nieuw Amerongen GP, van Bezu J, Paul MA, Groeneveld AB, et al. (2011) Opposing effects of the angiopoietins on the thrombin-induced permeability of human pulmonary microvascular endothelial cells. PloS one 6: e23448.
16. Koblizek TI, Weiss C, Yancopoulos GD, Deutsch U, Risau W (1998) Angiopoietin-1 induces sprouting angiogenesis in vitro. Current biology : CB 8: 529–532.
17. Mammoto T, Parikh SM, Mammoto A, Gallagher D, Chan B, et al. (2007) Angiopoietin-1 requires p190 RhoGAP to protect against vascular leakage in vivo. The Journal of biological chemistry 282: 23910–23918.
18. Cascone I, Napione L, Maniero F, Serini G, Bussolino F (2005) Stable interaction between alpha5beta1 integrin and Tie2 tyrosine kinase receptor regulates endothelial cell response to Ang-1. The Journal of cell biology 170: 993–1004.
19. Camenisch G, Pisabarro MT, Sherman D, Kowalski J, Nagel M, et al. (2002) ANGPTL3 stimulates endothelial cell adhesion and migration via integrin alpha vbeta 3 and induces blood vessel formation in vivo. The Journal of biological chemistry 277: 17281–17290.
20. Hersel U, Dahmen C, Kessler H (2003) RGD modified polymers: biomaterials for stimulated cell adhesion and beyond. Biomaterials 24: 4385–4415.
21. Cho CH, Kammerer RA, Lee HJ, Steinmetz MO, Ryu YS, et al. (2004) COMP-Ang1: a designed angiopoietin-1 variant with nonleaky angiogenic activity. Proceedings of the National Academy of Sciences of the United States of America 101: 5547–5552.
22. Holland J, Hersh L, Bryhan M, Onyiriuka E, Ziegler L (1996) Culture of human vascular endothelial cells on an RGD-containing synthetic peptide attached to a starch-coated polystyrene surface: comparison with fibronectin-coated tissue grade polystyrene. Biomaterials 17: 2147–2156.
23. Sagnella S, Anderson E, Sanabria N, Marchant RE, Kottke-Marchant K (2005) Human endothelial cell interaction with biomimetic surfactant polymers containing Peptide ligands from the heparin binding domain of fibronectin. Tissue engineering 11: 226–236.
24. Sagnella SM, Kligman F, Anderson EH, King JE, Murugesan G, et al. (2004) Human microvascular endothelial cell growth and migration on biomimetic surfactant polymers. Biomaterials 25: 1249–1259.
25. Wang X, Heath DE, Cooper SL (2012) Endothelial cell adhesion and proliferation to PEGylated polymers with covalently linked RGD peptides. Journal of biomedical materials research Part A 100: 794–801.
26. Van Slyke P, Alami J, Martin D, Kuliszewski M, Leong-Poi H, et al. (2009) Acceleration of diabetic wound healing by an angiopoietin peptide mimetic. Tissue engineering Part A 15: 1269–1280.
27. Cho CH, Kammerer RA, Lee HJ, Yasunaga K, Kim KT, et al. (2004) Designed angiopoietin-1 variant, COMP-Ang1, protects against radiation-induced endothelial cell apoptosis. Proceedings of the National Academy of Sciences of the United States of America 101: 5553–5558.
28. Cho CH, Kim KE, Byun J, Jang HS, Kim DK, et al. (2005) Long-term and sustained COMP-Ang1 induces long-lasting vascular enlargement and enhanced blood flow. Circulation research 97: 86–94.
29. Augustin HG, Koh GY, Thurston G, Alitalo K (2009) Control of vascular morphogenesis and homeostasis through the angiopoietin-Tie system. Nature reviews Molecular cell biology 10: 165–177.
30. Fukuhara S, Sako K, Minami T, Noda K, Kim HZ, et al. (2008) Differential function of Tie2 at cell-cell contacts and cell-substratum contacts regulated by angiopoietin-1. Nature cell biology 10: 513–526.
31. Saharinen P, Eklund L, Miettinen J, Wirkkala R, Anisimov A, et al. (2008) Angiopoietins assemble distinct Tie2 signalling complexes in endothelial cell-cell and cell-matrix contacts. Nature cell biology 10: 527–537.
32. Fachinger G, Deutsch U, Risau W (1999) Functional interaction of vascular endothelial-protein-tyrosine phosphatase with the angiopoietin receptor Tie-2. Oncogene 18: 5948–5953.
33. Papapetropoulos A, Fulton D, Mahboubi K, Kalb RG, O'Connor DS, et al. (2000) Angiopoietin-1 inhibits endothelial cell apoptosis via the Akt/survivin pathway. The Journal of biological chemistry 275: 9102–9105.
34. Gavard J, Patel V, Gutkind JS (2008) Angiopoietin-1 prevents VEGF-induced endothelial permeability by sequestering Src through mDia. Developmental cell 14: 25–36.
35. Bredt DS, Snyder SH (1994) Nitric oxide: a physiologic messenger molecule. Annual review of biochemistry 63: 175–195.
36. Harrison DG (1997) Cellular and molecular mechanisms of endothelial cell dysfunction. The Journal of clinical investigation 100: 2153–2157.
37. Noll G, Luscher TF (1998) The endothelium in acute coronary syndromes. European heart journal 19 Suppl C: C30–38.
38. DeBusk LM, Hallahan DE, Lin PC (2004) Akt is a major angiogenic mediator downstream of the Ang1/Tie2 signaling pathway. Experimental cell research 298: 167–177.
39. Rask F, Mihic A, Reis L, Dallabrida SM, Ismail NS, et al. (2010) Hydrogels modified with QHREDGS peptide support cardiomyocyte survival in vitro and after sub-cutaneous implantation. Soft Matter 6: 5089–5099.
40. Ismail NS, Pravda EA, Li D, Shih SC, Dallabrida SM (2010) Angiopoietin-1 reduces H(2)O(2)-induced increases in reactive oxygen species and oxidative damage to skin cells. The Journal of investigative dermatology 130: 1307–1317.
41. Oubaha M, Gratton JP (2009) Phosphorylation of endothelial nitric oxide synthase by atypical PKC zeta contributes to angiopoietin-1-dependent inhibition of VEGF-induced endothelial permeability in vitro. Blood 114: 3343–3351.
42. Ilan N, Madri JA (2003) PECAM-1: old friend, new partners. Current opinion in cell biology 15: 515–524.
43. Newman PJ, Newman DK (2003) Signal transduction pathways mediated by PECAM-1: new roles for an old molecule in platelet and vascular cell biology. Arteriosclerosis, thrombosis, and vascular biology 23: 953–964.
44. David S, Ghosh CC, Mukherjee A, Parikh SM (2011) Angiopoietin-1 requires IQ domain GTPase-activating protein 1 to activate Rac1 and promote endothelial barrier defense. Arteriosclerosis, thrombosis, and vascular biology 31: 2643–2652.
45. Kwak HJ, So JN, Lee SJ, Kim I, Koh GY (1999) Angiopoietin-1 is an apoptosis survival factor for endothelial cells. FEBS letters 448: 249–253.
46. Kim I, Kim HG, So JN, Kim JH, Kwak HJ, et al. (2000) Angiopoietin-1 regulates endothelial cell survival through the phosphatidylinositol 3′-Kinase/Akt signal transduction pathway. Circulation research 86: 24–29.
47. Carlson TR, Feng Y, Maisonpierre PC, Mrksich M, Morla AO (2001) Direct cell adhesion to the angiopoietins mediated by integrins. The Journal of biological chemistry 276: 26516–26525.
48. Davis GE, Koh W, Stratman AN (2007) Mechanisms controlling human endothelial lumen formation and tube assembly in three-dimensional extracellular matrices. Birth defects research Part C, Embryo today : reviews 81: 270–285.
49. Bayless KJ, Salazar R, Davis GE (2000) RGD-dependent vacuolation and lumen formation observed during endothelial cell morphogenesis in three-dimensional fibrin matrices involves the alpha(v)beta(3) and alpha(5)beta(1) integrins. The American journal of pathology 156: 1673–1683.
50. Folkman J, Klagsbrun M (1987) Angiogenic factors. Science 235: 442–447.
51. Coultas L, Chawengsaksophak K, Rossant J (2005) Endothelial cells and VEGF in vascular development. Nature 438: 937–945.
52. Topper JN, Gimbrone MA Jr (1999) Blood flow and vascular gene expression: fluid shear stress as a modulator of endothelial phenotype. Molecular medicine today 5: 40–46.
53. Davis GE, Camarillo CW (1996) An alpha 2 beta 1 integrin-dependent pinocytic mechanism involving intracellular vacuole formation and coalescence regulates

capillary lumen and tube formation in three-dimensional collagen matrix. Experimental cell research 224: 39–51.
54. Zhang C, Hekmatfer S, Karuri NW (2013) A comparative study of polyethylene glycol hydrogels derivatized with the RGD peptide and the cell-binding domain of fibronectin. Journal of biomedical materials research Part A.
55. Guarnieri D, De Capua A, Ventre M, Borzacchiello A, Pedone C, et al. (2010) Covalently immobilized RGD gradient on PEG hydrogel scaffold influences cell migration parameters. Acta biomaterialia 6: 2532–2539.

Is the Poly (L- Lactide- *Co*– Caprolactone) Nanofibrous Membrane Suitable for Urinary Bladder Regeneration?

Marta Pokrywczynska[1*], Arkadiusz Jundzill[1], Jan Adamowicz[1], Tomasz Kowalczyk[2], Karolina Warda[1], Marta Rasmus[1], Lukasz Buchholz[1], Sandra Krzyzanowska[1], Pawel Nakielski[3], Tomasz Chmielewski[3], Magdalena Bodnar[4], Andrzej Marszalek[4,5], Robert Debski[6], Malgorzata Frontczak-Baniewicz[7], Grzegorz Mikułowski[8], Maciej Nowacki[1], Tomasz A. Kowalewski[3], Tomasz Drewa[1,9]

1 Chair of Regenerative Medicine, Department of Tissue Engineering, Nicolaus Copernicus University in Torun, Ludwik Rydygier Medical College in Bydgoszcz, Bydgoszcz, Poland, **2** Department of Theory of Continuous Media, Institute of Fundamental Technological Research, Polish Academy of Sciences, Warsaw, Poland, **3** Department of Mechanics and Physics of Fluids, Institute of Fundamental Technological Research, Polish Academy of Sciences, Warsaw, Poland, **4** Department of Clinical Pathomorphology, Nicolaus Copernicus University in Torun, Ludwik Rydygier Medical College in Bydgoszcz, Bydgoszcz, Poland, **5** Department of Tumor Pathology, Center of Oncology, Poznan University of Medical Sciences, Poznan, Poland, **6** Department of Pediatrics, Hematology and Oncology, Nicolaus Copernicus University in Torun, Ludwik Rydygier Medical College in Bydgoszcz, Bydgoszcz, Poland, **7** Electron Microscopy Platform, Mossakowski Medical Research Centre, Polish Academy of Sciences, Warsaw, Poland, **8** Department of Intelligent Technologies, Institute of Fundamental Technological Research, Polish Academy of Sciences, Warsaw, Poland, **9** Department of Urology, Nicolaus Copernicus Hospital, Torun, Poland

Abstract

The purpose of this study was to compare: a new five-layered poly (L–lactide–*co*–caprolactone) (PLC) membrane and small intestinal submucosa (SIS) as a control in rat urinary bladder wall regeneration. The five-layered poly (L–lactide–*co*–caprolactone) membrane was prepared by an electrospinning process. Adipose tissue was harvested from five 8-week old male Wistar rats. Adipose derived stem cells (ADSCs) were seeded in a density of 3×10^6 cells/cm^2 onto PLC membrane and SIS scaffolds, and cultured for 5-7 days in the stem cell culture medium. Twenty male Wistar rats were randomly divided into five equal groups. Augmentation cystoplasty was performed in a previously created dome defect. Groups: (I) PLC+ 3×10^6ADSCs; (II) SIS+ 3×10^6ADSCs; (III) PLC; (IV) SIS; (V) control. Cystography was performed after three months. The reconstructed urinary bladders were evaluated in H&E and Masson's trichrome staining. Regeneration of all components of the normal urinary bladder wall was observed in bladders augmented with cell-seeded SIS matrices. The urinary bladders augmented with SIS matrices without cells showed fibrosis and graft contraction. Bladder augmentation with the PLC membrane led to numerous undesirable events including: bladder wall perforation, fistula or diverticula formation, and incorporation of the reconstructed wall into the bladder lumen. The new five-layered poly (L–lactide–*co*–caprolactone) membrane possesses poorer potential for regenerating the urinary bladder wall compared with SIS scaffold.

Editor: Xiaohua Liu, Texas A&M University Baylor College of Dentistry, United States of America

Funding: The authors have no support or funding to report.

Competing Interests: The authors have declared that no competing interests exist.

* Email: marta.pokrywczynska@interia.pl

These authors contributed equally to this work.

Introduction

The small intestine is commonly used for urinary tract reconstruction, either for bladder augmentation, replacement, or urinary diversion. Unfortunately, these techniques are associated with numerous complications [1]. Many natural and synthetic biomaterials such as plastic mold, gelatin sponge, Japanese paper, preserved dog bladder, lyophilized human dura, bovine pericardium, small intestinal submucosa, bladder acellular matrix, or composites of collagen and polyglycolic acid were used for human urinary bladder regeneration with a wide range of outcomes [2]. However, the ideal biomaterial for urinary bladder reconstruction has not been found thus far. The scaffolds used in tissue engineering should mimic the ability of the extracellular matrix (ECM) to regulate cell functions such as cell division, differentiation, and apoptosis [3]. The biomaterial intended for urinary bladder reconstruction should be not only biocompatible and biodegradable but also waterproof, flexible, elastic and able to provide good mechanical strength [4].

Decellularized extracellular matrices such as small intestinal submucosa (SIS) or bladder acellular matrix (BAM) have been widely used for urinary bladder reconstruction in various models [5–8]. However, acellular matrices have some disadvantages including immunogenicity, lot- to- lot variability, and inadequate biomechanical properties [9]. Consequently, there is a growing interest in developing new synthetic biomaterials for urinary bladder regeneration [10,11]. Aliphatic polyesters such as poly(-glycolide) (PGA), poly(lactide) (PLA), poly(ε-caprolactone) (PCL), and their copolymers have been widely used in tissue engineering with favorable results [12–16]. Several *in vitro* studies revealed the significant potential of PLC membranes for tissue engineering in urology [17,18]. This data has encouraged us to produce the PLC membrane for urinary bladder wall regeneration. The electro-

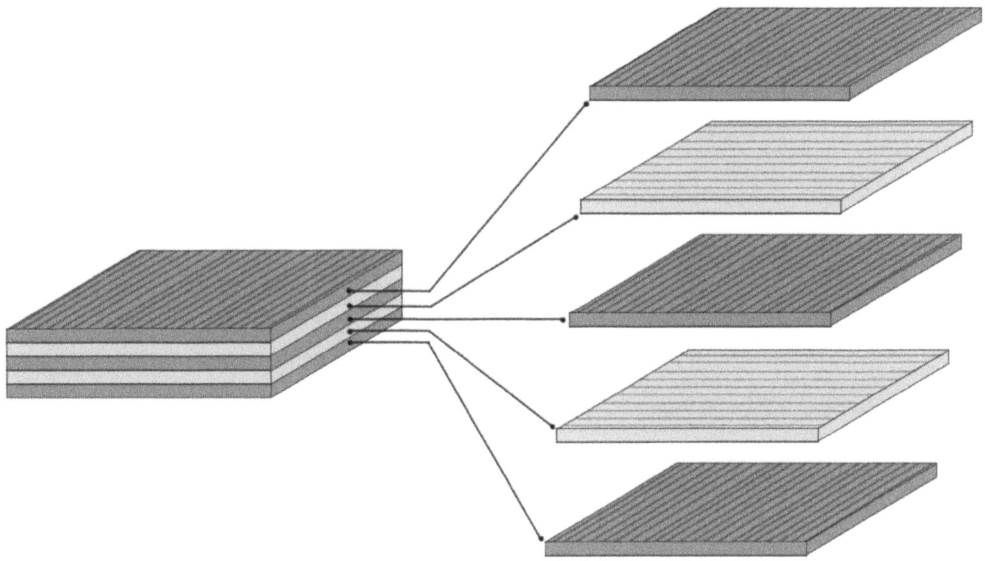

Figure 1. Scheme of five-layered electrospun membrane structure.

spinning process is a promising way for providing nanofibrous scaffolds, which closely mimic the natural extracellular matrix structure [19]. Matrices made of electrospun nanofibers when compared to natural collagen matrices have certain advantages: they are non-allergenic, do not involve risk of transferring diseases (e.g. prions), and are easy to prepare, handle, and store. Urine is extremely cytotoxic for stem cells [20,21]. Therefore, the separation of stem cells from the toxic environment of urine seems to be crucial for urinary bladder tissue engineering. In this experiment, we evaluate poly (L–lactide–*co*–caprolactone), a new nanofibrous and durable membrane, designed to isolate cells from urine in tissue engineered urinary bladders.

Materials and Methods

Synthesis of five-layered poly (L-lactide-co-caprolactone) membrane via an electrospinning technique

Poly (L–lactide–*co*–caprolactone) (PLC) (Purac, Netherlands), chloroform ($CHCl_3$) (POCh, Poland), N,N-dimethylformamide (DMF)(POCh, Poland), sodium bicarbonate ($NaHCO_3$) (POCh, Poland) were used as supplied, without further purification. Nanomaterials were prepared using an electrospinning process described by Li et al [22]. Briefly, PLC was dissolved in a mixture of $CHCl_3$ and DMF (mass proportion 16:1) to form 9% solution and left overnight. Electrospinning was conducted in a custom-made polycarbonate chamber of approximate volume of 1 m^3. The electrospinning setup consisted of custom-made voltage power supply (HVPS) with adjustable output voltage. The electrospun solution was transferred from a metering syringe pump (New Era Pumping Systems, Great Britain) through a hydraulic system to a nozzle made of blunt needle (outer volume 0.4 mm) connected to HVPS output. Each element of the setup has a commercial market counterpart. Electrically grounded aluminum foil attached to a fast rotating polypropylene mandrel served as a target. After collection of each consecutive layer of directed nanofibrous material, the foil was detached from the mandrel, positioned orthogonally to a previous direction, and attached back to the mandrel. Material of a plywood-like structure was formed after 5 repetitions of the fiber collection process (Fig. 1). Electrospinning parameters were as follows: voltage- 20 kV, polymer solution flow- 800 µl/h, spinneret to collector distance- 20 cm, mandrel speed- 4100 rpm, mandrel diameter- 60 mm.

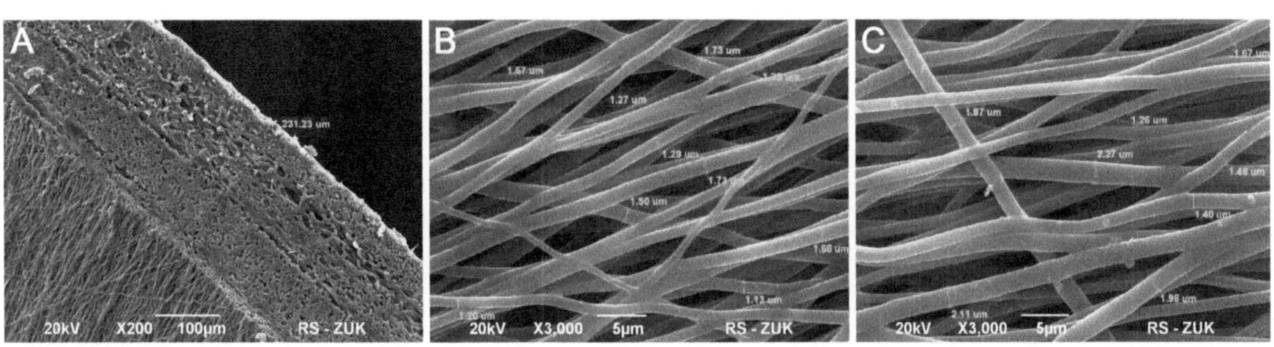

Figure 2. Ultrastructural analysis of the electrospun poly (lactydo-*co*-caprolactone) (PLC) membrane: thickness of five- layered PLC membrane (A), the surface of unmodified PLC membrane (B) the surface of PLC membrane modified with sodium bicarbonate (C). Scanning Electron Microscopy, bar 5 µm, 100 µm.

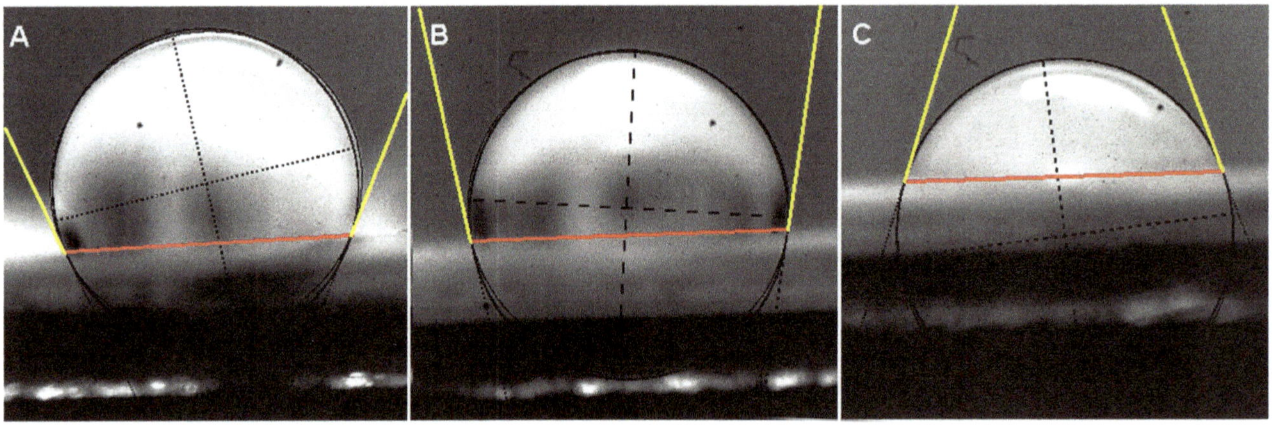

Figure 3. Contact angle measurements: untreated poly (lactydo-*co*-caprolactone) (PLC) membrane, 110° (A), $NaHCO_3$ treated PLC membrane: unmodified side- 97° (C) and modified side- 72° (B). Contact angle is the angle between the baseline of the drop (marked in red) and the tangent at the drop boundary (marked in yellow).

Table 1. Tensile properties of the five-layered poly (lactydo-*co*-caprolactone) (PLC) membrane and bladder specimens.

Material	Ultimate tensile strain (mm/mm)	Ultimate tensile strength (MPa)	Elastic modulus R1 (MPa)	Elastic modulus R2 (MPa)
PLC	4.35	11.4	16.0	2.25
Rat bladder (27)	2.03	0.72	-	0.76
Pig bladder (27)	1.66	0.32	-	0.26
Human bladder (27)	0.69	0.27	-	0.25

The elastic range 1 (R1) was demonstrated up to 24% of strain and the elastic range 2 (R2) between 70 % and 429 %.

Modification of five-layered poly (L-lactide-co-caprolactone) membrane

The surface of nanomaterial was modified by treatment with a 0.5 M solution of sodium bicarbonate dissolved in deionized water. Filter paper was soaked with a treatment solution and placed on top of the membrane for 30 min. The results were measured using a goniometric method by measuring contact angle of 2 μl droplet of water placed on a membrane before and after modification (OCA 20, DataPhysics Instruments GmbH, Filderstad, Germany). ImageJ software (NIST, USA) was used to evaluate data.

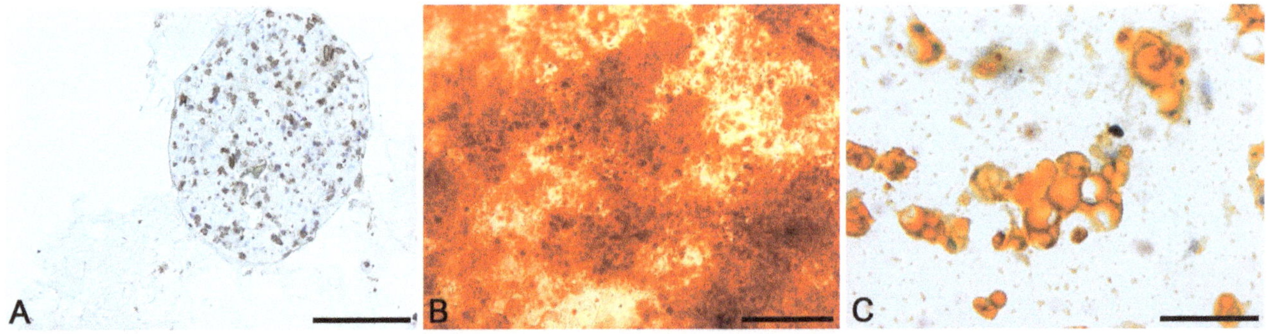

Figure 4. Differentiation potential of adipose derived stem cells: a positive anti-collagen II staining after chondrogenic induction, bar 200 μm (A), Alizarin Red staining after osteogenic induction, bar 500 μm (B), Oil Red O staining after adipogenic induction, bar 20 μm (C).

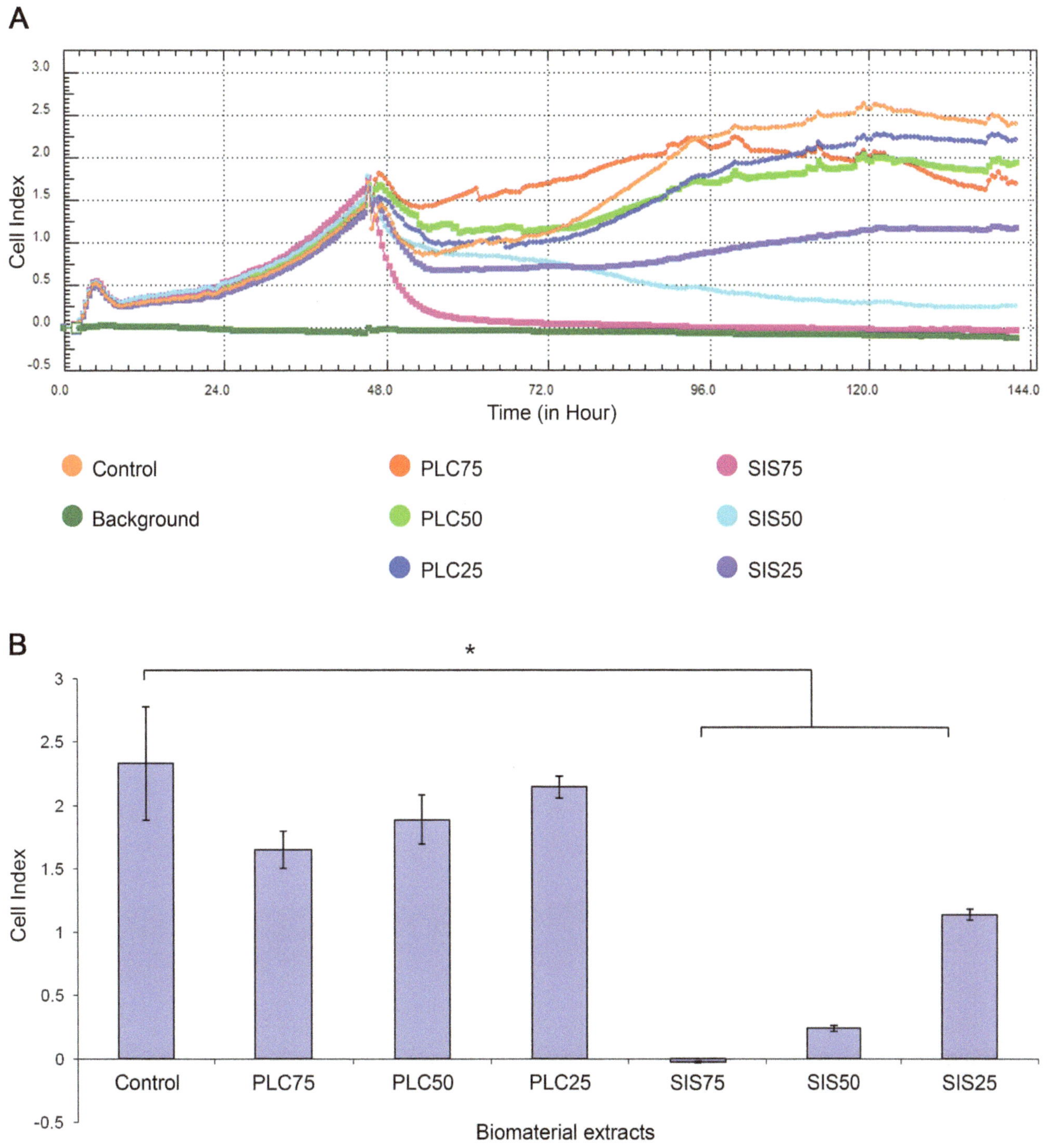

Figure 5. Analysis of poly (lactydo-*co*-caprolactone) (PLC) and small intestinal submucosa (SIS) cytotoxicity using Real Time Cell Analyzer (RTCA). Adipose derived stem cells were treated with 75%, 50% and 25% extracts of PLC (PLC75, PLC50, PLC25 respectively) and SIS (SIS75, SIS50, SIS25 respectively). The results are presented as: cell growth curves (A), mean cell index ± standard deviation after 96 hours of cell incubation with extracts. The statistical significance is shown as * $p<0.05$ (B)

Poly (L-lactide-co-caprolactone) membrane nanofiber thickness and pore size

The ImageJ software was used to assess fiber diameters. For each sample of nanofiber material, three SEM images were acquired and 33 measurements were taken from each image. The measurement was made perpendicular to the fiber's axis. Scaffold porosity (ε) was calculated by the method given by Hu [23] using density of the scaffold ($\rho_{scafflold}$) and density of neat PLC (ρ_{PLC}): $\varepsilon = 1 - (\rho_{scafflold}/\rho_{PLC})$. Characteristic 3-dimensional pore diameter (d_{3D}) was calculated by the method given by Tomadakis and Robertson [24] on the basis of scaffold porosity (ε) and mean fiber diameter (ω): $d_{3D} = -\ \omega/\ln(\varepsilon)$

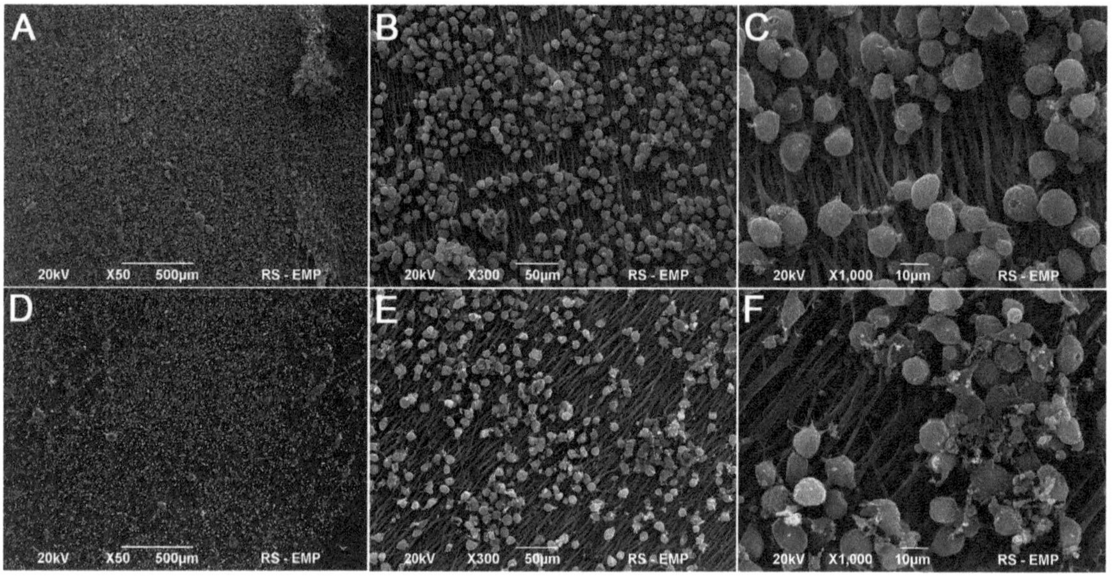

Figure 6. Poly (lactydo-*co*-caprolactone) membrane seeded with adipose derived stem cells (ADSCs), 7th (A,B,C) and 14th day of culture (D,E,F), Scanning Electron Microscopy, bar 500 µm, 50 µm, 10 µm.

Analysis of the mechanical properties of five-layered poly (L-lactide-co-caprolactone) membrane

The tensile tests were conducted on a load frame of a servohydraulic material testing machine (MTS 242.01 actuator, Eden Prairie, USA). The specimen (25 mm length, 12.5 mm width) was mounted into flat grips with gauge base 12.5 mm. During the test, the specimen was longitudinally extracted at a rate of 0.3 mm/s to failure. The grip travel and specimen load were continuously measured over the test procedure duration with a precision force transducer (Interface, model 1500, measuring range 125 N, resolution 0.0625 N) and MTS system linear variable differential transformer (measuring range 100 mm, resolution 0.01 mm). Stress/strain curve for the specimen was generated and the ultimate tensile strength, maximum strain and elastic modulus were determined. The strength (MPa) was calculated by dividing the value of failure load by initial cross-sectional area of the specimen. The maximum strain (the strain value corresponding to the ultimate stress) was calculated as the elongation of the specimen divided by the initial gauge length (mm/mm). The elastic modulus (MPa) was defined as the slope of the linear regions of the stress/strain curve.

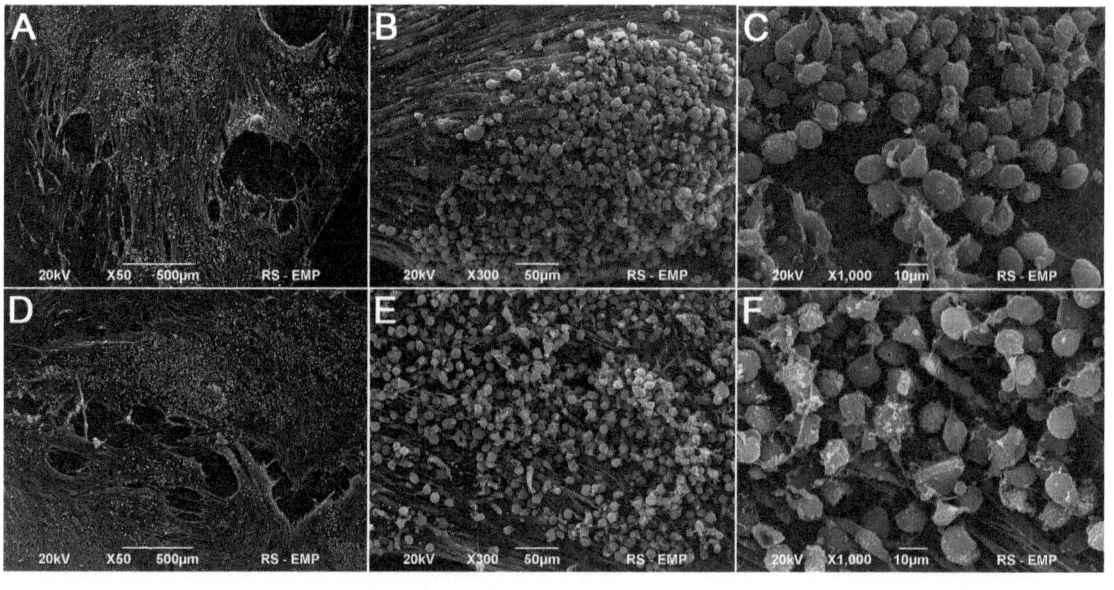

Figure 7. Small intestinal submucosa seeded with adipose derived stem cells (ADSCs), 7th (A,B,C) and 14th day of culture (D,E,F), Scanning Electron Microscopy, bar 500 µm, 50 µm, 10 µm.

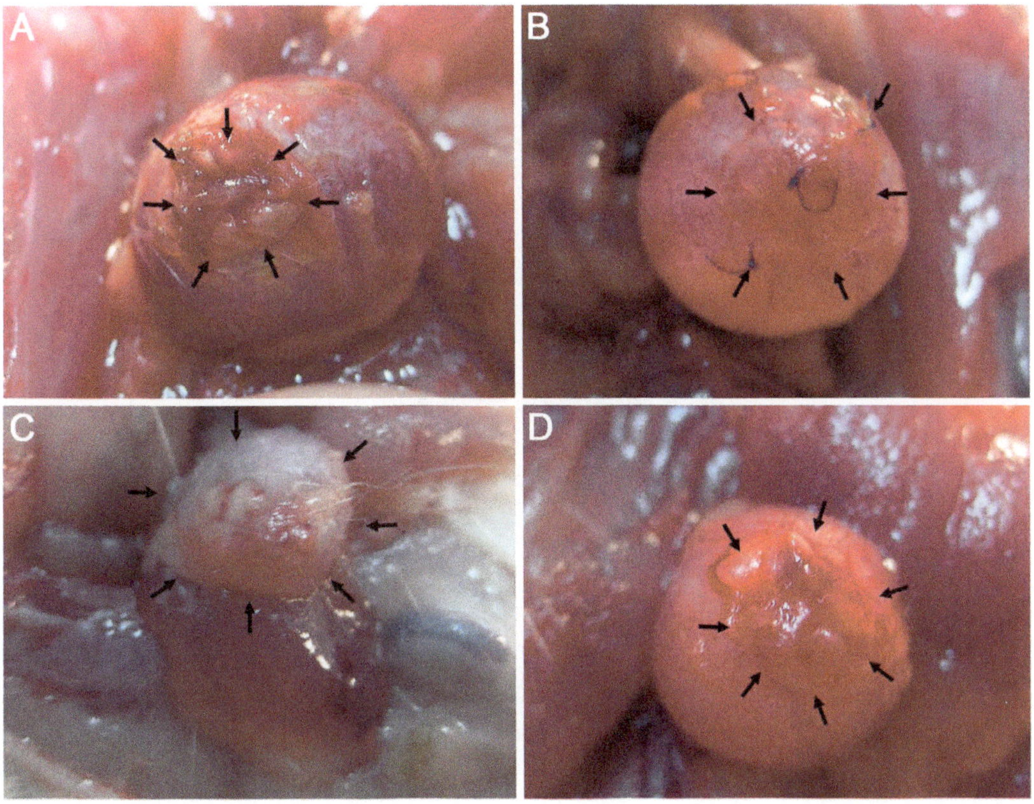

Figure 8. Macroscopic analysis of reconstructed bladders. Bladder augmented with adipose derived stem cells (ADSCs) seeded poly (lactydo-*co*-caprolactone) (PLC) (A), ADSCs seeded small intestinal submucosa (SIS) (B), unseeded PLC (C) and unseeded SIS (D). The arrows point out the reconstructed area.

Ethics Statement

This study was carried out in strict accordance with the recommendations in the Guide for the Care and Use of Laboratory Animals of the National Institutes of Health [25]. The protocol was approved by the Nicolaus Copernicus University Ethics Committee (no. 4/2012).

Adipose Derived Stem Cell Isolation and Culture

Adipose tissue was harvested from the retroperitoneal space from five syngeneic 8-week male Wistar rats. The animals were euthanized with an overdose of ketamine 75 mg/kg (Biowet, Poland). ADSCs were isolated according to the method described by Safford et al [26]. Briefly, adipose tissue (1 g) was digested in collagenase type I solution (1 mg/ml, Sigma- Aldrich, Germany) for 30 min at 37°C with shaking. The reaction was stopped by adding Dulbecco's Modified Eagle's Medium (DMEM; PAA, Austria), supplemented with 10% fetal bovine serum (FBS; PAA, Austria) and antibiotics (PAA, Austria). The cell suspension was filtered by a 100 μm cell strainer (Becton Dickinson, USA) and centrifuged at 350×g for 5 min. Viable cells were counted via trypan blue staining and seeded on 25 cm^2 flask in a density of 15×10^3 cells/cm^2. ADSCs were cultured in a medium consisting of DMEM supplemented with 10% FBS, fibroblast growth factor (FGF, 10 ng/ml; Sigma- Aldrich, Germany), penicillin/streptomycin (100 U/100 μg/ml) and amphotericin B (100 μg/ml) (PAA, Austria) at 37°C, 5%CO_2 and 95% humidity until 3rd passage.

Adipose Derived Stem Cells Characterization

To confirm the ADSCs phenotype, cells were subjected to antigen analysis by flow cytometry. Detached cells from the 3rd passage were washed and re-suspended with phosphate buffered saline (PBS). Approximately 1×10^6 cells were incubated with monoclonal primary antibodies conjugated with PE or FITC against CD34 (Santa Cruz Biotechnology, Inc, USA; 20 μl/sample), CD44 (Millipore, USA; 10 μl/sample), CD45 (BD, Pharmingen, USA; 0.06 μg/sample) and CD90 (Millipore, USA; 10 μl/sample) for 30 min. Expression level of each surface marker was quantified using an EPICS XL flow cytometer (Beckman Coulter, USA).

The differentiation of ADSCs into adipogenic, osteogenic and chondrogenic lineages was induced by culture in appropriate differentiation media, according to manufacturer's instruction (Invitrogen, USA). Negative control cells were maintained in DMEM/Ham's F-12 supplemented with 10% FBS and antibiotics. Adipogenesis was measured by the accumulation of neutral lipids in fat vacuoles, stained with Oil-Red-O (Sigma- Aldrich, Germany). Osteogenesis was confirmed using Alizarin Red staining (Millipore, USA). Chondrogenic differentiation was evaluated by anti-collagen type II immunocytochemical staining (anti-collagen II clone 6B3, Millipore, USA, 1:100, 16 h, 4°C).

Biomaterial Cytotoxicity Assay

Biomaterial cytotoxicity was determined using extract toxicity assay according to ISO-10993 norm. A piece of biomaterial (~6 cm^2) was extracted in DMEM supplemented with 10% FBS

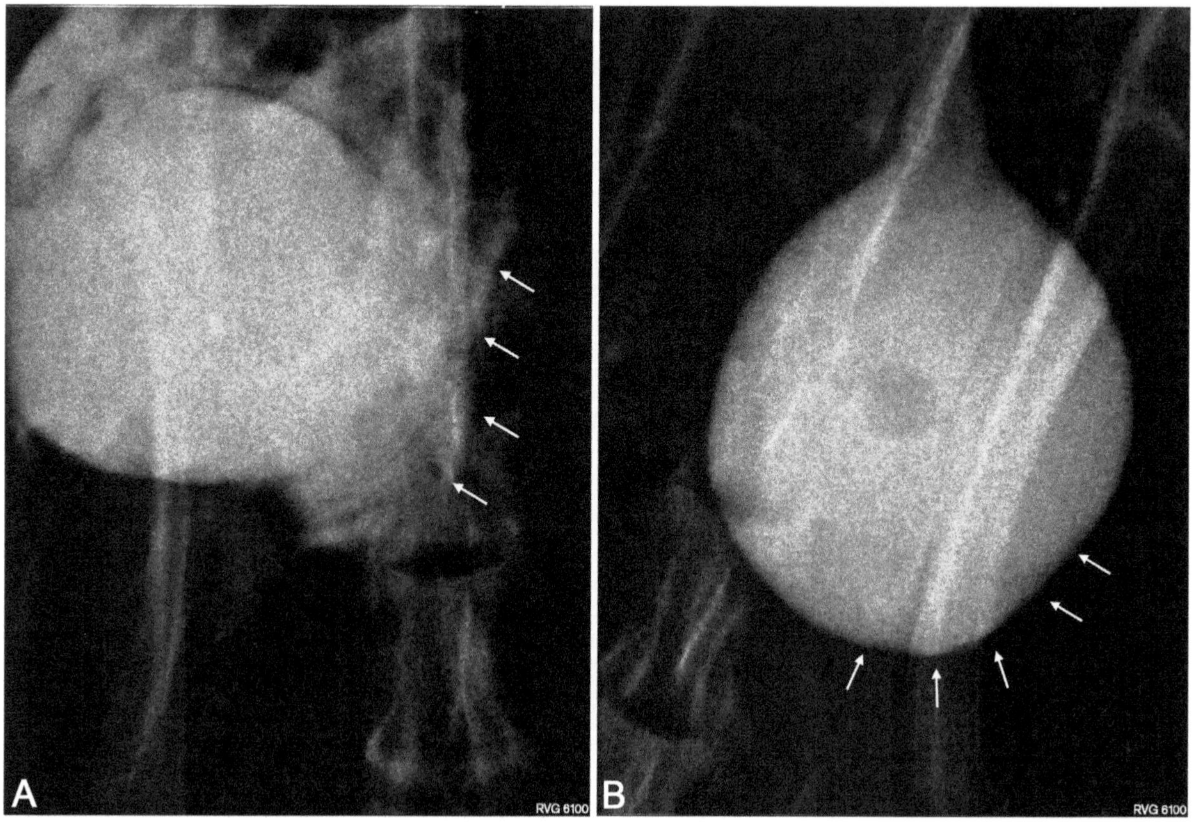

Figure 9. Cystography. Urinary bladders augmented with poly (lactydo-*co*-caprolactone) membrane (A) and small intestinal submucosa seeded with adipose derived stem cells (B). The arrows point out the reconstructed area.

and antibiotics (1 ml) at 37°C for 120 h. The extract was filtered and stored at 4°C for up to one week. Extract cytotoxicity was determined using the xCELLigence system (Real-Time Cell Analyzer Dual Plate, RTCA DP, Roche Applied Science, Germany). The xCELLigence system is a unique, cellular impedance- based systems that allows the real- time monitoring of cell growth. When adherent cells attach and spread on the sensor surface of an electrode, increases in impedance are recorded. The changes in impedance are expressed as cell index (CI). For this purpose, ADSCs were seeded on E-Plates (4×10^3 cells/well) and cultured until reaching a log- phase growth in DMEM supplemented with 10% FBS and antibiotics. Then the medium was changed for a fresh one (negative control) or medium supplemented with 25%, 50% and 75% biomaterial extract (PLC25, PLC50, PLC75, SIS25, SIS50, SIS75 respectively). The experiment was continued until ADSCs achieved the plateau growth phase.

Analysis of Adipose Derived Stem Cells Growth on the poly (L-lactide-co-caprolactone) membrane

ADSCs were seeded on the 1 cm^2 of poly (L-lactide-*co*-caprolactone) electrospun membrane or on small intestinal submucosa (SIS) (Surgisis, Biodesign, USA) mounted on cell crowns in a density of 3×10^6 cells/cm^2, and cultured for 7 or 14 days. ADSCs growth on PLC and SIS scaffolds was assessed by scanning electron microscopy. For this purpose, the specimens were fixed in 2% paraformaldehyde and 2,5% glutaraldehyde in phosphate buffer for 2 h, post-fixed in 1% OsO4 and dehydrated with grades series of ethyl alcohol followed by acetone. Next the specimens were critically-dried and coated with gold particles before observation in scanning electron microscope (JEOL JSM-6390LV, Japan).

Graft preparation

To prepare the grafts for bladder augmentation, the ADSCs were seeded on PLC or SIS scaffolds and cultured for 5-7 days.

Augmentation cystoplasty

Twenty syngeneic male Wistar rats weighing between 250 and 300 g were randomly divided into five equal groups. Sixteen rats, anesthetized with sodium pentobarbital (15 mg/kg, i.p., Biowet, Poland) and lidocaine (20 mg/kg, i.m., Polfa, Poland), underwent hemicystectomy and bladder augmentation with approximately 1 cm^2 of graft. The anastomosis line was marked by 8.0 monofilament non-absorbable marker sutures to identify the graft borders. In the first group, bladders were reconstructed using PLC membrane seeded with 3×10^6 ADSCs. In the second group, bladders were augmented using SIS seeded with 3×10^6 ADSCs. In the third and fourth groups, bladders were reconstructed with unseeded PLC and SIS membranes, respectively. In order to achieve a good analgesic effect, the lidocaine was additionally injected (20 mg/kg, i.m.) after the procedure. The fifth group (control) was left intact. Cystography, morphological and histological studies were performed after three months follow-up.

Cystography

Under general anesthesia with sodium pentobarbital (15 mg/kg, i.p.) and lidocaine (20 mg/kg, i.m.), the animal was placed in

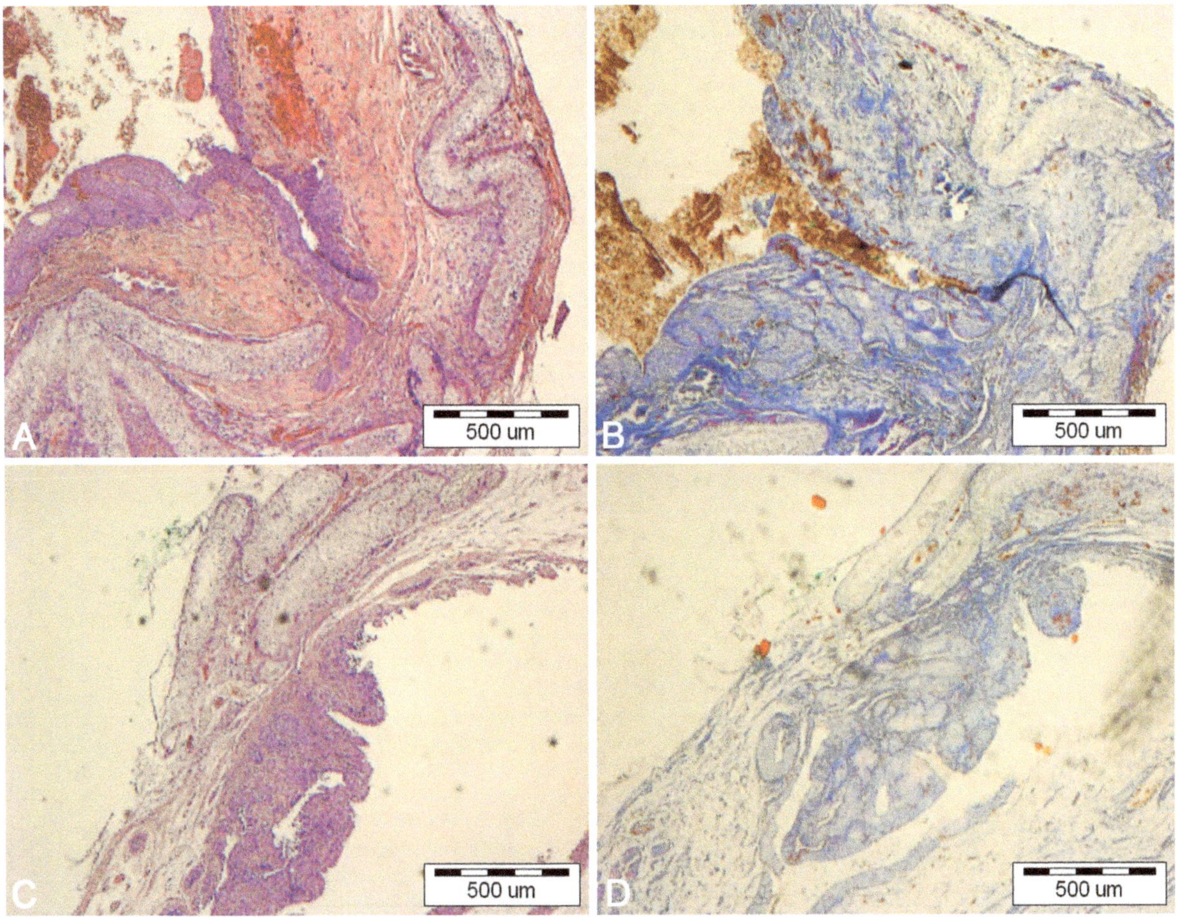

Figure 10. Bladders augmented with poly (lactydo-*co*-caprolactone) membrane seeded (A,B) and unseeded (C,D) with adipose derived stem cells. Light microscope, H&E (A,C) and Trichrome Masson staining (B,D), bar = 500 μm.

supine position with knees flexed. A small, flexible catheter (3 Fr, Galmed, Poland) was gently inserted into the rat's urethra and the urinary bladder was injected with radiocontrast (0.5 ml of 20% iopromide, Bayer Pharma AG, Germany). The cystograms were performed by fluoroscopy (Actube Dental 5D2, exposition 60 kV and 6mAm). The animals were euthanized with an overdose of ketamine (75 mg/kg).

Histological staining

The bladder samples were fixed in 10% buffered formaldehyde, using routine procedure of tissue processing, and embedded in paraffin. Cross-sections of whole bladders were made. The 4 μm thick paraffin sections were stained with hematoxylin and eosin. The connective tissue components and muscle layer were stained according to Masson staining. The smooth muscle abundance was assessed using the ImageJ program according to the method described previously [5]. The analysis was repeated for nine areas from each of the specimens.

Statistical analysis

Statistical differences between groups were determined by one-way ANOVA followed by Tamhane' *post hoc* multiple-comparison test (IBM SPSS Statistics, Predictive Solutions, Poland). Statistically significant differences were defined as having $p<0.05$.

Results

Synthesis and characterization of five-layered poly (L-lactide-co-caprolactone) membrane

The electrospun *five-layered* poly (L-lactide-*co*-caprolactone) membrane thickness was approximately 230 μm (Fig. 2A). The membrane had a micro-porous structure (Fig. 2B). The mean 3D pore diameter was 2.69±0.62 μm. The fiber diameter ranged between 740 nm and 2.27 μm (mean 1.39±0.32 μm). The $NaHCO_3$ treatment did not significantly change the thickness of PLC fibers (mean 1.51±0.36 μm) and 3D pore diameter (mean 2,91±0. μm)($p>0.05$) (Fig. 2C). The contact angle of PLC membrane before modification was 110 degrees. Sodium bicarbonate ($NaHCO_3$) treatment led to a decrease of the contact angle of the PLC membrane by 25°C (Fig. 3). The mechanical properties of the PLC membrane are given in Table 1.

Analysis of Phenotype and Multipotent Character of Adipose Derived Stem Cells

Flow cytometry confirmed the ADSCs phenotype. ADSCs derived from the third passage were positive for the $CD44^+$ (50% of cells) and $CD90^+$ (87% of cells) markers and negative for typical endothelial and hematopoietic markers $CD34^+$ (16% of cells) and $CD45^+$ (13% of cells). ADSCs were able to differentiate into adipocytes, osteoblasts and chondrocytes after cultivation in respective media (Fig. 4). Controls showed negative results.

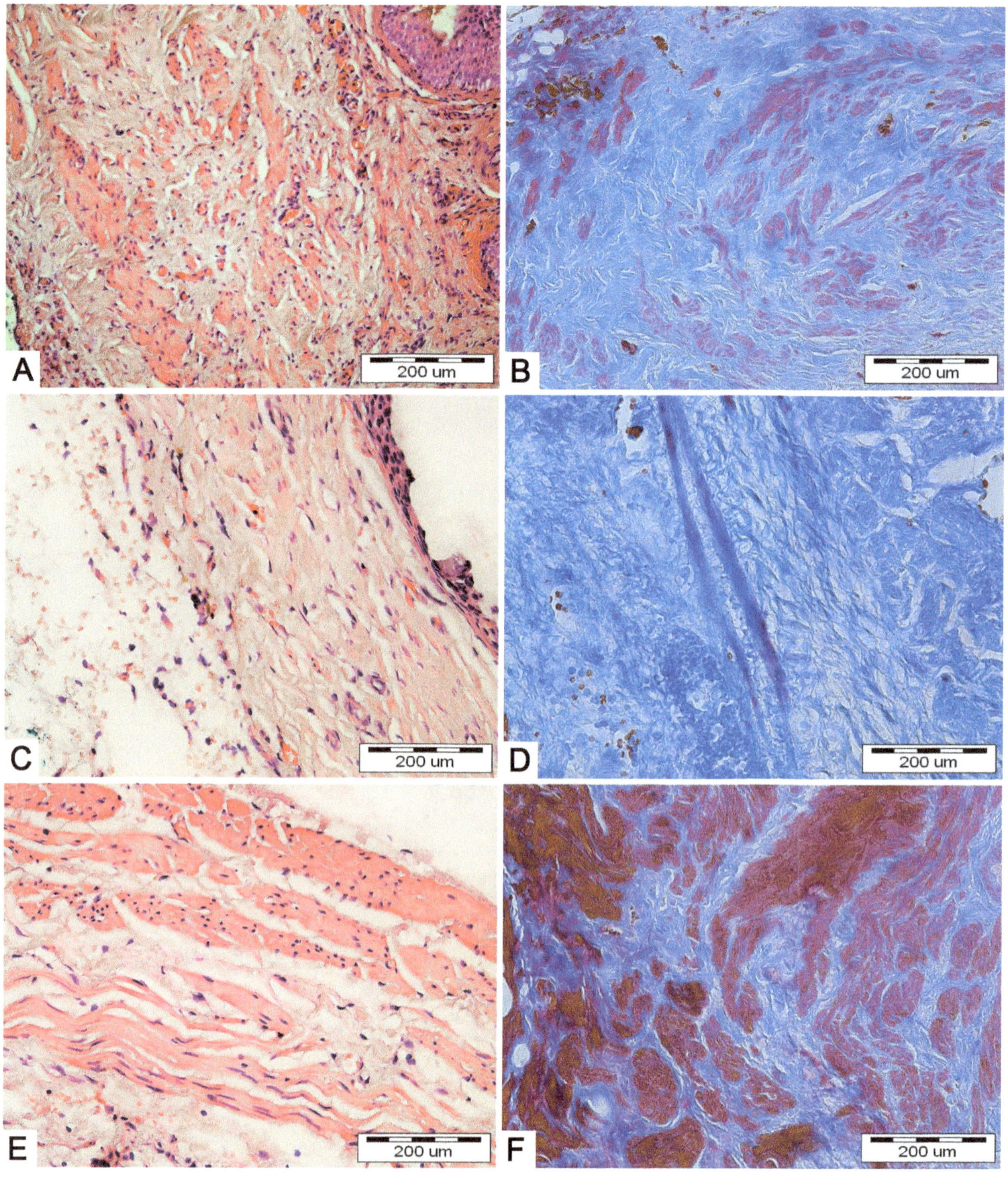

Figure 11. Bladders augmented with small intestinal submucosa seeded (A,B) and unseeded (C,D) with adipose derived stem cells. Native bladders in control group (E,F). Light microscope, H&E (A,C,E) and Trichrome Masson staining (B,D,F), bar = 200 μm.

Analysis of Biomaterial Cytotoxicity in vitro

Real- time cell analysis of the cytotoxicity of the PLC and SIS extracts is presented in Fig. 5A. There was no significant decrease in cell indexes following 96 hours of incubation of ADSCs with PLC25, PLC50 and PLC75 extracts compared to the negative control ($p > 0.05$). However, there was significant reduction in cell indexes following incubation of ADSCs with SIS25, SIS50 and SIS75 extracts compared to negative control ($p < 0.05$) (Fig. 5B).

Analysis of Growth of Adipose Derived Stem Cells on Five- layered Poly (L-lactide-co-caprolactone) Membrane and Small Intestinal Submucosa

ADSCs evenly covered the surfaces of both SIS and PLC membranes (Fig. 6, 7). After 7 days of culturing, the morphology of ADSCs cultured on SIS was normal but only single cells had a flattened shape and elongated cellular processes (Fig. 7A–C). The number of ADSCs adhering to the SIS surface increased after 14 days of culture (Fig. 7D–E). In contrast, a significant number of

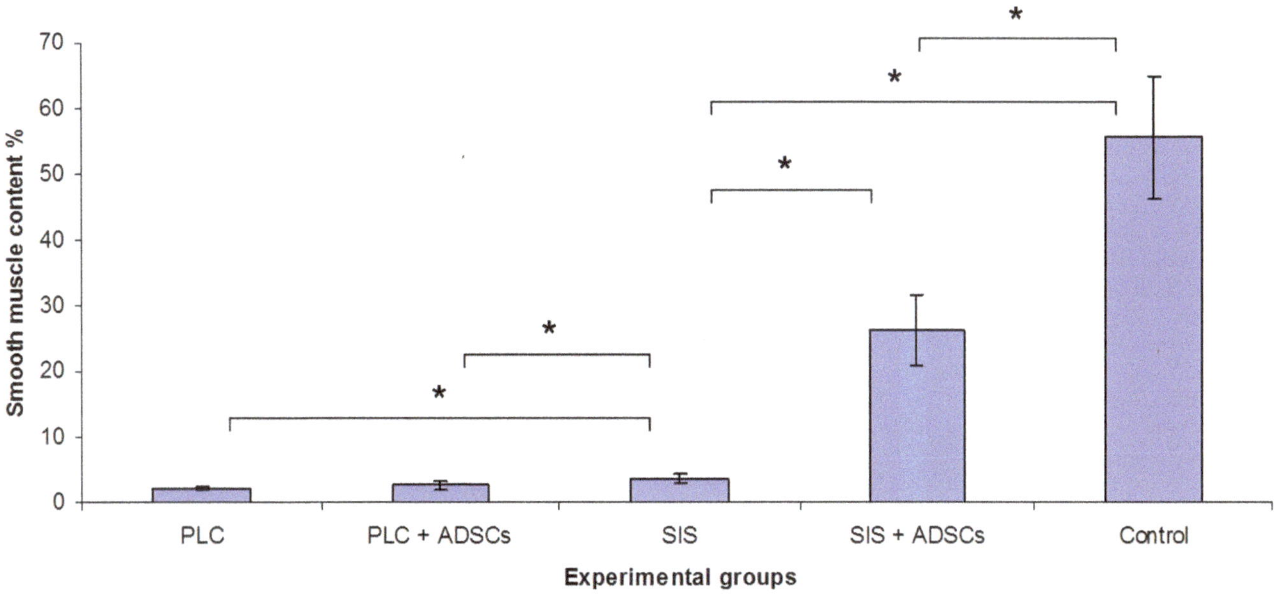

Figure 12. Smooth muscle content in bladders augmented with: poly (lactydo-*co*-caprolactone) (PLC), PLC seeded with adipose derived stem cells (ADSCs), small intestinal submucosa (SIS), SIS seeded with ADSCs and native bladder wall (control), respectively. Values are expressed as mean± standard deviation. The statistical significance is shown as * $p<0.05$.

ADSCs cultured on PLC membrane attached to the biomaterial surface after only 7 days of culture (Fig. 6A–C). The ADSCs had the normal morphology of living cells. The cell divisions were observed. After 14 days of culture, the cells formed a dense layer adhering well to the PLC membrane (Fig. 6D–E).

Analysis of Urinary Bladder Augmentation

Twelve from sixteen rats survived the 3 months observation time. The highest mortality rate (n = 50%) was observed in PLC groups. Surprisingly, deaths were observed between 2nd and 3rd months of observation. The cause of death was urinary bladder perforation, which occurred within the implanted biomaterial. We found that in many cases the biomaterial implant had twisted, formed a coil and detached from the augmented urinary bladder.

Urinary bladder augmentation with the PLC membrane (1st and 3rd groups) led to numerous undesirable events including: perforation, fistula or diverticula formation, and incorporation of the reconstructed wall into the urinary bladder lumen (Fig. 8AC; Fig. 9A). Histological analysis of reconstructed urinary bladders revealed no integration of PLC with bladder tissues (Fig. 10).

Bladders augmented with SIS seeded with ADSCs (2nd group) mimicked native bladders in gross examination (Fig. 8B). Cystography showed proper, regular shape of reconstructed bladders (Fig. 9B). Regeneration of all components of the normal urinary bladder wall was observed. However, smooth muscle fibers were less abundant in the reconstructed wall compared with native bladder wall (26.1±5.4% vs. 55.6±9.3%, $p<0.05$) (Fig. 11AB, Fig. 11EF, Fig. 12). In the bladders augmented with SIS matrices without cells (4th group), fibrosis and graft contraction occurred (Fig. 8D; Fig. 11CD).

Discussion

In the present study, the five- layered PLC nanofibrous membrane was produced with the use of a fast-rotating mandrel as a target. Such a type of target ensures a high degree of nanofibers orientation in a specified axis. In order to obtain a membrane of increased strength, a set of 5 orthogonally directed layers of nanofibers was collected. Such a membrane is supposed to have a reinforced structure preventing stretching in all directions. The ultimate tensile strain of PLC membrane was higher compared to rat (2.03 mm/mm, 0.72 MPa) and human (0,69 mm/mm, 0.27 MPa) urinary bladders. Similarly, the elastic modulus of PLC membrane was higher compared to human (0.25 MPa) and rat urinary bladder (0.76 MPa) [27]. These results indicate that the newly synthesized material is elastic and resistant to mechanical treatment. The material was designed to isolate stem cells from the toxic environment of urine. This was achieved because of the high hydrophobicity of the material (both comonomers: L-lactide and ε-caprolactone as well as their copolymer are hydrophobic) and very small air bubbles retained in its structure. However, highly hydrophobic surfaces are not colonized by cells and they cannot sustain cell proliferation. In order to decrease the material's surface hydrophobicity we selected chemical modification of nanofibers [28–30]. Other methods are also applied [31], among them are e. g. plasma treatment [32], adhesive protein treatment [33)], silk fibroin modification [34], arginine-glycine-aspartate (RGD) treatment, protein immobilization [35], or surface mineralization with apatite [36]. Our selection was based on the fact that only the surface of the nanofibrous membrane was to be modified while the bulk of the material was left unmodified in order to retain its waterproof properties. A too drastic decrease of hydrophobicity could lead to the loss of isolative properties of the bulk material against the toxic environment of urine, and consequently cause death of stem cells seeded on the nanomaterial surface. The surface of modified electrospun PLC membrane showed improved ADSCs adhesive properties compared with commercially available collagen matrix (SIS). ADSCs adhered to the electrospun PLC membrane and formed a dense layer. Just as in the case of SIS, the PLC membrane pore size was not large enough for cell migration into the scaffold. Such migration was only possible during a subsequent nanomaterial *in vivo* degradation. Further modifications of the porosity of the biomaterial are required. The biomaterial could be degraded *in vitro* before implantation. Another option is synthesis of a biomaterial comprised of layers of different porosity and hydrophobicity. The layer with larger pore sizes and decreased hydrophobicity would allow for directed cell adhesion and migration into the scaffold, while the layer with smaller pore sizes and increased hydrophobicity would isolate the cells from the toxic environment of urine.

We found that the PLC extract did not have a cytotoxic effect on ADSCs in all tested concentrations, as opposed to commercially available SIS extract, which was also confirmed by other authors [37]. In this study, SIS was used as a control because it is the reference material for urinary bladder tissue engineering. To exclude the variables arising from SIS self- preparation we used the commercially available and clinically applicable porcine SIS (Surgisis, Biodesign, USA).

Our *in vitro* analyses showed that the new PLC membrane has better *in vitro* properties compared to SIS, therefore, the use of new PLC membrane in bladder tissue engineering was rational.

Unfortunately, contrary to our expectations the electrospun PLC membrane did not provide an appropriate environment for urinary bladder wall regeneration. There were many differences in regeneration effects between PLC (1st and 3rd) and SIS (2nd and 4th) groups. The use of adipose derived stem cells did not affect the results of bladder augmentation with PLC membrane but enhanced the smooth muscle regeneration in bladders reconstructed with SIS. Numerous side effects such as: bladder wall perforation, fistula or diverticula formation, or incorporation of the reconstructed wall into the bladder lumen were observed in both PLC groups (1st and 3rd). The PLC membrane did not integrate with surrounding bladder tissues. Three months after surgery, PLC membrane degradation was observed, however no signs of remodeling occurred.

Probably, the numerous complications observed in this study result from the multi-layer construction of the membrane. We speculate that a membrane synthesized from non-oriented nanofibrous structure with only a thin layer of directed nanofibers on both surfaces would avoid complications, however it requires future investigation.

In conclusion, the poly L-lactide-co-caprolactone five-layered membrane is not suitable for urinary bladder wall regeneration.

Author Contributions

Conceived and designed the experiments: MP AJ JA. Performed the experiments: MP AJ JA KW MR LB TK PN TC MN RD MB MFB SK GM. Analyzed the data: MP AJ JA TK TAK MFB AM TD. Contributed reagents/materials/analysis tools: TK PN TC TAK. Wrote the paper: MP TK TD.

References

1. Baka-Ostrowska M (2011) Bladder augmentation and continent urinary diversion in boys with posterior urethral valves. *Cent Eur J Urol* 64: 237–239.
2. Pokrywczynska M, Adamowicz J, Sharma A, Drewa T (2014) Human urinary bladder regeneration through tissue engineering-an analysis of 131 clinical cases. *Exp Biol Med* 239: 264–71.
3. Adamowicz J, Kowalczyk T, Drewa T (2013) Tissue engineering of urinary bladder – current state of art and future perspectives. *Cent Eur J Urol* 66: 202–206.
4. Pokrywczynska M, Jundziłł A, Adamowicz J, Drewa T (2013) Tissue engineering-experimental method of urinary bladder regeneration. Postepy Hig Med Dosw 67: 790–9.
5. Pokrywczynska M, Jundzill A, Bodnar M, Adamowicz J, Tworkiewicz J, et al. (2013) Do mesenchymal stem cells modulate the milieu of reconstructed bladder wall? *Arch Immunol Ther Exp* 61:483–93.
6. Sharma AK, Bury MI, Marks AJ, Fuller NJ, Meisner JW, et al. (2011) A nonhuman primate model for urinary bladder regeneration using autologous sources of bone marrow-derived mesenchymal stem cells. *Stem Cells* 29: 241–50.
7. Landman J, Olweny E, Sundaram CP, Andreoni C, Collyer WC, et al. (2004) Laparoscopic mid sagittal hemicystectomy and bladder reconstruction with small intestinal submucosa and reimplantation of ureter into small intestinal submucosa: 1-year followup. *J Urol* 171: 2450–5.
8. Paterson RF, Lifshitz DA, Beck SD, Siqueira TM Jr, Cheng L, et al. (2002) Multilayered small intestinal submucosa is inferior to autologous bowel for laparoscopic bladder augmentation. *J Urol* 168: 2253–7.
9. Kropp BP, Cheng EY, Lin HK, Zhang Y (2004) Reliable and reproducible bladder regeneration using unseeded distal small intestinal submucosa. *J Urol* 172: 1710–3.
10. Shakhssalim N, Rasouli J, Moghadasali R, Aghdas FS, Naji M (2013) Bladder smooth muscle cells interaction and proliferation on PCL/PLLA electrospun nanofibrous scaffold. *Int J Artif Organ* 36: 113–20.
11. Maya H, Vincent M, Nötzli S, Madduri S, Tullio S, et al. (2013) Increased porosity of electrospun hybrid scaffolds improved bladder tissue regeneration. *J Biomed Mater Res A* DOI 10.1002/jbm.a.34889.
12. Fernández J, Etxeberria A, Sarasua JR (2012) Synthesis, structure and properties of poly(L-lactide-co-ε-caprolactone) statistical copolymers. *J Mech Behav Biomed Mater*; 9: 100–12.
13. Jeong SI, Kim BS, Lee YM, Ihn KJ, Kim SH, et al. (2004) Morphology of elastic poly(L-lactide-co-epsilon-caprolactone) copolymers and in vitro and in vivo degradation behavior of their scaffolds. *Biomacromolecules* 5:1303–9.
14. Knight TA, Payne RG (2013) Characterization of a PGA-based scaffold for use in a tissue-engineered neo-urinary conduit. *Methods Mol Biol* 1001: 179–88.
15. Andrychowski J, Frontczak-Baniewicz M, Sulejczak D, Kowalczyk T, Chmielewski T, et al. (2013) Nanofiber nets in prevention of cicatrisation in spinal procedures. Experimental study. *Folia Neuropathol*; 51: 147–57.
16. Bretcanu O, Misra SK, Yunos DM, Boccaccini AR, Roy I, et al. (2009) Electrospun nanofibrous biodegradable polyester coatings on Bioglass-based glass-ceramics for tissue engineering. *Materials Chemistry and Physics* 118: 420–426.
17. Sartoneva R, Haimi S, Miettinen S, Mannerström B, Haaparanta AM, et al. (2011) Comparison of a poly-L-lactide-co-ε-caprolactone and human amniotic membrane for urothelium tissue engineering applications. *J R Soc Interface* 8: 671–7.
18. Sartoneva R, Haaparanta AM, Lahdes-Vasama T, Mannerström B, Kellomäki M, et al. (2012) Characterizing and optimizing poly-L-lactide co-ε-caprolactone membranes for urothelial tissue engineering. *J R Soc Interface* 9: 3444–54.
19. Chen L, Bai Y, Liao G, Peng E, Wu B, et al. (2013) Electrospun Poly(L-lactide)/Poly(ε-caprolactone) Blend Nanofibrous Scaffold: Characterization and Biocompatibility with Human Adipose-Derived Stem Cells. *PLoS One* 8: e71265.
20. Adamowicz J, Kloskowski T, Tworkiewicz J, Pokrywczynska M, Drewa T (2012) Urine is a highly cytotoxic agent: does it influence stem cell therapies in urology? *Transplant Proc* 44: 1439–41.
21. Davis NF, Callanan A, McGuire BB, Flood HD, McGloughlin TM (2011) Evaluation of viability and proliferative activity of human urothelial cells cultured onto xenogenic tissue-engineered extracellular matrices. *Urology* 77: 1007.e1–7.
22. Li WJ, Cooper JA, Mauck RL, Tuan RS (2006) Fabrication And Characterization Of Six Electrospun Poly(α,ω-Hydroxy Ester)-Based Fibrous Scaffolds For Tissue Engineering Applications. *Acta Biomaterialia* 4: 377–385.
23. Hu Y, Grainger DW, Winn SR, Hollinger JO (2002) Fabrication of poly(alpha-hydroxy acid) foam scaffolds using multiple solvent systems, J. Biomed.Mater. Res. A 59, 563–572.
24. Tomadakis MM, Robertson TJ (2003) Pore size distribution, survival probability, and relaxation time in random and ordered arrays of fibers, J. Chem. Phys. 119, 1741–1749.
25. National Research Council (2011) Guide for the Care and Use of Laboratory Animals: Eighth Edition. Washington, DC: The National Academies Press.
26. Safford KM, Hicok KC, Safford SD (2002) Neurogenic differentiation of murine and human adipose derived stromal cells. *Biochem Biophys Res Commun* 294: 371–379.
27. Dahms SE, Piechota HJ, Dahiya R, Lue TF, Tanagho EA (1998) Composition and biomechanical properties of the bladder acellular matrix graft: comparative analysis in rat, pig and human. Br J Urol 82: 411–9.
28. Chen F, Lee CN, Teoh SH (2007) Nanofibrous modification on ultra-thin poly (e-caprolactone) membrane via electrospinning. Materials Science and Engineering: C 27: 325–332.
29. Park JS, Kim JM, Jun LS, Geun LS, Young-Keun J (2007) Surface hydrolysis of fibrous poly (ε-caprolactone) scaffolds for enhanced osteoblast adhesion and proliferation. Macromol Res 15: 424–429.
30. Ang LP, Cheng ZY, Beuerman RW, Teoh SH, Zhu X, et al (2006) The development of a serum-free derived bioengineered conjunctival epithelial equivalent using an ultrathin poly (ε-caprolactone) membrane substrate. Invest Ophthalmol & Vis Sci 47: 105–112.
31. Goddard JM, Hotchkiss JH (2007) Polymer surface modification for the attachment of bioactive compounds. Prog Polym Sci 32: 698–725.
32. Yamaguchi M, Shinbo T, Kanamori T, Wang PC, Niwa M, et al. (2004) Hirakawa K, Kamiya M. Surface modification of poly (L-lactic acid) affects

initial cell attachment, cell morphology, and cell growth. J Artif Organs 7: 187–193.

33. Zhu Y, Leong MF, Ong WF, Chan-Park MB, Chian KS (2007) Esophageal epithelium regeneration on fibronectin grafted poly (L-lactide-co-caprolactone)(PLLC) nanofiber scaffold. Biomaterials 28: 861–868.
34. Cai K, Yao K, Cui Y, Yang Z, Li X, et al. (2002) Influence of different surface modification treatments on poly (D, L-lactic acid) with silk fibroin and their effects on the culture of osteoblast in vitro. Biomaterials 23: 1603–1611.
35. Yoon JJ, Song SH, Lee DS, Park TG (2004) Immobilization of cell adhesive RGD peptide onto the surface of highly porous biodegradable polymer scaffolds fabricated by a gas foaming/salt leaching method. Biomaterials 25: 5613–5620.
36. Yu HS, Jang JH, Kim TI, Lee HH, Kim HW (2009) Apatite-mineralized polycaprolactone nanofibrous web as a bone tissue regeneration substrate. J Biomed Mater Res Part A 88: 747–754.
37. Feil G, Christ-Adler M, Maurer S, Corvin S, Rennekampff HO, et al. (2006) Investigations of urothelial cells seeded on commercially available small intestine submucosa. Eur Urol. 50, 1330–7.

19

Prefabrication of 3D Cartilage Contructs: Towards a Tissue Engineered Auricle – A Model Tested in Rabbits

Achim von Bomhard[1]*, Johannes Veit[2], Christian Bermueller[2], Nicole Rotter[2], Rainer Staudenmaier[3], Katharina Storck[4], Hoang Nguyen The[5]

1 Department of Maxillofacial Surgery, Technical University of Munich, Munich, Germany, **2** Department of Otorhinolaryngology, Ulm University Medical Center, Ulm, Germany, **3** HNO Praxis im Zentrum, ENT and Plastic Surgery, Munich, Germany, **4** Department of Otorhinolaryngology, Technical University of Munich, Munich, Germany, **5** Department of Hand- and Microsurgery, Military Hospital 108, Hanoi, Vietnam

Abstract

The reconstruction of an auricle for congenital deformity or following trauma remains one of the greatest challenges in reconstructive surgery. Tissue-engineered (TE) three-dimensional (3D) cartilage constructs have proven to be a promising option, but problems remain with regard to cell vitality in large cell constructs. The supply of nutrients and oxygen is limited because cultured cartilage is not vascular integrated due to missing perichondrium. The consequence is necrosis and thus a loss of form stability. The micro-surgical implantation of an arteriovenous loop represents a reliable technology for neovascularization, and thus vascular integration, of three-dimensional (3D) cultivated cell constructs. Auricular cartilage biopsies were obtained from 15 rabbits and seeded in 3D scaffolds made from polycaprolactone-based polyurethane in the shape and size of a human auricle. These cartilage cell constructs were implanted subcutaneously into a skin flap (15×8 cm) and neovascularized by means of vascular loops implanted micro-surgically. They were then totally enhanced as 3D tissue and freely re-implanted in-situ through microsurgery. Neovascularization in the prefabricated flap and cultured cartilage construct was analyzed by microangiography. After explantation, the specimens were examined by histological and immunohistochemical methods. Cultivated 3D cartilage cell constructs with implanted vascular pedicle promoted the formation of engineered cartilaginous tissue within the scaffold in vivo. The auricles contained cartilage-specific extracellular matrix (ECM) components, such as GAGs and collagen even in the center oft the constructs. In contrast, in cultivated 3D cartilage cell constructs without vascular pedicle, ECM distribution was only detectable on the surface compared to constructs with vascular pedicle. We demonstrated, that the 3D flaps could be freely transplanted. On a microangiographic level it was evident that all the skin flaps and the implanted cultivated constructs were well neovascularized. The presented method is suggested as a promising alternative towards clinical application of engineered cartilaginous tissue for plastic and reconstructive surgery.

Editor: Nuno M. Neves, University of Minho, Portugal

Funding: Grant sponsor: Bayerische Forschungsstiftung (FORTEPRO); Grant number: Az. 442/01. The funders had no role in study design, data collection and analysis, decision to publish, or preparation of the manuscript.

Competing Interests: The authors have declared that no competing interests exist.

* E-mail: achimvonbomhard@gmail.com

Introduction

Tissue engineering is a promising method for repair or replacement of any tissues in the human body that is injured or damaged as a result of disease or trauma. The first cultivated cartilage cell constructs have already been described in the early 70's by W.T. Green [1]. Nowadays, after 40 years of research, there continue to be problems in this field with regard to cell expansion, requirements regarding cell vitality in large cell constructs and homogenous seeding in cell carriers [2]. Therefore optimal cell carriers must be identified and implemented. Features with regard to biocompatibility, degradation, elasticity and mechanical stability, sufficient diffusion of nutrients into the center of the scaffold and suitable qualities for cell adhesion, growth, differentiation and proliferation play a decisive role.

Because cartilage is an avascular tissue consisting of chondrocytes only, which are embedded in a matrix composed of collagen and proteoglycans, the survival and growth of transferred native cartilage grafts depends in part on the maintenance of well vascularized perichondrium [3]. Engineered cartilage constructs with a large volume of expended cells usually require a hyperoxic environment for their growth and proliferation, and have no perichondrium [4]. So their blood and nutrient supply is clearly limited. This results in cell death and unavoidable necrosis of the engineered construct with subsequent loss of shape and function [2,5].

With respect to clinical applications, we hypothesized that the survival and transplantation of tissue-engineered constructs should be more effective if they, or their immediate surrounding, are neovascularized and transferred as an axial free flap based on a reliable vascular pedicle. Our previous experimental studies in chinchilla-bastard rabbits have revealed the reliable ability of neovacularization and free microsurgical transplantation of small engineered cartilage constructs using prefabricated flaps [6]. The present study was designed to investigate the ability of combining flap prefabrication and large engineered constructs with free microsurgical transplantation to improve cell survival especially in the center of the engineered cartilage constructs. Prefabrication is a clinically established approach from reconstructive surgery for the neovascularization of skin flaps. The principle of prefabricated flaps is based on the transformation of a previously non-axial perfused tissue area into an axially supplied flap by means of the implantation of an arteriovenous vascular pedicle [7]. From the implanted vascular pedicle, new vessels sprout into the skin flaps and connect with the originally existing vascular system. The new axially supplied skin flap formed in this way can – based on the implanted vascular pedicle – be well neovascularised and can then be freely transplanted with a secure blood supply [8].

Neovascularization of native cartilage was first described by Hirase et. al. in 1988 [3]. They dissected the central vessels out of the pinna in rabbits, and implanted them in a subcutaneous (extraperichondrial) pocket of a random pattern, distally based, full-thickness ear flap. After three weeks the neovascularized chondrocutaneous flaps (2×3 cm) could be freely transferred to the contralateral pinna. This procedure was performed to prevent resorption, which is present in free cartilage grafts. Morrison et. al. implanted the femoral artery and vein of rabbits in various vascular configurations, directly into the subdermal layer of the skin, in an attempt to fabricate a thin axial-pattern flap, containing only skin and immediately underlying connective tissue, which is significantly thinner than any free flap previously available [9]. In 1993, Walton et. al. demonstrated the feasibility to engineer a fibrovascular stroma to a biomaterial (expanded polytetrauoroethylene 2×2×0.2 cm) using an isolated vascular pedicle and produce a synthetic ap capable of supporting an overlying skin graft [10]. Kataras et al. prefabricated composite arteriovenous aps with implantation of an autologous graft (cartilage) or an alloplastic material (porous polyethylene) in rabbits, and indicated that arteriovenous perfusion can nourish a prefabricated ap containing an implanted material [11].

Unlike other authors, we have developed a new model of large tissue-engineered cartilage in shape and size of a human auricle, associated with large prefabricated aps (dimension of 8×15 cm) in chinchilla-bastard rabbits, based on the anatomic similarities between humans and rabbits. In order to design the study to be as close to clinical application as possible, we decided to adopt an autologous animal model and transplanted the cultivated 3D cartilage tissue in the form of a human auricle within a closed system. This had the advantage that no immunosuppression was required.

Materials and Methods

All reagents, cell culture media and supplements were purchased from Sigma Aldrich (Munich, Germany), Invitrogen (Darmstadt, Germany) and Biochrom (Berlin, Germany) unless indicated otherwise.

The trials were conducted on female chinchilla-bastard rabbits (Charles River Laboratories, Sulzfeld, Germany). This study was carried out in strict accordance with European guidelines for the Care and Use of Laboratory Animals. The protocol was approved by the Committee on the Ethics of Animal Experiments of the Technical University of Munich (Permit Number: Reg. Obb. AZ 55.2-1-54-2532-20-07). All surgical procedures were performed under general anesthesia with infusion of ketamine 40 mg/kg and xylazine 4 mg/kg i.v., and all efforts were made to minimize suffering. All operations were carried out under sterile conditions. A total of five operations were carried out on each animal (only step (1) and (2) in the cartilage construct control group):

(1) Auricular cartilage biopsy harvesting for the in vitro production of cultivated cartilage constructs using porous constructs made of polycaprolactone in the form of an auricle.
(2) In vivo construct implantation in an abdominal flap measuring 8×15 cm, three to four weeks after cartilage biopsy.
(3) In vivo neovascularization by means of the implantation of an arteriovenous vascular pedicle, one week after construct implantation.
(4) 6 weeks after implantation of the vascular pedicle, the abdominal skin flap and immediately underlying connective tissue, 8×15 cm in size, centered over the course of the transposed vessels, was raised as a rectangular island flap to proof the ability of the neovascularized cultivated 3D cartilage cell construct to survive solely perfused by the newly implanted axial vascular pedicle. After elevation the flap was resutured into position and observed daily for circulatory viability and the presence of necrosis.
(5) After wound healing the flap was raised again, with disconnection of the vascular pedivle, and was reanastomosed microsurgically to A. saphena and V. saphena magna to proof the possibility of free transplantation.

Total experimental time was 9 to 11 months.

Harvesting of Cartilage, Cell Culture, and Seeding of Biomaterials

In the first operation an auricular cartilage biopsy 5×30 mm in size from the auricle of each rabbit was harvested in order to obtain autologous chondrocytes. The auricular cartilage biopsy was cleaned and minced into 1×1 mm pieces under sterile conditions. Chondrocytes were enzymatically isolated during 16 hours incubation at 37°C in collagenase II (1,108 U/ml) and resuspended with Dulbecco Modified Eagle medium, foetal calf serum 100 ml/L and Antibiotic/Antimycotic 24 ml/L. For amplification chondrocytes were cultured at 37°C and 5.5% CO_2. Following the attachment of the cells, culture medium was changed twice to three times per week.

At subconfluence cells were removed after being washed in Phosphate Buffered Saline with trypsin 0.05% and then re-seeded at a concentration of 30,000 cells/cm^2. After about two weeks, and, on average, two to three passages, it was possible to achieve a sufficient cell count of approx. 20×10^6 cells/cm^3.

Biomaterial constructs. Polycaprolactone-based polyurethane scaffolds in the shape of human auricles were kindly manufactured by PolyMaterials, Kaufbeuren, Germany, as previously described [12].

Poly(caprolactone)diol (PCL diol) ($M_n = 1250$), poly(caprolactone)triol (PCL triol) ($M_n = 900$), DABCO DC-3042 (Air Products, Hattingen, Germany), dextrose, methylal, and diazabicycloundecene (Acros St. Augustin, Germany) were premixed and poured into one chamber of a two-chamber syringe equipped with a static mixer. The other chamber was filled with isophorone diisocyanate and methylal, and the syringe was heated to 37 to 38°C. The mixture was slowly injected into a silicone mold in the shape and size of a large (4×6 cm, 13 samples) and a small (3×4 cm, 2 samples) human auricle - created through rapid prototyping directly from computer aided design (CAD) data sources, via an STL file - and heated to 67 to 68°C for 2 h. Afterward, constructs were washed in boiling distilled water for 1 h and rinsed extensively with PBS, and autoclaved in a ask containing excess of PBS. The scaffold consisted mainly of PCL diol and PCL triol (39% each) as well as isophorone diisocyanate. The theoretical

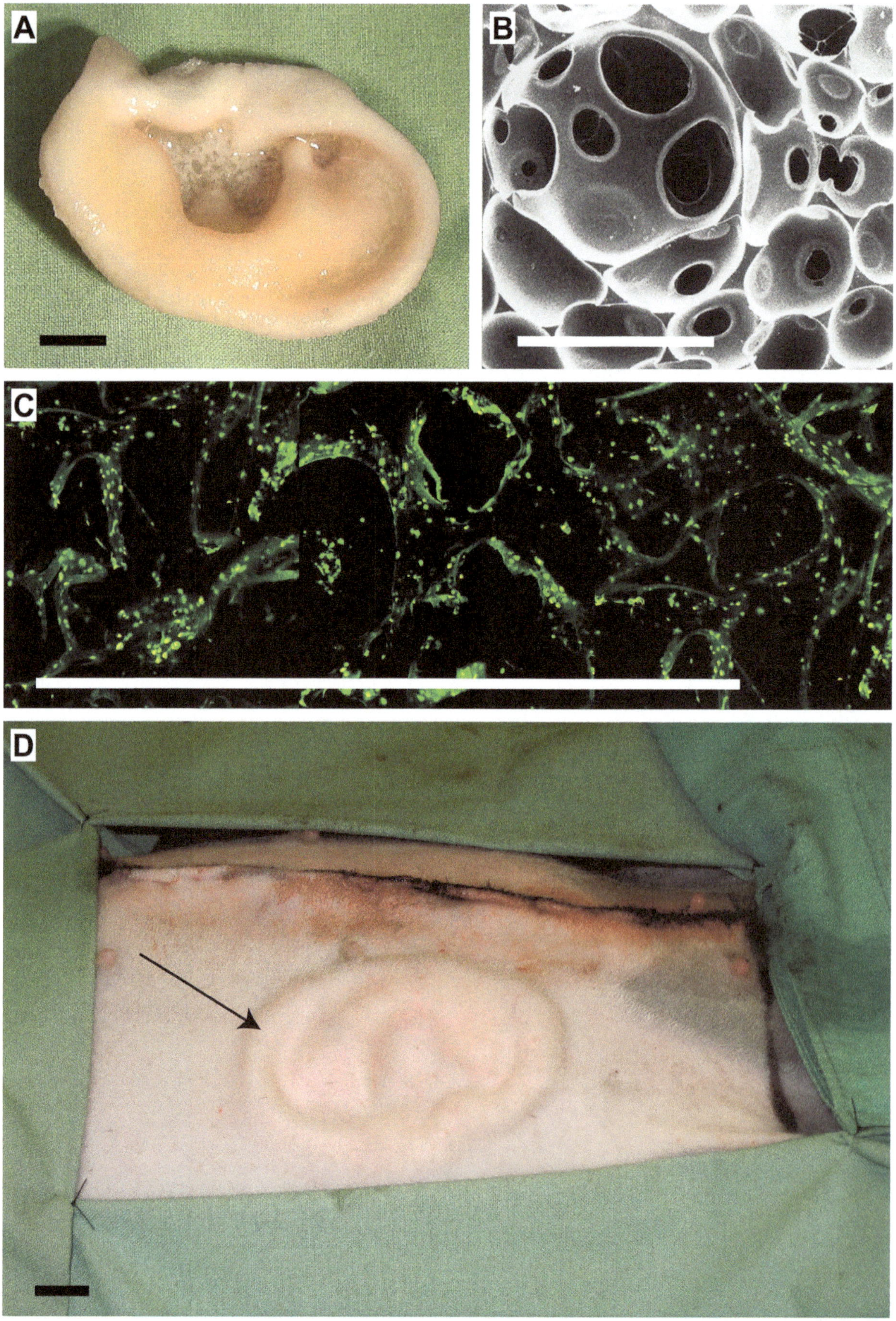
A
B
C
D

Figure 1. Biomaterial construct prior to implantation. A: Polyurethane based polycaprolactone construct seeded with chondrocytes. B: SEM picture of porous PCL construct, magnification 48×. C: Live/dead staining was performed with fluoresceindiacetate and propidium jodid using confocal laser scanning microscopy. D: Cultivated cartilage cell biomaterial construct (black arrow) implanted subcutaneously above the panniculus carnosus; day 0 postoperative. Scale bars = 1 cm.

cross-linking percentage of the repeating units of polyol was 3%. Pore size was 180–250 µm with a porosity of ca. 85% *(fig. 1B)*. Characterization of the material in detail was published by Eyrich et. al. in 2007 [13].

Cell seeding. Biomaterial constructs were autoclaved in a flask containing excess of PBS. In order to orientate the polar segments of the polymer to the surface of the scaffold to improve cell adhesion ability of the biomaterial constructs and to disinfect, they were placed in EtOH 70% prior to cell seeding and then washed in Phosphate Buffered Saline. The cultivated cells were dissolved in thrombin (Baxter, Unterschleißheim, Germany) and resuspended in a thrombin buffer (5 U/ml pharmacy, University Hospital Rechts der Isar, Munich, Germany). The thrombin quantities constituted half of the absorption volumes of the biomaterial construct. In addition, the same quantity of fibrinogen – dissolved in Trasylol 100 mg/ml (Bayer, Leverkusen, Germany) was added to the mixture. The resulting cell-glue was, with the help of a syringe, (Becton Dickson, Heidelberg, Germany) and a cannula (B. Braun, Melsungen, Germany) injected into the biomaterial construct (30×10^6 cells/cm^3). The seeded construct *(fig. 1A)* was left in an incubator until it hardened (approx. 30 mins) and then covered in a cell culture vessel with DMEM

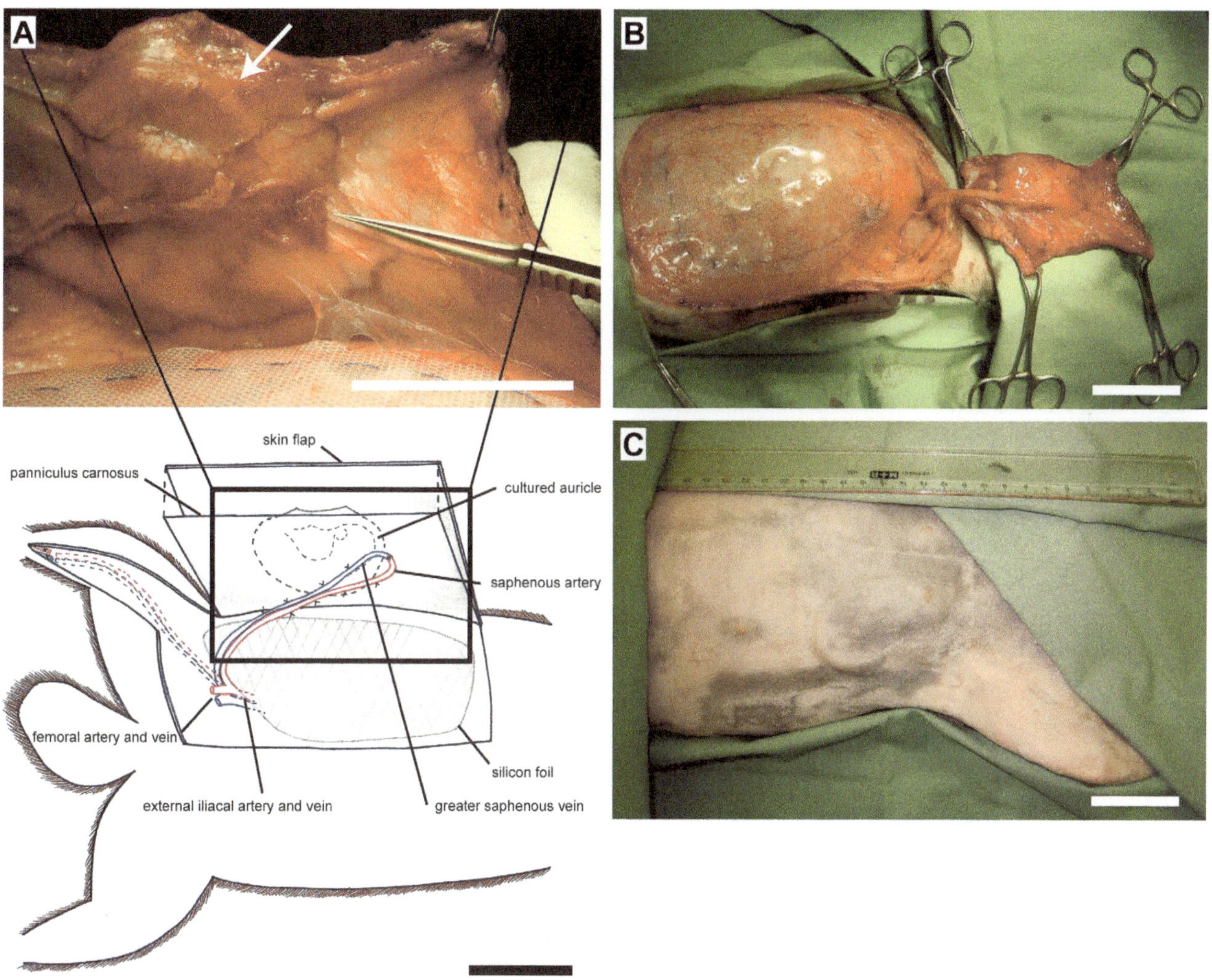

Figure 2. Surgical procedures. A: The implantation of the cultured auricle (white arrow) took place on the right side of the abdomen above the panniculus carnosus in a skin flap measuring 150×80 mm. A silicon foil was sewn onto the abdominal fascia to prevent neovascularization from the abdominal wall. Saphenous artery and greater saphenous vein were distally anastomosed turned in the cranial direction and attached below the abdominal skin flap. *indicates the original position of A. saphena and V. saphena magna. B: Four-sided flap excision. To proof the ability of the neovascularized cultivated 3D cartilage cell construct (black arrow) to survive solely perfused by the newly implanted axial vascular pedicle the abdominal flap was excised four-sided, leaving the vascular pedicle (white arrow) intact. C: Macroscopic view after free microsurgical transplantation. Healed skin flap three weeks after free microsurgical transplantation. Black arrow indicates the integrated neovascularized cultured auricle. Scale bars = 5 cm.

Table 1. Harvesting of cartilage, cell culture, and seeding of biomaterials.

	weight of cartilage biopsy (g)	Number of cells after isolation	Number of cells after amplification
construct 1	0.63	8.1×10^4	9.6×10^7
construct 2	0.477	32.4×10^4	9.6×10^7
construct 3	0.33	25.8×10^4	14.3×10^7
construct 4	0.791	66.2×10^4	15.4×10^7
construct 5	0.45	39.9×10^4	10.6×10^7
construct 6	0.75	42.7×10^4	11.1×10^7
construct 7	0.538	3×10^4	11.9×10^7
construct 8	0.79	4.6×10^4	13×10^7
construct 9	0.79	4.2×10^4	11.8×10^7
construct 10	0.218	4.2×10^4	9.7×10^7
construct 11	0.37	5.5×10^4	8.1×10^7
construct 12	0.8	9.7×10^4	13×10^7
construct 13	0.701	9×10^4	10.5×10^7
construct 14	0.28	4.9×10^4	9.7×10^7
construct 15	0.48	6.5×10^4	13×10^7
control 1	0.67	16.1×10^4	14×10^7
control 2	0.51	6.7×10^4	9.7×10^7
average	**0.56**	**17.0×10^4**	**11.5×10^7**

The auricular cartilage biopsies were weighed prior to cell isolation after thorough cleaning and sectioning. After cell isolation cell number was determined. After amplification cell number was increased 676× after passage two or three. There were no significant differences between test and control group; $p=0.881$ (weight of cartilage biopsy), $p=0.766$ (cells isolated), $p=0.655$ (amplificated cell number); Mann-Whitney Test (two-tailed).

culture medium. Foetal calf serum (100 ml/L), antibiotic/antimycotic (24 ml/L), insulin (2 μl/ml), proline (4 μl/ml) and vitamin C (5 μl/ml) were added and pre-cultivated in the incubator for 6 days prior to implantation. The medium was changed every other day. Before implantation, the biomaterial constructs were rinsed using an isotonic NaCl solution (B. Braun, Melsungen, Germany) to remove medium residues.

Neovascularization and free transplantation of cultivated cartilage cell constructs. Chondrocyte biomaterial cell constructs were first neovascularized in skin flaps by means of vascular pedicle implantation and then, based on the implanted axial vascular pedicle, were freely transplanted microsurgically back into position.

Implantation of Chondrocyte Biomaterial Constructs

In the second operation implantation was performed on the right side of the abdomen, in a skin flap measuring 150×80 mm *(fig. 1D)*. Working on the basis of a medial craniocaudal incision of 150 mm, a pocket of approx. 80 mm in depth was prepared above the panniculus carnosus where the seeded construct was attached using 3–4 stitches Vicryl 4/0. In order to provide protection against manipulation of the rabbits, the wound suture for was carried out intracutaneously.

Prefabrication of 3D Flaps by Vascular Pedicle Implantation

After wound healing in a third operation the planned skin flaps with implanted cartilage constructs were lifted laterally. Then, using ethilon 3/0, a silicon foil (LPI, Perouse, France, Bornel) with the dimensions 80×150×0.25 mm was sewn onto the abdominal fascia to prevent neovascularization from the abdominal wall. By means of an incision from the inguinal ligament to the calcaneus, saphenous artery and greater saphenous vein were exposed. The distal ends were anastomosed using 10/0 ethilon to achieve maximum blood flow through the newly created arteriovenous shunt [4]. The thus formed vascular pedicle was turned cranially and attached below the abdominal skin flap using Vicryl 8/0. The distance required to move the vascular pedicle was approximately 15 cm. *Figure 2A* gives an anatomical overview.

Four-sided Flap Excision (Rectangular Island Flap)

After a neovascularization period of 21 days, prior to free transplantation, incisions were made around the flap in the same dimension, while keeping the newly implanted vascular pedicle intact *(fig. 2B)*, and were then sewn back in place to analyse the ability of the neovascularized 3D chondrocyte construct to survive solely by perfusion of the newly implanted axial vascular pedicle.

Free Microsurgical Transplantation

After wound healing, the entire abdominal flap, with the integrated, cultivated auricular cartilage tissue was detached completely, vessels were disconnected, and then re-anastomosed in the same place using Prolene 10–0. The distal and proximal perivascular tissue was anastomosed using Prolene 7–0, in order to achieve strain relief for the anastomosis at the vascular pedicle. This prevented tearing of the vascular pedicle caused by movements of the rabbits.

Control Groups

In the two control groups, a total of 8 animals were operated on. All incisions were hereby carried out in the same manner as in the test group. In the first control group (cartilage construct control group) autologous cartilage constructs as in the study group were implanted subcutaneously in the same position in two rabbits

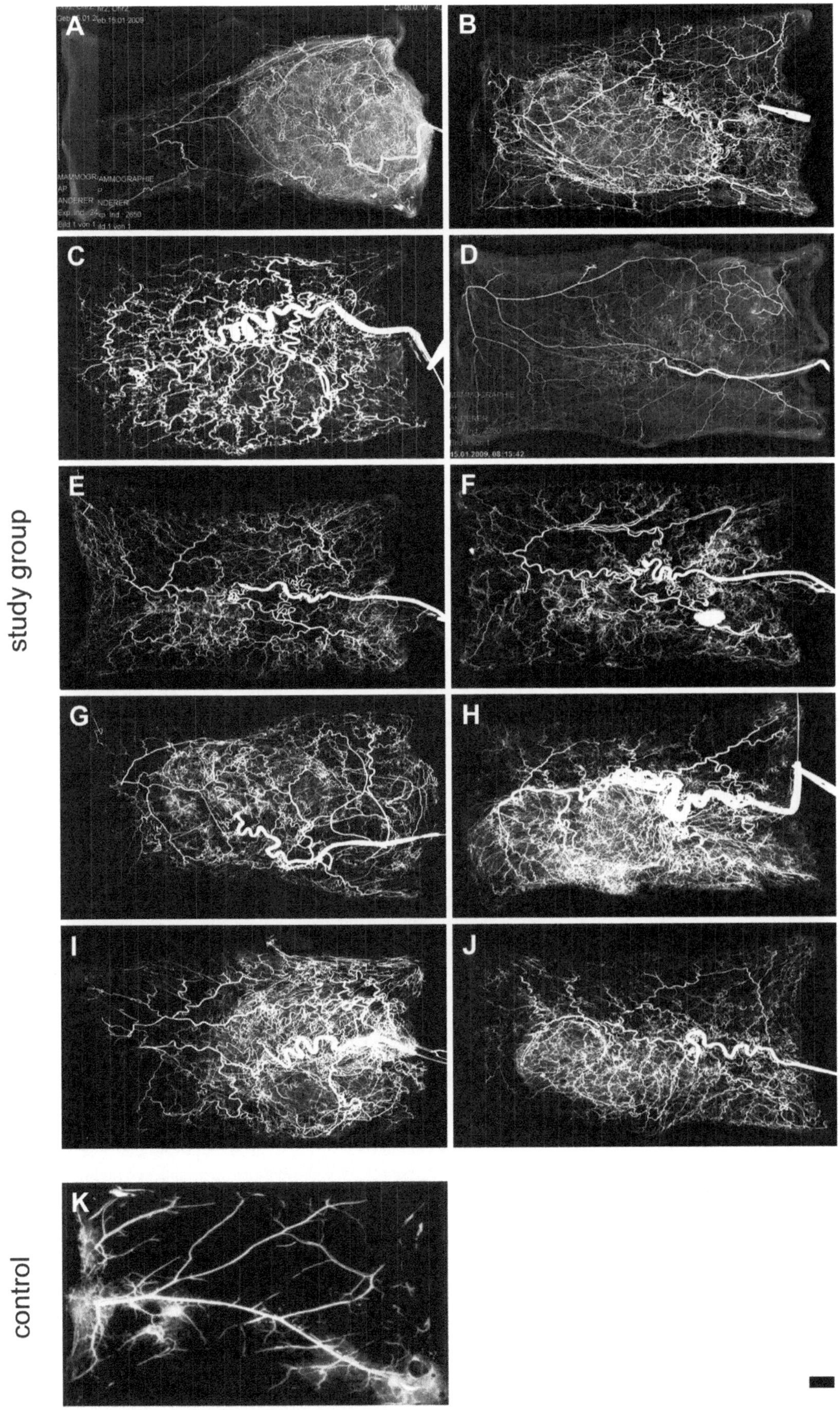
A
B
C
D
E
F
G
H
I
J
K
study group
control

Figure 3. Microangiographic images of prefabricated 3D skinflaps after neovascularisation. A–J: Images show vasculature of abdominal skin flaps, with integrated 3D cartilage constructs, perfused by Micropaque ® through the implanted vascular pedicle. K: The micro-angiographic control was performed using a Radifluor 120– TORR (Philips Electronic Instruments, Los Angeles, USA). Scale bars = 1 cm.

without the implantation of a vascular pedicle. In the second control group (microangiographic control) 6 animals served to determine the existing pattern of vessels and the morphology in the native abdominal skin flap.

Evaluation Parameters

Flaps and construct vitality were assessed macroscopically, before and after free transplantation, on a daily basis over an observation period of two weeks. Neovascularization was evaluated histologically and using selective microangiography. Formation of newly synthesized cartilage ECM was evaluated immunohistochemically.

Selective Microangiography and quantification of neovascularization. In order to carry out selective microangiography, animals were anaesthetised and i.v. heparinised (5.000 IU). The artery of the implanted vascular pedicle was cannulated with a catheter (0.9×25 cm) and the vascular system of the prefabricated skin flap was cleaned for 20 mins with heparinised saline solution (37°C). Furthermore, a suspension of Micropaque 30% and Rheomacrodex 10% in a ratio of 2:1 was injected with a constant pressure of 110 mmHg, for 45 mins into the vascular system via the artery. Once the animals had been sacrificed using Narcoren 160 mg/kg (Merial, Hallbergmoos, Germany) the flaps were removed and pinned to cork mats using cannula in order to prepare them for microangiographic photography using a mammography device (Mammomat 3000 Nova, Siemens Healthcare AG, Erlangen, Germany). In the microangiographic control group 6 animals served to determine the existing pattern of vessels and the morphology in the abdominal skin flap using a Radifluor 120– TORR (Philips Electronic Instruments, LA, California). 25 evenly spaced standard integral lines were placed corresponding to the 15-cm flap length. Vessel quantity was determined by counting the total number of

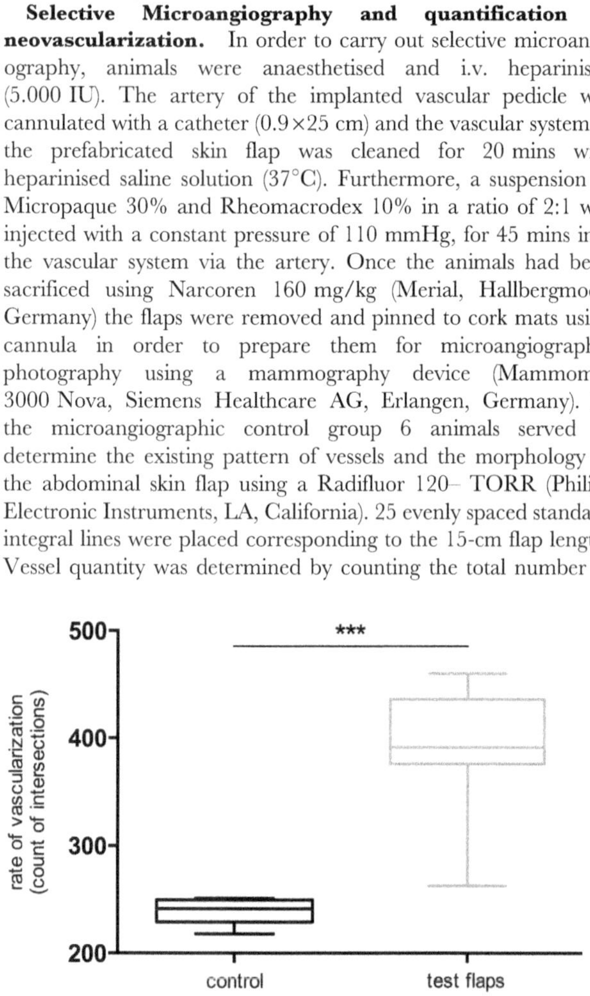

Figure 4. Angiographically quantified neovascularization in prefabricated 3D flaps. This boxplot shows the quantitative neovascularization of the 10 test flaps in comparison to 6 native abdominal skin flaps (control). To quantify the neovascularization in prefabricated 3D flaps 25 evenly spaced standard integral lines were placed corresponding to the 15-cm flap length. Vessel quantity was determined by counting the total number of times that vessels perfused by Micropaque ® intersected the integral lines over the entire surface of the flap under 2× magnification as described previously [14]. There was a significant difference between experimental and control group, ***p = 0.0002, Mann-Whitney Test (two-tailed).

times that vessels perfused by Micropaque ® intersected the integral lines over the entire surface of the flap under 2× magnification as described previously [14]. The contrast of the microangiographic recordings of the 3D flaps was reduced using an image processing program (Photoshop CS3, Adobe, US), until the resolution of the small vessels matched that of the microangiographic control group.

Histology. After microangiographic photography, conventional histological samples were prepared. The samples were prepared with cuts orthogonal to the vascular pedicle. The staining was done using Hematoxylin Eosin (HE) Alcianblue (Ab) and Elastica van Gieson (EvG).

Immunohistochemistry. For immunohistochemical staining for the detection of collagen II Zytomed KIT (Broad Spectrum Kit, Zytomed, Berlin, Germany) was used according to manufacturer's instructions.

Statistical analysis. Two-tailed Mann-Whitney test was performed using GraphPad Prism (version 6.0b for Mac OS X, GraphPad Software, La Jolla, USA).

Results

15 animals underwent surgery during this study. Three animals died around the time of implantation of the vascular pedicle. The remaining 12 flaps, with integrated auricular cartilage constructs could be successfully neovascularized. Four-sided flap excision was performed in every case and all skin flaps survived. After wound healing all flaps could be raised, and reanastomosed microsurgically to saphenous artery and greater saphenous vein. In two animals, however, the vascular pedicle tore on the second day following free transplantation and in one animal there was a thrombosis in the vascular pedicle and the flap was thus necrotic on the 8th day post-operation and the animal was euthanized.

By means of an auricular cartilage biopsy, it was possible to obtain, on average, 170,260 cells following cell isolation. On average 3×10^5 cells have been isolated per gram cartilage biopsy. Biomaterials of experimental and control group were seeded with 96 to 153.9 million cells following cell amplification. This corresponded, on average, to 28.5 million cells per cm^3 of the biomaterial construct. It was possible to achieve an adequate cell number following a maximum of 3 passages. *Table 1* shows the cartilage and cellular data in monolayer culture.

Cell Shape and vitality during in Vitro Culture

During amplification chondrocytes exhibited the typical pattern of dedifferentiation. Following the inoculation of chondrocytes in the biomaterial construct and further cultivation for one week, the cells underwent redifferentiation regaining an almost round, cartilage-specific shape. Using live/dead staining, vitality of almost 100% could be determined consistently *(fig. 1C)*.

Flap Viability

In all experimental animals and in the control group, the postoperative appearance of the flap was inconspicuous. The growth of hair was normal throughout the entire observation period. The edges of the wounds consistently healed uneventful *(fig. 2C)*. However, in some animals there was evidence of small, local infections at the edges of the wounds one to two weeks after the operation measuring a maximally 1.5×3.5 cm. Most likely they

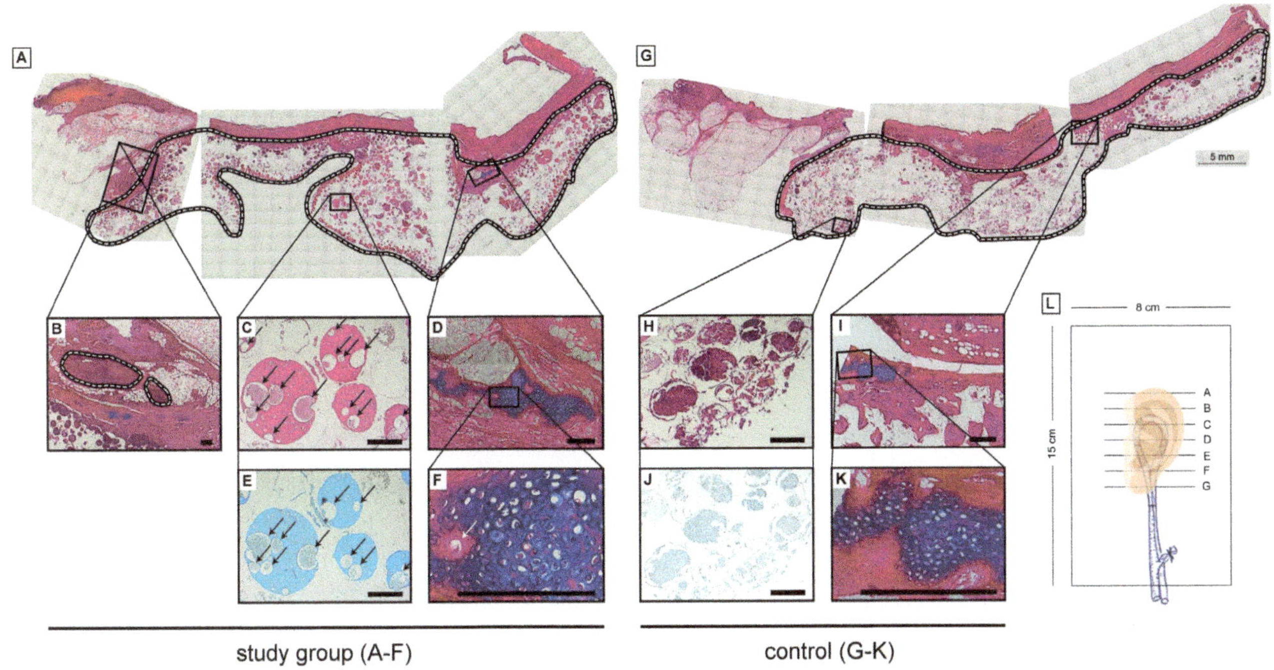

Figure 5. Histological results. A (study group): Stitched full cross section of the cultured auricle (black+white line). B: Implanted vascular pedicle (saphenous artery and greater saphenous vein), original magnification, x5. C+E: Low power view of the center of the cultivated cartilage cell biomaterial construct shows GAG, black arrows indicate ingrown vessels (C: Alcian blue counterstained with Kernechtred, original magnification, x50; E: Alcian blue staining, original magnification, x50). D+F: Periphery of cultivated cartilage cell biomaterial construct shows vital cartilage and good integration in the skin flap. White arrow indicates an ingrown vessel (Alcian blue counterstained with Kernechtred, original magnification, D×10; C×200). G (control group): Stitched full cross section of the cultured auricle (black+white line). H+J: Low power view of the center of the cultivated cartilage cell biomaterial construct shows necrosis with no GAG (H: Alcian blue counterstained with Kernechtred, original magnification, ×50; J: Alcian blue staining, original magnification, ×50). I+K: Periphery of cultivated cartilage cell biomaterial construct shows small areas of vital cartilage and good integration in the skin flap. (Alcian blue counterstained with Kernechtred, original magnification, I×10, K×200). L: Image indicates the performed sectional planes within the biomaterial construct. All showed histological images are from section plane E. Scale bars = 400 µm.

were caused by manipulation of the animals, scratching and consecutive wound-edge infection. This always occurred in the areas at the edge of the wounds and had no effect whatsoever on the overall vitality of the flap. In five of the ten freely transplanted skin flaps with integrated cultivated cartilage tissue there was evidence of venous congestion, beginning from the first day following the operation. This had generally abated by the second or third day after the operation at the latest. There were no signs of necrosis.

Shape and Size of Tissue Engineered Auricles Remain Largely Unaltered

The cartilage constructs were integrated well into the flaps. Directly after implantation, the auricle was clearly and distinctly evident beneath the skin. The structures of the seeded cell carrier were clearly visible and palpable and were only covered by a thin layer of skin. The formation of a capsule around the auricle led to a more plump appearance of the initially visible filigree structures of the helix, antihelix and the tragus. Even after free transplantation, the implanted large auricles (13 pieces) were stable in terms of structure, palpable in the skin flap and remained well integrated. The two small implanted auricles, however, had a considerably altered form due to a lack of stability. Changes in form caused by the degradation of the biomaterial construct were not observed during the observation period of a maximum of 12 months. The cultivated auricular cartilage biomaterial constructs were of firm consistency similar to that of native cartilage. The control constructs were as well integrated into the flaps, and over time the formation of a capsule around the auricle led to a plump appearance. In contrast to the constructs with prefabricated skin flap they revealed reduced mechanical stability.

In four animals a seroma (25–50 ml) formed around the implanted construct after implantation. Bacteriological tests showed no evidence of germs.

Selective Microangiography Demonstrates Neovascularization in the Complete Flap Area

A selective microangiography was carried out on all experimental animals after free transplantation and healing of incisions using a Mammomat 3000 Nova (Siemens Healthcare AG, Erlangen, Germany). Ten skin flaps were analysed. Two necrotic flaps could not be evaluated.

The prefabricated, 3D skin flaps were all very well neovascularized. Starting with the vascular pedicle the vascular system showed complete coverage of all flaps, with the exception of one flap. In this animal, the vascular pedicle was torn. Thus the distal section of the flap was not completely filled with contrast agent – however, still this was shown macroscopically to be vital after the operation. In the area of the integrated auricular cartilage biomaterial construct the main area of neovascularization was evident through a fine and very thick network of vessels *(fig. 3)*. Vessel density in the flap had increased on average by 158% (range 110–193%) as a result of vascular pedicle implantation. The boxplots in *figure 4* illustrate the acquired quantitative values for the vessel count in prefabricated, 3D skin flaps with integrated, tissue engineered auricular cartilage in comparison to the control group.

Histology

The skin was intact in all specimens. The implanted vascular pedicles exhibited a regular morphological pattern. Around the cartilage constructs there was histological proof of good neovascularization in all specimens *(fig. 5C, E, F)*.

In the area of the implanted, tissue engineered cell constructs, cartilage-like cells were found on their own and grouped as part of small cell structures. Large vital areas with cartilage-like tissue containing a significant amount of collagen II, as shown both in the alcian staining and immunohistochemically, were found mostly in the areas at the periphery of the construct *(fig. 5D, F)*. Some central areas also contained cartilage-like cells and collagen II *(fig. 5C, E)*. However, these were not as redifferentiated as the areas located at the periphery of the constructs. The central cells were arranged in a more sparse fashion and were surrounded by many fibres and an extra-cellular matrix. In the control group only small vital areas with cartilage-like tissue were found at the periphery of the construct *(fig. 5I, K)*. The center of the scaffold showed no vital cells, but instead large areas of necrosis *(fig. 5H, J)*.

The biomaterial construct did not fully degrade during observation periods ranging between 9 and 11 months after implantation in any of the animals, either in the study group nor in the control group. Degradation took place mainly at the periphery of the constructs. Centrally, the construct material appeared almost unchanged. In the area of the biomaterial (both in the study and the control group) there was evidence, in some of the specimens, of accumulations of leucocytes.

Discussion

The ultimate objective of the study described here is the transplantation of neovascularized, cultivated and formed cartilage tissue for the 3D reconstruction of complex cartilage defects. The vascularization of large, 3D cultivated tissue volumes was realized within this work by means of the prefabrication of flaps using an autologous model. These vital, manufactured connective tissue flaps could, following sufficient neovascularization in the skin flaps and around the bioconstructs at the recipient location, be transplanted to the defective cartilage area through free transplantation. This also offers the possibility of implanting the cultivated tissue in suitable locations, which fulfill both the aesthetic and functional requirements. For a particularly demanding tissue with regard to sufficient circulation, this represents a technology for the neovascularization of the biomaterial construct prior to cell seeding.

If an engineered biomaterial construct is implanted into a prefabricated flap, in this way a neovascularized, defect-specific, 3D tissue flap can be created to meet individual requirements. In addition, it appears feasible to use the neovascularization of the biomaterial construct separately, in a targeted manner. It would be possible to neovascularize all of the tissue cultivated using the tissue engineering procedure.

In recent literature, various additional approaches to improving the supply of nutrients in cultivated cell constructs have been described. This includes the use of growth factors, bioactive materials, cell transplants, targeted cell cultivation and the manipulation of physiological neoangiogenesis under hypoxic conditions [15]. Cell carriers from the field of nanotechnology with already-integrated channels for the improvement of diffusion also produced positive results [16]. Overall, these approaches show some success, but are, mostly quite expensive and cannot be applied to all types of tissue.

A significant number of authors have reported that the deposition of the extracellular matrix takes place mainly at the periphery of the engineered cartilage. In contrast, the central region of the cultivated tissue contains fewer cells and matrix [2,4]. According to Matin this phenomenon can be traced back to the uneven cell seeding within the construct pores and to diffusion limitations in the internal area [17]. In our study an uneven cell seeding can be excluded, as many vital chondrocytes have been found in the center of the construct after seeding. Cultivated 3D chondrocyte constructs within neovascularized skin flap produced cartilage-specific extracellular matrix (ECM) components, such as GAGs and collagen even in the center of the constructs.

Experimental in-vivo tests regarding tissue engineering of cartilage have previously been conducted mostly in nude mice, which have a compromised immune system [18,19]. In order to lead experimental work closer to clinical application, autologous cartilage tissue was used in this immunocompetent animal study. In addition, because no immunosuppression was used, it was possible to evaluate the immunological potential of the chondrocyte cell biomaterial constructs.

In four animals, which were operated on at an early stage during the study, there was clear evidence of seroma formation around the implanted construct following construct implantation. We hypothesize that a component of the medium – most likely foetal calf serum – was responsible for this allergic/inflammatory reaction. Once the auricular cartilage construct had been rinsed with in saline prior to implantation, no more seroma were. A slight swelling and reddening was, however, observed in most of the animals, which abated after a few days. It can therefore be assumed that the foetal calf serum indeed contributed considerably to the immunological reaction, but was not solely responsible [17,20]. Other studies confirm the immunological potential of foetal calf serum. It is therefore recommendable to use serum-free culture media [21,22] or carry out sufficient cleaning before implantation. Alternatively, autologous serum might be suitable [21]. The potential of non-human serum for infection with prion diseases should be acknowledged. Moreover, the implanted silicon foil can be responsible for foreign body reactions [2,23–25].

The selective microangiography demonstrated that skin flaps with integrated cultivated cartilage cell constructs were successfully neovascularized. In particular, the area of the biomaterial construct was rich in fine vessels. The quantitative analysis, viewed in comparison with the microangiographic control group (the native existing anatomical vessel density in 6 animals), showed that the vessel density in the flap had increased by 110–193% as a result of the vascular pedicle implantation. There was a significant difference between experimental and control group, $p = 0.0002$, Wilcoxon-Mann-Whitney Test. The increase of vessels, achieved by flap prefabrication, leads to a better vascular integration and thus better supply of nutrients and oxygen in cultured cartilage. If one were to compare these values with the neovascularization in the flaps without implanted constructs, as described in the available literature [7], these values are somewhat higher. This indicates that the implanted scaffold has a positive influence on the neovascularization. There were numerous vessels found histologically, within a fibrous capsule on the surface of the implanted cartilage constructs. In contrast, in the area of cultured cartilage, there was no neovascularization found. This was most likely due to inhibitory components of cultured cartilage like troponin I as described by Moses et. al. [26]. Thus prefabrication, by means of the implantation of a vascular loop, improves vascular integration without negative influence or dedifferentiation of implanted cartilage.

In the control group without implanted vascular pedicle, only small vital areas with cartilage-like tissue were found at the periphery of the construct. The center of the scaffold did not contain vital cells, but instead large areas of necrosis. In the study group some central areas also contained cartilage-like cells and collagen II, but in none of the constructs cartilage-like tissue was found throughout. In order to resolve this problem, an improved nutrient supply to the centre of the construct, among other solutions, will be required, which has so far been limited by the existing diffusion limits in the construct [17].

Conclusions

Adopting an autologous model it was possible, to realise a combination of prefabricated skin flaps by means of vascular pedicle implantation and the implantation of *in vitro* cultivated chondrocytes bioconstructs for the manufacture of functional, 3D cartilage grafts.

In this study, cartilage biopsies were taken from the auricles of 15 chinchilla-bastard rabbits, multiplied in monolayer culture and then seeded in a 3D cell carrier made from polycaprolactone. The chondrocyte biomaterial constructs were implanted in the skin flaps and then neovascularized using vascular loops with terminal arteriovenous anastomosis. They were then grown as connective tissue flaps and reimplanted in the same position, microsurgically.

It was possible to show macroscopically, that the combined, prefabricated 3D flaps could be freely transplanted. The integrated 3D constructs were stably integrated within the skin flaps. It was shown microangiographically that all skin flaps with implanted, cultivated constructs were well neovascularized. There was evidence of cartilage-like tissue with the formation of cartilaginous ECM even in the center of large cultured cartilage constructs. It was possible to demonstrate the presence of newly formed collagen type II. In contrast, in the control group, without implanted vascular loop, only small vital areas with cartilage-like tissue were found at the periphery of the construct. The center of the scaffold did not contain vital cells, but instead large areas of necrosis. Results of the experiment demonstrated that cultivated 3D auricular cartilage constructs with prefabricated skin flaps could be combined by means of vascular pedicle implantation and could be transplanted. With respect to future clinical application, the procedure is a promising alternative for clinical practice because of favorable esthetic outcomes with minimal donor site morbidity.

Acknowledgments

We are grateful to Hinrich Wiese Polymaterials, Kaufbeuren and Anne Hills for support and advice.

Author Contributions

Conceived and designed the experiments: AvB RS KS HNT. Performed the experiments: AvB HNT. Analyzed the data: AvB JV CB NR. Contributed reagents/materials/analysis tools: AvB RS KS HNT. Wrote the paper: AvB JV CB NR.

References

1. Green WT Jr, Dickens DR (1972) Culture model for investigation of cartilage matrix formation. Surg Forum 23: 453–455.
2. Vacanti CA (2006) The history of tissue engineering. J Cell Mol Med 10: 569–576.
3. Hirase Y, Valauri FA, Buncke HJ (1988) Neovascularized free cutaneous cartilage flap transfer with microsurgical anastomosis: an experimental model in the rabbit. Ann Plast Surg 21: 342–347.
4. Staudenmaier R, Hoang TN, Kleinsasser N, Schurr C, Frolich K, et al. (2004) Flap prefabrication and prelamination with tissue-engineered cartilage. J Reconstr Microsurg 20: 555–564.
5. Neumeister MW, Wu T, Chambers C (2006) Vascularized tissue-engineered ears. Plast Reconstr Surg 117: 116–122.
6. Hoang NT, Hoehnke C, Hien PT, Mandlik V, Feucht A, et al. (2009) Neovascularization and free microsurgical transfer of in vitro cartilage-engineered constructs. Microsurgery 29: 52–61.
7. Nguyen TH, Kloppel M, Staudenmaier R, Biemer E (2004) [Experimental investigation of neovascularisation in large prefabricated flaps after arteriovenous pedicle implantation]. Handchir Mikrochir Plast Chir 36: 212–217.
8. Pribaz JJ, Fine N, Orgill DP (1999) Flap prefabrication in the head and neck: a 10-year experience. Plast Reconstr Surg 103: 808–820.
9. Morrison WA, Dvir E, Doi K, Hurley JV, Hickey MJ, et al. (1990) Prefabrication of thin transferable axial-pattern skin flaps: an experimental study in rabbits. Br J Plast Surg 43: 645–654.
10. Walton RL, Brown RE (1993) Tissue engineering of biomaterials for composite reconstruction: an experimental model. Ann Plast Surg 30: 105–110.
11. Karatas O, Atabey A, Demirdover C, Barutcu A (2000) Delayed prefabricated arterial composite venous flaps: an experimental study in rabbits. Ann Plast Surg 44: 44–52.
12. Wiese HDP, Landsberg am Lech, DE Maier, Gerhard (2008) Open-pored polyurethane foam without skin formation, formulation for the production thereof and use thereof as a carrier material for cell and tissue cultures or medicaments. Polymaterials AG (Innovapark 20, 87600 Kaufbeuren, DE).
13. Eyrich D, Wiese H, Maier G, Skodacek D, Appel B, et al. (2007) In vitro and in vivo cartilage engineering using a combination of chondrocyte-seeded long-term stable fibrin gels and polycaprolactone-based polyurethane scaffolds. Tissue Eng 13: 2207–2218.
14. Hoang NT, Kloeppel M, Werner J, Staudenmaier R, Biemer E (2005) Proposed new method for angiographically quantifying neovascularization in prefabricated flaps. Microsurgery 25: 220–226.
15. Hadjipanayi E, Brown RA, Mudera V, Deng D, Liu W, et al. (2010) Controlling physiological angiogenesis by hypoxia-induced signaling. J Control Release 146: 309–317.
16. Song YS, Lin RL, Montesano G, Durmus NG, Lee G, et al. (2009) Engineered 3D tissue models for cell-laden microfluidic channels. Anal Bioanal Chem 395: 185–193.
17. Martin I (2006) [Engineering 3D cartilage grafts]. Praxis (Bern 1994) 95: 841–844.
18. Vacanti CA, Langer R, Schloo B, Vacanti JP (1991) Synthetic polymers seeded with chondrocytes provide a template for new cartilage formation. Plast Reconstr Surg 88: 753–759.
19. Cao Y, Vacanti JP, Paige KT, Upton J, Vacanti CA (1997) Transplantation of chondrocytes utilizing a polymer-cell construct to produce tissue-engineered cartilage in the shape of a human ear. Plast Reconstr Surg 100: 297–302; discussion 303–294.
20. Rotter N, Sittinger M, Hammer C, Bujia J, Kastenbauer E (1997) [Transplantation of in vitro cultured cartilage materials: characterization of matrix synthesis]. Laryngorhinootologie 76: 241–247.
21. Mandl EW, van der Veen SW, Verhaar JA, van Osch GJ (2002) Serum-free medium supplemented with high-concentration FGF2 for cell expansion culture of human ear chondrocytes promotes redifferentiation capacity. Tissue Eng 8: 573–580.
22. Bouwer AL, Netter P, Kemp RA, McLellan AD (2010) A defined serum-free medium useful for monitoring anti-melanoma responses induced by dendritic cell immunotherapy. J Immunol Methods 352: 178–181.
23. Imber G, Schwager RG, Guthrie RH Jr, Gray GF (1974) Fibrous capsule formation after subcutaneous implanation of synthetic materials in experimental animals. Plast Reconstr Surg 54: 183–186.
24. Gayou RM (1979) A histological comparison of contracted and non-contracted capsules around silicone breast implants. Plast Reconstr Surg 63: 700–707.
25. Aptekar RG, Davie JM, Cattell HS (1974) Foreign body reaction to silicone rubber. Complication of a finger joint implant. Clin Orthop Relat Res: 231–232.
26. Moses MA, Wiederschain D, Wu I, Fernandez CA, Ghazizadeh V, et al. (1999) Troponin I is present in human cartilage and inhibits angiogenesis. Proc Natl Acad Sci U S A 96: 2645–2650.

20

Pullulan/dextran/nHA Macroporous Composite Beads for Bone Repair in a Femoral Condyle Defect in Rats

Silke Schlaubitz[1], Sidi Mohammed Derkaoui[2,3], Lydia Marosa[4], Sylvain Miraux[5], Martine Renard[1], Sylvain Catros[4,6], Catherine Le Visage[2,3], Didier Letourneur[2,3], Joëlle Amédée[4], Jean-Christophe Fricain[4,6]*

1 CIC 1401, University hospital of Bordeaux/Inserm, Bordeaux, France, **2** U1148, LVTS/Inserm, Paris, France, **3** Près Sorbonne Paris Cité, University of Paris Nord and University Paris Diderot, Paris, France, **4** U1026 Tissue Bioengineering, University of Bordeaux/Inserm, Bordeaux, France, **5** RMSB Center/UMR 5536, CNRS, Bordeaux, France, **6** Dental School, University of Bordeaux, Bordeaux, France

Abstract

The repair of bone defects is of particular interest for orthopedic, oral, maxillofacial, and dental surgery. Bone loss requiring reconstruction is conventionally addressed through bone grafting. Depending on the size and the location of the defect, this method has limits and risks. Biomaterials can offer an alternative and have features supporting bone repair. Here, we propose to evaluate the cellular penetration and bone formation of new macroporous beads based on pullulan/dextran that has been supplemented with nanocrystalline hydroxyapatite in a rat model. Cross-linked beads of 300–500 μm diameters were used in a lateral femoral condyle defect and analyzed by magnetic resonance imaging, micro-computed tomography, and histology in comparison to the empty defects 15, 30, and 70 days after implantation. Inflammation was absent for both conditions. For empty defects, cellularisation and mineralization started from the periphery of the defect. For the defects containing beads, cellular structures filling out the spaces between the scaffolds with increasing interconnectivity and trabecular-like organization were observed over time. The analysis of calcified sections showed increased mineralization over time for both conditions, but was more pronounced for the samples containing beads. Bone Mineral Density and Bone Mineral Content were both significantly higher at day 70 for the beads in comparison to empty defects as well as compared with earlier time points. Analysis of newly formed tissue around the beads showed an increase of osteoid tissue, measured as percentage of the defect surface. This study suggests that the use of beads for the repair of small size defects in bone may be expanded on to meet the clinical need for a ready-to-use fill-up material that can favor bone formation and mineralization, as well as promote vessel ingrowth into the defect site.

Editor: Dominique Heymann, Faculté de médecine de Nantes, France

Funding: This work is funded by ANR-07-TecSan-011-01 ITOV and ANR-10-EMMA-009-01 MATRI+ from Agence National de la Recherche: (http://www.agence-nationale-recherche.fr/) to JA and DL. The funders had no role in study design, data collection and analysis, decision to publish, or preparation of the manuscript.

Competing Interests: Although the authors Sidi Mohammed Derkaoui and Didier Letourneur are shareholders of the company IMMATIS for development and commercialization of the reported matrices, this does not alter the authors' adherence to PLOS ONE policies on sharing data and materials.

* Email: jean-christophe.fricain@inserm.fr

Introduction

The repair of bone defects caused by trauma, surgery or due to cancer treatment is of particular interest for surgeons. Conventional bone grafting is still the gold standard, but depending on the size and the location of the defect, this method has limits and risks regarding the efficiency, possible infections and the manageable pain [1]. Biomaterials can offer an alternative and have osteoconductive [2,3] and/or osteoinductive [4–6] properties supporting bone repair while being biocompatible and biodegradable [7–12].

Bone substitute biomaterials are often based on calcium phosphates like hydroxyapatite (HA) [13], tricalcium phosphate (TCP) [13–15], biphasic calcium phosphates (BCP) [13,16,17], or coral [18] because of their biocompatibility due to a composition similar to mature bone, and their osteoconductive properties [2]. The use of calcium phosphate itself is limited to small defects, where biomaterials with osteoconductive properties are sufficient for bone reconstruction [5], or as a coating of other biomaterials [19], because those materials allow only limited neo-angiogenesis and are therefore not suitable for larger size defects.

Other groups of bone substitute biomaterials that promote vessel ingrowth are mainly based on polymers. The properties of natural polymers like collagen [20,21], dextran [22], chitosan [23], alginate/fibrin [24] or synthetic polymers like polyglycolic acid (PGA), polylactic acid (PLA), and poly(lactic-*co*-glycolic) acid (PLGA) among others [25,26] have been extensively studied in *in vitro* bone tissue engineering applications, mainly as delivery systems for encapsulated stem cells or growth factors. Polymer scaffolds can easily be generated in various 3D structures such as macroporous scaffolds, biospun scaffolds, micro beads, hydrogels, micro-molded matrices and nanoparticles [27]. Some polymers have little affinity with bone, but can easily be associated with cells or growth factors to promote bone tissue regeneration with angiogenic potential *in vivo* [28–32]. To combine properties favoring bone affinity and angiogenesis, specific composites of polymers and ceramics are being developed that have shown osteoinductive and osteoconductive properties [33]. Composite materials in different shape and sizes of scaffolds have been used in numerous *in vivo* investigations to promote bone healing [34–39], and can easily be associated with cells or growth factors to enhance osteoinductive and/or osteoconductive properties [40,41]. However the use of composite beads seems preferable for some clinical applications, because they could be injectable and therefore easy to handle and adapt to the morphology of any defect [42], as seen previously in a clinical trial [43].

Injectable beads have been studied *in vitro* as composite materials such as alginate/β-tricalcium phosphate [44], Chitosan-collagen/nano-hydroxyapatite [45], or fibrin/β-tricalcium phosphate [46], but *in vivo* studies have not been performed with those materials yet.

In the present study, a biomaterial based on a mixture of pullulan and dextran in association with nanocrystalline hydroxyapatite designed as cross-linked micro-beads was used as a scaffold. Previous experiments have shown the osteoconductive properties of the biomaterial when used as a massive macroporous scaffold [36]. In this study, the objective was to demonstrate for the first time that beads of a new ready-to-use fill-up composite material made of dextran/pullulan/nanocrystalline hydroxyapatite are suitable materials to activate and promote bone regeneration and mineralization in rat femoral condyle defects.

Materials and Methods

Preparation and characterization of the biomaterial

A nanocrystalline hydroxyapatite (nHA) solution was prepared by wet chemical precipitation using a 0.6 M solution of phosphoric acid (Prolabo, Paris, France) and a 1 M solution of calcium hydroxide (Alfa Aesar, Germany) [47,48]. The nHA powder was composed of 50–100 nm long rod-like crystals as observed by transmission electron microscopy. X ray quantitative micro analysis revealed a Ca/P ratio of 1.67 and FITR analysis confirmed the presence of phosphate, hydroxide and carbonate groups, as described previously [47].

Nanocrystalline hydroxyapatite-polysaccharide microparticles composite were obtained using water-in-oil (w/o) emulsification process. Briefly, 75% pullulan (MW 200000, Hayashibara Inc., Japan) and 25% dextran (MW 500000, Sigma, France) were dissolved in nHA suspension and NaCl as a porogen. The mixture was then dispersed in canola oil under mechanical stirring and polysaccharide beads were cross-linked by sodium trimetaphosphate at 50°C for 20 min. Resulting microbeads were extensively washed with 0.025% NaCl solution, and sieved, using a vibrating shaker (AS 200, Retsch, France) to obtain beads of 300–500 μm diameter that were freeze-dried [49].

Hydrated scaffolds were observed using environmental scanning electron microscopy (ESEM) with a Philips XL 30 ESEM-FEG (Eindhoven, The Netherlands) at an accelerating voltage of 15 keV and at a pressure of 3.5 Torr. Dry beads were observed by Back

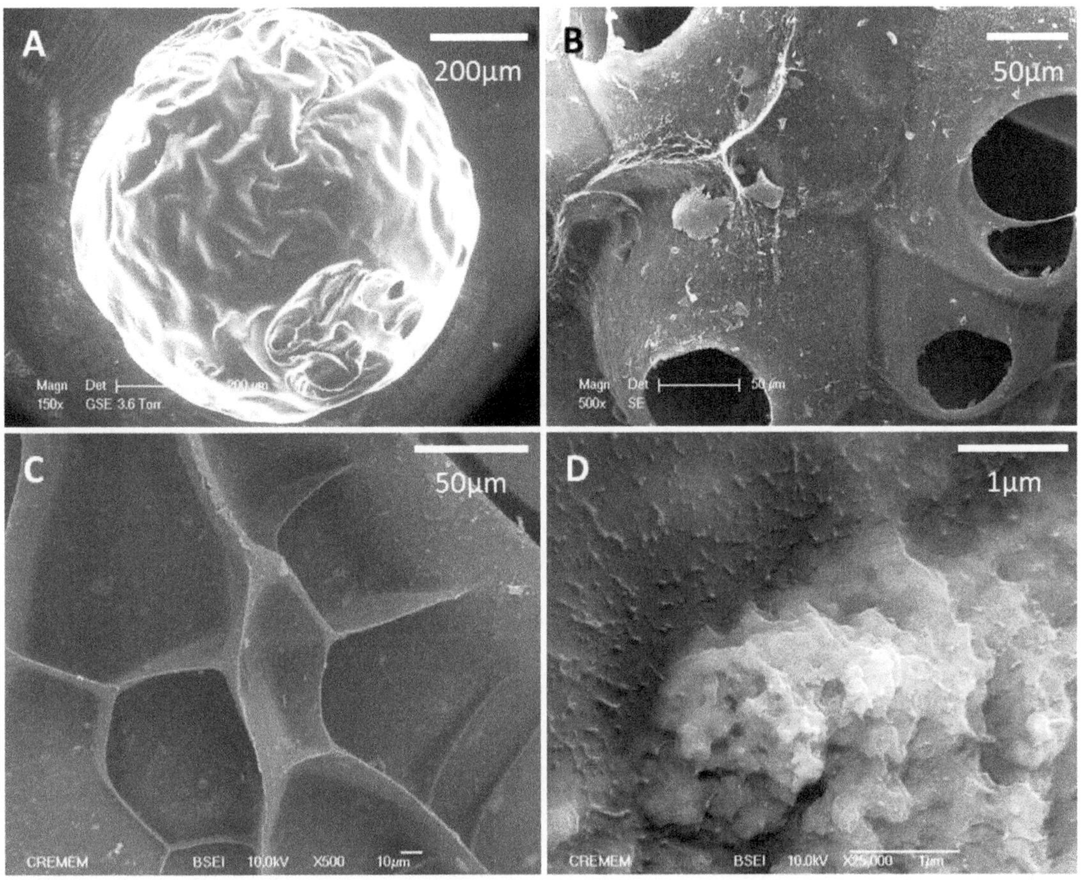

Figure 1. The morphology of microspheres was analyzed by scanning electron microscopy (A). After rehydration in PBS, porosity of hydrated scaffolds was observed with an Environmental Scanning Electron Microscopy (B). Back-scattered Electron Microscopy images of dry beads show pores and overall surface properties (C) and the distribution of nHA aggregates within the structure (D).

Scattered Electron Microscopy (BSEM) after gold sputtering on a 6700F electron microscope (JEOL, Peabody/MA, USA) working at 10 keV.

In vivo experiments

Animal experiments were performed in accordance with the "Principles of Laboratory Animal Care" recommended by the National Society for Biomedical Research in France. Interventions were carried out in an accredited animal facility (authorization n° A33-063-917) at the University of Bordeaux, under authorization n° B3310023 of the French Ministry of Agriculture and were approved by the Animal Ethic Committee of the University of Bordeaux.

Anesthesia was induced with a mixture of 4% isoflurane/Air 1–2 L/min (Baxter, Deerfield/IL, USA) and was maintained at 2% isoflurane/Air 0.8 L/min during the implantation. A stable body temperature was assured by the use of a heating device.

Medial defects of 3.5 mm diameter and 4 mm depth were introduced in left and right femoral condyles of 18 female Wistar RjHan rats at 10 weeks of age weighting 250–300 g (Janvier, Le Genest, France) using a dental microdrill (Thomas, France). Bone defects of about 38 mm^3 were rinsed with physiological solution NaCl 0.9% (w/v). The defect was either left empty or filled with 17±3 dry beads that were implanted into the site, leaving enough time for the beads to hydrate with the blood invading the defect and, as a consequence, to increase slightly in size. The scaffolds remained solid enough to stay inside the defect. Therefore, no membrane was needed to avoid leakage of the biomaterial. Absorbable Vicryl sutures were performed on lateral muscles before closing the cutaneous plan with non-absorbable prolene sutures and suture clips.

To assure analgesia, 0.3 mg/kg of buprenorphine (Vetergesic Multidose, Alstoe Veterinary, York, UK) was administered subcutaneously to each rat 30 min prior to the procedure as well as 24 hours afterwards. Additionally, animals were kept under free access to 0.2 mg/mL of Ibuprofen (Advil, Pfizer, Canada) 3 days prior to surgery and during 1 week after surgery through the drinking water. Twelve animals were euthanized by an overdose of CO_2 and the six animals that were followed longitudinally by MRI and therefore housed in a different facility, were euthanized by lethal injection (200 mg/animal) of sodium pentobarbital (Ceva, France) for the last time point when the regulator delivering the CO_2 was not available. The implants and the surrounding tissues were recovered 15, 30 and 70 days after surgery and fixed in 4% (w/v) paraformaldehyde for 4 days at 4°C. After rinsing in 0.1 M phosphate buffer pH 7.4, samples were processed for Micro-CT analysis and histology. For each time point, 6 condyles were left empty and 6 condyles were filled with the beads as matrices.

Magnetic Resonance Imaging

Immune response and degradation of the implanted scaffold was followed longitudinally in 6 rats by Magnetic Resonance Imaging (MRI). The rats were anaesthetized with 1–1.5% isoflurane/Air 0.8 L/min and maintained at a constant respiration rate of 75±15 respirations/min. Acquisitions at day 15, 30, and 70 after implantation were obtained with a 4.7T Biospec system (Bruker, Ettlingen, Germany) equipped with a 12 cm gradient system capable of 200 mT/m maximum strength. Measurements were performed with a Helmoltz coil (35/25 mm diameter) tuned to 200.3 MHz.

The animals were put in the prone position within the magnet, with the knee at the center of the NMR coil. A 3D TrueFISP imaging with alternating RF phase pulse method and sum of square reconstruction was used as described previously [50]. The aqueous volume was measured for the defect of each acquisition and the results are presented as the means of 6 samples with standard deviation.

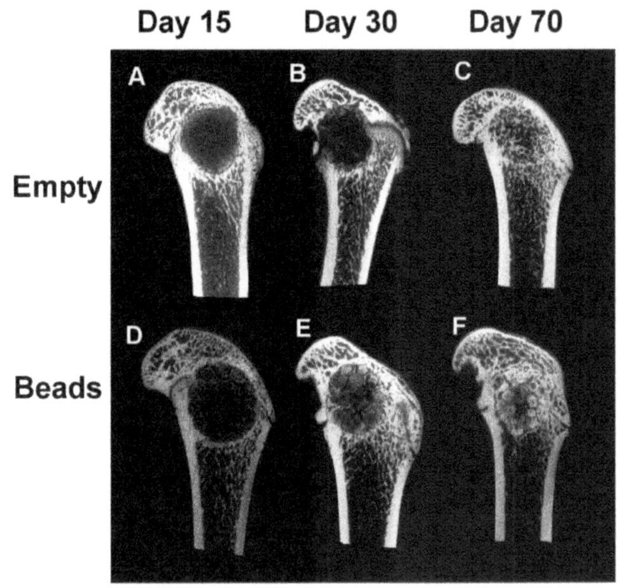

Figure 2. Representative 2D pictures of the median region of Micro-CT 3D reconstructed femoral condyles that were left empty (A–C) or received beads as implants (D–F). Mineralization around the beads is visible as early as 15 days after surgery (D). Empty defects show little mineralization within the defect 30 days after implantation (B), starting from the periphery of the defects. At day 70, mineralization is visible within the defect site for both conditions (C, F).

X-ray Micro-Computed Tomography (Micro-CT)

Three-dimensional images of the implants and surrounding tissue were produced by Micro-CT and used for quantification of the newly formed bone. *Ex vivo* acquisitions at 15 μm resolution were performed from 900 X-ray radiographs obtained on an Explore Locus SP X-Ray μCT device (General Electric, Milwaukee/WI, USA) with a source voltage of 80 kV, a current of 80 μA, and an exposure time of 3000 ms. Reconstruction of the region of interest was performed after correction of the center of rotation and calibration of mineral density. After calibration, three dimensional analyses were performed using Microview software (*GE Healthcare* Inc., Princeton/NJ, USA). The threshold for newly formed bone was determined for each scan individually. Quantification included Bone Mineral density (BMD), and Bone Mineral Content (BMC) as defined previously [51]. BMC values were analyzed from regions of interest (ROI) of a fixed volume. Six ROI throughout the defect were determined for each sample. The means of all ROI per sample were analyzed further to evaluate the means with standard deviation of six samples per condition at each time point. For the analysis of BMD, the entire defect site of six samples per condition was taken into account and the results of six samples per condition are shown as the means with standard deviation.

Histology

Each sample was cut median between the two femoral condyles. One half of each sample was decalcified for 7 hours in a commercially available decalcification solution (DC3, Labonord, France), dehydrated and embedded in paraffin, while the other

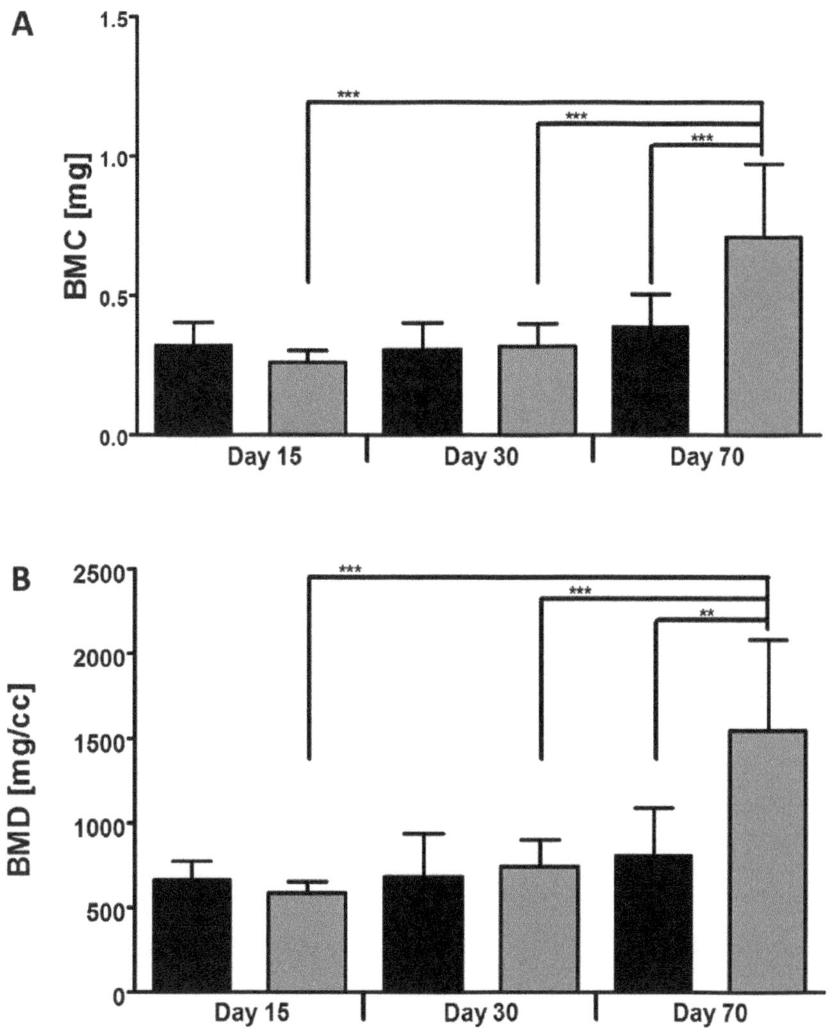

Figure 3. Ex *vivo* analysis of Bone mineral content (A) and Bone mineral density (B) by Micro-CT imaging. Regions of interest (ROI) were analysed in defects that were left empty (black columns) or received the beads (grey columns). Results are shown as means ± SD of 6 samples per condition. Significant differences are indicated with ** $p \leq 0.01$, and *** $p \leq 0.001$.

half was processed for resin inclusion in serial changes of 5% (w/v) Methyl benzoate (Sigma Aldrich), 35% (w/v) Butyl methacrylate (Sigma Aldrich), and 60% (w/v) Methyl methacrylate (VWR, Radnor/PA, USA) with increasing amounts of 4,N,N-Trimethylaniline (Sigma Aldrich, St. Louis/MO, USA).

Sections of 4–10 µm were obtained from each sample and stained with Masson's Trichrome (decalcified samples) or Von Kossa (calcified samples). Pictures were analyzed with an Eclipse 80i microscope light microscopy (Nikon, Japan) and captured using a DXM 1200C CCD camera (Nikon, Japan).

Immunohistochemistry

Infiltration of macrophages was evaluated by immunohistochemistry for transmembrane glycoprotein CD68, belonging to the lysosomal-associated membrane protein family. Vessel formation was investigated for the presence of the angiogenesis marker von Willebrand Factor, and the presence of osteoblast precursors was evaluated with immune-staining for BMP-2. Sections were deparaffinized in two changes of toluene for 5 min. Endogenous peroxidase was blocked by applying 3% (v/v) hydrogen peroxide in methanol for 10 min at −20°C. Antigen Retrieval was performed by applying Proteinase K (Invitrogen, Grand Island/NY, USA) for 10 min at room temperature. Endogenous biotin was blocked using an avidin/biotin blocking kit (Invitrogen, Grand Island/NY, USA), before blocking unspecific antibody binding using Histostain-SP kit (Invitrogen, Grand Island/NY, USA). Mouse monoclonal anti-rat Cluster of Differentiation 68 (CD68) antibody (AbD Serotec MCA341GA, Kidlington, UK; dilution 1:50), rabbit polyclonal anti rat Bone Morphogenetic Protein 2 (BMP2) antibody (Abcam 14933, Cambridge/MA, USA; dilution 1:250), and mouse monoclonal anti-rat von Willebrand Factor (vWF) antibody (DAKO A0082, Carpinteria/CA, USA; dilution 1:200) were all applied for 1 hour at room temperature. A biotinylated secondary antibody and the enzyme conjugate were applied for 10 minutes each, followed by 3 min with the AEC chromogen substrate (Histostain-SP kit, Invitrogen, Grand Island/NY, USA). Counterstaining was performed for 1 min with Mayer's hemalun (VWR, Radnor/PA, USA). Immune-staining was evaluated with an Eclipse 80i microscope light microscopy (Nikon, Tokyo, Japan).

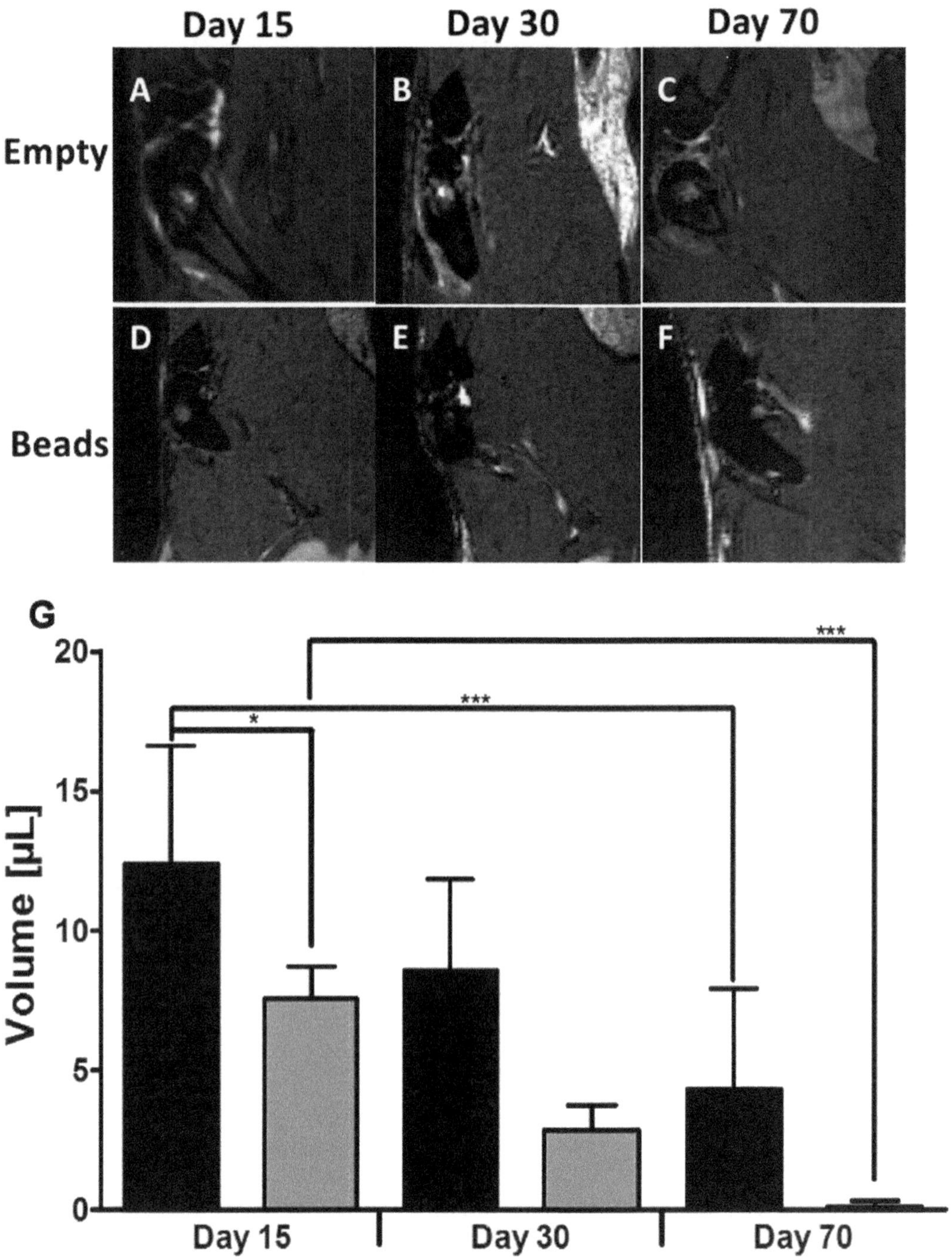

Figure 4. Representative images of MRI acquisition of condyles left empty (A–C) or after bead implantation (D–F) at 15 days (A, D), 30 days (B, E), and 70 days (C, F) after surgery. Quantification of the aqueous volume in μL of a fixed region of interest in defects that were left empty (black columns) or received the beads (grey columns), as obtained by longitudinal *in vivo* follow up by Magnetic Resonance Imaging (G). Results are presented as means ± SD of 6 samples and significant differences are indicated with * $p \leq 0.05$, and *** $p \leq 0.001$.

Histomorphometry

The percentage of newly formed osteoid tissue in defects filled with beads was determined from histological paraffin sections using NIS-Elements software (Nikon, Japan). Osteoid formation that was stained in dark red by Trichrome's Masson staining protocol using 2 volumes of 1% (v/v) Fushine acid and 3 volumes of 1% (v/v) Xylidine Ponceau, was detected semi-automatically by the software. For samples that were left empty, the surface of osteoid was expressed as a percentage of defect surface. For samples that received the beads, two analyses were performed. Once, we performed the same analysis and once we took into account the invading tissue only, subtracting the surface of the beads from the defect surface. Histologically stained slides from three samples per condition were processed for analysis, and 12 to 19 sections were analyzed per condition. Results are shown as means with standard deviation per condition.

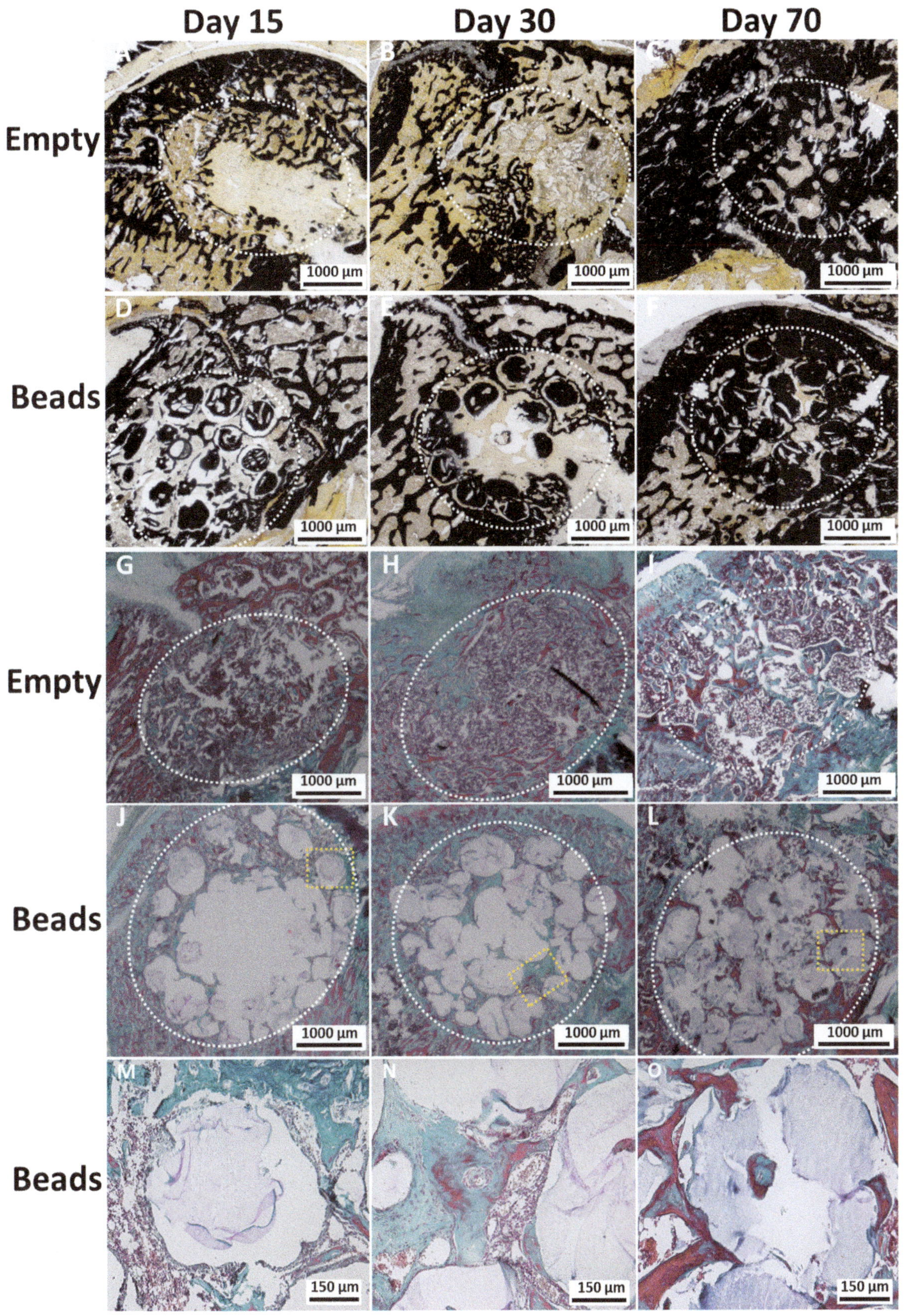

Figure 5. Representative images of histological staining of defects that were left empty (A–C, G–I) and for defects that received beads as implants (D–F, J–O). Calcified sections were stained with Von Kossa (A–F), while decalcified sections were stained with Masson's Trichrome (G–O). Increase of mineralization is visible over time when comparing sections obtained at day 15 (A, D, G, J) with sections of day 30 (B, E,

H, K) or those of day 70 (C, F, I, L). Areas of defects are marked with the dotted white line (A–L) and higher magnifications of the orange squares in J–L are shown in M–O.

Statistical analysis

Statistical significance was evaluated with a non-parametric "One-way Anova analysis of variance" followed by a Tukey's Multiple Comparison Test to compare all possible pairs of means independently, as provided by the GraphPad Prism Software (La Jolla/CA, USA). The significance level was set to alpha equal 0.05. Statistical significances are marked by stars with * indicating $p \leq 0.05$, ** $p \leq 0.01$, and *** $p \leq 0.001$.

Results

Chemical cross-linking with sodium trimetaphosphate of an alkaline solution containing NaCl as a porogen, nHA and a polysaccharide blend dispersed into canola oil under mechanical stirring allowed obtaining the beads. With this patented emulsification process [49], the resulting water-insoluble spherical beads were calibrated according to their size and then freeze-dried. The beads contained 2.8±0.1% (w/v) of nanocrystalline hydroxyapatite. Beads of 300–500 µm in diameter (Figure 1A) were observed by Back-scattered Electron Microscopy (BSEM), revealing micropores and a distribution of the nHA particles within the structure (Figure 1B–D).

2D images of 3D Micro-CT reconstructed femoral condyles after implantation of the beads in comparison with defects left empty revealed that mineralization occurred in both groups, empty defect and defect filled with beads. In the median of the defect, there was an increase in mineralized tissue over time of implantation (day 15 through day 70) (Figure 2) that seemed more pronounced and occurred faster for samples that received beads (Figure 2 D–F) than for defects that were left empty (Figure 2 A–C). At day 15, fine trabecular-like structures were only visible around the beads. These trabecular-like structures became more sophisticated over time. Empty defects showed little mineralization in the median of the defect after 30 days after implantation,

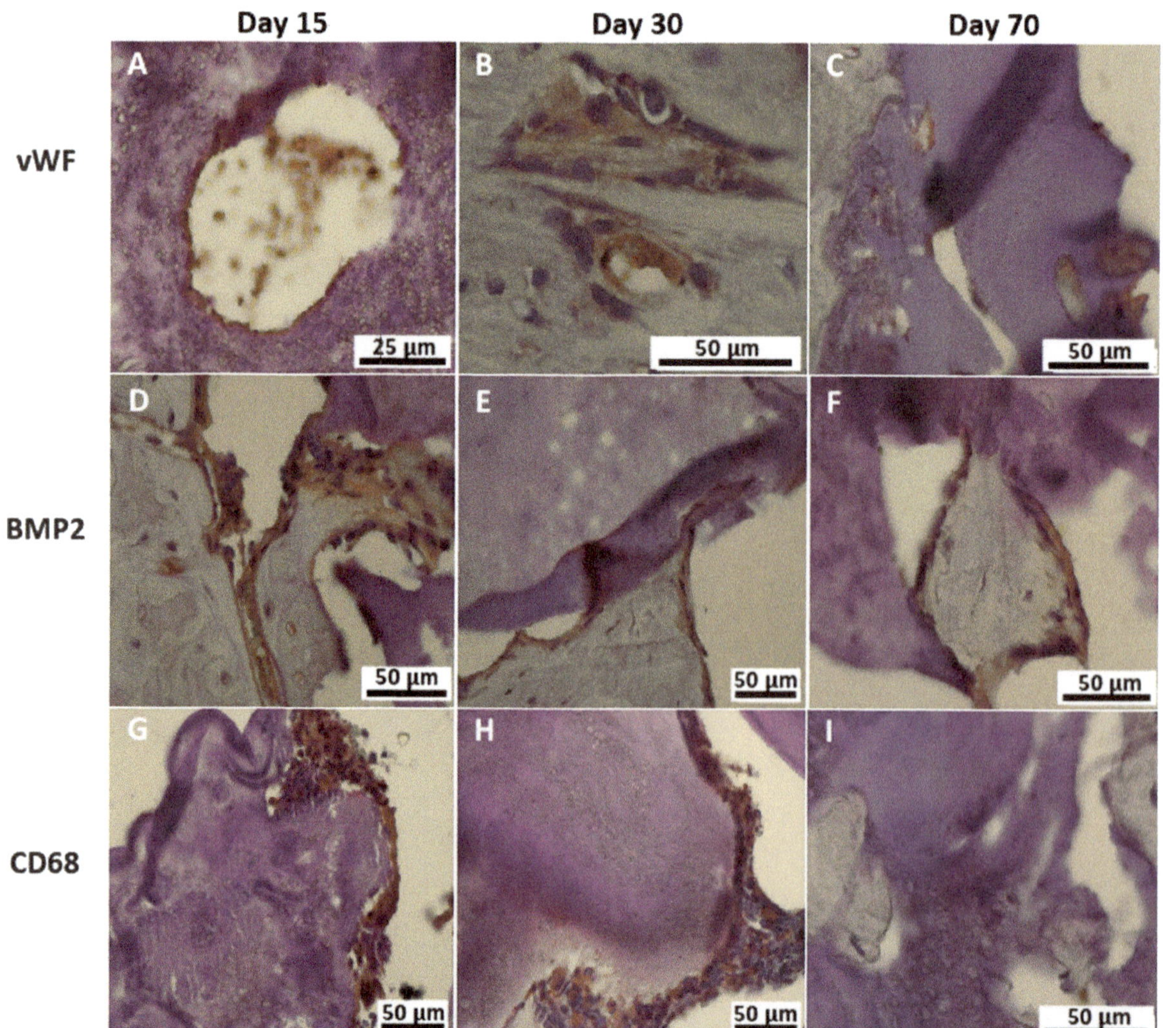

Figure 6. Representative images of immune-staining for cells in contact with the biomaterial using primary antibodies for von Willebrand Factor (A–C), Bone Morphogenetic Protein 2 (D–F) and CD 68 (G–I). Positive cells appear in brown and nuclei in dark blue.

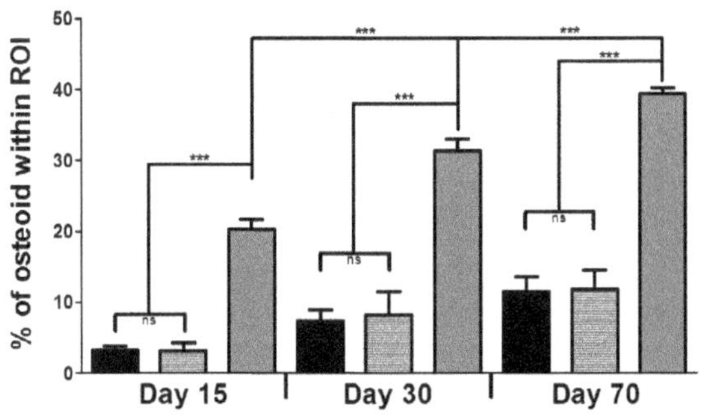

Figure 7. Histomorphometric analysis of newly formed osteoid in defects that were left empty (black columns) or received the beads (striped and grey columns). Osteoid surface over defect surface was expressed in percentage of osteoid within the region of interest (ROI), that was either the entire defect surface (black and striped columns) or the tissue surface around the beads (grey columns). Results are presented as means ± SD of 3 conditions and significant differences are indicated with *** $p \leq 0.001$.

starting from the periphery of the defects. At day 70, mineralization in form of trabecular structures was visible within the defect site.

When evaluating Bone Mineral density (BMD) and Bone Mineral Content (BMC), there were no statistically significant differences between the empty defects and the defects that received the beads at day 15 and day 30. Values were strongly increased at day 70 for samples treated with beads in comparison to the empty defects ($p \leq 0.01$) (Figure 3). When evaluating BMD and BMC for the samples that received beads, a significant increase was seen when comparing the two earlier with the latest time point.

Longitudinal *in vivo* follow-up by MRI did not reveal an inflammation in any of the conditions (Figure 4 A–F). The quantification of the aqueous volume showed a decrease for both conditions (Figure 4G). There were significant differences between the earliest and the latest time point for the two conditions ($p \leq 0.001$) and between the two conditions at day 15 ($p \leq 0.05$). Histological analysis did not reveal signs of inflammation at day 15 or at later time points (Figure 5). For defects that were left empty (Figure 5 A–C, G–I), cellularisation and mineralization started from the periphery of the defect. A similar process was observed with the beads, but cellular invasion also occurred within the defect (Figure 5 D–F, J–O).

Cellularisation of the material by endothelial cells forming vessels started within the first 15 days (Figure 6 A). Vessels were detected by vWF immune-staining in the trabecular-like structures (Figure 6 B) as well as within the material (Figure 6 A, C). BMP-2 positive cells secreting extracellular matrix were present at any time point, lining the trabecular-like structures, as well as in contact with the material (Figure 6 D–F). CD68 positive cells occurred mainly at the two earlier time points (Figure 6 G, H), while there was little presence of those cells at the latest time point (Figure 6 I). CD68 positive cells were seen in direct contact with the material.

Cellularisation was fully completed by day 70 for both conditions (Figure 5 I, L), but with different cellular structures. For the samples implanted with beads, cellular structures filled out all spaces between the beads with increasing interconnectivity and trabecular-like organization over time (Figure 5 D–F, J–L). Empty defects at day 15 after implantation showed few interconnected structures (Figure 5 A, G) and fibrosis that was still present at day 30 (Figure 5 H). At day 70, connective tissues were still visible, but most of the defect was filled with mineralized, trabecular structures (Figure 5 C, I). As it was seen in Von Kossa stained sections, the beads themselves, as well as the trabecular structures mineralized over time (Figure 5 D–F). Histomorphometric analysis of the newly formed tissue within the defect showed no significant difference at all time points between the two conditions when the percentage of osteoid surface was directly calculated within the surface of the entire defect (Figure 7). However, there was a significantly higher percentage of osteoid tissue formation at each time point when the surface of the beats was first subtracted from the defect surface in comparison to the empty defects (Figure 7). For the two calculations concerning the beads, there was a significant increase over time, while the increase in osteoid formation in empty defects was non-significant (not indicated in Figure 7).

Discussion

In the present study, macroporous cross-linked beads based on a mixture of pullulan and dextran [49,52] in association with nHA were used as a scaffold. This new scaffold, with injectable properties, was evaluated in a non-critical size bone defect, matching most clinical needs for small size defects.

Although most biomaterials are tested in critical-size defects, surgeons are often in need of materials to simply fill-up a small size defect that may not always be of a critical size [53]. The material tested in this experiment was previously implanted in the form of cylinders of 5 mm diameter and 6 mm depth in a critical size defect in rat femoral condyles and showed osteoconductive properties [36]. The goal of the present study was to investigate whether the material could modulate bone formation in a spontaneous bone regeneration model i.e. non-critical size defect to guide and activate bone mineralization and bone regeneration. The material was tested in the form of beads of 300–500 µm in diameter (Figure 1) to adapt to small size defect.

To fit specific needs for small size defects, the use of beads seems preferable. Scaffolds can be produced in any shape, but the advantage of small size beads is their capability to entirely fill up bone defects. The presented beads are sieved and can be produced at any desired diameter. In a previous study, the presented composite materials were used in the form of macroporous 3D cylinders [36], that showed limited access to invading cells and was therefore less cellularized than the material in the form of beads.

Although mineralization seemed to increase visually over time (Figure 2), it did not progress in the same way for the two conditions. It is known that defects left empty progress in bone repair from the periphery [54]. For that reason, the defects stayed rather empty in the median plan for the first 30 days (Figure 2 A, B). The presence of the beads seemed to allow a mineralization even within the defect; because trabecular-like structures were visible as early as 15 days after implantation (Figure 2 D). Thus, the material was used as a support for the invading cells, allowing a faster ingrowth of trabecular-like structures. Although those structures were less present within defects that were left empty, there was no significant difference for the BMC and the BMD in comparison with defects loaded with the material at day 15 and day 30. This is due to the fact that the entire defect was taken into account for the analysis and the defects left empty still repaired from the periphery. At day 70, bone repair was completed for the two conditions and BMC and BMD were significantly higher in the defect loaded with the beads. Overall, these results indicate that mineralization was delayed in empty defects in comparison with the condition that received the material. This indicates that beads of pullulan-dextran-nHA are suitable for a non-critical size defect and that mineralization was favored in the presence of the material.

As shown by histology, invading cells penetrated the scaffold within the first 15 days (Figure 5 J, M). Although pore sizes of the material were not investigated in this study, the beads allowed an attachment of invading cells, as cells, including vWF and BMP-2 positive cells, were visible in direct contact with the material at all time points (Figure 5 M–O, Figure 6). At day 30, Masson's Trichrome stained sections (Figure 5 H) showed the presence of fibrous tissue that was not seen at any time point with the beads, allowing a better cellularisation of the defect. Similar results concerning tissue preservation and bone formation were observed with Bio-Oss® collagen (Geistlich, Wolhausen, Switzerland) that contains demineralized bovine bone, considered as the gold standard biomaterial for alveolar sockets implantations to promote bone formation and ridge preservation [55]. In that study, non-critical-size defects in the extraction sockets of dogs were either filled with the material or left empty. Although the material did not enhance new bone formation, the profile of the ridge was better preserved with the graft.

As seen by immune-staining with a macrophage marker, CD68 positive cells were mainly seen at the two earlier time points (Figure 6 G, H). This is due to the cascade of events that is involved in inflammatory and wound healing responses following biomaterial implantation [56]. Directly after the implantation, polynuclear cells, monocytes and lymphocytes invade the defect site. The presence of a biomaterial as a foreign body provokes the infiltration with macrophages that differentiate and fuse to giant cells like osteoclasts at the end of the acute phase of the inflammation. At day 70, those processes are finished and CD68 positive cells are therefore no longer present.

MRI acquisition indicated a decrease of aqueous content over time for the two conditions (Figure 3 G), probably due to an increasing mineralization of the defect site over time, as seen by Micro-CT analysis. Although indicated in the graph, a direct comparison of the two conditions should be avoided, since the offset for the aqueous volume was not comparable for the two conditions. We did not measure the volume immediately after implantation because the animals were not transportable due to ethical reasons.

When analyzing Bone mineral density (BMD) and Bone mineral content (BMC), either the full defect or an area of interest, the distinction between newly formed bone and the biomaterial itself was difficult, since most micro-CT software choose thresholds representing the mineralized tissue. In the case of a biomaterial made of calcium phosphate containing hydroxyapatite, thresholding methods also capture the scaffold material in addition to the surrounding bone, as the densities of the bead material overlaps the heterogeneous densities found in bone and forming bone. We therefore decided to perform a histomorphometric analysis based on decalcified histological sections stained for osteoid tissue whose overall percentage was analyzed for the surface of the defect. While BMD and BMC values indicated an overall trend when using CaP based biomaterials, the histomorphometric analysis allows a more precise view on cellular processes. The main interest in performing this study was the evaluation of the increasing mineralized tissue surrounding the beads. Increasing osteoid formation, as analyzed by histomorphometric measurements, confirmed the increasing mineralization determined by Micro-CT analysis. As mentioned for the interpretation of the MRI results, a direct comparison of the two conditions seemed difficult, because the offset is not the same. Therefore, two series of measurements were performed. When the osteoid formation was expressed as percentage of the defect surface, there was no significant difference between the two conditions. But when the surface of the beads was first subtracted from the defect surface, the percentage of osteoid formation was significantly higher in samples that received beads (Figure 7). This result shows that beads have strong osteoconductive properties. The increasing presence of forming osteoid around the beads also indicates that the overall mineralization within the defect was not only due to the increased mineralization of the biomaterial itself as was shown before in an ectopic site [36].

While dextran is already used as a biomaterial in the form of hydrogels [57,58] to deliver proteins and its synthetic derivatives can control the biological activity of growth factors [59], pullulan-based scaffolds have been evaluated more recently and studied as a support for smooth muscle cells [60]. Cylinders, disks or electrospun fibers of pullulan and dextran were evaluated *in vitro* as a composite matrix for endothelial progenitor cells [61], human dermal fibroblast [62], human bone marrow stromal cells [63], and human adipose derived stem cells [64]. The composition of the scaffold studied in the form of cylinders has shown abilities for mineralization and formation of a dense mineral collagenous tissue in both an ectopic site and a critical-sized orthotopic defect [36]. Here, we demonstrated the *in vivo* potential of pullulan-dextran-nHA beads for bone repair. It is known that materials based on polymers often need to be associated with cells or growth factors to achieve osteoconductive properties [65]. In this study, potentially injectable cell-free beads have been tested in a rat femoral medial-epicondyle defect of 3.5 mm diameter and 4 mm depth, and showed osteoconductive properties as shown by µCT, histology and histomorphometric analysis.

It is known that materials based on polymers often have inflammatory properties.

Regarding *in vivo* applications, the host response to the material is therefore an important parameter to look at [66,67]. Polymers have been studied for their inflammatory properties and i.e. PLA or chitosan are long known to enhance the functions of inflammatory cells such as macrophages [68,69], and recent studies focused on the evaluation of the influence of different acetylation degrees [70], as well as the 3D structure [71] on the inflammatory response to chitosan. Similar studies have been performed for hydroxyapatite, suggesting an influence of the material shape, size and sintering temperature on the inflammatory response [72]. The composite beads did not show an obvious inflammatory response, as investigated by MRI and histology (Figure 4 and 5), but CD68 positive cells were present as a consequence of the biomaterial implantation. A loss of aqueous volume for the two conditions was observed over time (Figure 4G). In case of a chronic inflammation, the aqueous loss would not decrease, but stay stable or even increase.

Although the present study demonstrated non-inflammatory and osteoconductive properties of the new injectable macroporous composite, improvements regarding its biodegradability, bone tissue invasion, and injectable properties are needed and ongoing. Since a cellularisation of the defect was visible and happened quite fast after implantation, it would be preferable to have beads of an injectable size that are resorbing over this time frame, to allow the replacement of the biomaterial itself by cells and extracellular matrix secreting cells. It is likely that the abundant vascularization observed inside the defect would allow a continuous infiltration of the site and a progressive replacement of the scaffold to achieve complete bone regeneration.

Conclusion

Bone formation and mineralization for the defects filled with these beads were increased in comparison to empty defects. Cellular infiltration of the defect was well organized in trabecular-like structures, suggesting the use of beads for bone repair of small size defects. The beads are of an injectable size and can therefore be used to fill-up any type of bone defect of variable size and should become useful for surgeons in clinical applications.

Acknowledgments

The authors would like to thank the French Research National Agency and the institutions of Inserm (National Institute for Health and Medical Research), University of Bordeaux, Universities of Paris 7 and Paris 13 for financial support.

Thanks to the *Laboratory Animal Care Facility* for housing and taking care of the rats. The help of Letitia Medan, Myriam Touati and their co-workers is highly acknowledged.

The authors also would like to thank *The Resource Center of Electron Microscopy and Microanalysis* (CREMEM) at the University of Bordeaux for BSEM observations.

Special thanks to Brian Christopher Dawson for proofreading the manuscript.

Author Contributions

Conceived and designed the experiments: SS SMD SM CLV DL JA JCF. Performed the experiments: SS SMD LM SM MR SC. Analyzed the data: SS SMD SM. Contributed reagents/materials/analysis tools: SS SMD SM SC. Contributed to the writing of the manuscript: SS SMD SM SC CLV DL JA JCF.

References

1. Seiler JG, Johnson J (2000) Iliac crest autogenous bone grafting: donor site complications. J South Orthop Assoc 9: 91–97.
2. LeGeros RZ (2002) Properties of osteoconductive biomaterials: calcium phosphates. Clin Orthop Relat Res 395: 81–98.
3. Stodolak-Zych E, Szumera M, Blazewicz M (2012) Osteoconductive Nano-composite Materials for Bone Regeneration. Materials Science Forum 730–732: 38–43.
4. Barradas AMC, Yuan H, van Blitterswijk CA, Habibovic P (2011) Osteoinductive biomaterials: current knowledge of properties, experimental models and biological mechanisms. Eur Cell Mater 21: 407–429.
5. Yuan H, Fernandes H, Habibovic P, de Boer J, Barradas AMC, et al. (2010) Osteoinductive ceramics as a synthetic alternative to autologous bone grafting. Proc Natl Acad Sci U S A 107: 13614–13619.
6. Habibovic P, de Groot K (2007) Osteoinductive biomaterials-properties and relevance in bone repair. J Regen Med Tissue Eng 1: 25–32.
7. Ji W, Sun Y, Yang F, van den Beucken JJJP, Fan M, et al. (2011) Bioactive electrospun scaffolds delivering growth factors and genes for tissue engineering applications. Pharm Res 28: 1259–1272.
8. Liu J, Xu L, Li Y, Ma J (2011) Temporally controlled multiple-gene delivery in scaffolds: A promising strategy to enhance bone regeneration. Med Hypotheses 76: 173–175.
9. Porter JR, Ruckh TT, Popat KC (2009) Bone tissue engineering: a review in bone biomimetics and drug delivery strategies. Biotechnol Prog 25: 1539–1560.
10. Zhao X, Kim J, Cezar CA, Huebsch N, Lee K, et al. (2011) Active scaffolds for on-demand drug and cell delivery. Proc Natl Acad Sci U S A 108: 67–72.
11. Catros S, Guillemot F, Nandakumar A, Ziane S, Moroni L, et al. (2012) Layer-by-layer tissue microfabrication supports cell proliferation in vitro and in vivo. Tissue Eng Part C Methods 18: 62–70.
12. Garg T, Singh O, Arora S, Murthy R (2012) Scaffold: a novel carrier for cell and drug delivery. Crit Rev Ther Drug Carrier Syst 29: 1–63.
13. Bose S, Tarafder S (2012) Calcium phosphate ceramic systems in growth factor and drug delivery for bone tissue engineering: a review. Acta Biomater. 8: 1401–1421.
14. Uchida A, Nade S, McCartney E, Ching W (1985) Bone ingrowth into three different porous ceramics implanted into the tibia of rats and rabbits. J Orthop Res3: 65–77.
15. Uchida A, Nade SM, McCartney ER, Ching W (1984) The use of ceramics for bone replacement. A comparative study of three different porous ceramics. J Bone Joint Surg Br. 66: 269–275.
16. Castellani C, Zanoni G, Tangl S, van Griensven M, Redl H (2009) Biphasic calcium phosphate ceramics in small bone defects: potential influence of carrier substances and bone marrow on bone regeneration. Clin Oral Implants Res 20: 1367–1374.
17. Garrido C, Lobo SE, Turíbio FM, Legeros RZ (2011) Biphasic calcium phosphate bioceramics for orthopaedic reconstructions: clinical outcomes. Int J Biomater 2011: 129727.
18. Fricain JC, Bareille R, Ulysse F, Dupuy B, Amedee J (1998) Evaluation of proliferation and protein expression of human bone marrow cells cultured on coral crystallized in the aragonite of calcite form. J Biomed Mater Res 42: 96–102.
19. Junker R, Dimakis A, Thoneick M, Jansen J (2009) Effects of implant surface coatings and composition on bone integration: a systematic review. Clin Oral Implants Res 20 Suppl 4: 185–206.
20. Oh, Lee, Lee JH, Kim, Jang, et al. (2012) Collagen three-dimensional hydrogel matrix carrying basic fibroblast growth factor for the cultivation of mesenchymal stem cells and osteogenic differentiation. Tissue Eng Part A 18: 1087–1100.
21. Ferreira AM, Gentile P, Chiono V, Ciardelli G (2012) Collagen for bone tissue regeneration. Acta Biomater 8: 3191–3200.
22. Chen F, Wu Z, Wang Q, Wu H, Zhang Y, et al. (2005) Preparation of recombinant human bone morphogenetic protein-2 loaded dextran-based microspheres and their characteristics. Acta Pharmacol Sin 26: 1093–1103.
23. Costa-Pinto AR, Reis RL, Neves NM (2011) Scaffolds based bone tissue engineering: the role of chitosan. Tissue Eng Part B Rev 17: 331–347.
24. Zhou H, Xu HHK (2011) The fast release of stem cells from alginate-fibrin microbeads in injectable scaffolds for bone tissue engineering. Biomaterials 32: 7503–7513.
25. Kretlow JD, Mikos AG (2007) Review: mineralization of synthetic polymer scaffolds for bone tissue engineering. Tissue Eng 13: 927–938.
26. Dhandayuthapani B, Yoshida Y, Maekawa T, Kumar DS (2011) Polymeric Scaffolds in Tissue Engineering Application: A Review. Int J Polym Sc 2011: 1–19.
27. Mallick KK, Cox SC (2013) Biomaterial scaffolds for tissue engineering. Front Biosci 5: 341–360.
28. Borden M, Attawia M, Khan Y, El-Amin SF, Laurencin CT (2004) Tissue-engineered bone formation in vivo using a novel sintered polymeric microsphere matrix. J Bone Joint Surg Br. 86: 1200–1208.
29. Dragoo JL, Choi JY, Lieberman JR, Huang J, Zuk P, et al. (2003) Bone induction by BMP-2 transduced stem cells derived from human fat. J Orthop Res 21: 622–629.
30. Cho MH, Kim KS, Ahn HH, Kim MS, Kim SH, et al. (2008) Chitosan gel as an in situ-forming scaffold for rat bone marrow mesenchymal stem cells in vivo. Tissue Eng Part A 14: 1099–1108.
31. Grellier M, Granja PL, Fricain JC, Bidarra SJ, Renard M, et al. (2009) The effect of the co-immobilization of human osteoprogenitors and endothelial cells within alginate microspheres on mineralization in a bone defect. Biomaterials 30: 3271–3278.
32. Kolambkar YM, Dupont KM, Boerckel JD, Huebsch N, Mooney DJ, et al. (2011) An alginate-based hybrid system for growth factor delivery in the functional repair of large bone defects. Biomaterials 32: 65–74.
33. Swetha M, Sahithi K, Moorthi A, Srinivasan N, Ramasamy K, et al. (2010) Biocomposites containing natural polymers and hydroxyapatite for bone tissue engineering. Int J Biol Macromol 47: 1–4.
34. Cao H, Kuboyama N (2010) A biodegradable porous composite scaffold of PGA/beta-TCP for bone tissue engineering. Bone 46: 386–395.
35. Winter GD, Simpson BJ (1969) Heterotopic bone formed in a synthetic sponge in the skin of young pigs. Nature 223: 88–90.
36. Fricain JC, Schlaubitz S, Le Visage C, Arnault I, Derkaoui SM, et al. (2013) A nano-hydroxyapatite–pullulan/dextran polysaccharide composite macroporous material for bone tissue engineering. Biomaterials 34: 2947–2959.
37. Zheng H, Bai Y, Shih, Hoffmann C, Peters F, et al. (2014) Effect of a β-TCP collagen composite bone substitute on healing of drilled bone voids in the distal femoral condyle of rabbits. J Biomed Mater Res B Appl Biomater 102: 376–383.
38. Kim, Ahn, Park MS, Lee, Choi CY, et al. (2007) A poly(lactide-co-glycolide)/hydroxyapatite composite scaffold with enhanced osteoconductivity. J Biomed Mater Res A 80: 206–215.
39. Barbieri D, Renard JS, de Bruijn JD, Yuan H (2010) Heterotopic bone formation by nano-apatite containing poly(D,L-lactide) composites. Eur Cell Mater 19: 252–261.
40. Tsai C, Chou M, Jonas M, Tien Y, Chi EY (2002) A composite graft material containing bone particles and collagen in osteoinduction in mouse. J Biomed Mater Res 63: 65–70.
41. Wu D, Hao A, Zhang C, Cui F, Wang X, et al. (2012) [Promoting of angiogenesis and osteogenesis in radial critical bone defect regions of rabbits with nano-hydroxyapatite/collagen/PLA scaffolds plus endothelial progenitor cells]. Zhonghua Yi Xue Za Zhi 92: 1630–1634.
42. Daculsi G, Uzel AP, Weiss P, Goyenvalle E, Aguado E (2010) Developments in injectable multiphasic biomaterials. The performance of microporous biphasic calcium phosphate granules and hydrogels. J Mater Sci Mater Med 21: 855–861.

43. Weiss P, Layrolle P, Clergeau LP, Enckel B, Pilet P, et al. (2007) The safety and efficacy of an injectable bone substitute in dental sockets demonstrated in a human clinical trial. Biomaterials 28: 3295–3305.
44. Matsuno T, Hashimoto Y, Adachi S, Omata K, Yoshitaka Y, et al. (2008) Preparation of injectable 3D-formed beta-tricalcium phosphate bead/alginate composite for bone tissue engineering. Dent Mater J 27: 827–834.
45. Chen Y, Huang Z, Li X, Li S, Zhou Z, et al. (2012) In Vitro Biocompatibility and Osteoblast Differentiation of an Injectable Chitosan/Nano-Hydroxyapatite/Collagen Scaffold. J Nanomater 2012: 1–6. doi:10.1155/2012/401084.
46. Zhao H, Ma L, Gao C, Wang J, Shen J (2009) Fabrication and properties of injectable β-tricalcium phosphate particles/fibrin gel composite scaffolds for bone tissue engineering. Mater Sci Eng C Mater Biol Appl 29: 836–842.
47. Catros S, Guillemot F, Lebraud E, Chanseau C, Perez S, et al. (2010) Physico-chemical and biological properties of a nano-hydroxyapatite powder synthesized at room temperature. IRBM 31: 226–233.
48. Afshar a, Ghorbani M, Ehsani N, Saeri M, Sorrell C (2003) Some important factors in the wet precipitation process of hydroxyapatite. Mater Des 24: 197–202.
49. Letourneur D, Le Visage C, Derkaoui SM (2012) Crosslinked Polysaccharide Beads and their biomedical Uses. Available: http://patentscope.wipo.int/search/en/detail.jsf?docId=WO2012028623&recNum=195&maxRec=19887&office=&prevFilter=&sortOption=&queryString=calcitonin*+or+parathyroid*+or+pth+or+thrombomodulin*+or+milnacipran*+or+naftopidil*+or+fusudil*+or+mizoribin*+or+bredinin*&tab=PCT+Biblio. Accessed: 2012 March 08.
50. Miraux S, Massot P, Ribot EJ, Franconi, Thiaudiere E (2008) 3D TrueFISP imaging of mouse brain at 4.7T and 9.4 T. J Magn Reson Imaging 28: 497–503.
51. Parfitt AM, Drezner MK, Glorieux FH, Kanis JA, Malluche H, et al. (1987) Bone histomorphometry: standardization of nomenclature, symbols, and units. Report of the ASBMR Histomorphometry Nomenclature Committee. J Bone Miner Res 2: 595–610.
52. Abed A, Assoul N, Ba M, Derkaoui SM, Portes P, et al. (2011) Influence of polysaccharide composition on the biocompatibility of pullulan/dextran-based hydrogels. J Biomed Mater Res A 96: 535–542.
53. Bongio M, van den Beucken JJJP, Leeuwenburgh SCG, Jansen JA (2012) Preclinical evaluation of injectable bone substitute materials. J Tissue Eng Regen Med doi: 10.1002/term.1637.
54. Monfoulet L, Rabier B, Chassande O, Fricain JC (2010) Drilled hole defects in mouse femur as models of intramembranous cortical and cancellous bone regeneration. Calcif Tissue Int 86: 72–81.
55. Araújo MG, Lindhe J (2009) Ridge preservation with the use of Bio-Oss collagen: A 6-month study in the dog. Clin Oral Implants Res 20: 433–440.
56. Anderson JM, Rodriguez A, Chang DT (2008) Foreign body reaction to biomaterials. Semin Immunol 20: 86–100.
57. Van Tomme SR, Hennink WE (2007) Biodegradable dextran hydrogels for protein delivery applications. Expert Rev Med Devices 4: 147–164.
58. Rekha MR, Sharma CP (2007) Pullulan as a Promising Biomaterial for Biomedical Applications: A Perspective. Trends in Biomater Artif Organs 20: 111–116.
59. Degat, Dubreucq G, Meunier A, Dahri-Correia L, Sedel L, et al. (2009) Enhancement of the biological activity of BMP-2 by synthetic dextran derivatives. J Biomed Mater Res A 88: 174–183.
60. Autissier A, Letourneur D, Le Visage C (2007) Pullulan-based hydrogel for smooth muscle cell culture. J Biomed Mater Res A 82: 336–342.
61. Thébaud, Pierron D, Bareille R, Le Visage C, Letourneur D, et al. (2007) Human endothelial progenitor cell attachment to polysaccharide-based hydrogels: a pre-requisite for vascular tissue engineering. J Mater Sci Mater Med 18: 339–345.
62. Shi L, Le Visage C, Chew SY (2011) Long-term stabilization of polysaccharide electrospun fibres by in situ cross-linking. J Biomater Sci Polym Ed 22: 1459–1472.
63. Guerrero J, Catros S, Derkaoui SM, Lalande C, Siadous R, et al. (2013) Cell interactions between human progenitor-derived endothelial cells and human mesenchymal stem cells in a three-dimensional macroporous polysaccharide-based scaffold promote osteogenesis. Acta biomater 9: 8200–8213.
64. Lalande C, Miraux S, Derkaoui SM, Mornet S, Bareille R, et al. (2011) Magnetic resonance imaging tracking of human adipose derived stromal cells within three-dimensional scaffolds for bone tissue engineering. Eur Cell Mater 21: 341–354.
65. Janicki P, Schmidmaier G (2011) What should be the characteristics of the ideal bone graft substitute? Combining scaffolds with growth factors and/or stem cells. Injury 42 Suppl 2: S77–81.
66. Franz S, Rammelt S, Scharnweber D, Simon JC (2011) Immune responses to implants - a review of the implications for the design of immunomodulatory biomaterials. Biomaterials 32: 6692–6709.
67. Babensee J, Anderson J, McIntire L, Mikos A (1998) Host response to tissue engineered devices. Adv Drug Deliv Rev 33: 111–139.
68. Porporatto C, Bianco ID, Riera CM, Correa SG (2003) Chitosan induces different l-arginine metabolic pathways in resting and inflammatory macrophages. Biochem Biophys Res Commun 304: 266–272.
69. Vasconcelos DP, Fonseca AC, Costa M, Amaral IF, Barbosa M, et al. (2013) Macrophage polarization following chitosan implantation. Biomaterials 34: 9952–9959.
70. Barbosa JN, Amaral IF, Aguas AP, Barbosa MA (2010) Evaluation of the effect of the degree of acetylation on the inflammatory response to 3D porous chitosan scaffolds. J Biomed Mater Res A 93: 20–28.
71. Almeida CR, Serra T, Oliveira MI, Planell J, Barbosa M, et al. (2014) Impact of 3-D printed PLA- and chitosan-based scaffolds on human monocyte/macrophage responses: Unraveling the effect of 3-D structures on inflammation. Acta biomater 10: 613–622.
72. Laquerriere P, Grandjean-Laquerriere A, Addadi-Rebbah S, Jallot E, Laurent-Maquin D, et al. (2004) MMP-2, MMP-9 and their inhibitors TIMP-2 and TIMP-1 production by human monocytes in vitro in the presence of different forms of hydroxyapatite particles. Biomaterials 25: 2515–2524.

21

Geometry-Driven Cell Organization Determines Tissue Growths in Scaffold Pores: Consequences for Fibronectin Organization

Pascal Joly[1,2,4]*, Georg N. Duda[1,2,3,4], Martin Schöne[1,2,4], Petra B. Welzel[5,6], Uwe Freudenberg[5,6], Carsten Werner[5,6], Ansgar Petersen[1,3,4]

1 Julius Wolff Institute, Charité - Universitätsmedizin Berlin, Berlin, Germany, **2** Berlin-Brandenburg School for Regenerative Therapies, Charité - Universitätsmedizin Berlin, Berlin, Germany, **3** Berlin-Brandenburg Center for Regenerative Therapies, Charité - Universitätsmedizin Berlin, Berlin, Germany, **4** Center for Musculoskeletal Surgery, Charité - Universitätsmedizin Berlin, Berlin, Germany, **5** Max Bergmann Center of Biomaterials Dresden, Leibniz Institute of Polymer Research Dresden, Dresden, Germany, **6** Center for Regenerative Therapies Dresden, Technische Universität Dresden, Dresden, Germany

Abstract

To heal tissue defects, cells have to bridge gaps and generate new extracellular matrix (ECM). Macroporous scaffolds are frequently used to support the process of defect filling and thus foster tissue regeneration. Such biomaterials contain micro-voids (pores) that the cells fill with their own ECM over time. There is only limited knowledge on how pore geometry influences cell organization and matrix production, even though it is highly relevant for scaffold design. This study hypothesized that 1) a simple geometric description predicts cellular organization during pore filling at the cell level and that 2) pore closure results in a reorganization of ECM. Scaffolds with a broad distribution of pore sizes (macroporous starPEG-heparin cryogel) were used as a model system and seeded with primary fibroblasts. The strategies of cells to fill pores could be explained by a simple geometrical model considering cells as tensioned chords. The model matched qualitatively as well as quantitatively by means of cell number vs. open cross-sectional area for all pore sizes. The correlation between ECM location and cell position was higher when the pores were not filled with tissue (Pearson's coefficient $\rho = 0.45 \pm 0.01$) and reduced once the pores were closed ($\rho = 0.26 \pm 0.04$) indicating a reorganization of the cell/ECM network. Scaffold pore size directed the time required for pore closure and furthermore impacted the organization of the fibronectin matrix. Understanding how cells fill micro-voids will help to design biomaterial scaffolds that support the endogenous healing process and thus allow a fast filling of tissue defects.

Editor: Christophe Egles, Université de Technologie de Compiègne, France

Funding: This work has been funded by the Berlin-Brandenburg Center for Regenerative Therapies (BCRT) and the Berlin-Brandenburg School for Regenerative Therapies (BSRT). The funders had no role in study design, data collection and analysis, decision to publish, or preparation of the manuscript.

Competing Interests: The authors have declared that no competing interests exist.

* E-mail: pascal.joly@charite.de

Introduction

During the last 60 years, the design of biomaterials has evolved from bioinert (first generation) to bioactive and biodegradable materials (second generation) and finally to materials that will stimulate a cell specific response at the molecular level (third generation) [1,2]. This last generation aims at stimulating the cells, supporting them in their task to regenerate and at enabling them to accomplish healing in a fast and effective way. From endogenous mechanisms of bone healing, a tissue with the capacity to completely restore its function and structure, it is known that fibroblasts are among the first cells to enter the regenerative scenery. For healing of e.g. bone defects, matrix formation by fibroblasts paves the way for more complex tissue formation and maturation [3]. In order to form an organized extracellular matrix structure, it is necessary for the cells to cooperate and communicate by using a combination of their "sense of smell" (biochemical cues) but also their "sense of touch" (mechanical cues) [4]. 2D environments lack topological information present in the initial ECM tissue samples and therefore limit the cell possibility for 3D matrix adhesion, more specifically, spatial distribution of receptors anchored to the ECM [5,6]. Indeed, in 2D and 3D environments, cells will exhibit differences in terms of morphology, migration and signaling with their behavior strongly relating to the kind of 3D conditions provided [7]. Therefore, in the case of in-vitro experiments, three-dimensional matrices are recognized to be closer to physiological conditions. In scaffold design, substrate properties such as hydrophilicity have been shown to correlate with the capacity of the cells to adhere and to spread on pore walls [8]. Stiffness [9], pore size [10] as well as geometry [11] have also been gaining more attention lately as they have been shown to influence cell behavior. Local geometry was shown to have an impact on cell growth and viability by controlling the ability of the cell to expand [12] as well as a different number of cell functions [13]. However, larger scale geometry has also been observed to have an influence on tissue growth [11].

Macroporous scaffolds have the advantage of an enhanced nutrient supply to the cells and removal of metabolic waste out of the scaffold [14] due to convective transport through the pores.

They also promote cell migration when compared to nanoporous hydrogels. Hypoxia, nutrient transport as well as cell spreading are constant issues in the latter as the cells not only have to actively degrade their environment before they are able to migrate or spread to their full size but diffusion is hindered due to their nanoporous structure.

At the early stage of dermal wound healing, the existing matrix is mainly composed of fibronectin, type I and type III collagen as well as fibrin/fibrinogen and proteoglycans [3,15]. Early fibronectin deposition is essential as polymerization of type I and type III collagen depends on fibronectin [16] which will then further modulate the cell response to its environment [17]. Inhibition of fibronectin polymerization has also been linked with an increase in levels of endocytosed collagen I [18]. The localization of fibronectin within a scaffold will affect the localization of the collagen deposition.

In this work, we hypothesized that a simple geometrical model adapted from Bidan et al. [11] could accurately predict the number of cells involved in the process of pore filling in a macroporous scaffold and that the event of pore closure would impact the ECM distribution. Macroporous starPEG-heparin cryogels provide a broad range of pore sizes. Such scaffolds, used as a model system, offer human primary dermal fibroblasts a broad spectrum of voids to be filled. Cell spreading and subsequently pore filling by proliferation and production of the extracellular matrix component fibronectin was investigated in a timeframe between one and seven days. Pore filling was analyzed by immunostainings and confocal/multiphoton microscopy together with a custom made Matlab based code to gain information about cell location and fibronectin production. Time dependent changes in the open cross-sectional area of the pores were then compared to the simple geometrical model of pore filling. The aim of this study was to understand cell organization inside macroporous scaffolds during pore filling as well as its impact on ECM (here fibronectin) distribution. Such an understanding is fundamental for the design of biomaterial scaffolds controlling tissue growth for regenerative purposes.

Materials and Methods

Ethics Statement

Primary dermal human fibroblasts were used for this study. Ethical approval was obtained from the ethics committee of Charité – Universitätsmedizin Berlin and donor's written informed consent was given.

Macroporous starPEG-heparin Cryogel Scaffolds

The fabrication of well defined starPEG-heparin cryogels with large interconnected macropores suitable for the ingrowth of mammalian cells was already described elsewhere [19]. Briefly, we combined the well-established network formation via chemical crosslinking (EDC/sulfo-NHS chemistry) of amino terminated starPEG and heparin [20,21,22] with the cryogelation technology [23].

The reaction mixture containing the gel precursors and the activation agents was prepared as previously reported in [19,22] and pipetted into the cavities of a 96-well plate. A molar ratio of starPEG to heparin of $\gamma = 3$ and the total precursor concentration (starPEG and heparin) of 11.6% (w/w) was used for the present study. After freezing at $-20°C$ overnight, the samples were lyophilized for 24 hours. The resulting dry cylindrical cryogel scaffolds were cut into discs, which were punched to the desired diameter by means of a punching tool (Hoffmann GmbH, Qualitätswerkzeuge, München, Germany). The scaffolds used in the present study were about 3 mm in height and 5 mm in diameter after swelling in PBS (phosphate buffered saline, pH 7.4).

Prior to seeding cells, cryogel discs were sterilized with UV-light for 1 hour. To improve cell adhesion, the starPEG-heparin hydrogel matrix was biofunctionalized with an RGD containing peptide (cyclo(Arg-Gly-Asp-D-Tyr-Lys), Peptides International, Louisville, KY, USA) as described in [19]. Briefly, heparin carboxylic acid groups of swollen cryogels were activated with s-NHS/EDC solution (25 mM s-NHS, 50 mM EDC in 67 mM phosphate buffer (pH 5)) for 60 min at 4°C followed by flushing in borate buffer (100 mM, pH 8.0, 4°C). Next, the scaffolds were incubated in RGD-solution (200 µg/ml; dissolved in 100 mM borate buffer, pH 8) for 3 hours at room temperature. Finally, all samples were washed in PBS three times.

StarPEG-heparin cryogels were analyzed for mechanical properties, swelling, and 3-D architecture as reported elsewhere [19].

Cells and Culture Medium

Primary dermal human fibroblasts isolated from skin samples were used at passage 4–6. Cells were expanded in culture medium consisting of Dulbecco's modified Eagle's medium (DMEM, Invitrogen) supplemented with 10 vol-% fetal bovine serum (# S 0115, Biochrom AG), 1 vol-% penicillin/streptomycin (# A 2213, Biochrom AG) and 1 vol-% nonessential amino acids (# K 0293, Biochrom AG) in a humid incubator with 5% CO_2. Additionally ascorbic acid in a final concentration of 1.36 mM was added to the medium for cell seeded constructs to enhance ECM formation. Indeed, ascorbic acid has been shown to influence dermal fibroblasts proliferation and increase in collagen secretion. Other studies using trabecular meshwork cells also suggest an effect of ascorbic acid on fibronectin and laminin [24,25].

Cell Seeding into Scaffolds

Cells were trypsinized and brought into suspension in a concentration of 7.5×10^3 cells/µl in culture medium. The high seeding concentration was chosen to drive cells into pore filling without a prolonged phase of proliferation. The latter might itself be influenced by pores size and thus make the starting situation inhomogeneous. The actual concentration was chosen to result in a cell density that is close to the saturation cell density on the pore wall determined in preliminary seeding experiments. Scaffolds were placed on sterilized filter papers for 20 seconds to remove excess PBS from the pores. Scaffolds were dipped into the cell suspension, which they immediately absorbed. They were subsequently transferred into a 12-well culture plate without additional medium. Cell seeded constructs were kept in the humid incubator for 60 min to allow cell adhesion. Cell constructs were subsequently washed in culture medium to remove unattached cells.

Immunostainings

At the end of each experiment, cell-seeded constructs were fixed in 4% paraformaldehyde (PFA) at 4°C overnight. Then, they were stained for actin filaments (Alexa fluor 488 Phalloidin, # A12379, Invitrogen), fibronectin (anti-fibronectin antibody, # ab23750, Abcam plc) and cell nuclei (DAPI, # D3571, Invitrogen). Briefly, for confocal microscopy, scaffolds were washed in PBS three times for 5 minutes. Permeabilization was then performed using 0.2% Triton-X100 in PBS at room temperature. After washing the scaffolds twice in PBS for 5 minutes, blocking in 1%BSA/PBS at room temperature for 30 minutes was realized. Scaffolds were then incubated overnight at 4°C with primary antibody for fibronectin in Dako antibody diluent (dilution 1:150, # S3022, Dako).

Subsequently, washing in 0.2% Triton-X100 in PBS at room temperature was realized four times for 5 min. Scaffolds were then incubated for one hour in an incubation chamber at room temperature in 2% normal serum Donkey and 1% BSA in TBS with the Cy3 conjugated Anti-Rabbit together with Phalloidin Alexa 488. Subsequently, three washing steps in 0.2% Triton-X100 in PBS for 5 minutes were performed, as well as one washing step in distilled water for 3 min. Scaffolds were then stained for cell nuclei by DAPI (1:1500) in distilled water for 5 min at room temperature. Two final washing steps were performed in distilled water for 3 minutes.

Imaging

All images were taken using a LEICA SP5 confocal microscope equipped with a Mai Tai HP multiphoton laser and a 25× water immersion objective. Emitted light from the samples was detected using two non descanned detectors (NDD) as well as internal photomultiplier detector (PMT) between 555 nm and 625 nm to detect Cy3 secondary antibody emission with a resolution of 512×512 pixels. For the 20 μm z-stack, 6 slices were recorded resulting in voxel dimensions of $z=4$ μm and $x=y=400$ to 700 nm depending on the size of the region of interest. A HCX IRAPO L 25.0×0.95 water objective was used having a numerical aperture of 0.95. Laser power and detectors parameters were kept constant throughout the imaging of the different pores. Twelve pores out of two scaffolds were analyzed for each time point.

Analysis of Cell Location, Maximal Spreading Length and Minimum Pore Wall to Wall Distance

Acquired images were analyzed using a Matlab based code (Supporting information S1) which allowed the loading of a 20 μm z-stack, contouring the pore for each z-stack picture, selection of nuclei position through the z-stack as well as marking the points of connection between cells and scaffold. Maximal spreading length $d_{cell,spread}$ of each cell was then extracted as well as closest distance between the wall of the pore and the cell nucleus $d_{nucleus\text{-}wall}$. Maximal spreading length mean value $d^*_{cell,spread}$ was then calculated. From the drawing of the individual pore contours, an elliptic fit was performed and wall to wall distance $d_{pore,min}$ was defined as the small axis of the elliptical fit. The distributions of $d_{cell,spread}$ and $d_{pore,min}$ were used to define different categories of pore sizes: small, medium and large.

Analysis of Actin and Fibronectin within Pores and their Correlation

From the elliptic fit that was obtained from each of the contours drawn, a 360 degrees profile plot of 200 points resolution of intensity between the pore wall and the center of the elliptic fit was obtained in steps of two degrees. For both green (actin) and red (fibronectin) channels, an average over the z-stack was plotted to get an indication of actin and fibronectin distribution in the pores. Also, to study the colocalization between actin and fibronectin, the Pearson correlation coefficient was extracted from the maximum projection of the z-stack after having drawn the pore contour using the software Volocity (PerkinElmer) to define the region of interest. Developed by Manders et al. in 1992, the Pearson correlation coefficient aims at measuring the strength of linear relationship between two fluorescent intensities images. It is based on the comparison between pixel intensity of the same coordinates of different channels and yield values between 1 (perfect correlation) and −1, with 0 representing a random distribution [26,27]. The statistical analysis performed to compare the different groups was done using a paired samples T-Test.

Open Cross-sectional Area of the Pore

The fibronectin channel from the z-stack was used to get an idea of how fast and effective the filling was for the different categories. The projections of the z-stack were used to calculate the remaining empty area within the pore. A ratio was then made between the space to be filled (i.e. remaining open cross-sectional area) and the initial cross-sectional area of the pores and plotted as a function of the number of cells within that pore for the different pore size categories and time.

Modeling of Pore Closure by Cell Proliferation

To model the pore filling behavior of the fibroblasts, the principle of the chord model [11] was used. Initially the pore was modeled as a 2D elliptical area, which was filled step-by-step by another cell. Each cell was modeled as a line attached at two points to the current border. The area framed by the cell and the border was assumed as filled by the cell and the line was set as new border. The different cell positions available and the corresponding covered areas were then calculated. From this spectrum of potentially covered areas a certain range was selected (changing the lower cut-off point) represented by a factor R between $R=0$ (full spectrum) and $R=1$ (only the maximum area of the spectrum). The area covered by the cell and the according cell position within this variation range were then picked randomly. The simulation was stopped once the cells had filled the pore area.

The input parameters for the simulation were measured from the experiments and were the initial area A, the average elliptical ratio Φ_{ellips}, and the range of cell spreading length $d^*_{cell,spread} \pm \sigma_{dcell,spread}$. The elliptical fit enabled the calculation of the elliptical ratio of each pore (Φ_{ellips} = length of long axis/length of short axis). For each step, $d_{cell,spread}$ was chosen randomly in the range of cell spreading length observed ($d^*_{cell,spread} \pm \sigma_{d\ cell,spread}$).

Results

Key Characteristics of the starPEG-heparin Cryogel Scaffolds

Subzero temperature treatment of the gel forming reaction mixtures and subsequent lyophilization resulted in macroporous starPEG-heparin scaffolds as previously reported in [19]: the spongy materials exhibit a porosity of about 90% and undergo rapid swelling in aqueous solution. The swollen bulk materials are soft (bulk Youngs modulus about 6 kPa [19]) but very tough as demonstrated by uniaxial compression experiments [19]. They consist of interconnected irregular macropores surrounded by rather low hydrated polymer regions (pore walls) of only about 10 μm in width. Due to their high polymer concentration (cryoconcentration effect [23]) the pore walls exhibit a microscale Youngs modulus of about 1500 kPa [Welzel/Friedrichs, in prep.]. For a more detailed discussion of the cryogel properties the reader is referred to [19] and [Welzel/Friedrichs, in prep].

Cell Maximum Spreading Length

The maximal spreading length of the cells within the scaffold as well as the minimum wall to wall distance of the pores $d_{pore,min}$ analyzed in the scaffold are shown in Figure 1. Fibroblasts had a maximal spreading length of $d_{cell,spread} = 73.6 \pm 22.7$ μm. The distributions of $d_{cell,spread}$ and $d_{pore,min}$ were used to define different categories of pore sizes: Small pores ($60 < d_{pore,min} < 120$ μm), where most cells measured had the possibility to spread entirely across the pores. Medium pores ($120 < d_{pore,min} < 180$ μm) where only a marginal number of cells (3.5%) were measured to be able to reach such a size. Large pores were defined by $d_{pore,min} > 180$ μm where no cell was observed to cross the pores at day 1.

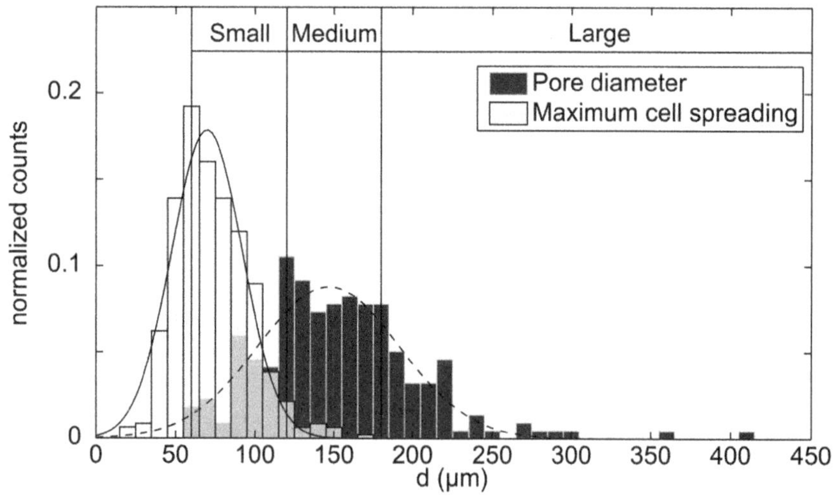

Figure 1. Maximal spreading length of fibroblasts and pore wall to wall distance within StarPEG-heparin cryogel. Maximal spreading length $d_{cell,spread}$ of the different human dermal fibroblasts are represented by white bars and wall to wall distance $d_{pore,min}$ of the pore analyzed in the scaffold by dark grey bars. The overlapping area is displayed in light grey. Gaussian fit was performed for maximum cell spreading (continuous line) and pore diameter (dashed line) distribution. The data was obtained by combining all the imaging performed on the starPEG-heparin cryogel (N_{pores} >100) and cells contained in pores (N_{cells} >200)."

Cell Spreading Behavior Inside Pores and Pore Filling

Figure 2 provides an insight into the general behavior of the pore filling process by human fibroblasts depending on incubation time and the scaffold pore size. As mentioned above, three categories were distinguished: small, medium and large pores. After 24 hours, human fibroblasts were fully spread on the scaffold walls. After 3 days, fibroblasts were observed to spread across the pores. Full pore filling was not observed in the medium and large pore size categories at day 3. After 7 days, all pores were closed and completely filled with cells.

The quantification of $d_{nucleus\text{-}wall}$ is shown in Figure 3a. The vertical line at 20 μm indicates an arbitrary limit corresponding to cells located primarily on the wall of the scaffold pores. This limit was chosen due to the fact that, at day 1, almost all cells were located on the wall and data analysis of nuclei position gave values of $d_{nucleus\text{-}wall}$ <20 μm. Figure 3b displays how the ratio of the cells having crossed the 20 μm limit changed over time in the three categories. As seen in the histology results, this quantitative data indicates that most cells were located on the walls of the pores at day 1. After 3 days, the cells were able to migrate or stretch further away from the wall and spread across the pores in the small category size. Interestingly, even if the cells have the possibility to fully cross over the pore in the small category, most of the cells were observed to be on the wall. Between 3 and 7 days, for the smaller pore, even though the total cell number was still increased, the ratio of cells more than 20 μm away from the wall divided by the total cell number showed a saturation behavior at a value of $\#cells_{in\_pore}/\#cells_{total} \approx 0.3$. For pores of medium and large category, $\#cells_{in\_pore}/\#cells_{total}$ increased strongly from day 3 to day 7 from 0.0 up to 0.4 and 0.5 respectively indicating a delayed pore closure compared to the small pores.

Pore Remaining Area

The simulation of the filling process illustrates the dependency of the open cross-sectional area on the total number of cells and gave an indication of how many cells were required to fill the pore to a certain degree (Fig. 4).

For the modeling, the initial mean cross-sectional area of the pores (area = 10.3, 22.5 and 44.9 × 10^3 μm^2 for small, medium and large categories, respectively) and $d^*_{cell,spread}$ served as input parameters (both parameters were measured from experiments). The pore cross-sectional geometry was simplified to an elliptic shape taking into account the average elliptic ratios Φ_{ellips} = length of long axis/length of short axis for the different pore size categories (Φ_{ellips} = 1.31, 1.29 and 1.47 for small, medium and large categories, respectively). A variety of simulations on the same geometry with different values for the random factor R were performed. Comparison between experimental data and simulation suggested a trend towards cells being positioned to cover a large area rather than taking a random position.

Fibronectin Formation Inside Scaffold Pores

Figure 5 shows fibronectin staining additional to the DAPI and actin staining from figure 2 to illustrate how the early ECM is deposited in the scaffold pores as a function of time and pore size category. At day 1, fibronectin was observed only at discrete locations on the walls of the pores. No individual fibers and no network formation could be identified. After 3 days, fibronectin had been deposited by the cells in a fibrillar form and the comparison with the actin cytoskeleton location indicates that its distribution followed cellular organization: fibers of fibronectin were observed across the pores only in the small category size. After 3 days, a higher amount of fibronectin fibrils were visible. This was potentially due to an additional recruitment of fibronectin by the cells through FN - integrin interaction and subsequent conformational changes during polymerization as reported in the literature [28]. After 7 days, fibrillar fibronectin was observed to fill pores of all size categories and was visible across the pores as well as on the pore walls forming a dense network.

To quantify cell and ECM distribution the average channel intensity for the green channel (actin) and the red channel (fibronectin) were analyzed as shown in Figure 6a. An overlayed intensity profile was plotted representing the radial distribution from the pore wall to the center of the pore (for details see method

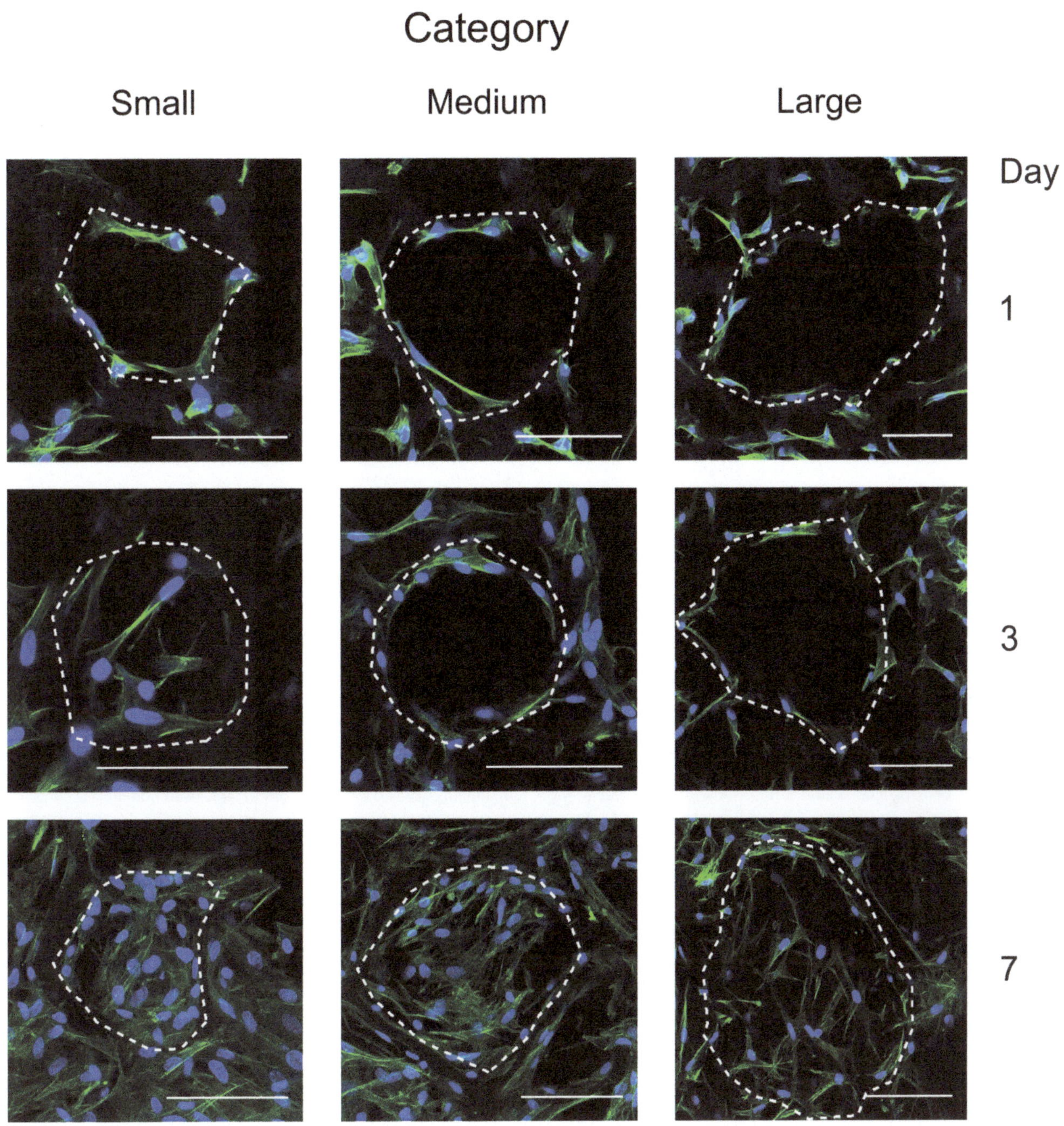

Figure 2. Filling of scaffold pores by fibroblasts for three pore size categories and three time points. Confocal microscopy for visualization of actin cytoskeleton (green) and cell nuclei (blue) to illustrate the filling of scaffold pores by human fibroblasts at day 1, 3, and 7 for the three pore size categories (scale bar equals 100 µm). Data shown corresponds to the maximum projection images of the 3D image stacks recorded inside scaffolds. During the imaging, scaffold pores could be identified due to background staining of the scaffold material and are highlighted by dashed lines.

section). At day 1, a high peak was observed in the close vicinity of the walls for all pore size categories indicating that actin and fibronectin, if any, were primarily located on the surface of the pores. Only for small pores an additional peak was observed closer to the pore center. In this case, cell crossing over the pores had already started in some pores accompanied by fibronectin deposition.

After 3 days, for all categories, the main peak for actin remained located next to the pore wall, indicating that most of the cells were still on the pore wall. For the small category pore size, the peak for actin and fibronectin at the pore center had grown strongly: the pore was being filled with cells and ECM. For the medium and large size category, actin and fibronectin were still mainly located next to the wall.

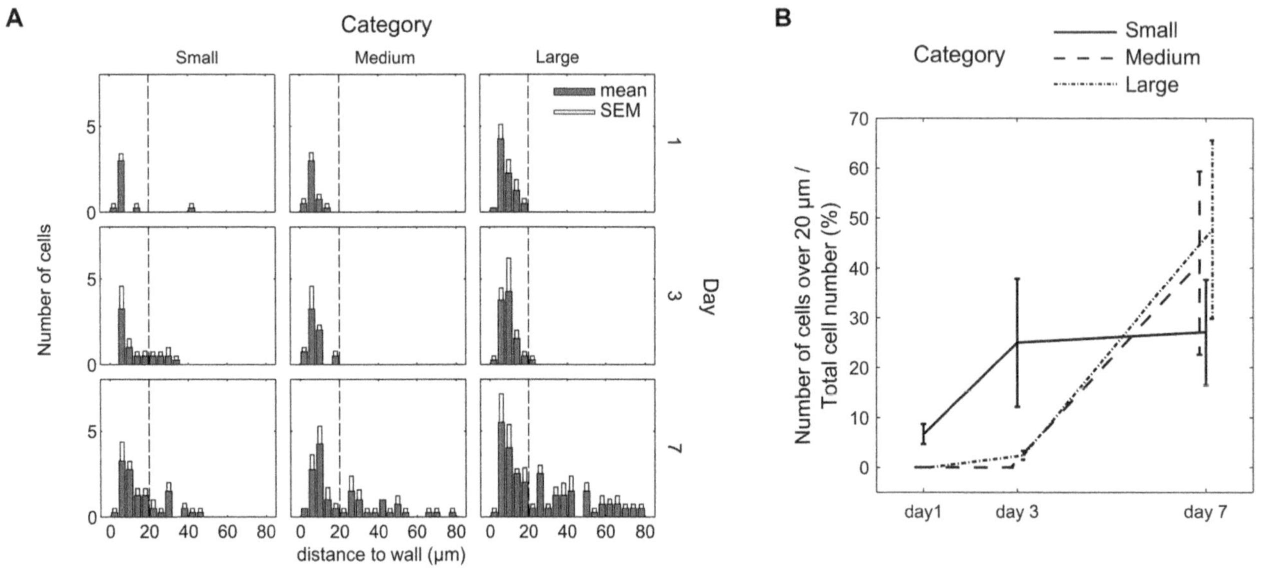

Figure 3. Quantification of cell number during pore filling. Evolution of cell position (cell nuclei) as a function of time and pore size for human fibroblasts (mean values of $n_{pores} = 4$ for each category, SEM = standard error of the mean) (a), the derived ratio between the number of cells that are located within the pores (#$cells_{in\_pore}$), and the total number of cells (#$cells_{total}$) in the pore (b).

The intensity distribution for day 7 indicated an almost homogeneous distribution of cellular actin inside the pores. This was confirmed by the histological images (Figure 2). For fibronectin, the curves showed higher intensity values in the middle of the pores indicating that a denser network had formed in the center of the pores. To further analyze this observation, the colocalization between actin and fibronectin was quantified by calculating the Pearson coefficient. The results for each category and points in time are shown in Fig. 6b. Values were observed to be comparable for the different categories at day 1. At day 3, a drop in correlation value was observed only in the small category, coinciding with the filling of the pore by fibronectin and actin, indicating that cells did not stay located to the place where they produced the fibronectin. This drop was not seen for medium and large categories at day 3. However, a clear drop in the Pearson coefficient was observed at day 7 for both categories. This is

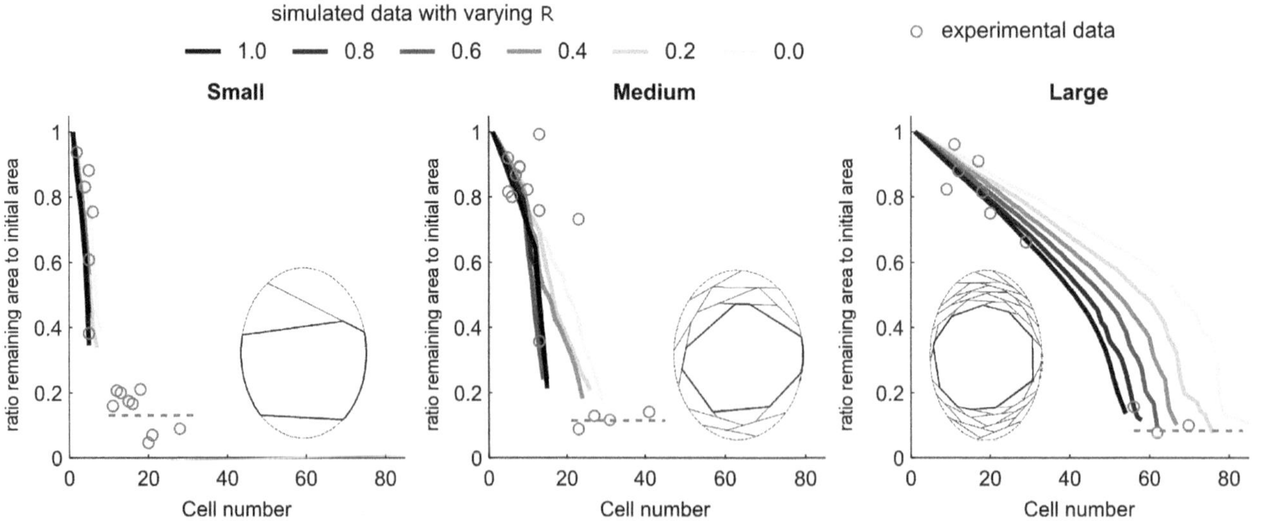

Figure 4. Remaining open cross-sectional area as a function of the number of cells. The ratio between the space to be filled (i.e. remaining open cross-sectional area) and the initial cross-sectional area of the pores was plotted as a function of the number of cells within that pore for the different categories. Experimental data is shown as circles for the different categories. Simulation data is displayed as lines with different shadings from black to light grey for different variation ranges for the covered area (R = 1: new cells positioned at locations of largest covered area per cell; R = 0: random position of new cells on the remaining inner pore surface). Simulation was based on a 2D model. Experimental data points were extracted from z-stack maximum projection for 2D analysis. Lines represent the average of ten simulations. Simulation stopped when the remaining inner diameter of the pore was smaller than the cell size. Within each graph the cell organization for a remaining area between 0.5 and 0.6 of the initial pore area according to the simulation is given. Dashed lines represent pore closure extracted from experimental data. Due to the porosity of the fibronectin network the remaining open area did not reach a zero value.

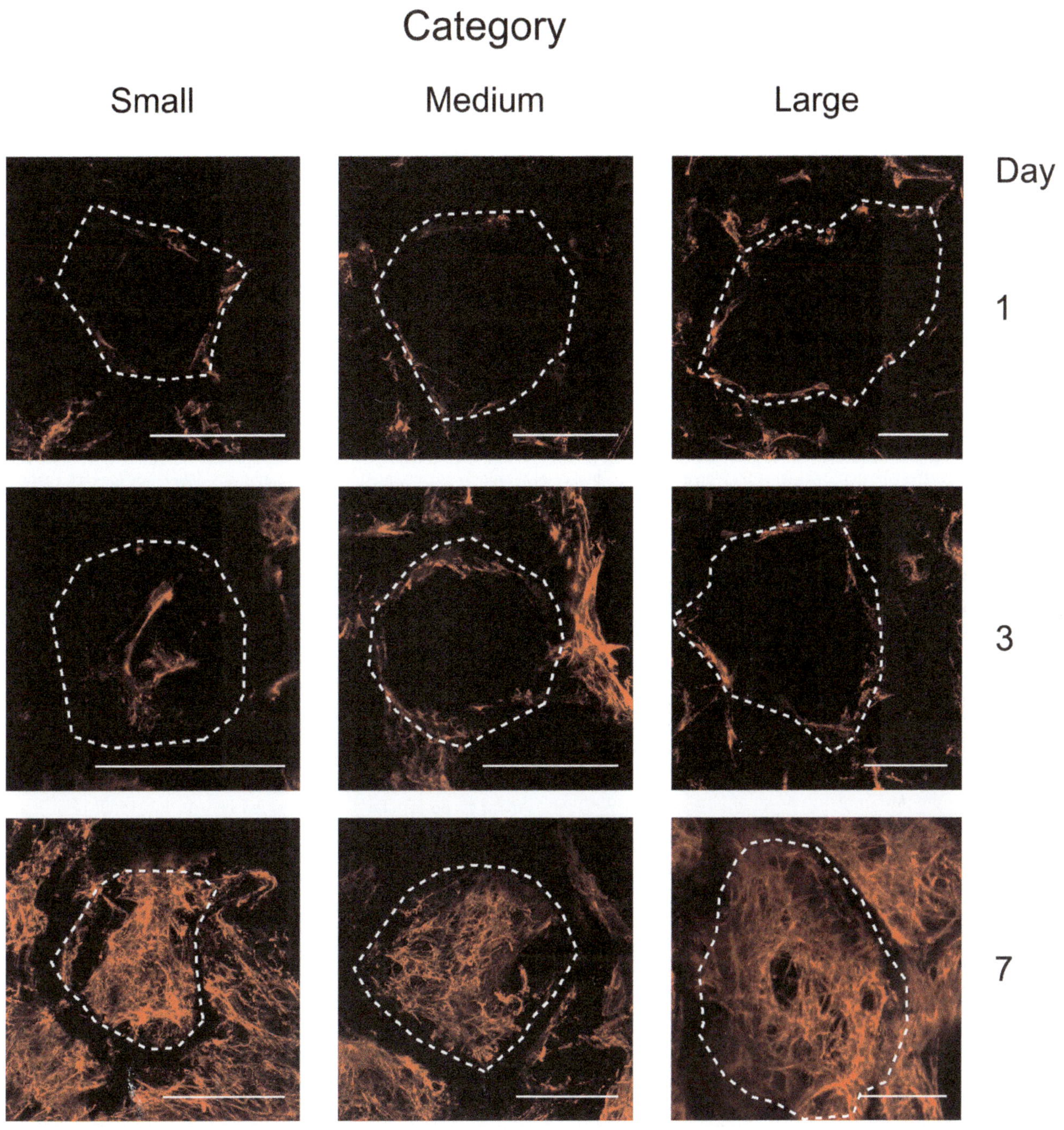

Figure 5. Fibronectin organization for three pore size categories and three time points. Confocal microscopy for visualization of fibronectin network (red) to illustrate the filling of scaffold pores until day 1, 3, and 7 and the three pore size categories (scale bar equals 100 μm). Data shown corresponds to the maximum projection images of the 3D image stacks recorded inside scaffolds. During the imaging, scaffold pores were observed due to some background staining and where delimited here by dashed lines.

interesting since it is the point in time when the pores had been filled with cells and fibronectin comparable to the situation at day 3 for the small pore category. By putting together those groups i.e. group A where the filling of the pore has not occurred (i.e. day 1 for all categories and day 3 for medium and large) and group B where the pore filling has happened (i.e. day 3 for small pores and day 7 for all categories) a significant difference ($p<0.05$) was observed between $\text{PearsonCoeff}_{\text{group A}} = 0.45 \pm 0.01$ and $\text{PearsonCoeff}_{\text{group B}} = 0.26 \pm 0.04$. Fibronectin and actin are only located on the pore wall when the pore is not yet filled. However, once the pore is filled (group B), although fibronectin still seem to be present in the pore wall vicinity, it appears that a change in distribution has occurred, as more fibronectin was observed in the middle of the pores. This process is summarized in figure 7.

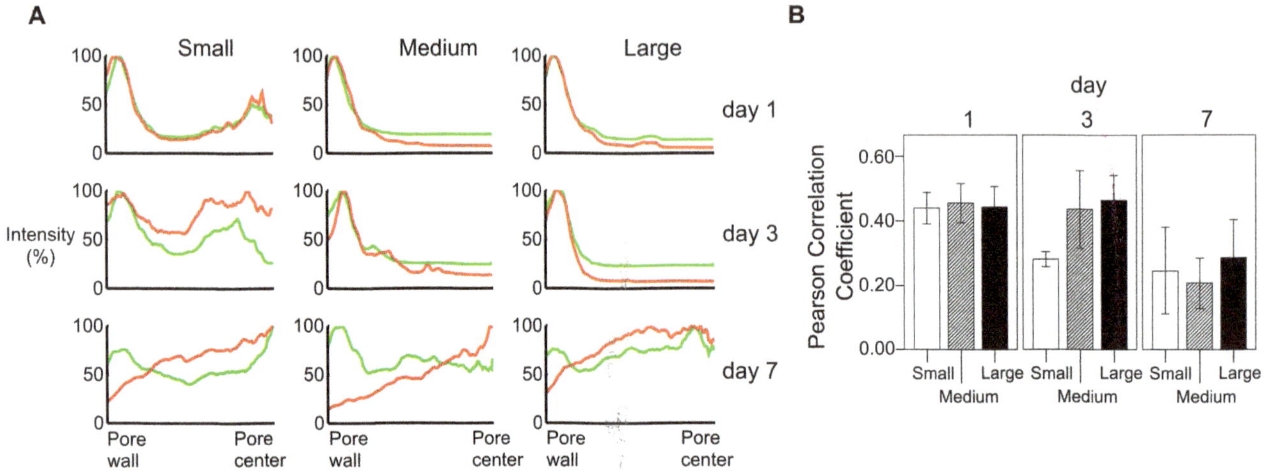

Figure 6. Distribution of fibronectin and actin inside scaffold pores. Actin (green) and fibronectin (red) intensity distribution from the pore center to the pore wall as extracted from histological images (Fig. 6a, $n_{pores}=4$ for each category). From the maximum projection of the z-stack and the contoured pore, the Pearson coefficient for green and red channel colocalization was calculated. The whole intensity data range for each channel (0–255) was considered here (Fig. 6b, n = 5 for each category).

Discussion

The goals of this study were to 1) qualitatively and quantitatively assess the strategy of cells to fill pores and 2) observe the variation in ECM localization and density before and after pore closure.

Macroporous scaffold pore filling proved to be a gradual process driven by the scaffold pore size. Initially, spreading of the cells on the walls was observed and correlated with the presence of fibronectin. When the pore size was in the range of maximal cell spreading length (73.6±22.7 μm), fibroblasts had the possibility of spreading across the pore, which resulted in fibronectin and actin being present already in the middle of some pores after 3 days (Fig. 6a). Assuming cells to behave like tensioned chords allowed to quantitatively describe the process of scaffold pore filling. This model implies that pore filling accelerates towards the end as a consequence of higher curvature in the inner remaining pore diameter (space that was not yet filled by cells) and an accordingly larger area covered by each cell (Fig. 4). The input parameters needed for a simulation of pore closure are only the initial pore geometry and the mean spreading length of the cells and can be measured experimentally. 20-μm stacks were chosen in order to visualize a full interconnected network of cells. Furthermore, going for higher z-values would result in an overlap of cells from different layers when performing the maximum projection. Since the experimental data was compared to a 2D simulation it was important to limit z to a single cell layer Additionally, loss of fibronectin intensity was quickly observed when going further into the pores. Finally, a small z-stack reduced variations in pore shape through the different images of the z-stack.

The geometrical model to simulate scaffold pore filling was based on a chord model developed to understand geometry dependent tissue growth at larger length scales [11]. Cell organization was considered from a geometrical point of view only. For a simulation of the complex cell-cell and cell-matrix interaction cellular mechanical properties like stiffness and traction force generation would have to be taken into account as done by other authors [29]. Realistic models for cell growth would additionally have to consider many parameters such as cell migration, nutrient transport, cell proliferation, and matrix deposition [30–32]. Here however, we were interested in the relation between the cell number and the according degree of pore closure. Thus, cell number was the only relevant variable and cell proliferation or nutrient transport issues did not have to be taken into account. Cells were considered as static and, aside from geometry, no interaction between cells was considered. The good agreement between the model and the experimental observations for all pore size categories suggests that the organization of fibroblasts as a consequence of a complex process is defined by simple geometrical constraints. Fibroblasts efficiently fill pores by advancing as far into the open space as their spreading size enables them to do (similar to a tensioned chord). For small pore sizes, spreading across the pores leads to a fast closure. For larger pores, the model predicted an initially slow decrease of the open pore area and an accelerated growth towards the end as a consequence of smaller remaining pore diameter and accordingly higher curvatures. The transition from the process of layer on layer growth to a phase where the cell are able to spread over the remaining pore might have consequences on the ECM structure as a consequence of altered cell arrangement and force distribution in the cellular network.

The model also allowed the introduction of a random factor R considering that the cells might locate randomly or choose positions where they can cover an area as large as possible. In order to further gain insight into the strategy of the cell behavior, a well defined void geometry would be useful. Using the biomaterial, each pore analyzed had indeed a different geometry and initial area to be filled. However, qualitatively, for larger pore, the data suggests that the cells preferentially selected locations of high curvature inside the remaining pore. This could be due to either curvature-dependent migration or proliferation [33–34]. Live cell imaging inside the pores would help to further elucidate this aspect. It should be underlined at this point that the model stops when the pore diameter decreased below the cell spreading size. Experimentally, at this final stage of pore filling, cells were observed to behave differently. They adopted a different shape with multiple spatially separated connection points rather than having the typical elongated spindle like appearance as they could not spread anymore to their maximal size.

Fibronectin is one of the early components of the ECM being synthesized and will strongly interact with and regulate the

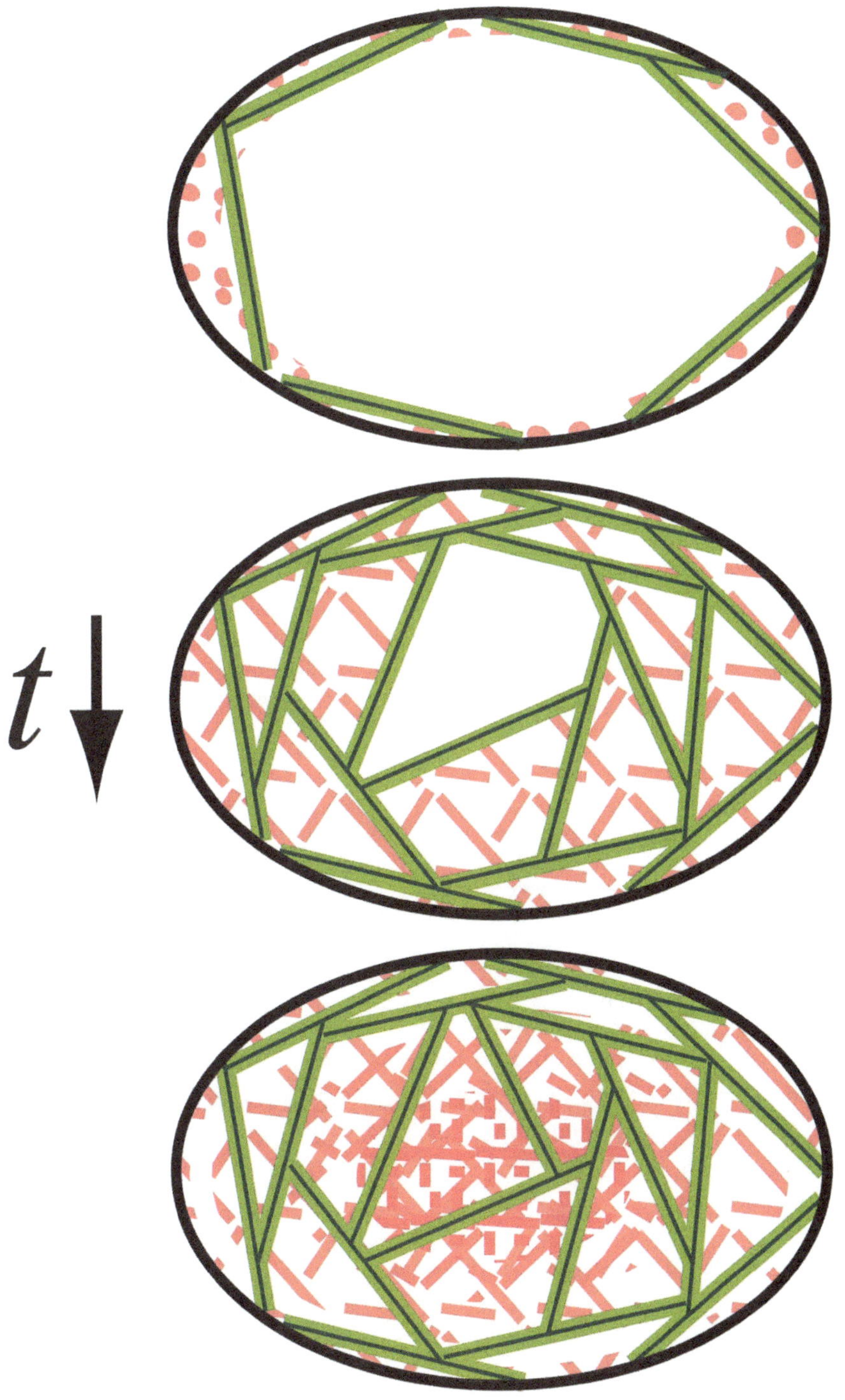

Figure 7. Schematic representation of pore filling as a function of time. Cells are represented by tensioned chords in green and fibronectin is in red. Initially cells are spreading on the pore wall and fibronectin is being produced but not visible as a connected network yet by the cells (top). The pore is subsequently filled by cells with time according to the model and produce fibronectin fibers being unfolded due to cell forces (middle). Finally the pore closes completely and at the same time fibronectin is further compacted towards the center (bottom).

formation and degradation of collagen type I and type III [16–18]. Collagen is the most abundant protein in the human body and one of the main components of the ECM structure that, together with elastin, is responsible for the structural framework and mechanical properties of the tissue being formed. Due to the way fibronectin interacts with collagen, the individual mechanical properties of the tissue formed inside the pore will be strongly influenced by the initial fibronectin deposition. It was shown that cells mechanically feel their environment and react to local mechanical stimuli like substrate stiffness and to extrinsic mechanical loading signals [35]. Concerning our study, the behavior of the cells might be potentially affected by the fact that inside the pores, where new tissue has been formed, cells sense a different mechanical environment from what they have been sensed initially when spreading on the walls [36–37]. Furthermore the increase in fibronectin network density present in the middle of the pore as suggested by the data in Figure 6a might be a result of ECM compaction due to cellular traction forces. The more tissue forms inside the pore, the higher are the forces on the interface layer with the comparably stiff biomaterial. This might lead to a fibronectin/ECM transport from the periphery to the center of the pore. One has however to be careful as even though the samples were stained in the same way and pictures were taken with the same microscopic settings, antibody diffusion might have been affected by the dense cell and ECM structures inside the pores influencing signal intensity distribution inside the pores. Surprisingly, even though we observed fibronectin to be present everywhere inside the pores of the scaffold after 7 days most of the cells nuclei remain located next to the wall of the scaffold. Further proliferation might have been triggered by changes in local fibronectin density as fibronectin is being pulled from the wall towards the center of the pores. The whole process, from initial spreading of the cells on the pore wall to pore closure and subsequent compaction of the fibronectin network is summarized by a schematic representation in Fig. 7.

For the current study, primary dermal fibroblasts were used as they are known to be key players in the early phase of wound or bone healing [38–39]. Their target is to synthesize a specific environment, composed mainly of fibronectin, collagen and elastin that provide first manifestations of structure and mechanical integrity. That first composition is the basis for further tissue maturation and specification by other cells such as MSCs to finally give specific structural properties and mechanical stability to the tissue.

The model presented here has the potential to be adapted to any kind of initial geometry and could be used to predict pore filling for the purpose of scaffold design [40]. Fast tissue growth inside areas of small pores (pores with diameter ≈ cell spreading length) providing early mechanical stabilization could be combined with slow growth inside supply channels (larger pores) that might be developed further into vascular structures. Chiu et al. reported that porous PEG hydrogels could support vascularization, depending also on the pore size [41]. This might have a significant impact in the design of biomaterials [42] to overcome the lack of sufficient vascularization in void healing. A bicomponent macroporous scaffold might be an advantageous solution as suggested by Feng et al. using porous bioceramics [43].

This study was performed in-vitro and under specific conditions of cell and scaffold choice. Consequently, several limitations have to be taken into account and could be the base for further experiments. First of all, tissue deposition is not only realized by fibroblasts. Mesenchymal stromal cells and tissue specific cells (osteoblasts, myoblasts, chondroblasts) are also known to create and remodel their environment. For example MSCs have been shown to stimulate wound healing [44]. Nonetheless, this study is justified by considering that the ratio of MSCs to fibroblasts is quite small and early tissue formation is assumed to be governed by fibroblasts. Also, the initial model and experiments by Bidan et al. were realized using MC3T3-E1 pre-osteoblasts, indicating that this model is not limited to fibroblasts [11]. A second point is that having performed the study under static conditions, the possibility of an insufficient nutrient and oxygen supply might have an impact on the way the pores are being filled even though the scaffold was macroporous [45]. This issue could be addressed by the usage of a bioreactor system to provide a better nutrient supply and compare it to the static case studied here [35]. Another point is that the study was only done using a single type of biohybrid scaffold. The resulting situation might be completely different when using scaffolds having a different surface chemistry, topography or pore architecture. Nevertheless we assume that the general geometric description of pore filling presented here is mostly governed by pore geometry and thus valid for different materials. A lot of evidence has accumulated showing that at least for large pored systems, the generalized chord model and curvature driven growth approach is valid [14,46]. A premise however is a sufficient mechanical stability of the pore against deformation through cell forces and a proper cell attachment to the biomaterial surface.

Conclusion

We used here a simple geometric description of the cell organization in scaffold pores that could correctly describe experimental observations of scaffold pore filling by fibroblasts. The model was originally developed to predict tissue growth at a larger scale. We showed here that the number of chords predicted by the adapted model agreed with the number of cells required to fill a scaffold pore. This knowledge might help to understand and finally predict tissue growth in other macroporous biomaterial structures. Only the spreading length of the cell and the initial pore geometry needed to be extracted experimentally to be able to predict void filling. Fibronectin as one component of the extracellular matrix was observed to follow cell organization until pore closure was reached. Once closed, reorganization of fibronectin inside the pore and a compaction of material in the pore center were observed. This might impact subsequent collagen network formation of the tissue inside the pores. These findings allow a better understanding of cell driven pore closure and ECM organization, which is needed for the optimized design of biomaterial scaffold pores for regenerative purposes.

Supporting Information

Supporting Information S1 MATLAB code used for the simulations. The settings for the simulation are given in the structure variable "Settings" (see lines 14–22). The shape of the void is defined by nodes with x- and y-coordinates and saved as complex numbers in P. Iterative adding of "cells" was done in the loop from line 64 to 188. The nodes covered by a "cell" where removed from P for the next step.

Acknowledgments

The authors would like to thank Dr. John Dunlop and Dr. Cécile Bidan from the Max Planck Institute of Colloids and Interfaces Department of Biomaterials for their constructive feedback.

Author Contributions

Conceived and designed the experiments: PJ GND PBW UF CW AP. Performed the experiments: PJ MS. Analyzed the data: PJ MS AP GND. Contributed reagents/materials/analysis tools: PJ MS PBW UF AP. Wrote the paper: PJ GND MS PBW UF CW AP.

References

1. Navarro M, Michiardi A, Castano O, Planell JA (2008) Biomaterials in orthopaedics. Journal of the Royal Society Interface 5: 1137–1158.
2. Hench LL, Polak JM (2002) Third-generation biomedical materials. Science 295: 1014–1017.

3. Midwood KS, Williams LV, Schwarzbauer JE (2004) Tissue repair and the dynamics of the extracellular matrix. International Journal of Biochemistry & Cell Biology 36: 1031–1037.
4. Kollmannsberger P, Bidan CM, Dunlop JWC, Fratzl P (2011) The physics of tissue patterning and extracellular matrix organisation: how cells join forces. Soft Matter 7: 9549–9560.
5. Beningo KA, Dembo M, Wang YI (2004) Responses of fibroblasts to anchorage of dorsal extracellular matrix receptors. Proceedings of the National Academy of Sciences of the United States of America 101: 18024–18029.
6. Schindler M, Nur-E-Kamal A, Ahmed I, Kamal J, Liu HY, et al. (2006) Living in three dimensions - 3D nanostructured environments for cell culture and regenerative medicine. Cell Biochemistry and Biophysics 45: 215–227.
7. Hakkinen KM, Harunaga JS, Doyle AD, Yamada KM (2011) Direct Comparisons of the Morphology, Migration, Cell Adhesions, and Actin Cytoskeleton of Fibroblasts in Four Different Three-Dimensional Extracellular Matrices. Tissue Engineering Part A 17: 713–724.
8. Vidal G, Delord B, Neri W, Gounel S, Roubeau O, et al. (2011) The effect of surface energy, adsorbed RGD peptides and fibronectin on the attachment and spreading of cells on multiwalled carbon nanotube papers. Carbon 49: 2318–2333.
9. Engler AJ, Sen S, Sweeney HL, Discher DE (2006) Matrix elasticity directs stem cell lineage specification. Cell 126: 677–689.
10. Salem AK, Stevens R, Pearson RG, Davies MC, Tendler SJB, et al. (2002) Interactions of 3T3 fibroblasts and endothelial cells with defined pore features. Journal of Biomedical Materials Research 61: 212–217.
11. Bidan CM, Kommareddy KP, Rumpler M, Kollmannsberger P, Brechet YJM, et al. (2012) How Linear Tension Converts to Curvature: Geometric Control of Bone Tissue Growth. Plos One 7.
12. Chen CS, Mrksich M, Huang S, Whitesides GM, Ingber DE (1997) Geometric control of cell life and death. Science 276: 1425–1428.
13. Vogel V, Sheetz M (2006) Local force and geometry sensing regulate cell functions. Nature Reviews Molecular Cell Biology 7: 265–275.
14. Edwards SL, Church JS, Alexander DLJ, Russell SJ, Ingham E, et al. (2011) Modeling Tissue Growth Within Nonwoven Scaffolds Pores. Tissue Engineering Part C-Methods 17: 123–130.
15. Shimizu M, Minakuchi K, Moon M, Koga J (1997) Difference in interaction of fibronectin with type I collagen and type IV collagen. Biochimica Et Biophysica Acta-Protein Structure and Molecular Enzymology 1339: 53–61.
16. Velling T, Risteli J, Wennerberg K, Mosher DF, Johansson S (2002) Polymerization of type I and III collagens is dependent on fibronectin and enhanced by integrins alpha(11)beta(1) and alpha(2)beta(1). Journal of Biological Chemistry 277: 37377–37381.
17. Sottile J, Shi F, Rublyevska I, Chiang H-Y, Lust J, et al. (2007) Fibronectin-dependent collagen I deposition modulates the cell response to fibronectin. American Journal of Physiology-Cell Physiology 293: C1934–C1946.
18. Shi F, Harman J, Fujiwara K, Sottile J (2010) Collagen I matrix turnover is regulated by fibronectin polymerization. American Journal of Physiology-Cell Physiology 298: C1265–C1275.
19. Welzel PB, Grimmer M, Renneberg C, Naujox L, Zschoche S, et al. (2012) Macroporous StarPEG-Heparin Cryogels. Biomacromolecules 13: 2349–2358.
20. Freudenberg U, Hermann A, Welzel PB, Stirl K, Schwarz SC, et al. (2009) A star-PEG-heparin hydrogel platform to aid cell replacement therapies for neurodegenerative diseases. Biomaterials 30: 5049–5060.
21. Welzel PB, Prokoph S, Zieris A, Grimmer M, Zschoche S, et al. (2011) Modulating Biofunctional starPEG Heparin Hydrogels by Varying Size and Ratio of the Constituents. Polymers 3: 602–620.
22. Freudenberg U, Sommer J-U, Levental KR, Welzel PB, Zieris A, et al. (2012) Using Mean Field Theory to Guide Biofunctional Materials Design. Advanced Functional Materials 22: 1391–1398.
23. Lozinsky VI (2002) Cryogels on the basis of natural and synthetic polymers: Preparation, properties and application. Uspekhi Khimii 71: 559–585.
24. Zhou LL, Higginbotham EJ, Yue B (1998) Effects of ascorbic acid on levels of fibronectin, laminin and collagen type 1 in bovine trabecular meshwork in organ culture. Current Eye Research 17: 211–217.
25. Hata RI, Senoo H (1989) L-ASCORBIC-ACID 2-PHOSPHATE STIMULATES COLLAGEN ACCUMULATION, CELL-PROLIFERATION, AND FORMATION OF A 3-DIMENSIONAL TISSUE-LIKE SUBSTANCE BY SKIN FIBROBLASTS. Journal of Cellular Physiology 138: 8–16.
26. Manders EMM, Stap J, Brakenhoff GJ, Vandriel R, Aten JA (1992) DYNAMICS OF 3-DIMENSIONAL REPLICATION PATTERNS DURING THE S-PHASE, ANALYZED BY DOUBLE LABELING OF DNA AND CONFOCAL MICROSCOPY. Journal of Cell Science 103: 857–862.
27. Barlow AL, MacLeod A, Noppen S, Sanderson J, Guerin CJ (2010) Colocalization Analysis in Fluorescence Micrographs: Verification of a More Accurate Calculation of Pearson's Correlation Coefficient. Microscopy and Microanalysis 16: 710–724.
28. Mao Y, Schwarzbauer JE (2005) Fibronectin fibrillogenesis, a cell-mediated matrix assembly process. Matrix Biology 24: 389–399.
29. Ramtani S (2004) Mechanical modelling of cell/ECM and cell/cell interactions during the contraction of a fibroblast-populated collagen microsphere: theory and model simulation. Journal of Biomechanics 37: 1709–1718.
30. Cheng G, Markenscoff P, Zygourakis K (2009) A 3D Hybrid Model for Tissue Growth: The Interplay between Cell Population and Mass Transport Dynamics. Biophysical Journal 97: 401–414.
31. Stolarska MA, Kim Y, Othmer HG (2009) Multi-scale models of cell and tissue dynamics. Philosophical Transactions of the Royal Society a-Mathematical Physical and Engineering Sciences 367: 3525–3553.
32. O'Dea RD, Osborne JM, El Haj AJ, Byrne HM, Waters SL (2012) The interplay between tissue growth and scaffold degradation in engineered tissue constructs. J Math Biol.
33. Dunn GA, Heath JP (1976) NEW HYPOTHESIS OF CONTACT GUIDANCE IN TISSUE-CELLS. Experimental Cell Research 101: 1–14.
34. Nelson CM, Jean RP, Tan JL, Liu WF, Sniadecki NJ, et al. (2005) Emergent patterns of growth controlled by multicellular form and mechanics. Proceedings of the National Academy of Sciences of the United States of America 102: 11594–11599.
35. Petersen A, Joly P, Bergmann C, Korus G, Duda GN (2012) The Impact of Substrate Stiffness and Mechanical Loading on Fibroblast-Induced Scaffold Remodeling. Tissue Engineering Part A 18: 1804–1817.
36. Discher DE, Janmey P, Wang YL (2005) Tissue cells feel and respond to the stiffness of their substrate. Science 310: 1139–1143.
37. Fujie T, Ricotti L, Desii A, Menciassi A, Dario P, et al. (2011) Evaluation of Substrata Effect on Cell Adhesion Properties Using Freestanding Poly(L-lactic acid) Nanosheets. Langmuir 27: 13173–13182.
38. Reinke JM, Sorg H (2012) Wound Repair and Regeneration. European Surgical Research 49: 35–43.
39. Carlson MA, Longaker MT (2004) The fibroblast-populated collagen matrix as a model of wound healing: a review of the evidence. Wound Repair and Regeneration 12: 134–147.
40. Bidan CM, Kommareddy KP, Rumpler M, Kollmannsberger P, Fratzl P, et al. (2013) Geometry as a Factor for Tissue Growth: Towards Shape Optimization of Tissue Engineering Scaffolds. Advanced Healthcare Materials 2: 186–194.
41. Chiu Y-C, Cheng M-H, Engel H, Kao S-W, Larson JC, et al. (2011) The role of pore size on vascularization and tissue remodeling in PEG hydrogels. Biomaterials 32: 6045–6051.
42. Hollister SJ (2005) Porous scaffold design for tissue engineering. Nat Mater 4: 518–524.
43. Bai F, Zhang J, Wang Z, Lu J, Chang J, et al. (2011) The effect of pore size on tissue ingrowth and neovascularization in porous bioceramics of controlled architecture in vivo. Biomedical Materials 6.
44. Jeon YK, Jang YH, Yoo DR, Kim SN, Lee SK, et al. (2010) Mesenchymal stem cells' interaction with skin: Wound-healing effect on fibroblast cells and skin tissue. Wound Repair and Regeneration 18: 655–661.
45. Rumpler M, Woesz A, Dunlop JWC, van Dongen JT, Fratzl P (2008) The effect of geometry on three-dimensional tissue growth. Journal of the Royal Society Interface 5: 1173–1180.
46. Knychala J, Bouropoulos N, Catt CJ, Katsamenis OL, Please CP, et al. (2013) Pore Geometry Regulates Early Stage Human Bone Marrow Cell Tissue Formation and Organisation. Annals of Biomedical Engineering 41: 917–930.

Permissions

All chapters in this book were first published in PLOS ONE, by The Public Library of Science; hereby published with permission under the Creative Commons Attribution License or equivalent. Every chapter published in this book has been scrutinized by our experts. Their significance has been extensively debated. The topics covered herein carry significant findings which will fuel the growth of the discipline. They may even be implemented as practical applications or may be referred to as a beginning point for another development.

The contributors of this book come from diverse backgrounds, making this book a truly international effort. This book will bring forth new frontiers with its revolutionizing research information and detailed analysis of the nascent developments around the world.

We would like to thank all the contributing authors for lending their expertise to make the book truly unique. They have played a crucial role in the development of this book. Without their invaluable contributions this book wouldn't have been possible. They have made vital efforts to compile up to date information on the varied aspects of this subject to make this book a valuable addition to the collection of many professionals and students.

This book was conceptualized with the vision of imparting up-to-date information and advanced data in this field. To ensure the same, a matchless editorial board was set up. Every individual on the board went through rigorous rounds of assessment to prove their worth. After which they invested a large part of their time researching and compiling the most relevant data for our readers.

The editorial board has been involved in producing this book since its inception. They have spent rigorous hours researching and exploring the diverse topics which have resulted in the successful publishing of this book. They have passed on their knowledge of decades through this book. To expedite this challenging task, the publisher supported the team at every step. A small team of assistant editors was also appointed to further simplify the editing procedure and attain best results for the readers.

Apart from the editorial board, the designing team has also invested a significant amount of their time in understanding the subject and creating the most relevant covers. They scrutinized every image to scout for the most suitable representation of the subject and create an appropriate cover for the book.

The publishing team has been an ardent support to the editorial, designing and production team. Their endless efforts to recruit the best for this project, has resulted in the accomplishment of this book. They are a veteran in the field of academics and their pool of knowledge is as vast as their experience in printing. Their expertise and guidance has proved useful at every step. Their uncompromising quality standards have made this book an exceptional effort. Their encouragement from time to time has been an inspiration for everyone.

The publisher and the editorial board hope that this book will prove to be a valuable piece of knowledge for researchers, students, practitioners and scholars across the globe.

List of Contributors

Xiao-Sheng Cheng
Department of Mathematics, Huizhou University, Huizhou, Guangdong, R China
School of Mathematical Sciences, Xiamen University, Xiamen, Fujian, P. R. China

Xian'an Jin
School of Mathematical Sciences, Xiamen University, Xiamen, Fujian, P. R. China

Ping Zhou, Yi Deng and Shicheng Wei
Department of Oral and Maxillofacial Surgery, School and Hospital of Stomatology, Peking University, Beijing, China
Center for Biomedical Materials and Tissue Engineering, Academy for Advanced Interdisciplinary Studies, Peking University, Beijing, China

Beier Lyu and Hongwei Ma
Suzhou Institute of Nano-Tech and Nano-Bionics, Chinese Academy of
Sciences, Suzhou, China

Ranran Zhang and Yalin Lyu
Department of Stomatology, Beijing Anzhen Hospital, Capital Medical University, Beijing, China

Hai Zhang
Department of Stomatology, Beijing Anzhen Hospital, Capital Medical University, Beijing, China
Department of Restorative Dentistry, School of Dentistry, University of Washington, Washington, United States of America

Karam Eldesoqi, Caroline Seebach, Christina Nguyen Ngoc, Simon Meier, Christoph Nau, Alexander Schaible, Ingo Marzi and Dirk Henrich
Department of Trauma-, Hand- and Reconstructive Surgery, Hospital of the Goethe- University, Frankfurt/Main, Germany

Nan Zhao and Donghui Zhu
Department of Chemical, Biological and Bio-Engineering, North Carolina Agricultural and Technical State University, Greensboro, North Carolina, United States of America
NSF Engineering Research Center-Revolutionizing Metallic Biomaterials, North Carolina Agricultural and Technical State University, Greensboro, North Carolina, United States of America

Beatriz Veleirinho
QOPNA Research Unit, Department of Chemistry, University of Aveiro, Aveiro, Portugal
Biotechnology and Biosciences Post-Graduation Program, Federal University of Santa Catarina, Florianópolis, Brazil

Daniela S. Coelho and Paulo F. Dias
Department of Cell Biology, Embryology, and Genetics, Federal University of Santa Catarina, Florianópolis, Brazil

Marcelo Maraschin
Plant Morphogenesis and Biochemistry Laboratory, Federal University of Santa Catarina, Florianópolis, Brazil

Rúbia Pinto
Epagri, Florianopolis, Brazil

Eduardo Cargnin-Ferreira
6 Federal Institute of Education, Science, and Technology of Santa Catarina, Garopaba, Brazil

Ana Peixoto and José A. Souza
Department of Pediatrics, Federal University of Santa Catarina, Florianópolis, Brazil

Rosa M. Ribeiro-do-Valle
Biotechnology and Biosciences Post-Graduation Program, Federal University of Santa Catarina, Florianópolis, Brazil

José A. Lopes-da-Silva
QOPNA Research Unit, Department of Chemistry, University of Aveiro, Aveiro, Portugal

Hélène Marie, Amélie Barrere and Laurence Mora
Université Paris 13, Sorbonne Paris Cité, BPC, Villetaneuse, France
Univ Paris Diderot, Hemostasis Bio-engineering and Cardiovascular Remodelling, Paris, France; INSERM, U698, Paris, France

Frédérique Schoentstein and Marie-Hélène Chavanne
Université Paris 13, Sorbonne Paris Cité, Laboratoire des Sciences des Procédés et des Matériaux, CNRS (UPR 3407), Villetaneuse, France

Brigitte Grosgogeat
UMR CNRS 5615 Laboratoire des Multimatériaux et des Interfaces, UFR d'Odontologie, Université Lyon 1, Lyon SCTD, Hospices Civils de Lyon, Lyon, France

Pierre Corre
INSERM (Institut National de la Santé et de la Recherche Médicale), UMR (Unité Mixte de Recherche) 791, center for osteoarticular and dental tissue engineering, Université de Nantes, Nantes, France
Centre Hospitalier Universitaire de Nantes, Clinique de Stomatologie et de Chirurgie maxillo-faciale, Nantes, France
Centre Hospitalier Universitaire de Nantes, Pôle Hospitalo-Universitaire 4, Nantes, France

Christophe Merceron, Caroline Vignes, Sophie Sourice, Martial Masson, Paul Pilet, Thomas Cordonnier, Jérôme Guicheux and Pierre Weiss
INSERM (Institut National de la Santé et de la Recherche Médicale), UMR (Unité Mixte de Recherche) 791, center for osteoarticular and dental tissue engineering, Université de Nantes, Nantes, France
Centre Hospitalier Universitaire de Nantes, Pôle Hospitalo-Universitaire 4, Nantes, France

Nicolas Durand and Florent Espitalier
INSERM (Institut National de la Santé et de la Recherche Médicale), UMR (Unité Mixte de Recherche) 791, center for osteoarticular and dental tissue engineering, Université de Nantes, Nantes, France
Centre Hospitalier Universitaire de Nantes, Clinique d'Oto-Rhino-Laryngologie et de Chirurgie cervico-faciale, Nantes, France
Centre Hospitalier Universitaire de Nantes, Pôle Hospitalo-Universitaire 4, Nantes, France

Jacques Mercier
Centre Hospitalier Universitaire de Nantes, Clinique de Stomatologie et de Chirurgie maxillo-faciale, Nantes, France
Centre Hospitalier Universitaire de Nantes, Pôle Hospitalo-Universitaire 4, Nantes, France

Séverine Remy and Ignacio Anegon
INSERM, UMR 1064, Centre pour la recherche en transplantation et immunologie et Plate-forme Transgenic Rats Nantes, Institut de Transplantation Urologie-Néphrologie (ITUN), Nantes, France

Alexandra Kroll, Renata Behra and Ralf Kaegi
Environmental Toxicology, Swiss Federal Institute of Aquatic Science and Technology (Eawag), Dübendorf, Switzerland

Laura Sigg
Environmental Toxicology, Swiss Federal Institute of Aquatic Science and Technology (Eawag), Dübendorf, Switzerland
Institute of Biogeochemistry and Pollutant Dynamics IBP, Swiss Federal Institute of Technology in Zurich (ETHZ), Zurich, Switzerland

Cristiano Rodrigues and Márcia Rosângela Wink
Laboratórios de Pesquisa em Biologia Celular, Departamento de Ciências Básicas da Saúde, Universidade Federal de Ciências da Saúde de Porto Alegre, Porto Alegre/RS, Brazil

Adriano M. de Assis
Laboratórios de Pesquisa em Biologia Celular, Departamento de Ciências Básicas da Saúde, Universidade Federal de Ciências da Saúde de Porto Alegre, Porto Alegre/RS, Brazil
Programa de Pós-Graduação em Ciências Biológicas: Bioquímica, Departamento de Bioquímica, Universidade Federal do Rio Grande do Sul, Porto Alegre/RS, Brazil

Dinara J. Moura and Jenifer Saffi
Genética Toxicológica, Departamento de Ciências Básicas da Saúde, Universidade Federal de Ciências da Saúde de Porto Alegre, Porto Alegre/RS, Brazil

Graziele Halmenschlager
Estresse Oxidativo e Poluição Atmosférica, Departamento de Ciências Básicas da Saúde, Universidade Federal de Ciências da Saúde de Porto Alegre, Porto Alegre/RS, Brazil

Léder Leal Xavier
Laboratório de Biologia Celular e Tecidual, Departamento de Ciências Morfofisiológicas, Pontifı´cia Universidade Católica do Rio Grande do Sul, Porto Alegre/RS, Brazil

Marilda da Cruz Fernandes
Laboratório de Pesquisa em Patologia, Departamento de Ciências Básicas da Saúde, Universidade Federal de Ciências da Saúde de Porto Alegre, Porto Alegre/RS, Brazil

Firouzeh Sabri, Thien-Chuong N. Phung and George-Rudolph M. Tamula
Department of Physics, University of Memphis, Memphis, Tennessee, United States of America

John D. Boughter Jr
Department of Anatomy and Neurobiology, UTHSC, Memphis, Tennessee, United States of America

David Gerth
Department of Otolaryngology Head and Neck Surgery, UTHSC, Memphis, Tennessee, United States of America

Omar Skalli
Department of Biological Sciences, University of Memphis, Memphis, Tennessee, United States of America

Nicholas Leventis
Department of Chemistry, Missouri University of Science and Technology,
Rolla, Missouri, United States of America

Kazuya Doi, Hiroshi Oue, Koji Morita, Shiho Kajihara, Takayasu Kubo, Katsunori Koretake and Yasumasa Akagawa
Department of Advanced Prosthodontics, Hiroshima University Graduate School of Biomedical Sciences, Hiroshima, Japan

Vittoria Perrotti, Giovanna Iezzi and Adriano Piattelli
Department of Medical, Oral and Biotechnological Sciences, Chieti-Pescara University, Chieti, Italy

Liao Wang, Chongyun Bao and Jing Hu
State Key Laboratory of Oral Diseases, West China Hospital of Stomatology, Sichuan University, Chengdu, China

Bi Zhang
The Affiliated Hospital of Stomatology, Chongqing Medical University, Chongqing, China

Pamela Habibovic
Department of Tissue Regeneration, MIRA-Institute for Biomedical Technology and Technical Medicine, University of Twente, Enschede, the Netherlands

Xingdong Zhang
National Engineering Research Center for Biomaterials, Sichuan University, Chengdu, China

Barbara Barboni, Luca Valbonetti and Aurelio Muttini
Department of Comparative Biomedical Science, University of Teramo, Teramo, Italy
Stem TeCh Group, Chieti, Italy

Carlo Mangano
Department of Surgical and Morphological Science, University of Insubria, Varese, Italy

Giuseppe Marruchella, Paolo Berardinelli, Alessandra Martelli,
Annunziata Mauro, Maura Turriani, Delia Nardinocchi and Mauro Mattioli
Department of Comparative Biomedical Science, University of Teramo, Teramo, Italy

Rossella Bedini and Raffaella Pecci
Department of Technologies and Health, Istituto Superiore di Sanità, Rome, Italy

Vincenzo Luca Zizzari and Adriano Piattelli
Department of Medical, Oral and Biotechnological Science, University "G.
d'Annunzio", Chieti, Italy, 5 Stem TeCh Group, Chieti, Italy

Stefano Tetè
Department of Medical, Oral and Biotechnological Science, University "G. d'Annunzio", Chieti, Italy
Stem TeCh Group, Chieti, Italy

Gemma Pascual, Sandra Sotomayor and Bárbara Pérez-Köhler,
Department of Medical Specialties, Networking Research Center on Bioengineering, Biomaterials and Nanomedicine (CIBER-BBN), Faculty of Medicine, Alcalá University, Alcalá de Henares, Madrid, Spain

Juan M. Bellón and Marta Rodríguez
Department of Surgery, Networking Research Center on Bioengineering, Biomaterials and Nanomedicine (CIBER-BBN), Faculty of Medicine, Alcalá University, Alcaláde Henares, Madrid, Spain

Marije Sloff, Paul Geutjes, Egbert Oosterwijk and Wout Feitz
Department of Urology, Nijmegen Centre for Molecular Life Sciences, Radboud University Medical Center, Nijmegen, The Netherlands

Rob de Vries and Merel Ritskes-Hoitinga
SYRCLE (SYstematic Review Centre for Laboratory animal experimentation), Central Animal Laboratory, Radboud University Medical Center, Nijmegen, The Netherlands
Joanna IntHout
Department for Health Evidence, Radboud University Medical Center, Nijmegen, The Netherlands

Ying Li, Zhuo Chen and Xin Tang
Clinical College of Ophthalmology, Tianjin Medical University, Tianjin Eye Hospital, Tianjin, China

Jiaxing Wang
Department of Ophthalmology, Tianjin Medical University General
Hospital, Tianjin, China

Jason W. Miklas and Lewis A. Reis
The Institute of Biomaterials and Biomedical Engineering, University of Toronto, Toronto, Ontario, Canada

Susan M. Dallabrida and Nesreen Ismail
Center for Vascular Biology Research, Beth Israel Deaconess Medical Center, Boston, Massachusetts, United States of America

Milica Radisic
The Institute of Biomaterials and Biomedical Engineering, University of Toronto, Toronto, Ontario, Canada
Department of Chemical Engineering and Applied Chemistry, University
of Toronto, Toronto, Ontario, Canada

Maria Rupnick
Brigham and Women's Hospital, Cardiovascular Division, Boston, assachusetts, United States of America (Affiliates of Harvard Medical School, Boston, Massachusetts, United States of America)

Marta Pokrywczynska, Arkadiusz Jundzill, Jan Adamowicz, Tomasz Karolina Warda, Marta Rasmus, Lukasz Buchholz, Sandra Krzyzanowska and Maciej Nowacki
Chair of Regenerative Medicine, Department of Tissue Engineering, Nicolaus Copernicus University in Torun, Ludwik Rydygier Medical College in Bydgoszcz, Bydgoszcz, Poland

Kowalczyk
Department of Theory of Continuous Media, Institute of Fundamental Technological Research, Polish Academy of Sciences, Warsaw, Poland

Pawel Nakielski, Tomasz Chmielewski and Tomasz A. Kowalewski
Department of Mechanics and Physics of Fluids, Institute of Fundamental Technological Research, Polish Academy of Sciences, Warsaw, Poland

Magdalena Bodnar
Department of Clinical Pathomorphology, Nicolaus Copernicus University in Torun, Ludwik Rydygier Medical College in Bydgoszcz, Bydgoszcz, Poland

Andrzej Marszalek
Department of Clinical Pathomorphology, Nicolaus Copernicus University in Torun, Ludwik Rydygier Medical College in Bydgoszcz, Bydgoszcz, Poland
Department of Tumor Pathology, Center of Oncology, Poznan University of Medical Sciences, Poznan, Poland

Robert Debski
Department of Pediatrics, Hematology and Oncology, Nicolaus Copernicus University in Torun, Ludwik Rydygier Medical College in Bydgoszcz, Bydgoszcz, Poland

Malgorzata Frontczak-Baniewicz
Electron Microscopy Platform, Mossakowski Medical Research Centre, Polish Academy of Sciences, Warsaw, Poland

Grzegorz Mikułowski
Department of Intelligent Technologies, Institute of Fundamental Technological Research, Polish Academy of Sciences, Warsaw, Poland

Tomasz Drewa
Chair of Regenerative Medicine, Department of Tissue Engineering, Nicolaus Copernicus University in Torun, Ludwik Rydygier Medical College in Bydgoszcz, Bydgoszcz, Poland
Department of Urology, Nicolaus Copernicus Hospital, Torun, Poland

Achim von Bomhard
Department of Maxillofacial Surgery, Technical University of Munich, Munich, Germany

Johannes Veit, Christian Bermueller and Nicole Rotter
Department of Otorhinolaryngology, Ulm University Medical Center, Ulm, Germany

Rainer Staudenmaier
HNO Praxis im Zentrum, ENT and Plastic Surgery, Munich, Germany

Katharina Storck
Department of Otorhinolaryngology, Technical University of Munich, Munich, Germany

Hoang Nguyen The
Department of Hand- and Microsurgery, Military Hospital 108, Hanoi, Vietnam

Silke Schlaubitz and Martine Renard
CIC 1401, University hospital of Bordeaux/Inserm, Bordeaux, France

Sidi Mohammed Derkaoui, Catherine Le Visage and Didier Letourneur
U1148, LVTS/Inserm, Paris, France
Prés Sorbonne Paris Cité, University of Paris Nord and University Paris Diderot, Paris, France

Lydia Marosa and Joëlle Amédée
U1026 Tissue Bioengineering, University of Bordeaux/Inserm, Bordeaux, France

Sylvain Miraux
RMSB Center/UMR 5536, CNRS, Bordeaux, France

Jean-Christophe Fricain and Sylvain Catros
Dental School, University of Bordeaux, Bordeaux, France
U1026 Tissue Bioengineering, University of Bordeaux/Inserm, Bordeaux, France

Pascal Joly and Martin Schöne
Julius Wolff Institute, Charité - Universitätsmedizin Berlin, Berlin, Germany
Berlin-Brandenburg School for Regenerative Therapies, Charité - Universitätsmedizin Berlin, Berlin, Germany
Center for Musculoskeletal Surgery, Charité - Universitätsmedizin Berlin, Berlin, Germany

Georg N. Duda
Julius Wolff Institute, Charité - Universitätsmedizin Berlin, Berlin, Germany
Berlin-Brandenburg School for Regenerative Therapies, Charité - Universitätsmedizin Berlin, Berlin, Germany
Berlin-Brandenburg Center for Regenerative Therapies, Charité - Universitätsmedizin Berlin, Berlin, Germany
Center for Musculoskeletal Surgery, Charité - Universitätsmedizin Berlin, Berlin, Germany

Petra B. Welzel, Uwe Freudenberg and Carsten Werner
Max Bergmann Center of Biomaterials Dresden, Leibniz Institute of Polymer Research Dresden, Dresden, Germany,
Center for Regenerative Therapies Dresden, Technische Universität Dresden, Dresden, Germany

Ansgar Petersen
Julius Wolff Institute, Charité - Universitätsmedizin Berlin, Berlin, Germany
Berlin-Brandenburg Center for Regenerative Therapies, Charité - Universitätsmedizin Berlin, Berlin, Germany
Center for Musculoskeletal Surgery, Charité - Universitätsmedizin Berlin, Berlin, Germany

Index